Bioactive Compounds in Nutraceutical and Functional Food for Good Human Health

Edited by Kavita Sharma, Kanchan Mishra, Kula Kamal Senapati and Corina Danciu

Published in London, United Kingdom

IntechOpen

Supporting open minds since 2005

Bioactive Compounds in Nutraceutical and Functional Food for Good Human Health
http://dx.doi.org/10.5772/intechopen.78846
Edited by Kavita Sharma, Kanchan Mishra, Kula Kamal Senapati and Corina Danciu

Contributors
Muzaffar Iqbal, Jose Otavio Carrera Silva Junior, Russany Silva Da Costa, Tais Vanessa Gabbay Alves, Raimundo Lopes Da Silva, Lindalva Maria De Meneses Costa Ferreira, Roseane Maria Ribeiro Costa, Hemant Sood, Somenath Ghosh, Doo-Gyung Moon, Nada Khazal Kadhim Hindi, Ruqaya Ewadh, Zainab Al-Mahdi, Kula Kamal Senapati, Uma Ranjan Lal, Snigdha Lal, Mahluga Jafarova Demirkapu, Hasan Raci Yananli, Nguyen Thi Hong Van, Nguyen Xuan Nhiem, Pham Minh Quan, Valentina Buda, Corina Danciu, Douglas Vieira Thomaz, Gayathri Mahalingam, Ashwini Devaraj, Telma Angelina Faraldo Correa, Adriana Rozembaum, Marcelo Rogero, Gabin Bitchagno, Vaderament-Alexe Nchiozem-Ngnitedem, Nadine Tseme Wandji, Guy Cedric Noulala Tchatchouang, Serge Alain Fobofou T., Bruno Lenta Ndjakou, Burhan Başaran, Lucía Ortega Cabello, Ashwini Venkatrao Mohite, Rajaram Gurav, Liza Meutia Sari, Minodora Andor, Antal Diana, Florina Ardelean, Ioana Zinuca Pavel, Cristina Dehelean, Codruta Soica, Roxana Folescu, Felicia Andrei

First published in London, United Kingdom, 2021 by IntechOpen
IntechOpen is the global imprint of INTECHOPEN LIMITED, registered in England and Wales, registration number: 11086078, 5 Princes Gate Court, London, SW7 2QJ, United Kingdom
Printed in Croatia

British Library Cataloguing-in-Publication Data
A catalogue record for this book is available from the British Library

Additional hard and PDF copies can be obtained from orders@intechopen.com

Bioactive Compounds in Nutraceutical and Functional Food for Good Human Health
Edited by Kavita Sharma, Kanchan Mishra, Kula Kamal Senapati and Corina Danciu
p. cm.
Print ISBN 978-1-83880-887-7
Online ISBN 978-1-83880-888-4
eBook (PDF) ISBN 978-1-83880-889-1

For EU product safety concerns:
IN TECH d.o.o., Prolaz Marije Krucifikse Kozulić 3, 51000 Rijeka, Croatia, info@intechopen.com or visit our website at intechopen.com.

Meet the editors

Dr. Kumari Kavita Sharma is a research assistant professor in the Department of Chemistry, Idaho State University (ISU), holding a joint appointment with Idaho National Laboratory. She earned her doctorate in Molecular Biotechnology from Konkuk University, Seoul, South Korea. She holds a master's in Analytical Chemistry and a bachelor's in Chemistry from the India's University of Pune. Before coming to ISU, she was an assistant professor in the School of Chemical Engineering, Yeungnam University, Gyeongsan, South Korea. Dr. Sharma is skilled in analytical instruments like HPLC, LC-MS, GC, NMR, UV-Visible spectroscopy, FTIR, and more. Her areas of research include life sciences, organic chemistry, and medicinal chemistry.

Kanchan Mishra is a postdoctoral associate at the School of Chemical Engineering, Yeungnam University, South Korea. He completed his Ph.D. on February 22, 2018, from the School of Chemical Engineering, Yeungnam University. He is conducting research on sustainable and green synthesis of metal and metal oxide nanoparticles for organic transformations. His research interests include nanocatalysts, mechanistic study, theoretical investigation, green synthesis, and recyclable and reusable catalysts.

Dr. Kula Kamal Senapati is a scientific officer at the Indian Institute of Technology Guwahati (IIT Guwahati). He received his MSc in Organic Chemistry and a Ph.D. in Chemistry from Gauhati University, India. He has been involved in characterizing materials and nanomaterials through highly sophisticated analytical instruments since 2005. He served as a scientist in Tocklai Tea Research Institute, India, and is involved in research of secondary metabolites in tea and their value addition. He has published his research works in many reputed international and national scientific journals (citations: 801; h-index: 13; and 1–10 index: 17) and one of his works was highlighted in Nature India (doi:10.1038/nindia.2017.49). Dr. Senapati has obtained awards and recognitions such as the National Green Innovation Award in 2011 (www.industrialgreenchem.com); top cited paper award for research papers published in the Journal of Molecular Catalysis A: Chemical in 2011-2012; a fellowship from National Science Council, Taiwan; and a post-doctoral offer from the Universidade da Madeira, Portugal. He is an editorial board member and reviewer for various international journals of repute.

Dr. Corina Danciu is the coordinator of the Discipline of Pharmacognosy, "Victor Babeş" University of Medicine and Pharmacy, Timişoara, Romania. She works in the field of natural compounds, her main research lines being phytochemical and biological investigations of different types of plant extracts as well as pure active phytocompounds. To date, Dr. Danciu has published more than eighty ISI papers with a cumulated impact factor above 190, a Web of Science h-index of 17, and a Google Scholar h-index of 20.

Contents

Preface

Bioactive compounds are widely used in the pharmaceutical industry, food processing, and drug, nutraceutical, and cosmetic manufacturing, among others. This book examines how to obtain bioactive compounds from different sources and their potential industrial uses. New and sophisticated techniques are used for obtaining bioactive compounds from different sources such as plants, marine life, and microorganisms. Currently, researchers are investigating the utilization of fruit and vegetable wastes as good and cheap sources of bioactive compounds. The bioactive compounds possess antioxidant, antimicrobial, anti-inflammatory, anticancer, and antidepressant properties. The biological activity of these compounds is based on the lead or privileged scaffolds present in their structure. These different scaffolds include indole, purine, chromone, coumarin, benzothiophene, lactone, and so on, and can be modified into multiple molecules for different bioactivities. Sesquiterpene lactones (SLs) are an example of containing lactone as a privileged scaffold. The γ-lactone ring, usually with an α-methylene group, is a significant characteristic of SLs. Their molecular structure may present hydroxyls, esterified hydroxyls, or epoxide groups. These bioactive molecules have direct potent effects on human health.

Kavita Sharma
Idaho State University,
United States of America

Kanchan Mishra
Yeungnam University,
South Korea

Kula Kamal Senapati
Indian Institute of Technology Guwahati,
India

Corina Danciu
Victor Babeş University of Medicine and Pharmacy,
Romania

Chapter 1

Bioactive Molecules from Medicinal Plants as Functional Foods (Biscuits) for the Benefit of Human Health as Antidiabetic Potential

Ashwini Devaraj and Gayathri Mahalingam

Abstract

Functional foods defined as "similar in appearance to a conventional food, and is demonstrated to have physiological benefits or reduce the risk of chronic disease beyond basic nutritional functions." The leading role in food industry plays ultimately by the functional food. In recent days, the designing of a functional food with the incorporation of medicinal plants, which is the natural product is the familiar one. The medicinal plants are scientifically proven, lesser side effects, and eco-friendly in nature. Many food types are chosen for the development of functional food with the incorporation of medicinal plants. Diabetes mellitus is a major chronic disease which affects the basic metabolism of insulin secretion and insulin functioning on glucose clearance from the blood stream. The modern inactive life style of the population leads to obesity and ultimately results in the major risk of diabetes mellitus and other risk factors alongside. The therapeutic alteration for DM is to minimize the burden of disease, and the targeted people were advised to follow proper physical activity and nutrient intake with healthy weight gain. The disease targeted people were recommended with the proper diet intake which aims at consuming the functional food with the incorporation of medicinal plants.

Keywords: functional food, medicinal plants, bioactive molecules, diabetes mellitus, nutrient intake

1. Introduction

In the recent days, the world has a strong belief in the field of food, which directly contributes to the human health. Each and every individual is aware of the food intake and its effects to the benefit of the health. The food which contains all vital nutrients not only to subside the hunger but also to provide essential nourishment apart from usual benefits. This will improve the physical and mental state of the human health in a disease free condition leads to a diet related therapeutic model for the lifestyle diseases [1]. Nowadays science and technology has its wide arm in every field because of the scientific evidence that proves the benefits and harmfulness of certain thing. In food technology also, the consumers see to the scientific evidence and proof for the beneficial effects of that food item to the health. Thus in the last few decades, the urge for the healthy combo of food demands the health promoting category in the food industry by the consumers [2]. Because of the modern life culture, the people are in hurry and fast, they were in need of fast foods and lead to change in their lifestyle. Thus, the emergence of lifestyle diseases started which is by the lack of physical activity, change in eating habits, taking unhealthy food in an unbalanced way and leads to hazardous health ill effects. The basic daily recommended dietary intake of macronutrients in gender wise is listed in **Table 1**.

Nutrients	Carbohydrates (g/d)	Fat (%K Cal)	Protein (g/d)	Total fiber (g/d)
Gender/category of age				
Male				
9–13 year	130		34	31
14–18 year	130		52	38
19–30 year	130	25–35	56	38
31–50 year	130		56	38
51–70 year	130		56	30
>70 year	130		56	30
Female				
9–13 year	130		34	26
14–18 year	130		46	26
19–30 year	130	20–35	46	25
31-50 year	130		46	25
51-70 year	130		46	21
>70 year	130		46	21

Table 1.
Recommended daily dietary intake of macronutrients for individuals – gender and age [3].

	Adult male	Adult female
Energy (kcal)	2400–3000	2200
Protein(g)	56	50
Vit. A (IU)	900	800
Vit. D (IU)	600	600
Vit. C (IU)	90	75
Vit. E (IU)	15	15
Folate (mcg)	400	400
Ca (mg)	1000	800–1000
Ph (mg)	700	700
Fe (mg)	18	8
Zn (mg)	11	8
I (mg)	18	8
Se (mg)	55	55

Table 2.
Daily nutritional requirement in gender based on recommended dietary intakes [7].

Nutrient is the term to be explained or mentioned before the definition of bioactive components in and as food. The food is a combination of vital components in a proper ratio called balanced diet. The nourishment provided by the food for the healthy functioning of the human body is said to be the nutrients [4]. About 2500 years ago, the tenet "Let food be thy medicine and medicine be thy food," espoused Hippocrates receives interest nowadays. Lifestyle and diet intake are common couple factors responsible for major chronic diseases. Major lifestyle diseases like cancer, diabetes, osteoporosis, respiratory diseases, cardiovascular diseases, gastrointestinal diseases, and obesity, account for 63% annual deaths [5]. Non-communicable diseases are coined as lifestyle diseases, which are linked with the people nature of living and behavior involved in diet, lifestyle, and environment [6]. Thus, the food not only helps to promote health, physical development, and growth but also to prevent or treat various disorders and diseases. Some of the daily recommended nutritional elements gender wise is listed in **Table 2**.

2. Diabetes mellitus

Diabetes mellitus is one of the most deadly chronic diseases with metabolic disorder which is associated with major life threatening complications. The diabetes is a growing global problem which affects the metabolism and results in many complications leads tom death at early stages of life. Diabetes is of three types: type 1 – non-insulin dependent diabetes, type 2 – insulin dependent diabetes, type 3 – gestational diabetes. According to international diabetic federation (IDF), type 2 diabetes is accounting 90% of worldwide diabetes. The prevalence of diabetes in 2030 is 1.2 billion, where as in 2045, it increases to 1.3 billion unconditionally shown in **Figure 1**. It is mainly because of insulin insufficient secretion to lower the blood glucose level because of impairment of pancreatic beta cells. The elevated blood glucose level in the body is because of insulin impairment termed as insulin resistance. The major cause of insulin impaired diabetes is due to improper physical activity, unhealthy food habits, and increase of obese condition [8]. As defined by WHO, the obese condition and overweight of body are the major risk factors for the cause of diabetes [5]. The risk factors are directly correlated to the food intake, physical inactivity, and modern lifestyle, thus the energy imbalance plays a vital role in the

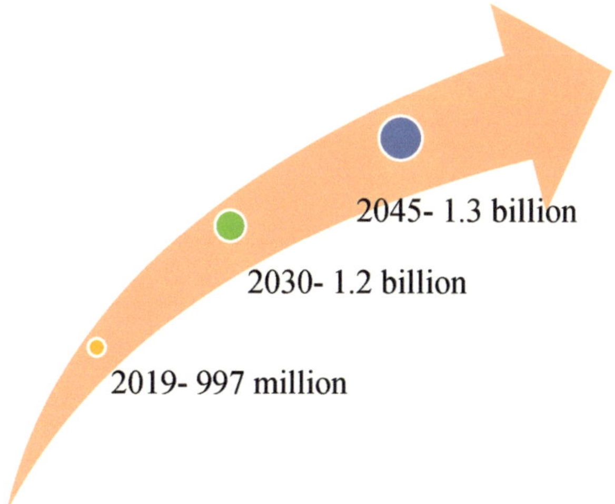

Figure 1.
Prevalence of diabetes worldwide, IDF.

prevalence of diabetes worldwide. Thus, the improper food habit and inactiveness of modern lifestyle inculcate the human life in a dangerous route where the people are unaware of diabetes emergence in their life. Prevalence of diabetes is estimated by sex among 65–99 years old adult population by IDF in **Figure 2** [9].

The major risk factors of type 2 diabetes mellitus are unhealthy eating, lack of physical exercise, obesity, and family epigenetics. Thus, the overeating of unhealthy food leads to increased body weight because of lack of exercise, results in obese condition. This obese condition portrays the beta cell destruction results in insulin impairment. It takes many years to exhibit the hyperglycemic condition in the body [10]. Various studies show that the high intake of fatty food gradually results in the lack of glucose resistant. The intake of unsaturated fatty acid is good for diabetic patients while the saturated and the Trans fats are associated with diabetic risk in a very high ratio. Reduced insulin secretion in the pancreas associated with decreased glucose uptake due to insulin excitation in muscles, also increased fundamental liver glucose production, thus glucose absorption in gastrointestinal tract is increased. This is the pathophysiological structure of type 2 diabetes mellitus. It is shown in **Figure 3**.

2.1 Medicinal plants and DM

As earlier defined by WHO, the cure for diabetes mellitus is possibly from the usage of medicinal plants in the form of herbal medicine whose remedies and proportion are involved in the control of the same. The scientific evidence and therapeutic application of the medicinal plants is the ultimate goal for the researchers and healthcare systems in the management of diabetes [11]. From the olden days, the management of severe diseases has its loophole in the potential of some medicinal plants which acts as drugs. These are having high belief that the drugs prepared from the medicinal plants are lesser side effects, and risk of toxicity is minimal and free from harmful effects [12]. Thus, World Health Organization recommends the usage of medicinal plants of traditional methods for the management of diabetes mellitus because of its lesser side effects when compared to the synthetic drugs. The usage of medicinal plants is extensively benefited worldwide for the management of various diseases in the pharmaceutical industry. Since plants are the rich source of phytochemicals whose benefits are countless and endless. In the pharma industries for most of the chronic diseases, the drugs are synthesized by using 50% of the medicinal plants from the historical origin [13].

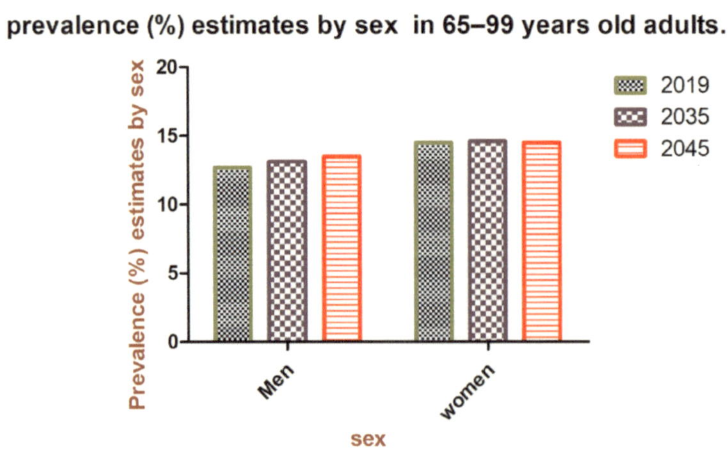

Figure 2.
Prevalence of diabetes estimated by sex among 65–99 years old adult population [7].

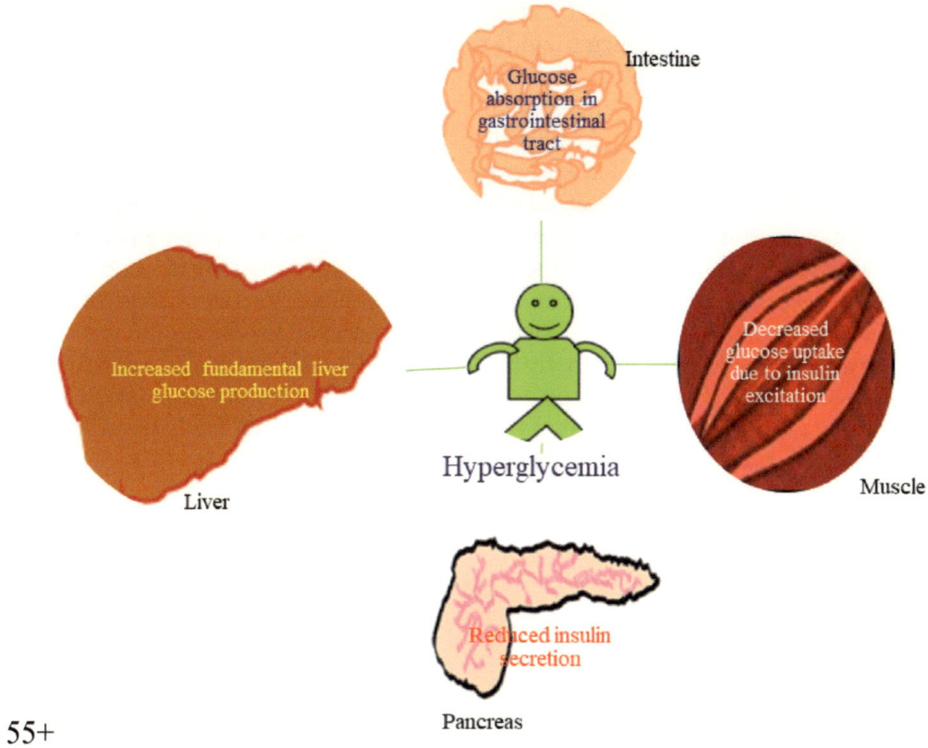

Figure 3.
Pathophysiology of type 2 diabetes mellitus.

The complications of diabetes mellitus are countless, which leads to gradual impairment of vital body parts one by one. The challenging part of diabetes is its management for maintaining the blood glucose level to normal range. The traditional usage of medicinal plants for the management of diabetes because of the bioactive molecules present in it which fight against the condition [14]. From the ancient history, the medicinal plants are having its main role for the curing of many diseases which paves the way for the utilization of the same to the management of diabetes mellitus. Thus, our ancestors commonly used the medicinal plants in the food itself, and the leaf, stem, fruits, flowers, and roots all are utilized in the management of various diseases. This enlightens the idea of using medicinal plants in the management of DM.

3. Bioactive molecules and functional food

The plants whose bioactive molecules are so called phytochemicals, which makes the plant to be said as medicinal plant. The medicinal plants contain some potential biological properties that help the human beings to get rid of some diseases and protect them from hazardous health issues [15]. The application of phytochemicals is in many fields such as food industry, health and nutrition, agriculture, and pharmaceuticals also. Thus, medicinal plants are having ancient origin for the home made treatment and desirable beneficial effects to human health. The bioactive molecules are called so because it is the biologically active compounds which give additional nutritional value than the primary nutrition. Thus the food contains some additional nutritional value because of the bioactive molecules present in it said to be known as functional food. The phytochemicals like alkaloids, phenols, flavonoids, terpenes and antioxidants, and polyunsaturated fatty acids like omega 3 fatty acids containing foods are called as functional foods [16].

According to American dietetic association, functional foods are said to be the food and its components which provide some additional nutrition than the basic nutrition. Examples for the functional food are the fortified food, enhanced or enriched foods which are having good health impact for the growth and development [17]. The pharmaceutical compounds than giving as drug form it can be provided as nutritional supplement with added advantage [12]. The functional food is similar to normal conventional food but beyond normal physiological role, it is having additional nutritional content that said to be as enhanced, enriched, or fortified food [18]. The functional properties are added in many food products like bread, biscuits, powders, mixes, decoction, suranams, also as various forms. The addition of nutrient rich food stuff in all the diets like legumes, grains, nuts, fruits, and vegetables will eventually result in less glycemic index foods and low fat substances which are good for type 2 DM [19]. Various probiotics, prebiotics, and also combination of both called symbiotic combo of functional foods are utilized [20]. The risk factors of DM and link to food habit are stated in **Figure 4**.

3.1 Functional food: biscuits

Among various functional foods, biscuits are of having wide range of advantages like a product with less moisture content, easily available, less cost, ready to consume at any time, can be prepared as varieties because of changing the proportion of the major ingredients. Biscuits can be made in wide functionality and nutritional value [21]. Biscuits are rich in cereal and best nutraceutical for delivery health benefits to the consumers in a right proportion. It can be used in the daily diet and improve the health state of human beings [22]. Biscuits can be available in different tastes, and variations also possess prolonged shelf life which suits all age group people [23]. In this review, especially biscuits are covered because of its efficacy and possess good functional properties in **Table 3**.

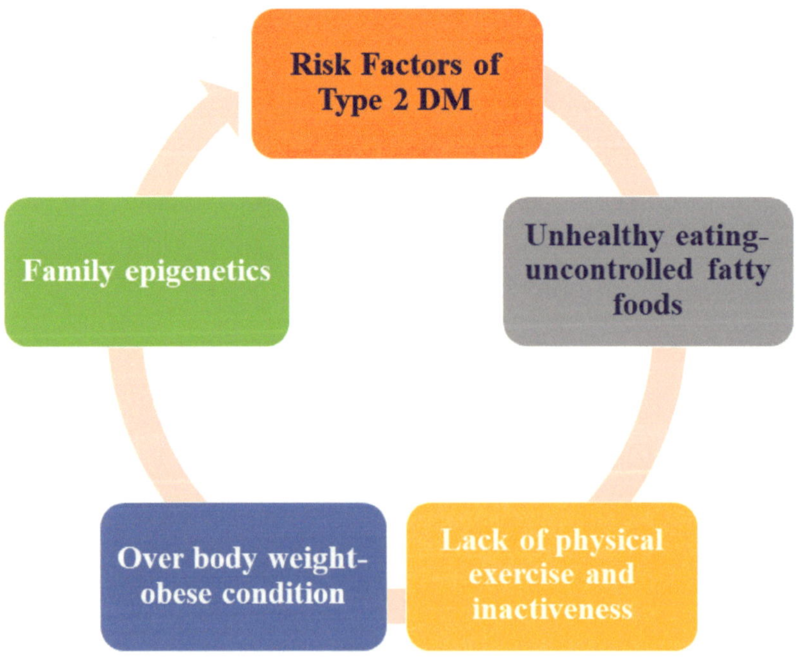

Figure 4.
Risk factors of DM and link of food intake.

S. No.	Medicinal plants	Common names	Bioactive molecules	Functional food	Reference
1	*Lupinus albus*	Sweet white lupin (SWL) grains	Dietary fiber and phenolic compounds	SWL biscuits	[24]
2	Cissus quadrangularis stem powder	Hadjor	Phytosterols, phenols, ascorbic acid, and calcite	Biscuits and cookies	[25]
3	*Musa paradisiaca*	Banana	Higher protein and phenol content	Biscuits products	[26]
4	Ipomoea batatas, Daucus carota, and *Musa paradisiaca*	Sweet potato, carrot and banana	Micro and macronutrients	Whey protein and banana incorporated biscuits	[27]
5	Hibiscus sabdariffa L. seed	Roselle seed	Protein content, dietary fiber and micronutrients	Biscuits and cookies	[28]
6	Ammannia baccifera L.	Monarch redstem	Phenols, flavonoids and terpenoids	Drug formulation	[29]
7	Lepidium sativum	Garden cress seed	Proteins, minerals and essential amino acids	Garden cress seed biscuit	[30]
8	(*Fragaria ananassa*) extract (Fisetin)	Strawberry	Fisetin – a flavonol	Biscuits	[31]
9	Beta vulgaris	Sugar beet molasses	Protein potassium, calcium, magnesium and iron content.	Ginger nut type biscuits	[21]
10	Phaseolus vulgaris	Common bean	Improved nutritional properties	Biscuits	[32]
11	Betifore-type	Egyptian butter cookie type	α-amylase added	Cookies	[21]
12	(Avena sativa), and maltitol	Oats	Inulin, a fructooligosachararide (FOS)	Biscuits	[33]
13	(*Musa species*), (Citrus sinensis), (Citrullus lanatus), (*Ananas cosmosus*) and (Carica papaya)	Banana, oranges, watermelon, pineapple, pawpaw	High fiber	Biscuits	[34]
14	Cinnamomum verum	Cinnamon powder	Protein and dietary fiber	Butter biscuits	[35]
15	De-oiled peanut meal flour (DPMF)		Nutritionally rich	Biscuits	[36]

S. No.	Medicinal plants	Common names	Bioactive molecules	Functional food	Reference
16	Amla (*Emblica officianalis*), (Moringa oleifera) and (Vitis vinifera)	Drumstick leaves, raisins.	Antioxidant effect	Biscuits	[37]
17	Hylocereus undatus	Pitaya	Nutritional quality	Biscuits	[38]
18	Fructo - ligosaccharide (FOS),	—	A prebiotic soluble fiber	Cookies	[39]
19	*Vitis*	Grape Seed Powder	Fatty acids and tocopherols	Iranian Sangak Bread	[40]
20	*Beta vulgaris* L.	Beetroot	Antioxidants	Mayonnaise	[41]
21	Curcuma longa L.	Turmeric flower	Formulating healthy cookie	Cookies	[42]
22	Brewer's spent grain (BSG)	—	Protein and fiber content	Cookies	[43]
23	*Carica papaya*	Papaya pulp flour (PPuF)	Protein and fiber content	Cookies	[44]
24	Sour cherry pomace extract		Polyphenols anthocyanins, antioxidant activity	Cookies	[45]
25	Linum usitatissimum flour	Flaxseed	Dietary fiber and linolenic acid	Cakes	[46]
26	Hordeum vulgare, Brassica nigra, *Linum usitatissimum*	Barley mustard, defatted mustard, flaxseed meal and flaxseed oil	Lowering blood lipids	Functional prebiotic biscuits	[47]
27	Citrus limetta	Sweet lime	Antioxidant potential	Herbal Juice	[48]
28	*Moringa olifera* leaves	Drumstick leaves	Beta-carotene	Biscuits	[49]
29	Holy Basil and *Moringa*	Thulasi	Protein and fiber enriched	Herbal biscuit	[50]
30	Inulin (Raftilin) (in combination with one of the following raw materials: soy flour, amaranth, carob apple fiber or oat fiber)	—	Essential mineral (Ca, Mg, Mn, Cu, and Fe) content and protein content	Biscuits	[51]

S. No.	Medicinal plants	Common names	Bioactive molecules	Functional food	Reference
31	Psyllium fiber content	—	Fiber incorporation	Biscuits	[23]
32	Green tea extract (GTE) was	—	Stability	Biscuits	[52]
33	(*Trigonella foenum graecum*) and (*Momorodica charantia*), Gudmar leaves	Fenugreek seeds, bitter gourd fruit.	Hypoglycemic properties	Idli and vegetable soup	[53]
34	*Moringa oleifera*	Drumstick	Nutritional value of food	Fortifying amala (stiff dough), ogi (maize gruel), bread, bis-cuits, yoghurt, cheese	[54]
35	Sesamum indicum powder	Black sesame	Antioxidant properties	Cookies	[55]
36	(*Musa species*), (*Citrus sinensis*), (*Citrullus lanatus*), (*Ananas cosmosus*), and (*Carica papaya*)	Banana, oranges, watermelon, pineapple, pawpaw	Dietary fiber, antidiabetic	Fiber-enriched cake	[56]
37	*Mangifera indica* L.	Mango	Phenolic content	Biscuits	[57]
38	Trigonella foenum graecum L	Fenugreek	Saponins	Biscuits	[58]
39	Soybean (*Glycine max*), Mushroom	Mushroom, soy bean	Protein supplemented cereal snack	Biscuits	[59]
40	Multigrain flour	Bengal gram flour (BGF),germinated pearl millet	Reduced -calorie	Biscuits	[60]
41	*Spinacia oleracea* L.	Spinach	Antioxidant properties	Biscuits	[61]
42	*Spirulina platensis*	Spirulina	Antioxidant activity	Biscuits	[62]

Table 3.
The incorporation of various bioactive molecules from the medicinal plants in the functional food – biscuits.

4. Conclusion

The functional food acts as a bridge between the nutritional diet and healthy living. The importance of medicinal plant incorporation is well known by the public from the ancient time but they were unaware of it. Nowadays its innovative way of incorporation in many kinds of food by means of scientific approach in order to avoid all kinds of queries and confusions is well understood. Overall, the maintenance of healthy living through nutritional approach is by the consumption of medicinal plant incorporated functional food also by having a physical activity to prevent the diseases from consuming life and have a healthy weight gain to avoid unnecessary health issues and sufferings. The potentiality and consequences of functional food suggest many inspiring opportunities on the whole. As a segment of a varied diet, on regular basis at effective levels, the consumption of functional food is recommended to major nutrition-related chronic diseases. For effective strategy in health claims, the functional food acts as one of the part of it, to maximize health and minimize the disease risk. The successful cyclic promo of functional food is depicted in **Figure 5**. From the novel idea in theory form have to cross many barriers like evidence, reviews, marketing, and publications results in the final destination the prototype, is product development.

Figure 5.
Functional food creation to success – a cyclic health promo. Source: https://www.slideshare.net/ektabelwal/development-of-nutraceuticals-functional-foods.

Acknowledgements

The authors wish to show their gratitude to the chancellor of Vellore Institute of Technology, Vellore, Tamilnadu, India for providing all the support to achieve the present research process.

Conflicts of interest

The authors declare no conflict of interest.

Funding

No external funding received for the research.

Author details

Ashwini Devaraj and Gayathri Mahalingam*
Department of Biotechnology, School of Bioscience and Technology, Vellore Institute of Technology, Vellore, India

*Address all correspondence to: gayathrigopinath@vit.ac.in

References

[1] Siro I, Kapolna E, Kapolna B, Lugasi A. Functional food. Product development, marketing and consumer acceptance — A review. Appetite. 2008;**51**:456-467

[2] Corbo MR, Bevilacqua A, Petruzzi L, Casanova FP, Sinigaglia M. Functional beverages: The emerging side of functional foods commercial trends, research, and health implications. Comprehensive Reviews in Food Science and Food Safety. 2014;**13**:1192-1206

[3] Institute of Medicine. Dietary Reference Intakes for Energy, Carbohydrate, Fiber, Fat, Fatty Acids, Cholesterol, Protein, and Amino Acids. Washington, DC: The National Academies Press; 2005. pp. 1-1331

[4] Skerrett PJ, Willett WC. Essentials of healthy eating: A guide. Journal of Midwifery & Women's Health. 2010;**55**(6):492-501

[5] WHO. Global report on diabetes. Vol. 58. World Health Organization; 2016

[6] Manjula K, Suneetha C. Designer foods-their role in preventing lifestyle disorders. International Journal of Science and Nature. 2011;**2**(4):878-882

[7] Institute of Medicine. Dietary Reference Intakes: The Essential Guide to Nutrient Requirements. Washington (DC): The National Academies Press; 2006. pp. 2-5

[8] IDF Diabetes Atlas. In: Karuranga S, Malanda B, Pouya Saeedi PS, editors. International Diabetes Federation. 9th ed. Brussels, Belgium: International Diabetes Federation; 2019

[9] Sinclair A, Saeedi P, Kaundal A, Karuranga S, Malanda B, Williams R. Diabetes and global ageing among 65 – 99-year-old adults: Findings from the international diabetes federation diabetes atlas, 9 th edition. Diabetes Research and Clinical Practice. 2020;**162**:108078

[10] Lindström J, Virtanen SM. Functional foods and prevention of diabetes. Functional Foods: Concept to Product: Second Edition. Woodhead Publishing Limited; 2011. 261-276 p.

[11] Ezuruike UF, Prieto JM. The use of plants in the traditional management of diabetes in Nigeria: Pharmacological and toxicological considerations. Journal of Ethnopharmacology. 2014;**155**(2):857-924

[12] Ramalingum N, Mahomoodally MF. The therapeutic potential of medicinal foods. Advances in Pharmacological Sciences. 2014;**2014**. Article ID: 354264

[13] Fehresti Sani M, Montasser Kouhsari S, Moradabadi L. Effects of three medicinal plants extracts in experimental diabetes: Antioxidant enzymes activities and plasma lipids profiles in comparison with metformin. Iranian Journal of Pharmaceutical Research. 2012;**11**(3):897-903

[14] Tiwari AK, Madhusudana RJ. Diabetes mellitus and multiple therapeutic approaches of phytochemicals: Present status and future prospects. Current Science. 2002;**83**(1):30-38

[15] Harborne JB. Phytochemical Methods. London New York: Chapman and Hall; 1973. pp. 1-278

[16] Monjiote DP, Leo EEM, Campos MRS. Functional and biological potential of bioactive compounds in foods for the dietary treatment of type 2 diabetes mellitus. In: Functional Food - Improve Health through Adequate Food. London: IntechOpen; 2017. pp. 143-163

[17] American Dietetic Association. Position of the American dietetic association: Functional foods. Journal of the American Dietetic Association. 2009;**109**(4):735-746

[18] Spence JT. Challenges related to the composition of functional foods. Journal of Food Composition and Analysis. 2006;**19**:2005-2007

[19] Mirmiran P, Bahadoran Z, Azizi F. Functional foods-based diet as a novel dietary approach for management of type 2 diabetes and its complications: A review. World Journal of Diabetes. 2014;**5**(3):267-281

[20] Cencic A, Chingwaru W. The role of functional foods, nutraceuticals, and food supplements in intestinal health. Nutrients. 2010;**2**:611-625

[21] Filipcev B, Simurina O, Bodroza-Solarov M. Quality of gingernut type biscuits as affected by varying fat content and partial replacement of honey with molasses. Journal of Food Science and Technology. 2012;**51**(11):3163-3171

[22] Kumar PKP, Manohar RS, Indiramma AR, Krishna AGG. Stability of oryzanol fortified biscuits on storage. Journal of Food Science and Technology. 2014;**51**(10):2552-2559

[23] Raymundo A, Fradinho P, Nunes MC. Effect of Psyllium fibre content on the textural and rheological characteristics of biscuit and biscuit dough. Bioactive Carbohydrates and Dietary Fibre. 2014;**3**(2):96-105

[24] Mousa R, Mousa A. Formulation of reduced calorie and trans-free fat biscuits using palm oil and sucralose: Study of their hypoglycemic activity on albino rats. American Journal of Food and Nutrition. 2015;**3**(6):131-140

[25] Karadbhajne SV, Kumbhare V, Bombaywala MA. Application of Cissus quadrangularis Linn. (Hadjor) in Nutraceutical food (bakery) products. International Journal of ChemTech Research. 2014;**6**(4):2263-2270

[26] Elaveniya E, Jayamuthunagai J. Functional, physicochemical and anti-oxidant properties of dehydrated banana blossom powder and its incorporation in biscuits. International Journal of ChemTech Research. 2014;**6**(9):4446-4454

[27] Devika I, Banu AT. Development of value added biscuit for people living with human immuno deficiency virus. International Journal of Pharma and Bio Sciences. 2016;**7**(2):390-394

[28] Nyam K, Leao S, Tan C, Long K. Functional properties of roselle (Hibiscus sabdariffa L.) seed and its application as bakery product. Journal of Food Science and Technology. 2014;**51**(12):3830-3837

[29] Poornima V, Jeyam M. Assessing the nutraceutical significance of the medicinal herb ammannia baccifera l. by proximate, mineral analysis and phytochemical screening. World Journal of Pharmaceutical and Medical Research. 2016;**2**(5):65-71

[30] Patil DD, Lal EA, Nandkule VD. Development and quality evaluation of garden cress seed biscuits. International Journal of Engineering, Science and Technology. 2015;**3**(3):770-774

[31] Surnis MSA, Patil PS, Jadhav RH. Extraction, isolation and quantification of bioactive compound (Fisetin) and its product formulation. International Journal of Engineering Research & Technology. 2016;**5**(08):56-58

[32] Sparvoli F, Laureati M, Pilu R, Pagliarini E, Toschi I, Giuberti G, et al. Exploitation of common bean flours with low antinutrient content for making nutritionally enhanced biscuits. Frontiers in Plant Science. 2016;**7**(6):928

[33] Stamataki NS, Nikolidaki EK, Yanni AE, Stoupaki M, Konstantopoulos P, Tsigkas A, et al. Evaluation of a High Nutritional Quality Snack Based on Oat Flakes and Inulin: Effects on Postprandial Glucose, Insulin and Ghrelin Responses of Healthy Subjects. Food & Function. 2016;**7**(7):3295-3303

[34] Erukainure OL, Ebuehi OAT, Adeboyejo FO, Okafor EN, Muhammad A, Elemo GN. Fiber-enriched biscuit enhances insulin secretion, modulates beta-cell function, improves insulin sensitivity, and attenuates hyperlipidemia in diabetic rats. PharmaNutrition. 2013;**1**(2):58-64

[35] Ng SH, Wan Rosli WI. Effect of cinnamon powder addition on nutritional composition, physical properties and sensory acceptability of butter biscuits. Malaysian Journal of Nutrition. 2014;**20**(2):245-253

[36] Yadav DN, Thakur N, Sunooj KV. Effect of de-oiled peanut meal flour (DPMF) on the textural, organoleptic and physico chemical properties of bread. Food and Nutrition Sciences. 2012;**20**(3):471-476

[37] Reddy V, Urooj A, Kumar A. Evaluation of antioxidant activity of some plant extracts and their application in biscuits. Food Chemistry. 2005;**90**(1-2):317-321

[38] Ho L-H, Abdul Latif NW. Binti. Nutritional composition, physical properties, and sensory evaluation of cookies prepared from wheat flour and pitaya (Hylocereus undatus) peel flour blends. Cogent Food & Agriculture. 2016;**2**(1):1-10

[39] Handa C, Goomer S, Siddhu A. Physicochemical properties and sensory evaluation of fructoligosaccharide enriched cookies. Journal of Food Science and Technology. 2012;**49**(2):192-199

[40] Peighambardoust S, Aghamirzaei M. Physicochemical , nutritional, shelf life and sensory properties of Iranian Sangak bread fortified with grape seed powder. Journal of Food Processing & Technology. 2014;**5**(10):1-5

[41] Raikos V, McDonagh A, Ranawana V, Duthie G. Processed beetroot (Beta vulgaris L.)as a natural antioxidant in mayonnaise: Effects on physical stability, texture and sensory attributes. Food Science and Human Wellness. 2016;**5**(4):191-198

[42] Azmi NS, Bhat R, Yeoh TK. Quality evaluation of novel cookies prepared by supplementing with fresh turmeric flower (Curcuma longa L.) extracts as a value added functional ingredient. International Food Research Journal. 2016;**23**(4):1514-1522

[43] Petrovic J, Pajin B, Tanackov-Kocic S, Pejin J, Fistes A, Bojanic N, et al. Quality properties of cookies supplemented with fresh brewer's spent grain. Food and Feed Research. 2017;**44**(1):57-63

[44] Varastegani B, Zzaman W, Yang TA. Investigation on physicochemical and sensory evaluation of cookies substituted with papaya pulp flour. Journal of Food Quality. 2015;**38**(3):175-183

[45] Tumbas Saponjac V, Cetkovic G, Canadanovic-Brunet J, Pajin B, Djilas S, Petrovic J, et al. Sour cherry pomace extract encapsulated in whey and soy proteins: Incorporation in cookies. Food Chemistry. 2016;**207**:27-33

[46] Moraes EA, de Souza Dantas MI, de Castro Morais D, da Silva CO, de Castro FAF, Martino HSD, et al. Aguiar SMRRE, Maria Ines de Souza Dantra SMRR. Sensory evaluation and nutritional value of cakes prepared with whole flaxseed flour. Food Science and Technology International: SAGE Journals. 2010;**30**(4):974-979

[47] Hassan AA, Rasmy NM, Foda MI, Bahgaat WK. Production of functional biscuits for lowering blood lipids. World Journal of Dairy & Food Sciences. 2012;**7**(1):1-20

[48] Aarti H. Quality evaluation of herbal juice developed from traditional Indian medicinal plants using Citrus limetta as base. Journal of Nutrition & Food Sciences. 2015;**5**(5):396-400

[49] Rathnayake ARMHA, Navarathna SB. Utilization of Moringa olifera leaves as a functional food ingredient in bakery industry. International Journal of Scientific Research. 2015;**6**(2):339-344

[50] Alam MA, Alam J, Hakim A, Huq AKO, Golam Moktadir SM. Development of fiber enriched herbal biscuits: A preliminary study on sensory evaluation and chemical composition. International Journal of Food Sciences and Nutrition. 2014;**3**(4):246-250

[51] Vitali D, Dragojević IV, Šebečić B. Effects of incorporation of integral raw materials and dietary fibre on the selected nutritional and functional properties of biscuits. Food Chemistry. 2009;**114**(4):1462-1469

[52] Sharma A, Zhou W. A stability study of green tea catechins during the biscuit making process. Food Chemistry. 2011;**126**(2):568-573

[53] Goel S, Kochar GK, Kaur T, Goel MS. Development and organoleptic evaluation of food preparations incorporated with selected antidiabetic medicinal plants. Ethno-Medicine. 2011;**5**(2):101-106

[54] Oyeyinka AT, Oyeyinka SA. Moringa oleifera as a food fortificant: Recent trends and prospects. Journal of the Saudi Society of Agricultural Sciences. 2016;**17**(2):127-136

[55] Lim JA, Lee JH. Quality and antioxidant properties of cookies supplemented with with black sesame powder. Journal of the Korean Society of Food Science and Nutrition. 2015;**44**(7):1058-1063

[56] Erukainure OL, Ebuehi OAT, Adeboyejo FO, Aliyu M, Elemo GN. Hematological and biochemical changes in diabetic rats fed with fiber-enriched cake. Journal of Acute Medicine. 2013;**3**(2):39-44

[57] Ashoush IS, Gadallah MGE. Utilization of mango peels and seed kernels powders as sources of phytochemicals in biscuit. World Journal of Dairy & Food Sciences. 2011;**6**(1):35-42

[58] El-azeem ASA, Hegazy AM, Afifi AA. Physiochemical, sensory and nutritional properties of corn-fenugreek flour composite biscuits. 2011;**5**(4):84-95

[59] Mohajan TF. Effect of incorporation of soy flour to wheat flour on nutritional and sensory quality of biscuits fortified with mushroom. Food Science & Nutrition. 2015;**3**(5):363-369

[60] Aggarwal D, Sabikhi L, Sathish Kumar MH. Formulation of reduced-calorie biscuits using artificial sweeteners and fat replacer with dairy-multigrain approach. NFS Journal. 2016;**2**:1-7

[61] Galla NR, Pamidighantam PR, Karakala B, Gurusiddaiah R, Akula S. Nutritional, textural and sensory quality of biscuits supplemented with spinach (Spinacia oleracea L.). International Journal of Gastronomy and Food Science. 2017;**7**(December 2016):20-26

[62] Singh P, Singh R, Jha A, Rasane P, Gautam AK. Optimization of a process for high fibre and high protein biscuit. Journal of Food Science and Technology. 2015;**52**(3):1394-1403

Chapter 2

Bioactive Molecules from Indian Medicinal Plants as Possible Candidates for the Management of Neurodegenerative Disorders

Uma Ranjan Lal and Snigdha Lal

Abstract

The present review gives an account of various bioactive molecules obtained from Indian medicinal plants for neurological degenerative disorders. Emphasis is laid on their correlation with the plants used in traditional system of medicine in India. The methodology involved in present review was enlisting of medicinal plants used for neurodegenerative disorders followed by their chemistry. A correlation with the chemical constituents and their recent findings has been done. Many medicinal plants such as *Aloe vera* and *Bacopa monnieri* have documented correlations and also need to be explored more. Molecules like garcinol **(34)**, which was originally an anti-cancer compound, have good correlation as neuroprotective agent. Likewise many plants that have not been explored but are used in traditional system of medicine have also been listed. *Jaggery* and honey, which are used in traditional formulations in large quantity, also have natural products that are used as neuroprotective agents. In conclusion, a lot more study is required to correlate the medicinal plants and herbal formulations to have much more natural products for neurodegenerative disorders.

Keywords: ayurvedic formulations, Indian medicinal plants, aloin, garcinol

1. Introduction

Neurodegeneration is defined as a complex process that involves progressive damage to the brain with neuronal loss leading to incurable and devitalizing consequences. The neurons, which get degenerated, are not restored, which result in impaired cognition and neurological disorders leading to depression, schizophrenia, Alzheimer's disease, dementia, epilepsy, cerebral ischemia, and parkinsonism [1, 2]. These diseases involve various characteristic pathological and molecular features affecting neurons in various regions of the brain. Neurodegenerative diseases are accelerated by the way we live our daily lives. As per recent report by Indian Council of Medical Research (ICMR), the proportion of deaths estimated due to lifestyle diseases has increased from 37.09% in 1990 to 61.8% in 2016 [3]. Thus, there is an immense need for the development of strategies for its prevention as a number of patients suffering from lifestyle diseases are increasing day by day. Ayurveda is considered an old age traditional medicine of Indian practice involving considerable usage of plants and herbal preparations, which are known to cure various disorders. Various classical formulations for neurodegenerative diseases are listed in Ayurvedic Formulary of India (AFI), which also provides a file of information regarding single drugs of plant, animal, and mineral origin, providing their official names and English equivalents for their easy identification [4]. In present review, the major plants listed in the formulations indicated as neuroprotective is discussed. Present review outlines the chemical constituents and pharmacology of the listed plants. The sole aim of present review is to extract the hidden potential ancient formulations supported by modern findings so as to enhance their acceptability in masses. Further, a systematic approach for their chemical standardization is a need as it will be supportive in identifying therapeutic molecules, stating the need for combining scientific interpretations and traditional knowledge.

2. Indian medicinal plants and neurodegenerative diseases

The present sections describe plant material along with their chemical constituents and related pharmacology pertaining to neurodegenerative disorders. Few molecules with prominent activity have also been discussed.

3. *Aloe vera* (L.) Burm.f.

A. vera is a succulent, evergreen, perennial herb, which is being cultivated all over the world mainly for the purpose of agricultural and medicinal uses. Anthraquinones (aloe-emodin (**1**), chrysophanol (**2**), and a bitter reddish-yellow exudate containing majorly anthrones (aloin A and B (**3**)), aloe-emodin (**4**) [5], and few aloe resins such as aloeresin (**5**) and aloenin (**6**)) [6] are the major constituents in aloe (**Figure 1**). Aloe-emodin **4** is reported to enhance cognition against scopolamine-induced cognitive impairment in mice [7]. Aloin **3** is shown to decrease intracellular ROS generation and reduced Ca^{2+} production, which is responsible for depolarization and death of the neuron, suggesting it as a useful and alternative therapy for cerebrovascular diseases [8]. *A. vera* gel demonstrated antioxidant and anti-inflammatory activities when given to the rats with sciatic nerve reperfusion injury [9].

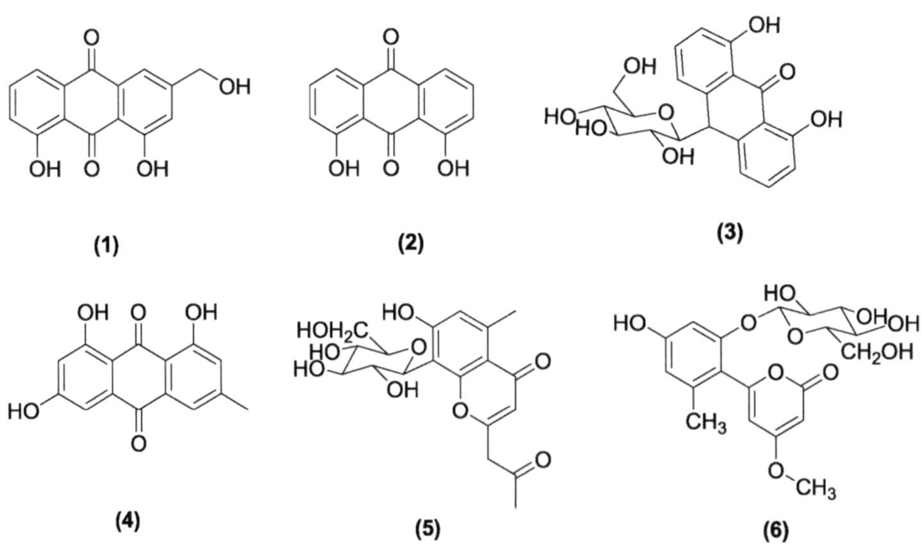

Figure 1.
Structures of compounds present in A. vera.

4. *Terminalia chebula* Retz

Terminalia chebula (Combretaceae) commonly called *Haritaki*, is used in ayurvedic formulations, also reported to demonstrate neuroprotective activities. The principle constituents in pericarp of *T. chebula* fruits are phenolics (chebulic acid (7), chebulagic acid (**8**), chebulinic acid (**9**), ellagic acid (**10**), gallic acid (**11**)) [10] and triterpenoid glycosides chebuloside II (**12**) (**Figure 2**) [11]. Pericarp of fruits of *T. chebula* constitutes *Triphala* (one of the popular formulations of ayurvedic system of medicine). The methanolic (70%) extract of the fruit *Fructus chebulae* has been shown to rescue cerebral ischemia by protecting neurons from degeneration. The in-vitro study and in-vivo studies have shown promising results [12]. Also, there is a promising decrease in the levels of malondialdehyde (MDA), NO, and microglial death stimulated by lipopolysaccharide (LPS) in the cells after treating with the extract [13]. The underlying mechanism might be the inhibition of inflammatory and oxidative processes. *T. chebula* constituents such as chebulagic acid, chebulinic acid, and ellagic acid have shown to be neuroprotective on various cell lines by showing its effect on various targets [14–16].

5. Honey and *Jaggery*

Both Honey and *Jaggery* are used in making of many ayurvedic preparations, and they constitute major part of many formulations [4]. Honey, a sweet and viscous substance, is produced by the honey bees (*Apis* spp.). It contains sugar and others such as minerals, proteins, essential oils, and flavonoids. Dietary polyphenols found in honey as well as other food and plant materials can prevent neurodegenerative disease in various ways [17–22]. These include oxidative protection of neurons [17], enhancement of neuronal function and regeneration [18], protection of neurons from Aβ-induced neuronal injury and neurotoxicity [19], protection of hippocampal cells against nitric oxide-induced toxicity [20], and modulation of neuronal and glial cell signaling pathways [21]. Luteolin (**13**) shows neuroprotective activity via prevention of microglia-associated inflammation in the hippocampus of aged rats [22]. Other flavonoids such as quercetin (**14**) and kaempferol (1) have also shown to be neuroprotective in various models. Kaempferol's (**15**) neuroprotective effect was confirmed by the histochemical findings where it prevented the loss of TH-positive neurons induced by MPTP [23–25]. Luteolin **13**, a flavone, has a significant role in treating CNS disorders through various mechanisms [26]. The activity of luteolin **13**

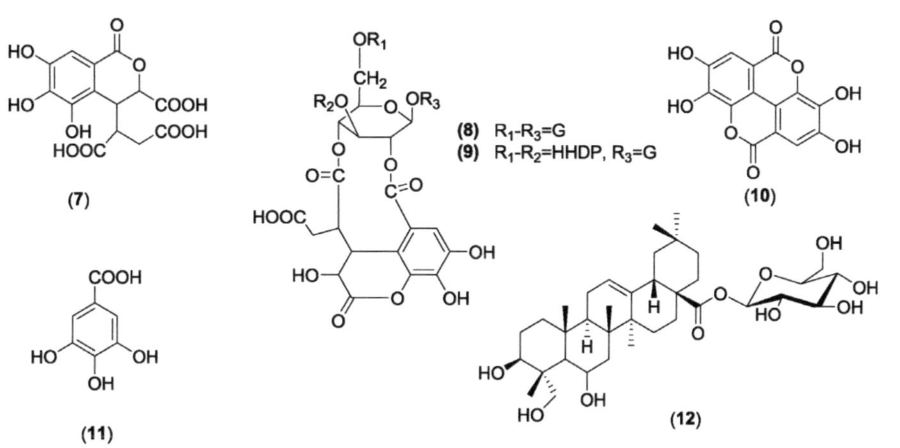

Figure 2.
Structures of compounds present in T. chebula.

was also reported at its microglial transcriptome level for the treatment of CNS disorders via anti-inflammatory activity [27]. Pretreatment of mice with luteolin **13** was expected to reduce the frequency of seizure in the PTZ mice model [28]. Luteolin **13** also has the potential to improve the activity of oxygen-glucose deprivation/reperfusion (OGD/R) induced neurons in a dose-dependent manner by enhancing the Na+/K+-ATPase activity suggesting a potential molecule to treat cerebral ischemia [29]. Apigenin (**16**), a trihydroxyflavone, was reported to inhibit the release of glutamate from hippocampal nerve terminals, which might be useful in treating epilepsy [30]. The role of apigenin (**16**) in the treatment of cognition was demonstrated by attenuating the Aβ-induced cytotoxicity in rat cortical neurons but having no intervention with oxidative stress. The biological activities of the flavonoids might be due to hydroxyl groups at R2 and R3 position influencing various cellular events eventually leading to apoptosis [27–30]. Other phenolics such as ferulic acid (**17**) exert neuroprotective effect through middle cerebral artery occlusion [31] and by decreasing the number of microglia/macrophages after cerebral ischemia/reperfusion injury in rats [32]. Chlorogenic acid (**18**) present in honey exerts a neuroprotective effect against methyl mercury-induced apoptosis in pheochromocytoma-12 (PC12) cell lines. It prevents the generation of reactive-oxygen species (ROS), suppressing the decreasing action of glutathione peroxidase (GPx) and GSH and attenuating apoptosis by the activation of caspase-3 [33] and also inhibits the activity of acetylcholine esterase and MDA in the hippocampus as well as in the frontal cortex in mice [34]. Honey as such has also shown to be neuroprotective in male Sprague-Dawley rats, and there was a decrease in the thiobarbituric acid reactive substance levels in the rats, which were pretreated with honey [35]. Chrysin (**19**) improved the morphological integrity of nigrostriatal neurons and increased the endogenous levels of BDNF, S100B, NGF, and GDNF in mice striatum improving behavioral and better muscular coordination [36]. Its lipophilic nature and bioflavonoid nucleus confers an added advantage to grow as a compelling therapeutic agent for neurological disorders. Pinocembrin

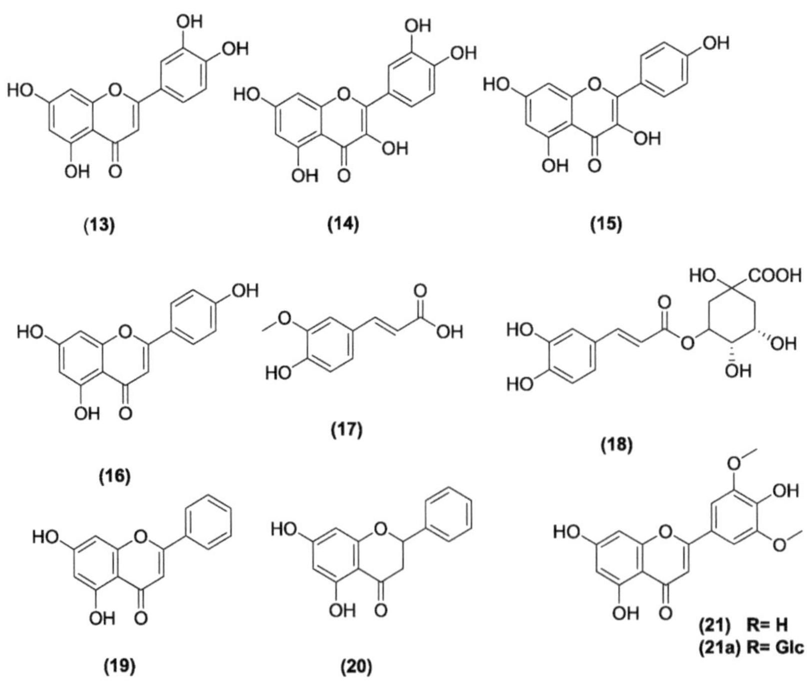

Figure 3.
Structures of compounds present in Jaggery *and honey.*

(20) elevated glutamate level, which was an important excitatory neurotransmitter produced by global cerebral I/R [37].

Jaggery is used in most of the ayurvedic formulations [4]. Traditionally, it is indicated for treating neurodegenerative disorders. The chemistry of sugar cane reveals that it is rich in phenolics and their glycosides [38], flavonoid glycosides, and flavones [39]. Tricin-7-glucoside **(21)** was reported to possess neuroprotective activity against cerebral ischemia via reduced expression of NF-κB and HMGB1. SH-SY5Y cells, which were pretreated with tricin **(21a)**, reduced the apoptosis induced by OGD. Attenuation of histopathological damage reduced brain edema was reported in animal models of ischemia along with reduced NF-κB and HMGB1 expression [40]. It also has potential for Alzheimer's disease [41] (**Figure 3**).

6. *Bacopa monnieri* L.

Bacopa monnieri L. (Scrophulariaceae) is notable ayurvedic medicinal plant and referred to have impact on brain function and is a neural tonic to improve intelligence and cognition [42]. Normal formulation Illumina® containing *B. monnieri* (leaf dry extract 43%, containing bacoside, 20%) decreased of 8-iso-PFG2α and ROS/RNS generation in the rat brain, consequently, decreasing the basal and H_2O_2 and amyloid β peptide-incited oxidative stress [43]. In another study, standardized extract of *B. monnieri* (Bacoside A **(22)** content 82 ± 0.5%) when given in the doses of 5 and 10 mg/kg, orally, shows a dose-related increment in superoxide dismutase, catalase, glutathione peroxidase activities in the frontal cortex, striatum, and hippocampus [44]. *B. monnieri*, when administered orally, improves cognitive impairment and neurodegeneration in rats. Immunohistological recognition of superoxide dismutase (Cu/Zn-SOD) and histopathological changes in the CA1 region of the hippocampus were observed [45]. An extract containing 5% (w/w) of the saponins, which contains bacoside A3, bacopasaponin X, bacopasaponin C **(23)**, bacopaside II **(24)**, and bacopaside I **(25)**, shows improvements in cognitive abilities and neuroprotective impacts in Alzheimer's disease model [46, 47]. When rats treated with bacoside A (22), huge changes were found in the degrees of both nonenzymatic and enzymatic antioxidants, which recommends that bacoside A **22** improves *B. monnieri* antioxidant status in rat brain [48]. Other saponins such as **(26-28)** reported from this plant have been shown in **Figure 4**. A survey of 10 years of research at Swinburne University done by Con Stough et al. recommends that an extract of *B. monnieri* (CDRI 08: KeenMind) is a safe and effectual cognitive enhancer [49]. *B. monnieri* extract suppressed the generation of free radical levels and indicated critical assurance against 3-nitropropionic acid (3-NPA)-mediated cytotoxicity in dopaminergic (N27 cell lines) [50]. *B. monnieri* standard extract containing 55.34% of bacosides indicated a defensive impact on ischemia-induced memory hindrance and diminished the infarct size in the ischemic brain. It additionally demonstrated a critical increment in catalase action and exhaustion in lipid peroxidation, nitrite, and nitrate activity [51]. Neuronal cell cultures when treated with *B. monnieri* extract shielded neurons from β-amyloid-induced cell toxicity. The extract gave protection to cell cultures against glutamate-induced excitotoxicity since it was not able to repress glutamate-mediated toxicity [52]. In Alzheimer's disease animal model (C57/Bl6 mice), *B. monnieri* extract has decreased lipoxygenase action and hydrogen peroxide-induced lipid peroxidation. *B. monnieri* has neuroprotective mechanisms, and it lessens β-amyloid deposits in the brain of C57/Bl6 mice Alzheimer's disease animal model [53]. *B. monnieri* extract have also shown to have protective effect against rotenone induced Parkinson's disease in PC-12 cell lines [54].

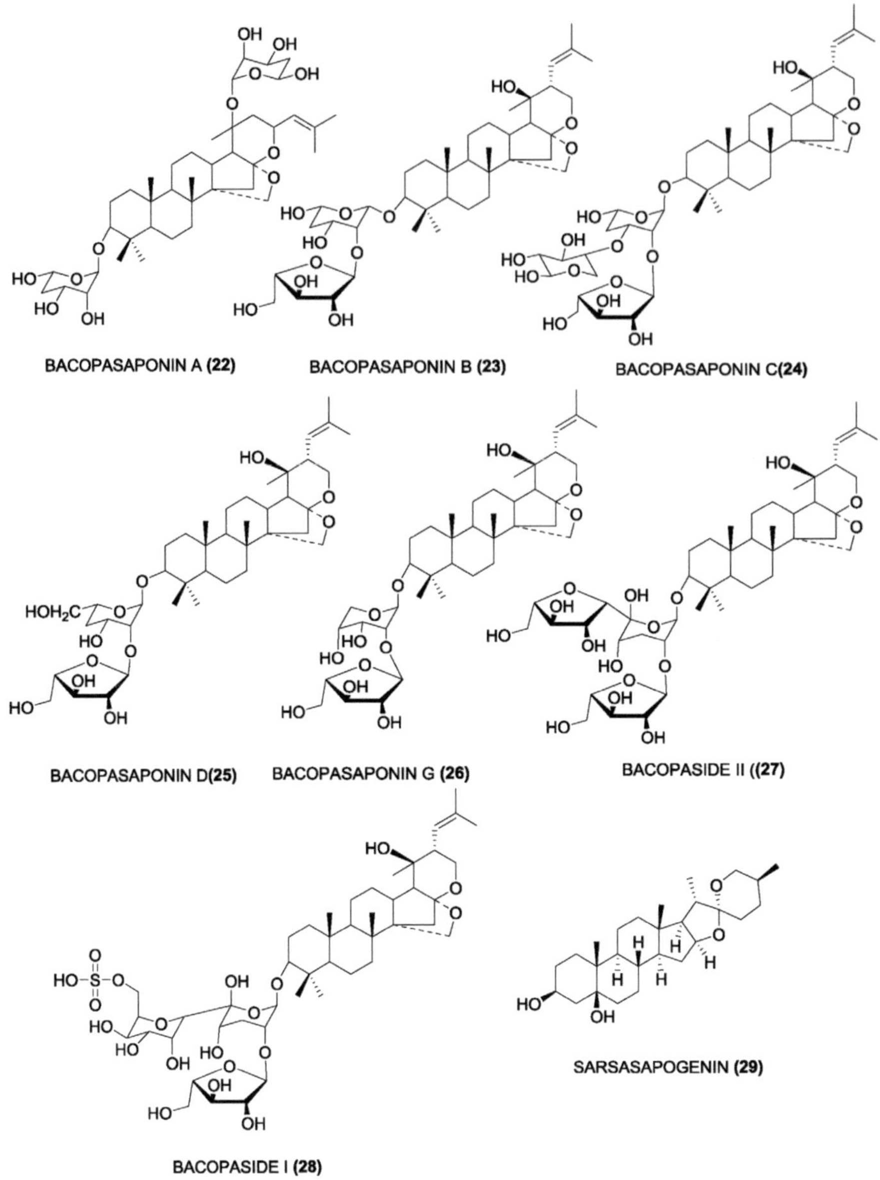

Figure 4.
Structures of compounds present in B. monnieri.

7. *Asparagus racemosus* Willd.

Methanolic extract of *Asparagus racemosus* decreased the degrees of cytokines, malondialdehyde (as a marker of lipid peroxides formation), and nitric oxide with a critical increment in the levels of catalase, superoxide dismutase, and glutathione, concluding its neuroprotective activity [55]. Its root extract (100 mg/kg) cured region-specific neurodegeneration created in Swiss albino mice [56] and also demonstrated dose-dependent improvement in memory after histochemical and behavioral studies. A noteworthy decrease in transfer latency time and a significant increment in acetylcholinesterase (AchE) staining in histochemical identification were observed suggesting antioxidant, cholinergic, and neuroprotective properties of *A. racemosus* [57]. EuMil, a formulation containing standardized extracts *A. racemosus* Willd., has been used for stress-related problems and has been found to re-establish modified degree of nor-adrenaline, 5-hydroxytryptamine, and dopamine normal (100 mg/kg, p.o. 14 days) [58]. *Withania somnifera* and *A. racemosus* together have shown noteworthy impacts in cell viability test, lactic dehydrogenase, malondialdehyde, glutathione disulfide, glutathione, nerve growth factor, pro-brain-derived growth factor levels, and reactive oxygen species generation [59]. Ovariectomized adult female Wistar rats showed noteworthy upregulation of estrogen receptors (ERα and ERβ) in hippocampus and frontal cortex area alongside enhancement in the levels of brain-derived neurotrophic factor. Upregulation of estrogen receptors and enhancement in the levels of brain-derived neurotrophic factor can be filled in as proof for neuroprotective impact of ethanolic concentrate of *A. racemosus* roots [60, 61]. Significant protection is seen after the supplementation of Mentat (BR-16A) is an herbal psychotropic preparation, containing *A. racemosus* against ethanol withdrawal-induced decrease of pentylenetetrazole threshold in rats and mice [62]. Sarsasapogenin, a steroidal saponin from *A. racemosus,* has been studied for neuroprotective impact in Alzheimer's disease. Sarsasapogenin **(29)** indicated noteworthy restraint of butyrylcholinesterase, monoamine oxidase-B, beta-secretase 1, and acetylcholinesterase, key enzymes related to the pathogenesis of Alzheimer's disease. At the point when tested against Aβ42 and H_2O_2-interceded cytotoxicity, sarsasapogenin showed a huge neuroprotective impact on PC12 cells. These discoveries recommended that sarsasapogenin can go about as a multi-target directed ligand and as a reasonable lead compound for treating different elements engaged with the pathogenesis of Alzheimer's disease [63]. Alterations in the normal levels of neurotransmitters, glutamate, acetylcholinesterase, dopamine, and protein because of worldwide cerebral ischemia were normalized by standardized *A. racemosus* Willd. root methanolic extract, appeared by the abatement in the degree of glutamate, acetylcholinesterase, and increment in dopamine levels and protein levels. Group treated with methanolic root extract of *A. racemosus* Willd. shown 85% neuronal protection in the CA1 area of the hippocampus. This demonstrated the cerebroprotective role of *A. racemosus* Willd. [64].

8. *Foeniculum vulgare* Mill.

Foeniculum vulgare extract reduced the amnesic effect and memory deficits in mice induced due to aging. *F. vulgare* extract demonstrated inhibition of acetylcholine, and in the exteroceptive behavioral model, it increased the step-down latency in mice significantly [65]. Fennel essential oil inhalation inhibits beta-amyloid (1-42)-induced depression and anxiety and also indicates that it may have further clinical applications [66]. There were improvements in Parkinson's disease in the animal model produced by *F. vulgare* Mill. essential oil [67]. Clinically fennel supplementation to obese middle-aged women decreased bodyweight, reduction in serum Aβ protein along with improvements in cognitive functions and metabolic profiling [68]. It normalized the expression levels of oxidative stress markers [Superoxide dismutase and Peroxiredoxin-6 (Prdx6)] and APP isoforms (APP common, 770 and 695) and also improved the Pb-induced morphological deterioration of cortical neurons [69].

9. *Azadirachta indica* A. Juss.

Commonly known as *Neem* in Indian subcontinent, *Azadirachta indica* has been shown to attenuate cisplatin-induced neurotoxicity in rats, and it also had neuroprotective effect on cerebral post-ischemic reperfusion and hypoperfusion [70, 71]. *A. indica* extracts have shown to be anti-oxidative and anti-apoptotic neuroprotective in Parkinson-induced functional damage [72]. *A. indica* standardized leaf extract (total bitters 4.3%) has shown to be neuroprotective in partial sciatic nerve injury in rats as evidenced from anti-inflammatory, antioxidant, and anti-apoptotic studies [73].

10. *Picrorhiza kurroa* Royle ex Benth

Picrorhiza kurroa, an ayurvedic herb, has been shown to potentiate photochemotherapy in vitiligo [74]. Apocynin **(30)** from *P. kurroa* has shown to be neuroprotective *in vivo* [75]. It also shows protective effect in a mouse model of chemically induced colitis [76]. It (4-hydroxy-3-methoxy-acetophenone, **30**) mediates long-lasting memory recovery, helps in hippocampal neuroprotection, and reduces glial cell activation after transient global cerebral ischemia in rats [77]. Interestingly, the concentration of picrosides I **(31)** and II **(32)** and apocynin **30** (iridoid rich fraction) in plasma (C_{max}) was found to be 244.9, 104.6, and 502 ng/ml with half-life ($t_{1/2}$) 14, 8, and 6 h, respectively [78]. *P. kurroa* also prevents memory deficits by inhibiting NLRP3 Inflammasome Activation and BACE1 Expression in 5xFAD Mice [79]. Acylated iridoid glycosides with hyaluronidase inhibitory activity from the rhizomes of *P. kurroa* Royle ex Benth have recently been isolated [80]. Therapeutic potentials of plant iridoids in Alzheimer's and Parkinson's diseases have separately been reviewed recently [81].

11. *Berberine* from *Berberis aristata* DC

Berberine **(33)** has been shown to be neuroprotective through the Nrf2 upregulation and also alleviates rotenone-induced cytotoxicity by antioxidation and activation of PI3K/Akt signaling pathway in SH-SY5Y cells [82]. Neuroprotective effects of berberine have also been confirmed in animal models of Alzheimer's disease [83]. Berberine nanoparticles have shown protective effect against LPS-induced neurodegenerative changes [84]. Berberine confers neuroprotection in coping with focal cerebral ischemia by targeting inflammatory cytokines [85]. Berberine had also shown protective effect against the altered intrinsic properties of the CA1 neurons induced by Aβ neurotoxicity [86]. There are many pathways by which berberine acts to protect neurons and has recently been reviewed. Authors have concluded that it being a potential candidate for combating neurodegenerative diseases [87, 88].

12. Garcinol from *Garcinia indica* Choisy

Garcinol **(34)** is one of the major constituents of *Garcinia indica* (a plant found in the region of Western Ghats of India). Garcinol can effectively restore the balance between the neurotransmitters glutamate and the γ-aminobutyric acid (GABA), rescue neural precursor cells, and promote their rapid growth. It regulates the expressions of glutamic acid decarboxylase 65 and GABAA receptors, preventing hyperactivation of NMDA receptor and the resultant excitotoxicity. It enhances memory and cognition in C57BL/6 mice, significantly lowering epileptic seizure scores [89, 90]. It also serves as a strong inhibitor of histone acetyltransferases (HAT), thus contributing protection against rapid neurodegeneration in parkinsonian brain [91, 92]. Garcinol helps in restoration of dopamine potency and has homocysteine lowering ability as well as have been shown to counter LDOPA-induced dyskinesia in PD model, demonstrate its worthiness as a potential drug candidate against Parkinson's Disease [93, 94]. Garcinol also exhibits desirable anti-cholinesterase properties by inhibiting the enzyme acetyl

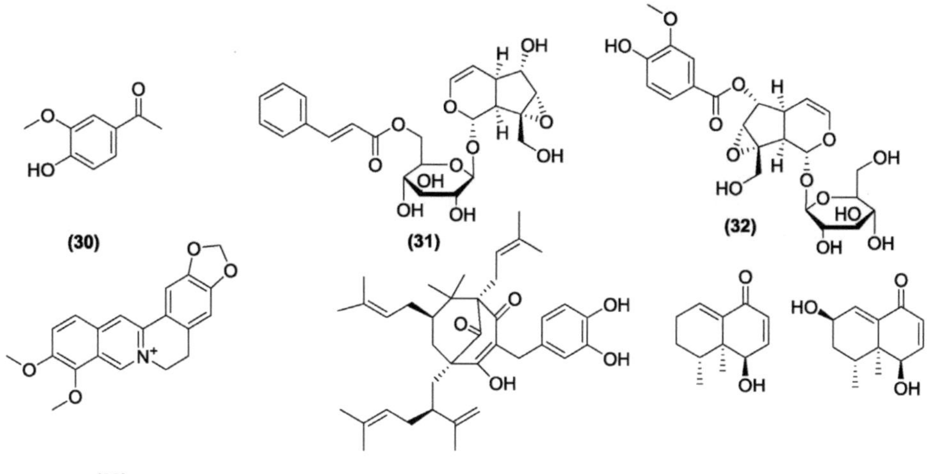

Figure 5.
Structures of compounds present in various plants for neuroprotective activity.

cholinesterase with an IC50 value of 0.66 μM, and it improves the neuronal count in hippocampal regions following administration of pentylenetetrazole (PTZ). In cultured rat cortical progenitor cells, garcinol can reduce cell death associated with growth factor deprivation. It also promotes neurite outgrowth in epidermal growth factor-responsive neural precursor cells and supports the survival of neurons [90]. In unilaterally 6-hydroxydopamine (6-OHDA)-lesioned hemi parkinsonian mice, 5 mg/kg of garcinol co-treatment with L-DOPA effectively controlled the axial, limb, and orofacial (ALO) score for dyskinesia analysis. Following the administration of garcinol, a decreased expression of c-Fos, FRA-2, and ARC genes has been visualized [94], which is usually over-activated in L-DOPA induced dyskinesia [95]. Moreover, methanolic extract of *G. indica*, which is known to contain garcinol, effectively elevates dopamine level in the striatum and confers neuroprotection to dopaminergic neurons in 6-OHDA lesioned experimental rats [96]. Even pharmacological inhibition of HATs by garcinol can notably suppress MPP+-induced cell death due to reduction in ATP content [97]. *In silico* studies on the molecular interaction between garcinol and the active sites of COMT and MAO-B revealed that garcinol can potentially inhibit the activity of the two enzymes, similar to their known inhibitors [93]. In conclusion, garcinol may prove to be a dependable remedial measure in PD therapeutics since its inhibition of MAO-B and COMT can be correlated to increase the availability of dopamine as well as prevent the generation of toxic dopamine metabolites including homocysteine, 3-omethyldopa, 3-methoxytyramine, and 3,4-dihydroxyphenylacetaldehyde [98] (**Figure 5**).

13. *Nardostachys jatamansi* (D. Don)DC.(Jatamansi)

The plant *Nardostachys jatamansi* is well exploited in Ayurvedic system of medicine for its role in neurological disorders. Various phytochemicals have been reported from this plant by various authors. Its sedative and antidepressant activity, its mechanism (inhibition of MAO and GABA), and its role in rat cerebral ischemia have been documented [99]. Anticonvulsant activity and neurotoxicity profile of the plant have been generated [100]. The extracts of *N. jatamansi* have been found to attenuate 6-hydroxydopamine-induced parkinsonism in rats, as proven by behavioral, neurochemical, and immunohistochemical studies [101]. It also improves learning and memory in mice. It also has stress modulating antioxidant effect. Its formulation was also found to enhance the learning and memory process in rats [102–104]. It has shown neuroprotective efficacy in conjunction with selenium in cognitive impairment [105]. *N. jatamansi* root extract was found to modulate the growth of IMR-32 and SK-N-MC neuroblastoma cell lines through MYCN-mediated regulation of MDM2 and p53 [106]. Novel Sesquiterpeniods and Anti-neuroinflammatory metabolites from *N. jatamansi* have also been isolated [107]. Compounds such as Desoxo-narchinol A **35** and Narchinol B **36** isolated from *N. jatamansi* have been shown to exert anti-neuroinflammatory effects by upregulating nuclear transcription factor erythroid-2-related factor 2/heme oxygenase-1 signaling [108].

14. Other plants

Many plants, which have been used in traditional formulations for neurological disorders, need to be phytochemically explored and correlated with the modern findings. **Table 1** gives the list of those plants with reported neuroprotective activity.

Plant name	Part used	Reports on neuroprotective activity	Reference
Fumaria indica (Hausskn.) Pugsley	Leaf	Significant activity of ethanolic extract on rat cognitive dysfunctions. Potential antianxiety activity of leaf extract; preclinical study	[109, 110]
Alhagi pseudalhagi (M. Bieb) Desv. ex B. Keller & Shap.	Whole plant	Traditionally used for neuroprotective disorders. Compounds having neuroprotective activity like flavanone glycosides and alkaloids like β-phenethylamine and tetrahydroisoquinoline have been reported	[111–114]
Pluchea lanceolata (Oliver & Hiern.)	Leaf	Protection of hippocampal neurons from endothelin-1 induced ischemic injury to ameliorate cognitive deficits; protective effect against aluminum chloride-induced neurotoxicity in Swiss Albino mice; protective effect on LPS-induced neuro-inflammation in C6 rat glial cells	[115–117]
Premna mucronata Roxb.	Whole plant	Luteolin and apigenin are reported, and they are reported to be neuroprotective	[118]
Semecarpus anacardium L.f.	Fruits	Stress induced neuroprotective activity	[119]
Sida cordifolia L.	Whole plant	Ameliorative effect in parkinsonism	[120]
Tinospora cordifolia (Thunb.) Miers.	Stems	Suppresses neuro-inflammation in Parkinsonian Mouse Model; potential neuro-regenerative candidate against glutamate induced excitotoxicity: an *in vitro* perspective	[121–123]
Trichosanthes dioica Roxb.	Rhizome	Neuropharmacological properties of root	[124]
Strobilianthes ciliatus (Nees.)		No such reports	—

Table 1.
List of plants whose exploration is required, as they are used tremendously in traditional system of medicine in India.

15. Conclusions

In present review we focus on evidence to prevent neurodegenerative disorders in various studies (*in vitro* and *in vivo*). The mentioned medicinal plants play their protective roles *via* increased SOD and catalase levels, restoration of GSH, and decreased MDA levels and also protect neurons against ROS as antioxidant activities. The neuroprotective effects of the mentioned plants occur *via* reduction of inflammatory cytokines as well as enhancement of anti-inflammatory cytokines, inhibition of the acetylcholinesterase activity, and decreased MDA levels in the neural system via modulating GABAergic and glutamatergic neurons and also increasing amount of amino acids and serotonin (5-HT) in the neurotransmitters systems. Based on the evidence produced, it is suggested that more exploration of traditional formulations is required, and also repurposing in natural products is also required to a greater extent, as is evident from recent reports on garcinol and berberine.

Conflict of interest

Authors do not have any conflict of interest.

Author details

Uma Ranjan Lal[1*] and Snigdha Lal[2]

1 Independent Researcher, Bariatu, Ranchi, Jharkhand, India

2 Ram Lakhan Singh Yadav College, Bettiah, Bihar, India

*Address all correspondence to: urlalvv@yahoo.com

References

[1] Mattson MP. Metal-catalyzed disruption of membrane protein and lipid signaling in the pathogenesis of neurodegenerative disorders. Annals of the New York Academy of Sciences. 2004;**1012**:37-50

[2] Saxena S, Caroni P. Selective neuronal vulnerability in neurodegenerative diseases: From stressor thresholds to degeneration. Neuron. 2011;**71**:35-48

[3] Available from: https://pib.gov.in/Pressreleaseshare [Accessed: 31 January 2020]

[4] Anonymous. Ayurvedic Formulary of India, (part I, II, III). Government of India, Ministry of Health and Family Welfare Department of Indian Systems of Medicine and Homoeopathy; 2011

[5] Saleem R, Faizi S, Deeba F, Siddiqui BS, Qazi MH. Anthrones from *Aloe barbadensis*. Phytochemistry. 1997;**45**:1279-1282. DOI: 10.1016/S0031-9422(97)00107-6

[6] Speranza G, Gramatica P, Dadá G, Manitto P. Aloeresin c, a bitter *C,O*-diglucoside from cape aloe. Phytochemistry. 1985;**24**:1571-1573. DOI: 10.1016/S0031-9422(00)81068-7

[7] Tao L, Xie J, Wang Y, Wang S, Wu S, Wang Q, et al. Protective effects of aloe-emodin on scopolamine-induced memory impairment in mice and H_2O_2-induced cytotoxicity in PC12 cells. Bioorganic & Medicinal Chemistry Letters. 2014;**24**:5385-5389. DOI: 10.1016/j.bmcl.2014.10.049

[8] Chang R, Zhou R, Qi X, Wang J, Wu F, Yang W, et al. Protective effects of aloin on oxygen and glucose deprivation-induced injury in PC12 cells. Brain Research Bulletin. 2016;(121):75-83. DOI: 10.1016/j.brainresbull.2016.01.001

[9] Guven M, Gölge UH, Aslan E, Sehitoglu MH, Aras AB, Akman T, et al. The effect of *Aloe vera* on ischemia-reperfusion injury of sciatic nerve in rats. Biomedicine & Pharmacotherapy. 2016;**79**:201-207. DOI: 10.1016/j.biopha.2016.02.023

[10] Lee DY, Kim HW, Yang H, Sung SH. Hydrolyzable tannins from the fruits of *Terminalia chebula* Retz and their α-glucosidase inhibitory activities. Phytochemistry. 2017;**137**:109-116. DOI: 10.1016/j.phytochem.2017.02.006

[11] Kundu AP, Mahato SB. Triterpenoids and their glycosides from *Terminalia chebula*. Phytochemistry. 1993;**32**:999-1002. DOI: 10.1016/0031-9422(93)85243-K

[12] Gaire BP, Kim H. Neuroprotective effects of Fructus Chebulae extracts on experimental models of cerebral ischemia. Journal of Traditional Chinese Medicine. 2014;**34**:69-75. DOI: 10.1016/s0254-6272(14)60057-1

[13] Gaire BP, Jamarkattel-Pandit N, Lee D, Song J, Kim JY, Park J, et al. *Terminalia chebula* extract protects OGD-R induced PC12 cell death and inhibits LPS induced microglia activation. Molecules. 2013:3529-3542. DOI: 10.3390/molecules18033529

[14] Kim HJ, Kim J, Kang KS, Lee KT, Yang HO. Neuroprotective effect of chebulagic acid via autophagy induction in SH-SY5Y cells. Biomolecules & Therapeutics. 2014;**22**:275-281. DOI: 10.4062/biomolther.2014.068

[15] Shen YC, Juan CW, Lin CS, Chen CC, Chang CL. Neuroprotective effect of *Terminalia chebula* extracts and ellagic acid in Pc12 cells. African Journal of Traditional, Complementary and Alternative Medicines (AJTCAM). 2017;**14**:22-30. DOI: 10.21010/ajtcam.v14i4.3

[16] Song JH, Shin MS, Hwang GS, Oh ST, Hwang JJ, Kang KS. Chebulinic acid attenuates glutamate-induced HT22 cell death by inhibiting oxidative stress, calcium influx and MAPKs phosphorylation. Bioorganic & Medicinal Chemistry Letters. 2018;**28**:249-253. DOI: 10.1016/j.bmcl.2017.12.062

[17] Cirmi S, Ferlazzo N, Lombardo GE, Ventura-Spagnolo E, Gangemi S, Calapai G, et al. Neurodegenerative diseases: Might *Citrus* flavonoids play a protective role? Molecules. 2016;**21**:1312

[18] Inanami O, Watanabe Y, Syuto B, Nakano M, Tsuji M, Kuwabara M. Oral administration of (−) catechin protects against ischemia-reperfusion-induced neuronal death in the gerbil. Free Radical Research. 1998;**29**:359-365

[19] Youdim KA, Joseph JA. A possible emerging role of phytochemicals in improving age-related neurological dysfunctions: A multiplicity of effects. Free Radical Biology & Medicine. 2001;**30**:583-594

[20] Vauzour D, Vafeiadou K, Rice-Evans C, Williams RJ, Spencer JP. Activation of pro-survival Akt and ERK1/2 signalling pathways underlie the anti-apoptotic effects of flavanones in cortical neurons. Journal of Neurochemistry. 2007;**103**:1355-1367

[21] Jang S, Dilger RN, Johnson RW. Luteolin inhibits microglia and alters hippocampal-dependent spatial working memory in aged mice. The Journal of Nutrition. 2010;**140**:1892-1898

[22] Xu B, Li X-X, He G-R, Hu J-J, Mu X, Tian S, et al. Luteolin promotes long-term potentiation and improves cognitive functions in chronic cerebral hypoperfused rats. European Journal of Pharmacology. 2010;**627**:99-105

[23] Hu P, Wang M, Chen W-H, Liu J, Chen L, Yin S-T, et al. Quercetin relieves chronic lead exposure-induced impairment of synaptic plasticity in rat dentate gyrus *in vivo*. Naunyn-Schmiedeberg's Archives of Pharmacology. 2008;**378**:43-51

[24] Kumar A, Sehgal N, Kumar P, Padi S, Naidu P. Protective effect of quercetin against ICV colchicine-induced cognitive dysfunctions and oxidative damage in rats. Phytotherapy Research. 2008;**22**:1563-1569

[25] Li S, Pu X-P. Neuroprotective effect of kaempferol against a 1-methyl-4-phenyl-1, 2, 3, 6-tetrahydropyridineinduced mouse model of Parkinson's disease. Biological & Pharmaceutical Bulletin. 2011;**34**:1291-1296

[26] Ashaari Z, Hadjzadeh MAR, Hassanzadeh G, Alizamir T, Yousefi B, Keshavarzi Z, et al. The flavone luteolin improves central nervous system disorders by different mechanisms: A review. Journal of Molecular Neuroscience. 2018;**65**:491-506. DOI: 10.1007/s12031-018-1094-2

[27] Dirscherl K, Karlstetter M, Ebert S, Kraus D, Hlawatsch J, Walczak Y, et al. Correction: Luteolin triggers global changes in the microglial transcriptome leading to a unique anti-inflammatory and neuroprotective phenotype. Journal of Neuroinflammation. 2010;**7**:1-2. DOI: 10.1186/1742-2094-7-3

[28] Hosseinzadeh H, Nassiri-Asl M. Neuroprotective effects of flavonoids in epilepsy. In: Brahmchari G, editor. Neuroprotective Natural Products Clinical Aspects Mode of Action, Chapter 10. Weinheim, Germany : Wiley-VCH; 2017. pp. 279-291. ISBN: 978-3-527-34186-3

[29] Fang L, Zhang M, Ding Y, Fang Y, Yao C, Zhang X. Protective effects of luteolin on neurons against oxygen-glucose deprivation/reperfusion injury via improving Na+/K+-ATPase activity. Zhongguo Zhongyao Zazhi.

2010;**35**:1051-1054. DOI: 10.4268/cjcmm20100825

[30] Kang SS, Lee JY, Choi YK, Kim GS, Han BH. Neuroprotective effects of flavones on hydrogen peroxide-induced apoptosis in SH-SY5Y neuroblostoma cells. Bioorganic & Medicinal Chemistry Letters. 2004;**14**:2261-2264. DOI: 10.1016/j.bmcl.2004.02.003

[31] Koh P-O. Ferulic acid prevents the cerebral ischemic injury-induced decrease of Akt and bad phosphorylation. Neuroscience Letters. 2012;**507**:156-160

[32] Cheng C-Y, Su S-Y, Tang N-Y, Ho T-Y, Chiang S-Y, Hsieh C-L. Ferulic acid provides neuroprotection against oxidative stress-related apoptosis after cerebral ischemia/reperfusion injury by inhibiting ICAM-1 mRNA expression in rats. Brain Research. 2008;**1209**:136-150

[33] Li Y, Shi W, Li Y, Zhou Y, Hu X, Song C, et al. Neuroprotective effects of chlorogenic acid against apoptosis of PC12 cells induced by methylmercury. Environmental Toxicology and Pharmacology. 2008;**26**:13-21

[34] Kwon S-H, Lee H-K, Kim J-A, Hong S-I, Kim H-C, Jo T-H, et al. Neuroprotective effects of chlorogenic acid on scopolamine-induced amnesia via anti-acetylcholinesterase and anti-oxidative activities in mice. European Journal of Pharmacology. 2010;**649**:210-217

[35] Shafika N, Sairazi M, Sirajudeen KNS, Asari MA, Mummedy S, Muzaimi M, et al. Effect of tualang honey against KA-induced oxidative stress and neurodegeneration in the cortex of rats. BMC Complementary and Alternative Medicine. 2017;**17**(1):31. DOI: 10.1186/s12906-016-1534-x

[36] Goes ATR, Jesse CR, Antunes MS, Lobo Ladd FV, Lobo Ladd AAB, Luchese C, et al. Protective role of chrysin on 6-hydroxydopamine-induced neurodegeneration a mouse model of Parkinson's disease: Involvement of neuroinflammation and neurotrophins. Chemico-Biological Interactions. 2018;**279**:111-120. DOI: 10.1016/j.cbi.2017.10.019

[37] Saad MA, Salam RMA, Kenawy SA, Attia AS. Pinocembrin attenuates hippocampal inflammation , oxidative perturbations and apoptosis in a rat model of global cerebral ischemia reperfusion. Pharmacological Reports. 2015;**67**:115-122. DOI: 10.1016/j.pharep.2014.08.014

[38] Palla G. Isolation and identification of phenolic glucosides in liquid sugars from cane molasses. Journal of Agricultural and Food Chemistry. 1982;**30**:764-766. DOI: 10.1021/jf00112a035

[39] Colombo R, Yariwake JH, Queiroz EF, Ndjoko K, Hostettmann K. On-line identification of minor flavones from sugarcane juice by LC/UV/MS and post-column derivatization. Journal of the Brazilian Chemical Society. 2009;**20**:1574-1579. DOI: 10.1590/S0103-50532009000900003

[40] Jiang W, Xu Y, Zhang S, Zhu H, Hou J. Tricin 7-glucoside protects against experimental cerebral ischemia by reduction of NF- j B and HMGB1 expression. European Journal of Pharmaceutical Sciences. 2012;**45**:50-57. DOI: 10.1016/j.ejps.2011.10.019

[41] Na CS, Hong SS, Choi YH, Lee YH, Hong SH, Lim JY, et al. Neuroprotective effects of constituents of *Eragrostis ferruginea* against Aβ-induced toxicity in PC12 cells. Archives of Pharmacal Research. 2010;**33**:999-1003. DOI: 10.1007/s12272-010-0704-5

[42] Sukumaran NP, Amalraj A, Gopi S. Neuropharmacological and cognitive effects of *Bacopa monnieri* (L.)

Wettst – A review on its mechanistic aspects. Complementary Therapies in Medicine. 2019;**44**:68-82. DOI: 10.1016/j.ctim.2019.03.016

[43] Menghini L. A natural formulation increases brain resistance to oxidative stress. European Journal of Medicinal Plants. 2014;**4**:171-182. DOI: 10.9734/ejmp/2014/7150

[44] Bhattacharya SK, Bhattacharya A, Kumar A, Ghosal S. Antioxidant activity of *Bacopa monniera* in rat frontal cortex, striatum and hippocampus. Physical Therapy Research. 2000;**14**:174-179. DOI: 10.1002/(SICI)1099-1573(200005)14:3<174::AID-PTR624>3.0.CO;2-O

[45] Khan MB, Ahmad M, Ahmad S, Ishrat T, Vaibhav K, Khuwaja G, et al. *Bacopa monniera* ameliorates cognitive impairment and neurodegeneration induced by intracerebroventricular-streptozotocin in rat: Behavioral, biochemical, immunohistochemical and histopathological evidences. Metabolic Brain Disease. 2014;**30**:115-127. DOI: 10.1007/s11011-014-9593-5

[46] Shinomol GK, Mythri RB, Srinivas Bharath MM. *Bacopa monnieri* extract offsets rotenone-induced cytotoxicity in dopaminergic cells and oxidative impairments in mice brain. Cellular and Molecular Neurobiology. 2012;**32**:455-465. DOI: 10.1007/s10571-011-9776-0

[47] Uabundit N, Wattanathorn J, Mucimapura S, Ingkaninan K. Cognitive enhancement and neuroprotective effects of *Bacopa monnieri* in Alzheimer's disease model. Journal of Ethnopharmacology. 2010;**127**:26-31. DOI: 10.1016/j.jep.2009.09.056

[48] Anbarasi K, Vani G, Balakrishna K, Devi CSS. Effect of bacoside A on brain antioxidant status in cigarette smoke exposed rats. Life Sciences. 2006;**78**:1378-1384. DOI: 10.1016/j.lfs.2005.07.030

[49] Stough C, Scholey A, Cropley V, Wesnes K, Zangara A, Pase M, et al. Examining the cognitive effects of a special extract of *Bacopa monniera* (CDRI 08: KeenMind): A review of ten years of research at Swinburne University. Journal of Pharmacy & Pharmaceutical Sciences. 2013;**16**:254-258. DOI: 10.18433/j35g6m

[50] Shinomol GK, Srinivas Bharath MM. Neuromodulatory propensity of *Bacopa monnieri* leaf extract against 3-nitropropionic acid-induced oxidative stress: *In vitro* and *in vivo* evidences. Neurotoxicity Research. 2012;**22**:102-114. DOI: 10.1007/s12640-011-9303-6

[51] Saraf MK, Prabhakar S, Anand A. Neuroprotective effect of *Bacopa monniera* on ischemia induced brain injury. Pharmacology, Biochemistry, and Behavior. 2010;**97**:192-197. DOI: 10.1016/j.pbb.2010.07.017

[52] Limpeanchob N, Jaipan S, Rattanakaruna S, Phrompittayarat W, Ingkaninan K. Neuroprotective effect of *Bacopa monnieri* on beta-amyloid-induced cell death in primary cortical culture. Journal of Ethnopharmacology. 2008;**120**:112-117. DOI: 10.1016/j.jep.2008.07.039

[53] Dhanasekaran M, Tharakan B, Holcomb LA, Hitt AR, Young KA, Manyam BV. Neuroprotective mechanisms of ayurvedic antidementia botanical *Bacopa monniera*. Physical Therapy Research. 2007;**21**:965-969. DOI: 10.1002/ptr.2195

[54] Swathi G, Ramaiah CV, Rajendra W. Protective role of *Bacopa monnieri* against Rotenone- induced Parkinson's disease in PC 12 cell lines. International Journal of Phytomedicine. 2017;**9**:219-222. DOI: 10.5138/09750185.2008

[55] Ahmad MP, Hussain A, Siddiqui HH, Wahab S, Adak M. Effect of methanolic extract of *Asparagus racemosus* Willd. On lipopolysaccharide

induced-oxidative stress in rats. Pakistan Journal of Pharmaceutical Sciences. 2015;**28**:509-513

[56] Saxena G, Singh M, Meena P, Barber S, Sharma D, Shukla S, et al. Neuroprotective effects of *Asparagus racemosus* Linn root extract: An experimental and clinical evidence. Annals of Neurosciences. 2007;**14**:57-63. DOI: 10.5214/ans.0972.7531.2007.140302

[57] Parihar MS, Hemnani T. Experimental excitotoxicity provokes oxidative damage in mice brain and attenuation by extract of *Asparagus racemosus*. Journal of Neural Transmission. 2004;**111**:1-12. DOI: 10.1007/s00702-003-0069-8

[58] Bhattacharya A, Muruganandam AV, Kumar V, Bhattacharya SK. Effect of poly herbal formulation, EuMil, on neurochemical perturbations induced by chronic stress. Indian Journal of Experimental Biology. 2002;**40**:1161-1163

[59] Bhatnagar M, Meena P, Barbar S, Joshi C. Neuroprotective response of the hippocampus region of the brain to *Withania somnifera* and *Asparagus racemosus* root extract: An *in vitro* study. Journal of Medicinal Plant Research. 2013;**7**:2259-2264. DOI: 10.5897/JMPR12.497

[60] Lalert L, Kruevaisayawan H, Amatyakul P, Ingkaninan K. Neuroprotective effect of *Asparagus racemosus* root extract via the enhancement of brain-derived neurotrophic factor and estrogen receptor in ovariectomized rats. Journal of Ethnopharmacology. 2018;**225**:336-341. DOI: 10.1016/j.jep.2018.07.014

[61] Pahwa P, Goel RK. Ameliorative effect of *Asparagus racemosus* root extract against pentylenetetrazol-induced kindling and associated depression and memory deficit. Epilepsy & Behavior. 2016;**57**:196, 201. DOI: 10.1016/j.yebeh.2016.01.025

[62] Kulkarni SK, Verma A. Protective effect of Mentat (BR-16A), a herbal preparation on alcohol abstinence-induced anxiety and convulsions. Indian Journal of Experimental Biology. 1993;**31**:435-439

[63] Kashyap P, Muthusamy K, Niranjan M, Trikha S, Kumar S. Sarsasapogenin: A steroidal saponin from *Asparagus racemosus* as multi target directed ligand in Alzheimer's disease. Steroids. 2020;**153**:108529. DOI: 10.1016/j.steroids.2019.108529

[64] Nandagopal M, Muralidharan P, Thirumurugan G. Cerebroprotective effect of root extract of *Asparagus racemosus* Willd. in global cerebral ischemia in rats. Journal of Pharmacology and Toxicology. 2011;**6**:49-61. DOI: 10.3923/jpt.2011.49.61

[65] Joshi H, Parle M. Cholinergic basis of memory-strengthening effect of *Foeniculum vulgare* Linn. Journal of Medicinal Food. 2006;**9**:413-417. DOI: 10.1089/jmf.2006.9.413

[66] Cioanca O, Hancianu M, Mircea C, Trifan A, Hritcu L. Essential oils from Apiaceae as valuable resources in neurological disorders: *Foeniculum vulgare* aetheroleum. Industrial Crops and Products. 2016;**88**:51-57. DOI: 10.1016/j.indcrop.2016.02.064

[67] Nemati M, Hemmati AA, Najafzadeh H, Mansouri MT, Khodayar MJ. Evaluation the effects of *Foeniculum vulgare* essence on behavioral-motor disorders of Parkinson's disease induced by reserpine in ovariectomized and non ovariectomized rats. Jundishapur Journal of Natural Pharmaceutical Products. 2018;**13**(1):e67391. DOI: 10.5812/jjnpp.67391

[68] El Shebini SM, Abdel-Moaty M, Kazem YI, Ahmed NH, Fouad S, Mohamed MS, et al. Relation between obesity, cognition and serum amyloid β protein level and potential role of *Foeniculum vulgare* in reducing weight and improving cognitive functions. Journal of Biological Sciences. 2017;**17**:202-212. DOI: 10.3923/jbs.2017.202.212

[69] Bhatti S, Ali Shah SA, Ahmed T, Zahid S. Neuroprotective effects of *Foeniculum vulgare* seeds extract on lead-induced neurotoxicity in mice brain. Drug and Chemical Toxicology. 2018;**41**:399-407. DOI: 10.1080/01480545.2018.1459669

[70] Xiang X, Wu L, Mao L, Liu Y. Anti-oxidative and anti-apoptotic neuroprotectiveeffects of *Azadirachta indica* in Parkinson-induced functional damage. Molecular Medicine Reports. 2018;**17**(6):7959-7965. DOI: 10.3892/mmr.2018.8815

[71] Yanpallewar S, Rai S, Kumar M, Chauhan S, Acharya SB. Neuroprotective effect of *Azadirachta indica* on cerebral post-ischemic reperfusion and hypoperfusion in rats. Life Sciences. 2005;**76**(12):1325-1338

[72] Kandhare AD, Mukherjee AA, Bodhankar SL. Neuroprotective effect of *Azadirachta indica* standardized extract in partial sciatic nerve injury in rats: Evidencefrom anti-inflammatory, antioxidant and anti-apoptotic studies. EXCLI Journal. 2017;**16**:546-565. DOI: 10.17179/excli2017-161

[73] Abdel Moneim AE. *Azadirachta indica* attenuates cisplatin-induced neurotoxicity in rats. Indian Journal of Pharmacology. 2014;**46**(3):316-321. DOI: 10.4103/0253-7613.132182

[74] Bedi KL, Zutshi U, Chopra CL, Amla V. *Picrorhiza kurroa*, an ayurvedic herb, may potentiate photochemotherapy in vitiligo. Journal of Ethnopharmacology. 1989;**27**(3):347-352

[75] Simonyi A, Serfozo P, Lehmidi TM, Cui J, Gu Z, Lubahn DB, et al. The neuroprotective effects of apocynin. Frontiers in Bioscience (Elite Edition). 2012;**4**:2183-2193

[76] Marín M, Giner RM, Ríos JL, Recio MC. Protective effect of apocynin in a mouse model of chemically-induced colitis. Planta Medica. 2013;**79**(15):1392-1400. DOI: 10.1055/s-0033-1350710

[77] Romanini CV, Ferreira ED, Soares LM, Santiago AN, Milani H, de Oliveira RM. 4-Hydroxy-3-methoxy-acetophenone-mediated long-lasting memory recovery, hippocampal neuroprotection, and reduction of glial cell activation after transient global cerebral ischemia in rats. Journal of Neuroscience Research. 2015;**93**(8):1240-1249. DOI: 10.1002/jnr.23575

[78] Zahiruddin S, Khan W, Nehra R, Alam MJ, Mallick MN, Parveen R, et al. Pharmacokinetics and comparative metabolic profiling of iridoid enriched fraction of *Picrorhiza kurroa* - An Ayurvedic herb. Journal of Ethnopharmacology. 2017;**197**:157-164. DOI: 10.1016/j.jep.2016.07.072

[79] Kim N, Do J, Ju IG, Jeon SH, Lee JK, Oh MS. *Picrorhiza kurroa* prevents memory deficits by inhibiting NLRP3 Inflammasome activation and BACE1 expression in 5xFAD mice. Neurotherapeutics. 2020;**17**(1):189-199. DOI: 10.1007/s13311-019-00792-7

[80] Morikawa T, Nakanishi Y, Inoue N, Manse Y, Matsuura H, Hamasaki S, et al. Acylated iridoid glycosides with hyaluronidase inhibitory activity from the rhizomes of *Picrorhiza kurroa* Royle ex. Benth. Phytochemistry.

2020;**169**:112185. DOI: 10.1016/j.phytochem.2019.112185

[81] Dinda B, Dinda M, Kulsi G, Chakraborty A, Dinda S. Therapeutic potentials of plant iridoids in alzeimer's and parkinson's diseases. A review. European Journal of Medicinal Chemistry. 2019;**169**:185-199

[82] Tavakkoli A, Iranshahi M, Hasheminezhad SH, Hayes AW, Karimi G. The neuroprotective activities of natural products through the Nrf2 upregulation. Phytotherapy Research. 2019;**33**(9):2256-2273. DOI: 10.1002/ptr.6427

[83] Yuan NN, Cai CZ, Wu MY, Su HX, Li M, Lu JH. Neuroprotective effects of berberine in animal models of Alzheimer's disease: A systematic review of pre-clinical studies. BMC Complementary and Alternative Medicine. 2019;**19**(1):109. DOI: 10.1186/s12906-019-2510-z

[84] Soudi SA, Nounou MI, Sheweita SA, Ghareeb DA, Younis LK, El-Khordagui LK. Protective effect of surface-modified berberine nanoparticles against LPS-induced neurodegenerative changes: A preclinical study. Drug Delivery and Translational Research. 2019;**9**(5):906-919. DOI: 10.1007/s13346-019-00626-1

[85] Maleki SN, Aboutaleb N, Souri F. Berberine confers neuroprotection in coping with focal cerebral ischemia by targeting inflammatory cytokines. Journal of Chemical Neuroanatomy. 2018;**87**:54-59. DOI: 10.1016/j.jchemneu.2017.04.008

[86] Haghani M, Shabani M, Tondar M. The therapeutic potential of berberine against the altered intrinsic properties of the CA1 neurons induced by Aβ neurotoxicity. European Journal of Pharmacology. 2015;**758**:82-88. DOI: 10.1016/j.ejphar.2015.03.016

[87] Lin X, Zhang N. Berberine: Pathways to protect neurons. Phytotherapy Research. 2018;**32**(8):1501-1510. DOI: 10.1002/ptr.6107

[88] Fan D, Liu L, Wu Z, Cao M. Combating neurodegenerative diseases with the plant alkaloid Berberine: Molecular mechanisms and therapeutic potential. Current Neuropharmacology. 2019;**17**(6):563-579. DOI: 10.2174/1570159X16666180419141613

[89] Hao F, Jia L-H, Li X-W, Zhang Y-R, Liu X-W. Garcinol upregulates GABAA and GAD65 expression, modulates BDNF-TrkB pathway to reduce seizures in pentylenetetrazole (PTZ)-induced epilepsy. Medical Science Monitor. 2016;**22**:4415-4425

[90] Weng M-S, Liao C-H, Yu S-Y, Lin J-K. Garcinol promotes neurogenesis in rat cortical progenitor cells through the duration of extracellular signal-regulated kinase signaling. Journal of Agricultural and Food Chemistry. 2011;**59**:1031-1040. DOI: 10.1021/jf104263s

[91] Balasubramanyam K, Altaf M, Varier RA, Swaminathan V, Ravindran A, Sadhale PP, et al. Polyisoprenylated benzophenone, garcinol, a natural histone acetyltransferase inhibitor, represses chromatin transcription and alters global gene expression. The Journal of Biological Chemistry. 2004;**279**:33716-33726. DOI: 10.1074/jbc

[92] Hegarty SV, Sullivan AM, O'Keeffe GW. The Epigenome as a therapeutic target for Parkinson's disease. Neural Regeneration Research. 2016;**11**:1735-1738. DOI: 10.4103/1673-5374.194803

[93] Mazumder MK, Paul R, Phukan BC, Dutta A, Chakrabarty J, Bhattacharya P, et al. Garcinol, an effective monoamine oxidase-B inhibitor for the treatment

of Parkinson's disease. Medical Hypotheses. 2018;**117**:54-58. DOI: 10.1016/j.mehy.2018.06.009

[94] Ryu Y-K, Park H-Y, Go J, Kim Y-H, Hwang JH, Choi D-H, et al. Effects of histone acetyltransferase inhibitors on l-DOPA-induced dyskinesia in a murine model of Parkinson's disease. Journal of Neural Transmission. 2018;**125**:1319-1331. DOI: 10.1007/s00702-018-1902-4

[95] Bastide MF, Dovero S, Charron G, Porras G, Gross CE, Fernagut P-O, et al. Immediate-early gene expression in structures outside the basal ganglia is associated to L-DOPA-induced dyskinesia. Neurobiology of Disease. 2014;**62**:179-192. DOI: 10.1016/j.nbd.2013.09.020

[96] Antala BV, Patel MS, Bhuva SV, Gupta S, Rabadiya S, Lahkar M. Protective effect of methanolic extract of *Garcinia indica* fruits in 6-OHDA rat model of Parkinson's disease. Indian Journal of Pharmacology. 2012;**44**:683-687. DOI: 10.4103/0253-7613.103242

[97] Park G, Tan J, Garcia G, Kang Y, Salvesen G, Zhang Z. Regulation of histone acetylation by autophagy in Parkinson disease. The Journal of Biological Chemistry. 2016;**291**:3531-3540. DOI: 10.1074/jbc.M115.675488

[98] Deb S, Banashree C, Phukan M, Mazumder K, Duttaa A, Paul R, et al. Garcinol, a multifaceted sword for the treatment of Parkinson's disease. Neurochemistry International. 2019;**128**:50-57

[99] Dhingra D, Goyal PK. Inhibition of MAO and GABA: Probable mechanisms for antidepressant-like activity of *Nardostachys jatamansi* DC. in mice. Indian Journal of Experimental Biology. 2008;**46**(4):212-2018

[100] Salim S, Ahmad M, Zafar KS, Ahmad AS, Islam F. Protective effect of *Nardostachys jatamansi* in rat cerebral ischemia. Pharmacology, Biochemistry, and Behavior. 2003;**74**(2):481-486

[101] Ahmad M, Yousuf S, Khan MB, Hoda MN, Ahmad AS, Ansari MA, et al. Attenuation by *Nardostachys jatamansi* of 6-hydroxydopamine-induced parkinsonism in rats: Behavioral, neurochemical, and immunohistochemical studies. Pharmacology, Biochemistry, and Behavior. 2006;**83**(1):150-160

[102] Joshi H, Parle M. *Nardostachys jatamansi* improves learning and memory in mice. Journal of Medicinal Food. 2006;**9**(1):113-118

[103] Lyle N, Bhattacharyya D, Sur TK, Munshi S, Paul S, Chatterjee S, et al. Stress modulating antioxidant effect of *Nardostachys jatamansi*. Indian Journal of Biochemistry & Biophysics. 2009;**46**(1):93-98

[104] Shah J, Goyal R. Investigation of neuropsychopharmacological effects of a polyherbal formulation on the learning and memory process in rats. Journal of Young Pharmacists. 2011;**3**(2):119-124. DOI: 10.4103/0975-1483.80296

[105] Khan MB, Hoda MN, Ishrat T, Ahmad S, Moshahid Khan M, Ahmad A, et al. Neuroprotective efficacy of *Nardostachys jatamansi* and crocetin in conjunction with selenium in cognitive impairment. Neurological Sciences. 2012;**33**(5):1011-1020. DOI: 10.1007/s10072-011-0880-1

[106] Suryavanshi S, Raina P, Deshpande R, Kaul-Ghanekar R. *Nardostachys jatamansi* root extract modulates the growth of IMR-32 and SK-N-MC neuroblastoma cell lines through MYCN mediated regulation of MDM2 and p53. Pharmacognosy Magazine. 2017;**13**(49):21-24. DOI: 10.4103/0973-1296.197645

[107] Yoon CS, Kim DC, Park JS, Kim KW, Kim YC, Oh H. Isolation

of novel Sesquiterpeniods and anti-neuroinflammatory metabolites from *Nardostachys jatamansi*. Molecules. 2018;**23**(9):E2367. DOI: 10.3390/molecules23092367

[108] Kim KW, Yoon CS, Kim YC, Oh H. Desoxo-narchinol a and Narchinol B isolated from *Nardostachys jatamansi* exert anti-neuroinflammatory effects by up-regulating of nuclear transcription factor Erythroid-2-related factor 2/Heme Oxygenase-1 Signaling. Neurotoxicity Research. 2019;**35**(1):230-243. DOI: 10.1007/s12640-018-9951-x

[109] Singh GK, Rai G, Chatterjee SS, Kumar V. Effects of ethanolic extract of *Fumaria indica* L. on rat cognitive dysfunctions. AYU. 2013;**34**(4):421-429. DOI: 10.4103/0974-8520.127727

[110] Singh GK, Chauhan SK, Rai G, Chatterjee SS, Kumar V. Potential antianxiety activity of *Fumaria indica*: A preclinical study. Pharmacognosy Magazine. 2013;**9**(33):14-22. DOI: 10.4103/0973-1296.108129

[111] Asghari MH, Fallah M, Moloudizargari M, Mehdikhani F, Sepehrnia P, Moradi B. A systematic and mechanistic review on the Phytopharmacological properties of *Alhagi* species. Ancient Science of Life. 2016;**36**(2):65-71. DOI: 10.4103/asl.ASL_37_16

[112] Muhammad G, Hussain MA, Anwar F, Ashraf M, Gilani AH. *Alhagi*: A plant genus rich in bioactives for pharmaceuticals. Phytotherapy Research. 2015;**29**(1):1-13. DOI: 10.1002/ptr.5222

[113] Singh VP, Yadav B, Pandey VB. Flavanone glycosides from *Alhagi pseudalhagi*. Phytochemistry. 1999;**51**(4):587-590

[114] Ghosal S, Srivastava RS, Bhattacharya SK, Debnath PK. The active principles of *Alhagi pseudalhagi*: Beta-phenethylamine and tetrahydroisoquinoline bases. Planta Medica. 1974;**26**(4):318-326

[115] Mundugaru R, Sivanesan S, Popa-Wagner A, Udaykumar P, Kirubagaran R, Kp G, et al. *Pluchea lanceolata* protects hippocampal neurons from endothelin-1 induced ischemic injury to ameliorate cognitive deficits. Journal of Chemical Neuroanatomy. 2018;**94**:75-85. DOI: 10.1016/j.jchemneu.2018.09.002

[116] Mundugaru R, Sivanesan S, Udaykumar P, Rao N, Chandra N. Protective effect of *Pluchea lanceolata* against aluminum chloride-induced neurotoxicity in Swiss albino mice. Pharmacognosy Magazine. 2017;**13**(Suppl 3):S567-S572. DOI: 10.4103/pm.pm_124_17

[117] Srivastava P, Mohanti S, Bawankule DU, Khan F, Shanker K. Effect of *Pluchea lanceolata* bioactives in LPS-induced neuroinflammation in C6 rat glial cells. Naunyn-Schmiedeberg's Archives of Pharmacology. 2014;**387**(2):119-127. DOI: 10.1007/s00210-013-0924-6

[118] Patel NG, Patel KG, Patel KV, Gandhi TR. Validated HPTLC method for quantification of luteolin and apigenin in *Premna mucronata* Roxb., Verbenaceae. Advances in Pharmacological Sciences. 2015;**2015**:682365. DOI: 10.1155/2015/682365

[119] Shukla SD, Jain S, Sharma K, Bhatnagar M. Stress induced neuron degeneration and protective effects of *Semecarpus anacardium* Linn. And *Withania somnifera* Dunn. in hippocampus of albino rats: An ultrastructural study. Indian Journal of Experimental Biology. 2000;**38**(10):1007-1013

[120] Devi SJR, Kumar BP. In silico screening for anti-inflammatory

bioactive molecules from ayurvedic decoction, *Balagulchyadi Kashayam*. Current Computer-Aided Drug Design. 2019. DOI: 10.2174/157340991566619101 5113753

[121] Birla H, Rai SN, Singh SS, Zahra W, Rawat A, Tiwari N, et al. *Tinospora cordifolia* suppresses neuroinflammation in parkinsonian mouse model. Neuromolecular Medicine. 2019;**21**(1):42-53. DOI: 10.1007/s12017-018-08521-7

[122] Sharma A, Kaur G. *Tinospora cordifolia* as a potential neuroregenerative candidate against glutamate induced excitotoxicity: An *in vitro* perspective. BMC Complementary and Alternative Medicine. 2018;**18**(1):268. DOI: 10.1186/s12906-018-2330-2336

[123] Singh H, Kaur T, Manchanda S, Kaur G. Intermittent fasting combined withsupplementation with Ayurvedic herbs reduces anxiety in middle aged female rats by anti-inflammatory pathways. Biogerontology. 2017;**18**(4):601-614. DOI: 10.1007/s10522-017-9706-8

[124] Bhattacharya S, Haldar PK. Neuropharmacological properties of *Trichosanthes dioica* root. Chinese Journal of Natural Medicines. 2013;**11**(2):158-163. DOI: 10.1016/S1875-5364(13)60043-6

Chapter 3

Phytochemical and Nutritional Studies in the Genus *Abelmoschus* Medik

Ashwini Venkatrao Mohite and Rajaram Vithoba Gurav

Abstract

Genus *Abelmoschus* Medik (family - Malvaceae Juss.) comprising about 11 species in India. Among which some are cultivated on an economic scale as important vegetables and the rest are wild. Apart from cultivated species, wild species of *Abelmoschus* also showed a wide range of phytochemicals and nutritional components. *Abelmoschus esculentus* (L.) Moench an economically important vegetable crop popularly known as okra which cultivated throughout the world. Nutritionally, okra plays an important role in the human diet because it has enormous value of nutritional factors like carbohydrates, protein, fibers, minerals and vitamins, including vitamin C. Mucilage obtained from *Abelmoschus* is natural and digestive in nature and used as a tablet binder. The plant shows various pharmacological activities like, antioxidants, antidiabetic, antiulcer and antimicrobial. Apart from cultivated species, the studies on wild species were carried out and concluded some important findings *viz. Abelmoschus manihot* (L.) Medik. is rich source of various secondary metabolites like; hyperin, isoquercetin, myricetin, hibifolin, adenosine and stigmasterol. *Abelmoschus ficulneus* (L.) Wight & Arn. and *A. manihot* are consumed by the local people worldwide. The present chapter is focused on the previous work done in genus *Abelmoschus* in the area of nutrition, phytochemical, genetic diversity and breeding.

Keywords: *Abelmoschus*, okra, phytochemical, nutritional, mucilage

1. Introduction

The word *Abelmoschus* has meaningful etymology Arabic language "Abu-al-misk", means "Father of Musk and Kaab-el-misk means source of musk in reference with its musky odor of seeds [1, 2]. Genus is native to Africa, South and South East Asia and distributed throughout tropical and subtropical regions of the world with center of diversity at South Asia and South-West Pacific region [3–5]. Indian subcontinent has been considered as the center of diversity of the genus *Abelmoschus* due to the presence of wide morphological diversity. In India *Abelmoschus* species are found in the dense evergreen forests to open wastelands as well as cultivated in gardens and commercial farm. In addition, *Abelmoschus* species are also distributed in various regions from a range of Himalayan Mountain [6] to the Southern Peninsular India [7, 8]. About 11 species, 3 sub-species and

4 varieties of *Abelmoschus* are known in India [8]. Due to potential nutritional importance okra gain attention to improve nutrition and health status of mankind. Sabitha et al. demonstrated anti-diabetic activity through *in vitro* α-glucosidase and α-amylase enzyme inhibitory effect of aqueous extract of the okra peel and seeds and concluded that the consumption of okra may help to maintain blood sugar [9]. Due to widely use in the arena of food and medicine, the area under cultivation of *Abelmoschus* has progressively increased during last few years [10]. Besides that, *Abelmoschus* species are used for various other purposes like fiber yielding, paper making, waste water treatment, substitute for jute and used in the textile industries [11]. It is the rich reservoir of essential micronutrients for food fortification process e.g.; bread fortification. In this chapter, we have interpreted phyto-constituents from *Abelmoschus* species and their potential roles in human diet and pharmaceutical applications. Along with that we are also focused on qualitative and quantitative assessment of mucilage and its application in various fields. Further chapter revealed genetic diversity, plant breeding and mutation studies of various species of *Abelmoschus*.

2. Phytochemical analysis of genus *Abelmoschus*

During regular metabolic activities of plant some chemical compounds are formed known as phytochemicals. These chemicals are produced by plants for their defense mechanism, but many research studies reveals the numerous phytochemicals can be used against many diseases of humans. Phytochemicals are often referred as "secondary metabolites" which includes alkaloids, flavonoids, phenols, tannins, terpenoids, gums and polysaccharides [12]. Adetuyi et al. analyzed Vitamin C, total phenolic content, iron chelating activity and reducing power of six varieties of okra cultivated in Nigeria and proved that during storage period from 0 to 10 days, the loss of antioxidants percentage were lowest in "Benin" okra variety [13]. A variety of phytochemicals and antioxidants have been isolated from different *Abelmoschus* species. The list of phytochemicals isolated from *Abelmoschus* species are depicted in **Table 1**. *A. moschatus* solvent extract contains higher level of polyphenols and flavonoids which are responsible for antioxidant and other cumulative activities, so it can be used as food and medicine mainly to improve insulin sensitivity [19]. The four quercetin derivatives and epigallocatechin were first time reported in okra [14]. *A. esculentus* fruit is rich in phenolics and flavonoids may be serve as good source of natural antioxidants [15, 16]. Along with cultivated some wild species also rich in antioxidants [20].

2.1 Phytochemical analysis of *Abelmoschus* seeds

Abelmoschus seeds are the potential source of various nutrients as well as having immense biological properties; therefore some researchers focused their attention towards the examination of chemical composition and their use in the area of nutrition and medicines. Rao reported amino acids, fats, heat labile proteins, dietary fibers from seeds and kernels of okra variety "Pusa Swani" [21]. Sami et al. carried-out studies on analysis of fatty acids and amino acid from the fruit of A. *esculentus* collected from four localities by GC-MS and amino acid analyzer [22]. Camiciuc et al. carried out the GC-MS and NMR spectroscopic analysis of essential oil obtained from okra and ambrette seeds and subsequently reported the 40 and 35 bio-active compounds respectively [23, 24]. *A. moschatus* species is known for its aromatic compound and it has been used in Chinese traditional medicine to cure depression and anxiety. The effect of decoction has

Name of species	Plant part	Analytical techniques used	Solvent	Phytochemicals	Ref.
A. esculentus Cv. Benin, Auchi, Ikaro, Akure, Okeneand Lokoja	Fruit	Spectroscopy	Acetone, Methanol, Water	Vitamin C, total phenolic content, iron chelating activity and reducing power	[13]
A. esculentus	Fruit	HPLC and NMR	70% ethanol, hexane, methanol	ABTS, Quercetin 3-O-xylosyl (1 → 2) glucoside, quercetin 3-Oglucosyl (1 → 6) glucoside, quercetin 3-O-glucoside, quercetin3-O-(6-O-malonyl)-glucosideand epigallocatechin	[14]
A. esculentus	Fruit	—	Ethanol, Water	Carbohydrate, mucilage, protein, amino acids, fat and oil, flavonoids, phenolic compounds, tannins, saponins, phytosterols, alkaloid, glycoside, hypoglycemic activity	[11]
A. esculentus (25 accessions)	Fruit	Spectroscopy	Distilled water	Total phenolics, flavonoids and antioxidant contents	[15]
A. esculentus (7 accessions)	Fruit	Spectroscopy	Methanol	Total phenolics, flavonoids	[16]
A. manihot	Flower	HPLC	Ethanol, Methanol	Hyperin, isoquercetin, hibifolin, myricetin, quercetin-3'-O-glucoside, and quercetin	[17]
A. manihot	Leaf	Spectroscopy and FTIR	Ethanol	Flavonoid and DPPH	[18]
A. moschatus	Fruit	—	70% Ethanol	Total phenolic, total flavonoids	[19]
A. esculentus, *A. esculentus* cv. Phule Utkarsha *A. ficulneus*, *A. manihot*	Fruit	Spectroscopy	Ethanol, Methanol, Distilled water	Total phenolic, total flavonoids, DPPH and FRAP	[20]

Table 1.
Phytoconstituents from Abelmoschus *taxa.*

been reported to exhibit hypotensive properties [25]. The volatile compounds identified from *A. moschatus* seeds having an odor similar to the musk therefore used in perfume and cosmetics formulations [26]. In addition, the cultivated as well as wild species of *Abelmoschus* were also used to study seed oil and fatty acid content using GC-MS and structural analysis using TD-NMR [27]. Thirty five compounds were identified using GC-MS along with their antibacterial properties were reported from ambrette seed oil [28].

2.2 Importance of mucilage in *Abelmoschus*

Mucilage is the water soluble polysaccharides found in various plant systems and in some microorganisms [29]. In present days, there is an immense interest have been seen in studying the mucilaginous compounds due to their viscosity and pharmaceutical applications like, excipient, tablet binder, thickeners in oral liquids, gelling agents, purifiers, protective colloids in suspension gum substitute and effluents in rheological engineering [30]. Most of the species from family Malvaceae are well-known for their mucilage content and it was studied by Ahmad et al. for their properties [31]. The whole plant of *Abelmoschus* species harbor considerable quantity of mucilage, associate mucilage is an acidic polysaccharide composed of galacturonic acid, rhamnose, and glucose with the ratio of 1.3:1.0:0.1, respectively [32]. Okra mucilage contains a significant amount of carbohydrate, neutral sugars, minerals and other complex polysaccharides [33] which medically confirmed that mucilage is associated with antimicrobial, antiulcer, hypoglycemic and anticancer activities [26].

Mucilage isolated from immature fruits and roots of okra showed significant anti-complementary activity and extensive hypoglycemic activities because root mucilage possesses side chains composed of D-galactopyranose residues and L-rhamnopyranosyl residues in the part of the backbone [34]. Okra mucilage was described as water soluble polysaccharide based material which can be further modified by grafting acrylamide for the synthesis of green polymeric material and it was issued as the biomaterial for waste water treatment as an environment cleaning approach [35]. Nair and Fasha analyzed mucilage of *A. esculentus* and *A. moschatus* and recommended for their used in preparation of pharmaceutical suspensions [36]. Apart from the environment cleaning and medicinal values, there are various applications discussed, okra mucilage can be utilized for pharmaceutical adjuvant, emulsifying agent [37] excipients and binding agent for the formulation of pharmaceutical dosage using Ibuprofen and Paracetamol tablet [38]. It is also used to prepare polyelectrolyte complex with the help of chitosan and used as coating material [39]. Okra mucilage is also used as the suspending agent and pharmaceutical excipients [40, 41]. Mucoadhesive gel was prepared from okra fruit mucilage and used for nasal drug delivery [42].

Abelmoschus stem and *Hibiscus* leaves mucilage was used to analyze the nutritive values such as moisture, fat, fiber, protein, carbohydrate, and energy value and further same mucilage powder was used to prepare idli, upma and roti and evaluate for consumer acceptability and nutritive values. These products have great acceptableness in terms of texture, color, taste and flavor [43]. Okra mucilage can be potentially utilized as the blending mediator in a food emulsion system [44] and having biodegradable and non-toxic coagulant property [45]. Apart from these it is natural source for edible film production [46] and has excellent potential in food packaging [47].

3. Nutritional potential of genus *Abelmoschus*

Production of nutritionally rich food is the major challenge in the fulfillment of healthy diet against tremendous explosion of population. Throughout the year, numerous vegetables have been basically examined for their nutritional parameters. Nutritionally, okra plays an important role in the human diet because it contains carbohydrates, protein, fibers, minerals and vitamins, including vitamin C which will fulfill dietary requirements of the body [27]. Some nutritional parameters with their quantity were depicted in **Table 2**.

Species name	Plant part used	Nutritional parameter	Quantity	Ref.
A. esculentus, A. esculentus cv. Phule Utkarsha *A. ficulneus, A. manihot*	Fruit	Proximate (Moisture, fat, ash, fiber, protein, carbohydrate)	2.52–48.47 g/100 g	[20]
		Mineral Composition (B, Ca, Cu, Fe, Mg, Mn, Mo, N, P, K, Na, S, Zn)	0.0005–3.51 g/100 g	
A. esculentus (6 varieties)	Fruit	Moisture	88.02–90.13%	[48]
		Protein	13.61–16.27%	
		Fiber	10.15–11.63%	
		Fat	9.03–10.57%	
		Ash	7.19–9.63%	
		Mineral (Zn, Fe, Mg, Cl and K)	0.87–62.17 mg/100 g	
A. esculentus (22 accessions)	Fruits	Macro elements (Ca, Cl, K, Mg and Na)	32.06–319 mg/kg	[49]
		Micro elements (Al, Cu and Mn)	17.8–42.45 mg/kg	
		Trace elements (As and Br)	2.84–34.41 mg/kg	
A. esculentus (4 accessions)	Fruits	Water soluble vitamins (B3, B6, B12, C)	1.42–91.20 μg/100 g	[50]
		Fat soluble vitamins (E, K3)	0.05–1.47 μg/100 g	
A. esculentus (8 accessions)	Fruit	Moisture	9.69–13.33 g/100 g	[51]
		Crud protein	10.25–26.16 g/100 g	
		Ash	5.62–11.30 g/100 g	
		Crude fiber	11.97–29.93 g/100 g	
		Crude fat	0.56–1.69 g/100 g	
		Carbohydrate	36.66–50.97 g/100 g	
		Mineral (Ca, Fe, K, Zn, P, Na)	3.33–318.20 mg/100 g	
A. manihot (23 accessions)	Leaves	Mineral composition (Cl, Fe, Mg, Mn, K, Na, Zn and Cu)	0.8–635 mg/100 g	[52]

Table 2.
Nutritional potential of Abelmoschus *taxa.*

Sun dried okra fruits were examined for the nutritional parameters like moisture, ash, crude fat, fiber, carbohydrate, protein and microbial composition. Dried okra with light deep green and light purple colored had highest carbohydrate (76.8%) and crude protein (23.2%) [53]. Effect of different processes like cooking, sun drying of okra fruit caused effects on proximate composition and some other parameters like loss of vitamin C and nutritional factor [54]. Fruits of cultivated as well as wild taxa are the rich in, proximate and mineral composition. The highest fiber content i.e. 23.49 and 22.90% were isolated from the wild species *A. ficulneus* and *A. manihot* respectively [20].

4. Genetic diversity study in *Abelmoschus*

The accessibility of the genetic diversity and its collection, maintenance and conservation is essential for the crop improvement program [55]. *Abelmoschus* is an important genus in terms of nutrition, pharmaceuticals point of view, due to it has high industrial and economic importance. Despite that very slight attention has been paid to assess genetic diversity of *Abelmoschus* species at molecular level. The conservation and distribution of the germplasm genetic diversity of *Abelmoschus* species has been studied with the various DNA markers. Techniques based on the genetic material such as Random Amplified Polymorphic DNA (RAPD) have been employed to assess the genetic diversity in the five species of *Abelmoschus* [56]. Similarly Martinello et al. also successfully employed RAPD marker to study of genetic diversity within 42 accessions of *Abelmoschus* species. The study revealed existence of the most important genetic heterogeneity in the tested germplasm [57]. RAPD makers were applied over the okra accessions which lead to 96% polymorphism [58] along with the phenotypic markers [59, 60]. Sequence related amplified polymorphism (SRAP) and phenotypic markers determined the genetic diversity of Turkish okra. These markers showed 50% polymorphism among the studied germplasm and useful for studying diversity and relationships among them and have potential marker aided selection, linkage mapping and evolutionary studies [61]. Cross species Simple sequence repeats (SSR) primers were used to study genetic diversity of 20 [62], 65 [63] and 24 [64] different accessions of okra by different scientists. SSR markers help to discrimination among okra accessions and provide vital information for use in the improvement of genomic resource in vegetable crop [64]. AFLP markers were separated the Greek landraces from other significant pool of variation [65]. Inter simple sequence repeat (ISSR) markers and morphological markers were effectively used to study 28 genotypes of West African okra [66].

5. Breeding studies

5.1 Mutation breeding in *Abelmoschus*

Induction of the mutation in plant breeding has become a well-known and important tool to supplement current germplasm and improvement of cultivars for the expression of specific traits. Several improved crop varieties have been released to farmers shows great economic value of the technology of mutation breeding [67]. From the past 70 years, near about 2252 mutant varieties from 175 crop plants including cereals, pulses, oilseeds, fibers, fruits, vegetables and ornamentals have been released in different countries throughout the world [68]. Induced mutations using different chemical and physical mutagens were studied by many breeders in *Abelmoschus*. Induction of mutation with the help of gamma radiations, ethyl methane sulphonate and diethyl sulphate in *Abelmoschus esculentus* cv. PusaSawani was reported [69] and 35 true breeding mutants were isolated [70]. The physical mutagen, Gamma radiations with different doses were useful to induced mutations for the screening of yellow vein mosaic disease resistant [71] and to develop agronomical and yield characters in okra [72–76]. Ethyl methane sulphonate was greatly affected most of the agronomic and yield characters in M_2 generation [77].

5.2 Crossability study among different species of *Abelmoschus*

Interspecific hybridization plays a vital role in the increasing genetic variation by interchanging genetic information in between different species, which is helpful to solve taxonomic relationship and also useful in preparation of genetic linkage map [78]. However artificial crossing methods are easy and simple in *Abelmoschus* but the rate of success is still an important constraint in the interspecific hybridization. Earlier reports revealed that crossing between cultivated species and wild species of *Abelmoschus* is more difficult. The breeding possibilities among the four species and one variety of *Abelmoschus* were studied [79] by Pal et al. The sterile hybrid obtained from the cross between *Abelmoschus esculentus* (2n = 130) and *A. tetraphyllus* (2n = 138) [80], partially fertile and resistant to yellow vein mosaic (YVM) virus was obtained from crossed between *A. manihot* and *A. manihot* ssp., *manihot* with the cultivated okra, *A. esculentus* cv. "PusaSawani" [81]. The interspecific crosses seems to be a major cause of variation perceived in cultivated species that are, *A. esculentus, A. manihot* and *A. moschatus* [82]. During reciprocal crosses between *A. caillei* and *A. tetraphyllus* 69–76 meiotic bivalents were obtained in F_1 hybrids [83]. The pollen germination and pollen tube growth behaviour with respect to seed set between four species of *Abelmoschus* revealed that, *A. caillei* may be served as potential connection parent for the transfer of alien gene for the okra breeding programmes [84].

6. Conclusion

Abelmoschus is an economically important genus which is distributed worldwide. Along with the cultivated species, some of wild species are traditionally used to cure many disorders. *A. manihot* is rich source of phytochemicals and antioxidant activities. *A. moschatus* has great number of phytoconstituents and contains aromatic compounds in its seeds which are utilized in perfume industries. *Abelmoschus* species contains mucilage which has several food and medicinal applications. It is useful in cleaning the sugar cane juice in jaggery preparation. Isolation and identification of many compounds from wild species proved to have diverse medicinal properties along with extraordinary nutritional potential. A large number of intervarietal combinations have been studied, but not much progress has been made in the improvement of this genus. As a rich source of phytochemicals and nutrition, wild species of *Abelmoschus* offers opportunities for the development as a substitute for cultivated species as a vegetable.

Author details

Ashwini Venkatrao Mohite and Rajaram Vithoba Gurav*
Department of Botany, Shivaji University, Kolhapur, Maharashtra, India

*Address all correspondence to: botanyraj@rediffmail.com

References

[1] Don G. General History of the Dichlamydeous Plants. Vol. 1. London: Malvaceae; 1831. pp. 430-505

[2] Quattrocchi U. CRC World Dictionary of Plant Names: Common Names, Scientific Names, Eponums, Synonyms, and Etymology. Boca Raton: CRC Press; 2000

[3] Borssum Waalkes JV. Malesian Malvaceae revised. Blumea. 1966;**14**:89-105

[4] Vredebregt JH. Taxonomic and ecological observations on species of *Abelmoschus* Medik. In: Report of an International Workshop on Okra Genetic Resources Held at NBPGR. New Delhi, India; 1991, 8-12 October 1990. pp. 60-76

[5] Bisht IS, Bhat KV. Okra (*Abelmoschus* ssp.). In: Singh RJ, editor. Genetic Resources, Chromosome Engineering, and Crop Improvement: Vegetable Crops. Vol. 3. New York: CRC Press; 2006. pp. 147-184

[6] Negi KS, Pant KC. Wild species of *Abelmoschus* Medik. (Malvaceae) from central Himalayan regions of India. Journal of the Bombay Natural History Society. 1998;**95**:148-150

[7] Sivarajan VV, Pradeep AK. Malvaceae of Southern Peninsular India. Delhi, India: Daya Pub. House; 1996. pp. 1-73

[8] Sutar S, Patil P, Aitawade M, John J, Malik S, Rao S, et al. A new species of *Abelmoschus* Medik.(Malvaceae) from Chhattisgarh, India. Genetic Resources and Crop Evolution. 2013;**60**:1953-1958

[9] Sabitha V, Panneerselvam K, Ramachandran S. In vitro α-glucosidase and α-amylase enzyme inhibitory effects in aqueous extracts of *Abelmoschus esculentus* (L.) Moench. Asian Pacific Journal of Tropical Biomedicine. 2012;**2**(1):S162–S164

[10] Anonymous. Fruits, Vegetable and Condiments Statistics 2006-07. Islamabad, Pakistan: MINFAL; 2008

[11] Saha D, Jain B, Jain VK. Phytochemical evaluation and characterization of hypoglycemic activity of various extracts of *Abelmoschus esculentus* Linn. Fruit. International Journal of Pharmacy and Pharmaceutical Sciences. 2011;**3**(2):183-185

[12] Harborne JB. Phytochemical Methods. London: Chapman and all, Ltd; 1973. pp. 49-188

[13] Adetuyi FO, Osagie AU, Adekunle AT. Antioxidant degradation in six indigenous okra *Abelmoschus esculentus* (L) Moench varieties during storage in Nigeria. Journal of Food Technology. 2008;**6**(5):227-230

[14] Shui G, Peng LL. An improved method for the analysis of major antioxidants of *Hibiscus esculentus* Linn. Journal of Chromatography A. 2004;**1048**:17-24

[15] Ahiakpa JK, Quartey EK, Amoatey HM, Klu GYP, Achel DG, Achoribo E, et al. Total flavonoid, phenolic contents and antioxidant scavenging activity in 25 accessions of okra (Abelmoschus spp L.). African Journal of Food Science and Technology. 2013;**4**(5):129-135

[16] Mohammed MI, Bayero AS, Shettima UI. Levels of total phenolic and flavonoids in *Abelmoschus esculentus* L. from some irrigation areas of Kano state- Nigeria. Bayero Journal of Pure and Applied Sciences. 2016;**9**(2):121-124

[17] Lai X, Liang H, Zhao Y, Wang B. Stimultaneous determination of seven active flavonols in the flower of *Abelmoschus manihot* HPLC. Journal

of Chromatographic Science. 2009;**47**:206-210

[18] Sudewi S, Lolo WA, Warongan M, Refai Y, Rante H. The ability of *A. manihot* L. leaf extract in scavenging of free radical DPPH and total flavonoid determination. IOP conf. Series. Materials Science and Engineering. 2017;**259**:12-20

[19] Liu IM, Tzeng TF, Liou SS. *Abelmoschus moschatus* (Malvaceae), an aromatic plant, suitable for medical or food uses to improve insulin sensitivity. Phytotherapy Research. 2010;**24**:233-239

[20] Mohite AV, Gurav RV. Nutraceutical and antioxidant evaluation of *Abelmoschus* taxa. International Journal of Vegetable Science. 2019;**25**(6):610-618

[21] Rao PU. Chemical composition and biological evaluation of okra (*Hibiscus esculentus*) seeds and their kernels. Plant Foods for Human Nutrition. 1985;**35**:389-396

[22] Sami R, Lianzhou J, Yang L, Ma Y, Jing J. Evaluation of fatty acid and amino acid compositions in okra (*Abelmoschus esculentus*) grown in different geographical locations. BioMed Research International. 2013;**2013**:1-6

[23] Camciuc M, Vilarem G, Gaset A. Volatile constituents of the seed teguments of *Abelmoschus esculentus* (L.) Moench. Journal of Essential Oil Research. 1999;**11**:545-552

[24] Dung NX, Khien PV, Nhuan DD, Hoi TM, Ban NK, Leclercq PA, et al. Composition of the seed oil of *Hibiscus abelmoschus* L. (Malvaceae) growing in Vietnam. Journal of Essential Oil Research. 1999;**11**(4):447-452

[25] Bown D. Encyclopaedia of Herbs and Their Uses. London: Dorling Kindersley; 1995

[26] Pieroni A, Ansari NM. Antioxidant activity of five vegetables traditionally consumed by South Asian migrants in Bradford, Yorkshire, UK. Phytotherapy Research. 2005;**19**:907-911

[27] Jarret RL, Wang ML, Levy IJ. Seed oil and fatty acid content in okra (*Abelmoschus esculentus*) and related species. Journal of Agricultural and Food Chemistry. 2011;**59**:4019-4024

[28] Arokiyaraj A, Choi SH, Lee Y, Bharanidharan R, Islam V, Vijaykumar B, et al. Characterization of ambrette seed oil and its mode of action in bacteria. Molecules. 2015;**20**:384-395

[29] Woolfe ML, Chaplin MF, Otchere G. Studies on mucilages extracted from okra fruits (*Hibiscus esculentus* L.) and baobab leaves (*Adansonia digitata* L.). Journal of the Science of Food and Agriculture. 1977;**28**:519-529

[30] Tomoda M, Shimizu N, Oshima Y, Takahashi M, Murakami M, Hikino H. Hypogycemic activity of twenty plant mucilages and three modified products. Planta Medica. 1987;**53**:8-12

[31] Ahmad MA, Bari HAA, Yunus RM. Studying the effect addition of okra-natural mucilage as drag reducing agent in different size of pipes in turbulent water flowing system. In: National Conference on Postgraduate Research (NCON-PGR), 1st October 2009. Malaysia: UMP Conference Hall, Centre for Graduate Studies, Universiti Malaysia Pahang; 2009. pp. 128-133

[32] Sengkhamparn N, Verhoef R, Schols HA, Sajjaanantakul T, Voragen AGJ. Characterisation of cell wall polysaccharides from okra (*Abelmoschus esculentus* (L.) Moench). Carbohydrate Research. 2009;**344**(14):1824-1832

[33] Ndjouenkeu R, Goycoolea FM, Morris ER, Akingbala JO. Rheology of okra (*Hibiscus esculentus* L.)

and dika nut (*Irvingia gabonensis*) polysaccharides. Carbohydrate Polymers. 1996;**29**:263-269

[34] Tomoda M, Shimizu N, Gonda R, Kanari M. Anticompementary and hypoglycemic activity of okra and hibiscus mucilages. Carbohydrate Research. 1989;**190**:323-328

[35] Mishra A, Clark JH, Pal S. Modification of okra mucilage with acrylamide: Synthesis, characterization and swelling behaviour. Carbohydrate Polymers. 2008;**72**:608-615

[36] Nair BR, Fasha KS. Isolation and characterization of mucilage from some selected species of *Abelmoschus* Medik. (Malvaceae) and their application in pharmaceutical suspension preparation. International Journal of Pharmacy and Pharmaceutical Sciences. 2013;**5**(1):398-402

[37] Gangurde AB. Preliminary characterization of *Abelmoschus esculentus* (L.) pod mucilage as o/w type emulsifier. International Journal of advances in Pharmacy, Biology and Chemistry. 2012;**1**(1):30-41

[38] Patel VI, Patel HA, Jani M, Shah A, Kumar S, Patel JA. Formulation and evaluation of okra fruit mucilage as a binder in Paracetamol and ibuprofen tablet. International Journal for Pharmaceutical Research Scholars. 2012;**1**(4):156-161

[39] Rajendra A, Bushetti SS, Giri A. Design and evaluation of compression coated formulations for an anti-inflammatory drug based on modified okra mucilage. Journal of Applied Pharmaceutical Science. 2012;**2**(7):238-245

[40] Jayan SC, Manoj NK. Isolation and evaluation of mucilage of *Abelmoschusesculentus* as a suspending agent. International Journal of Pharmaceutical Sciences and Chemistry. 2013;**2**(3):1311-1315

[41] Farooq U, Malviya R, Sharma PK. Extraction and characterization of okra mucilage as pharmaceutical excipient. Academic Journal of Plant Sciences. 2013;**6**(4):168-172

[42] Sharma N, Kulkarni GT, Sharma A. Development of *Abelmoschus esculentus* (okra) based mucoadhesive gel for nasal delivery of Rizatriptan benzoate. Tropical Journal of Pharmaceutical Research. 2013;**12**(2):149-153

[43] Nazni P, Vigneshwar P. Study on extraction and organoleptic evaluation of okra and hibiscus mucilage incorporated products. International Journal of Food and Nutritional Sciences. 2014;**3**(1):99-103

[44] Noorlaila A, Siti AA, Asmeda R, Norizzah AR. Emulsifying properties of extracted okra (*Abelmoschus esculentus* L.) mucilage of different maturity index and its application in coconut milk emulsion. International Food Research Journal. 2015;**22**(2):782-787

[45] Freitas TKFS, Oliveira VM, de Souza MTF, Geraldino HCL, Almeida VC, Favaro SL, et al. Optimization of coagulation-flocculation process for treatment of industrial textile waste water using okra (*A. esculentus*) mucilage as natural coagulant. Industrial Crops and Products. 2015;**76**:538-544

[46] de Cotrim MAP, Mottin AC, Ayres E. Preparation and characterization of okra mucilage (*Abelmoschus esculentus*) edible films. Macromolecular Symposia. 2016;**367**:90-100

[47] Araujo A, Galvão A, Filho CS, Mendes F, Oliveira M, Barbosa F, et al. Okra mucilage and corn starch bio-based film to be applied in food. Polymer Testing. 2018;**71**:352-361

[48] Adetuyi F. O., Osagie A.U. and Adekunle (2011) Nutrient, antinutrient, mineral and zinc bioavailability of okra *Abelmoschus esculentus* (L) Moench variety. American Journal of Food and Nutrition, 1(2): 49-54.

[49] Ahiakpa JK, Amoatey HM, Amenorpe G, Apatey J, Ayeh EA, Quartey EK, et al. Mucilage content of 21 accessions of okra (*Abelmoschus*spp L.). Scientia. 2014;**2**(2):96-101

[50] Sami R, Li Y, Qi B, Wang S, Zhang Q, Han F, et al. HPLC analysis of water-soluble vitamins (B2, B3, B6, B12 and C) and fat-soluble vitamins (E, K, D, A, and β-carotene) of okra (*Abelmoschus esculentus*). Journal of Chemistry. 2014;**2014**:1-6

[51] Gemede HF, Haki GD, Beyene F, Woldegiorgis AZ, Rakshit SK. Proximate, mineral, and antinutrient compositions of indigenous okra (*Abelmoschus esculentus*) pod accessions: Implications for mineral bioavailability. Food Science & Nutrition. 2015;**4**(2):223-233

[52] Yalambing L, Arcot J, Greenfield H, Holford P. Aibika (*Abelmoschus manihot* L.): Genetic variation, morphology and relationships to micronutrient composition. Food Chemistry. 2014;**193**:62-68

[53] Kolawole OM, Ajiboye AE, Aturu EE, Anibijuwon II. Effect of solar drying on the proximate and microbial composition of *Abelmoschus esculentus*. Journal of Microbiology and Biotechnology Research. 2011;**1**(1):71-81

[54] Tsado EK. Effect of method of drying okra fruits (*Abelmoschus esculentus*) on proximate composition of the dried product. Donnish Journal of Agricultural Research. 2015;**2**(8):66-70

[55] Poehlman JM, Sleper DA. Breeding Field Crops. 4th ed. Ames, Iowa, United States: Iowa State Press; 1995. pp. 75-76

[56] Bhat KV, Bisht IS, Mahajan RK, Rana RS. Analysis of the genetic relationship among *Abelmoschus* species using isozyme and RAPD markers. In: Abstract of Paper Presented in 2nd International Crop Science Congress. New Delhi, India; 1996. pp. 17-24

[57] Martinello GE, Leal NR, Amaral AT, Pereira MG, Daher RF. Genetic diversity in okra using RAPD markers. Horticultura Brasileira. 2003;**21**(1):20-25

[58] Haq I, Khan AA, Azmat MA. Assessment of genetic diversity in okra (*Abelmoschus esculentus* L.) using RAPD markers. Pakistan Journal of Agricultural Sciences. 2013;**50**(4):655-662

[59] Singh D, Dudi BS, Dhankhar SK, Rajkumar. Genetic diversity analysis of okra genotypes using morphological markers. International Journal of Current Microbiology and Applied Sciences. 2018;**7**(1):1667-1675

[60] Patel JS, Japda AR, Dhruve JJ. Assessment of genetic diversity of okra (*Abelmoschus esculentus* L.) for YVMV using RAPD and SSR markers. International Journal of Advanced Biological and Biomedical Research. 2018;**8**(2):2250-3579

[61] Gulsen O, Karagul S, Abak K. Diversity and relationships among Turkish okra germplasm by SRAP and phenotypic markers polymorphism. Biologia. 2007;**62**(1):41-45

[62] Sawadogo M, Ouedraogo JT, Balma D, Ouedraogo M, Gowda BS, Botanga C, et al. The use of cross species SSR primers to study genetic diversity of okra from Burkina Faso. African Journal of Biotechnology. 2009;**8**(11):2476-2482

[63] Schafleitner R, Kumar S, Lin C, Hegde SG, Ebert A. The okra (*Abelmoschus esculentus*) transcriptome as a source for gene sequence information and molecularmarkers for diversity analysis. Gene. 2013;**517**:27-36

[64] Fougat RS, Purohit AR, Kumar S, Parekh MJ, Kumar M. SSR based genetic diversity in *Abelmoschus* species. Indian Journal of Agricultural Sciences. 2015;**85**(9):99-104

[65] Kyriakopoulou OG, Arens P, Pelgrom KTB, Karapanos I, Bebeli P, Passam HC. Genetic and morphological diversity of okra [Abelmoschus esculentus (L.) Moench] genotypes and their possible relationships, with particular reference to Greek landraces. Scientia Horticulturae. 2014;**171**:58-70

[66] Ahiakpa JK, Magdy M, Werner O, Amoatey HM, Yeboah MA, Appiah AS, et al. Intra- specific variation in West African and Asian germplasm of okra (*Abelmoschus* spp. L.). Annals of Agricultural Science. 2017;**62**(2):131-138

[67] Micke A, Donini B, Maluszynski M. Induced mutations for crop improvement. Mutation Breeding Review. 1990;**7**:1-44

[68] Maluszynski M, Nichterlein K, Zanten LV, Ahloowalia BS. Officially released mutant varieties- the FAO/IAEA databasc. Mutation Breeding Review. 2000;**12**:1-84

[69] Yashvir. Induced quantitative mutations in okra (*Abelmoschus esculentus* Moench). Proceedings of the Indiana Academy of Sciences. 1975;**81**(4):181-185

[70] Abraham V. Inheritance of ten induced mutants in okra (*Abelmoschus esculentus* (Linn) Moench). Current Science. 1982;**54**(18):931-935

[71] Phadvibulya V, Boonsirichai K, Adthalungrong A, Srithongchai W. Selection for resistance to yellow vein mosaic virus disease of okra by induced mutation. In: Shu QY, editor. Induced Plant Mutations in the Genomics Era. Plant Breeding Section, Joint FAO/IAEA Division of Nuclear Techniques in Food and, International Atomic Energy Agency. Vienna, Austria, Rome: Food and Agriculture Organization of the United Nation; 2009. pp. 349-351

[72] Hegazi AZ, Hamideldin N. The effect of gamma irradiation on enhancement of growth and seed yield of okra [*Abelmoschus esculentus* (L.) Monech] and associated molecular changes. Journal of Horticulture Science and Forestry. 2010;**2**(3):38-51

[73] Warghat AR, Rampure NH, Wagh P. Effect of sodium azide and gamma rays treatment on percent germination, survival, morphological variation and chlorophyll mutation in musk okra (*Abelmoschus moschatus* L.). International Journal of Pharmacy and Pharmaceutical Sciences. 2011;**3**(5):483-486

[74] Ullah H, Khan RU, Khan SU, Mehmood S, Sherwani SK, Muhammad A, et al. Estimation of induced variability of yield contributing traits in M1 gamma irradiation germplasm of okra (*Abelmoschus esculentus* L.). South Asian Journal of Life Sciences. 2013;**2**(1):4-7

[75] Amir K, Hussain S, Shuaib M, Hussain F, Urooj Z, Khan WM, et al. Effect of gamma irradiation on okra (*Abelmoschus esculentus* L.). Acta Ecologica Sinica. 2018;**38**(5):368-373

[76] Mohite AV, Gurav RV. Induced mutation using gamma rays in okra (*Abelmoschus esculentus* (L.) Moench). Journal of Applied Horticulture. 2019;**21**(3):205-208

[77] Baghery MA, Kazemitabar SK. Effect of EMS induction on some morphological traits of okra (*Abelmoschus esculentus* L). International Journal of Biosciences. 2015;**6**(2):216-221

[78] Kaga A, Ishii T, Tsukimoto K, Tokoro E, Kamijima O. Comparative molecular mapping in *Ceratotropis* species using an interspecific cross between *Vignaumbellate* and *V. angularis*. Theoretical and Applied Genetics. 2000;**100**:207-213

[79] Pal BP, Singh HB, Swarup V. Taxonomic relationships and breeding possibilities of species of *Abelmoschus* related to okra (*A. esculentus*). Botanical Gazette. 1952;**113**:455-464

[80] Mamidwar RB, Nerkar YS, Jambhale ND. Cytogenetics of interspecific hybrids in the genus *Abelmoschus*. Indian Journal of Heredity. 1979;**11**:35-39

[81] Jambhale ND, Nerkar YS. Inheritance of resistance to okra yellow vein mosaic disease in interspecific crosses of *Abelmoschus*. Theoretical and Applied Genetics. 1981;**60**:313-316

[82] Siemonsma JS. West African okra- morphology and cytogenetical indications for the existence of a natural amphidiploid of *Abelmoschus esculentus* (L.) Moench and *A. manihot* (L.) Medikus. Euphytica. 1982;**31**:241-252

[83] Kondaiah GM, Kulkarni UG, Nerkar YS. The evolution of West Africa cultivated "Guinean" okra: Further evidence. Journal Of Maharashtra Agricultural Universities. 1990;**15**:65-68

[84] Patil P, Malik SK, Negi KS, John J, Yadav S, Chaudhari G, et al. Pollen germination characteristics, pollen-pistil interaction and reproductive behaviour in interspecific crosses among *Abelmoschus esculentus* Moench and its wild relatives. Grana. 2013;**52**(1):1-14

Chapter 4

Plant-Derived Compounds against Microbial Infections and Cancers

Gabin Thierry M. Bitchagno,
Vaderament-A. Nchiozem-Ngnitedem, Nadine Tseme Wandji,
Guy Cedric T. Noulala, Serge Alain T. Fobofou
and Bruno Ndjakou Lenta

Abstract

Plants synthesize and preserve a variety of metabolites known as natural products. Many of them are easily extractable and can be used as starting material or chemical scaffolds for various purposes, especially in drug discovery. Numbers of reports have listed valuable candidates with privilege scaffolds currently in active development as drugs. New compounds with anticancer and antiinfective activities have been discovered recently, some presented these backbones. The present book chapter aims to highlight these findings from plants which can be considered valuable for the development of new drugs against malignant cells and infective diseases. Interest in anti-infective agents is increasing due to the resistance of microorganisms to existing drugs and newly emerging infectious diseases. This resistance is also, nowadays, associated to some forms of cancers. In addition, the value of plants as essential part in the health care pipeline in low- and middle-income countries is under consideration even though these countries are almost all surrounded by a rich and untapped biodiversity. People are always relying on "modern drugs and treatment" which is unfortunately not affordable to all. Therefore, the present compilation of data on plant-derived compounds can inspire the formulation of ameliorated traditional medicines (ATM) against the targeted diseases and the conservation of species.

Keywords: phytoconstituents, anticancer, antimicrobial, biological cutoff points, sesquiterpenoid lactones, phenolic compounds

1. Introduction

1.1 General statement

As any other organisms on Earth, plants are said to possess multi-functional properties. They constitute feedstock materials to feed people and are reputed for their uses in medicines [1, 2]. History of plants has been always related to that of Human. Reports said Human have always insured their primary health care by using plants [3–5]. Even with the discovery of technology leading to synthetic drugs with sometimes more efficiency, plants still remain ubiquitous and safe for health concerns.

Research currently overflows in the literature related to the chemistry and biology of plants. Interests focus on experimental validation of ethnopharmacological uses of certain herb and formulation of plant extracts for a sustainable health care [1–5]. Therefore, plants are ground, exhausted and evaluated for various biological activity including properties to inhibit the growth of or to kill microorganisms and tumor cell lines. However, both microorganism and cancer cells become more and more resistant and remain serious threats for life. As an example, resistance to penicillin used for the treatment of lung infection ranged from 0 to 51% around the World and between 8 and 65% *Escherichia coli* associated with urinary tract infections presented resistance to ciprofloxacin, another antibiotic (https://www.who.int/health-topics/antimicrobial-resistance). WHO took some measures to diagnose and eradicate the issue but the problem is still actual and present.

More than half of existing antibiotics and anticancers are from synthesis of which almost a quarter takes its origin in natural substances isolated from plants, marine organisms and microorganisms [6]. Nevertheless, plant supply extracts continue to play a relevant role in human beings daily life. Up to date data show that plant extracts are reputed in food science where they are used as dietary supplements [6–8]. This practice is prevalent in Europe and North America where the interest in plants and related materials is rising up. Despite the progress made in the field of the synthesis of active principles for the formulation of medicaments, people still rely on natural occurring drugs due to their safety and uniqueness. The list of valuable substances from plants cannot be exhaustive.

In ancient time, the discovery of salicin, an *ortho-O*-glucopyranosylphenylethanol, from *Salix alba* led to the development of the reputed anti-inflammatory agent aspirin [9, 10]. Morphine, a benzylisoquinoline alkaloid isolated from *Papaver somniferum*, is a painkiller quite known in medicine and which also exist under its derivatives, heroin and codeine [9, 10]. Another alkaloid namely quinine isolated from *Cinchona succirubra* has been for long employed to cure malaria and fever related ailments but since 2004, almost all antimalarial drugs in the markets is made up of artemisinin isolated from the Chinese medicinal plant *Artemisia annua*. Artemisinin is commercialized under various acronyms including arteether, artesunate or artemether [10, 11]. In other hand, the chemotherapy of breast cancer uses the drug taxol which is the commercial name for paclitaxel, a diterpene isolated from *Taxus brevifolia*. Other compounds like ingenol-3-angelate, from *Euphorbia peplus*, known under the acronym Ingenol mebutate or the L-histidine-derived alkaloid pilocarpine found in *Pilocarpus jaborandi* are also some drugs used against other form of cancer [12–14].

Days after days, we keep discovering the deeply wealth of our surrounded nature. Reports abound in the literature especially on valuable natural compounds in drug development [6–8]. Most of them highlight natural product scaffolds as building blocks to the development of other compounds through synthesis [6–8]. A list of priority backbones has even been proposed to lead the development of new drugs [6].

However, interest in health care could also be to find out natural occurring compounds with considerable effects and low toxicity which can be introduced in the actual pipelines of treatment of a disease. That is, the sensibility of the found natural substance is not strong enough to compete commercial drugs but could be proposed alongside prescribed medicines because of its safety and availability. This can actually help in low- or middle-income countries to face certain diseases and build up a sustainable health care system. One can question how useful was the discovery of artemisinin for indigenous people if they have to wait years for pharmaceuticals companies to manufacture the drugs before its consumption. The same interrogation can also be valid to other discovery from plants, e.g. taxol, michellamine B or vinblastine. The plant sources of these active substances are known but still people from villages are waiting for "modern medicines" to take care of their respective health problems.

1.2 Problematic of microbial infections nowadays: drug resistance and climate change

According to WHO microbial infections are said to be the second cause of death globally, with low- and middle-income countries bearing the greatest burden. They include bacteria or fungi but viruses and protozoa diseases are also listed in this category (https://www.who.int/news-room/fact-sheets/detail/antimicrobial-resistance). Their origin preceded that of human life on earth [15]. In fact, human comes from successive mutations and evolution of bacteria [15]. Our body is made up of more than 100 trillion bacteria [16–18]. Some of them are useful in human life where they played a critical role in metabolism. However, greater percentage of them has been found to be harmful. By analogy to what is being said by believers, "you are dust and will return to dust," one can also argue that "you are bacteria and will return bacteria."

A lot of concerted effort has been put forward since the existence of mankind in trying to understand the biology of infective pathogens and their control. Some success has been achieved although there still more room for further research on this area. Through our constant manipulation and uses of these pathogens together with huge amount of chemicals, including drugs, we end up developing "new organisms" with different properties compared to their natural counterparts. In fact, the original pathogens start developing resistance to the drugs that were previously used for their eradication, making the problem worse [19–22].

The question of resistance of pathogens to commercialized drugs relies on the living environment of these small organisms. A misconception of bacteria considered that they exist as individual organisms [23]. However, things are different. Bacteria accumulated in colonics to live. They generally stick on a surface and gathered to survive together [23, 24]. Such a constitution known as biofilm is made up of bacteria somehow wrapped in a certain liquid (extracellular matrix) with strange properties. The entire constitution acts as a safety membrane for bacteria. The so-described making-up of this living organism constitutes the first barrier to bacteria and therefore the first stage of resistance [25, 26]. Biofilm are quite distributed in hospitals and nursing homes. They are claimed in household and industrial pipes, biomaterials such as contact lenses, medical devices including implants and urinary catheters, as well as plant and animal tissues. Cases of bacteria resistance have exploded this last decade especially in those zones [23–26]. This includes bacteria like *Acinetobacter*, *Pseudomonas* and various *Enterobacteriaceae* (*Klebsiella*, *E. coli*, *Serratia* and *Proteus*) which cause severe and often deadly infections such as bloodstream infections and pneumonia. Bacteria are carried over by devices such as ventilator systems and blood stream catheters. In high-income countries, 7% of all hospitalized people will contract some form of infection, including one in three people in intensive care units. In low- and middle-income countries, this figure rises to at least 10% of hospitalized people, and up to half of people in intensive care units, said the WHO (https://www.who.int/news-room/fact-sheets/detail/antimicrobial-resistance).

The World is currently facing an unprecedented rising of temperature, leading to certain change in habits and behavior. Discussion are mainly focused on how it could have an impact on human life, as it's getting colder and colder when it is the cold season or more and more hot when it is the hot season or even hot when it is supposed to be cold and vice versa. Politics in every country recommend adopting new habits to stop the rising up of the earth temperature. In the meantime, it seems that no one is caring about changes occurring at microorganism scale. It comes out that changes occurring in microorganisms due to climate change are not so important for us. And yet we should beware at least, the occurrences of the new viruses Ebola and Corona these last years should make us change our mind on how the World is changing. New microorganisms would be discovered, and the existing microorganisms could mutate to more harmful organisms. Fortunately, almost the same changes are also expected in plant kingdom even though the global warming and deforestation contribute also to plant extinctions. It is also awaited that new metabolites are being made as a result of new biosynthetical routes. These metabolites could either be directly active against pathogenic microorganisms or inspire new synthetical routes in laboratories to reach new drugs and medicines.

1.3 Standard antibacterial and anticancer cutoff points

Discussions are ongoing in the literature between scientists to clearly established standards for a substance to be considered for further steps in drugs development. Established standards are rare or inaccessible. The Kuete's group proposed standards of evaluating antimicrobial or anticancer properties of secondary metabolites derived from plants. They established that the antimicrobial activity of a crude extract can be considered significant when it MIC is below 100 μg/mL, moderate when between 100 and 625 μg/mL and low when more than 625 μg/mL. For pure compounds, the activity is considered significant when the MIC is below 10 μg/mL, moderate when between 10 μg/mL < MIC<100 μg/mL or low when greater than 100 μg/mL. One can notice that these standards are not considering the MIC of standard commercial drug references. We are not saying that the comparison of obtained MIC with those of the commercial drug is not necessary but the abovementioned cut-off points precise the ranges for a plant substance to be considered as valuable in drug discovery. Likewise, an extract is said to possess a good cytotoxicity if the IC_{50} values are below 4 μg/mL, modererate when 4 < IC_{50} < 20 μg/mL/10 < IC_{50} < 50 μM and low with IC_{50} above 100 μg/mL (250 μM) [14, 15]. While for cancer cell lines the activity of the pure compound is consider strong when IC_{50} < 10 μM [20, 21].

Another point of constant intensive discussion remains the relative low activity of most plant extracts and related constituents against microbial and cancer strains. Their activities are sometimes hundred- or thousand-fold less than the sensitivity of existing drugs. Some scientists find these activities not significant enough to be considered for clinical trials as a phytochemical substance should show comparable sensitivity with the commercial drugs. However, knowing that most cancer treatments are based on chemotherapy which is, as known, as harmful as the disease, can natural substances replace synthetic compounds? Likewise, one can also question infectious disease treatments in the same words. In what extent can we clearly consider a plant based products in drug development with respect to their activity against strains of bacteria or tumor cell lines? As mentioned above, the objective is not to compete with existing drugs developed with many expenditures but simply to select valuable extracts and phytoconstituents which could be used alongside actual treatments because of their safety and availability status especially for less developing countries.

1.4 World sustainable development goals related to the field

In September 2015, the UN adopted a list of 17 goals for a better life on the planet with emphasis on the quality of life for posterity (https://sustainabledevelopment.

un.org/?menu=1300). Globally it is recognized that ending poverty and other deprivations must go hand-in-hand with strategies that improve health and education, reduce inequality, and spur economic growth, by tackling climate change and working to preserve our oceans and forests are essential for our future as human beings. In the health sector, people should commit themselves to promote healthy lively hood and well-being for all at all ages. These are objectives stated in the Sustainable Development Goal number 3 of the list. Targets within this goal include ending the preventable deaths of newborns and children and ensuring access to effective medicines to all.

However, we are still living in a place where basic infections (malaria, typhoid, diarrhea, cholera and others) can cause death; a World where medicines are too expensive and inaccessible to everyone; a region where people have to walk more than 10 Km to expect treatment in a hospital or a World with increasing political and economic crisis. Nowadays, one should also highlight the increasing resistance of microbes and other pathogens to existing drugs and the occurrence of new strains of bacteria and virus. The former has been related to the overuse and misuse of drugs which modify the living pathogens environment making them used to it, thus developing tolerances to the used drugs.

One of the alternatives to tackle these challenging issues remain natural remedies and drugs. Many sources are being investigated but plants remain the most exploited. Substances from plants are quantitative, affordable, reachable and biologically recognized and easily metabolized by other organisms. They are environmentally friendly and can thus be promoted ever. Numbers of reports are available in the literature, highlighting the antimicrobial and anticancer properties of phytoextracts and products. Extracts can then be standardized and proposed to our fellow population to alleviate the cost of various and diverse drugs available in the markets.

1.5 Rationale of this survey

The present research literature aims to review recent plant compounds reported for their anticancer or antimicrobial properties which constitute valuable candidates to drug development. Our survey covers research reported from 2010. We only listed compounds with MICs or IC_{50s} < 10 μg/mL for a molarity scale ranging from 10^{-6}–20 μM. Activities of extracts were not highlighted herein. Both sensitive and resistant strains were checked out without restriction.

2. Plant-based secondary compounds with antimicrobial properties

Infective diseases are one of the most common illnesses in the World. They are currently the main concern on earth due to the ongoing Coronavirus (Covid-19) outbreak. Some pathogens spread out in animals and much of them are not known so far. But, at one moment or at another, due to our growing familiarization with wild animals, pathogens can spread within Human kingdom. Research are constantly been done to contain the diseases and come over the pathogens. Most of them are based on drug discovery, one of the oldest fields of Human concern so far. Plants constitute the main source of drugs although interests have moved to bioactive microbial constituents in the last decades mainly against microbial infectious. Owing to the rich biodiversity in our planet, the search for bioactive compounds from untapped natural resources is among the important ongoing projects.

One of the main constituents of plants with pronounced therapeutic interests against infective diseases are volatile oil. They are found in almost every organ of a plant but are said to be present in high extent in fruits and seeds. The composition of essential oil consists of monoterpenes and sesquiterpenes paired with aromatic compounds and lightweight esters, fatty acids, alcohols, ketones and aldehydes. Some examples include γ-terpinene, carvacrol, *p*-cymene, thymol, linalool, α-terpinene, limonene, eucalyptol, geranyl propionate and α- and β-pinene [27]. Owing to their high hydrophobicity, essential oil are said to impair the cell membrane of microbes, increase their membrane permeability and decrease their cytoplasmic pH [28]. The so-described abilities explained their significant activity against bacteria and fungi including resistant strains like *Staphylococcus sp.* and *Pseudomonas sp.* with MIC values approaching 0.01 µg/mL [27]. Volatile oil play also an essential role in protecting and even preventing biofilm development which is very important as presented above [29]. However, the same lipophilicity capacity of essential oil, relevant for their good antiinfective properties, constitutes also their

Family	Species (part)	Compound name	Test microorganisms (MIC in µg/mL)	Refs.
Guttiferae	*Garcinia mangostana*	Mangostin A (**1**)	MRSA (6.25 µg/mL), VRE (3.13 µg/mL)	[30]
	Garcinia cowa (fruits)		*B. cereus* (0.5 µg/mL), *B. subtilis* (0.25 µg/mL), *M. luteus* (1.0 µg/mL)	[31]
	Garcinia mangostana	Mangostin Y (**2**)	MSSA (6.25 µg/mL), MRSA (3.13 µg/mL), VRE (6.25 µg/mL), VSE (6.25 µg/mL)	[30]
	Garcinia cowa (stem barks)	Cowanol (**3**)	MRSA SK1 (2 µg/mL), *S. aureus* (8 µg/mL)	[32]
		Cowagarcinone E (**4**)	MRSA SK1 (8 µg/mL)	
		Garciniacowone (**5**)	MRSA SK1 (2 µg/mL), *S. aureus* (2 µg/mL)	
		Cowanin (**6**)	MRSA SK1 (4 µg/mL)	
	Garcinia cowa (fruits)		*B. subtilis* (4 µg/mL), *M. luteus* (4 µg/mL)	[31]
	Garcinia cowa (fruits)	Garcicowanone A (**7**)	*B. cereus* (0.25 µg/mL), *B. subtilis* (2 µg/mL), *M. luteus* (4 µg/mL),	
		9-Hydroxycalabaxanthone (**8**)	*B. cereus* (8 µg/mL), *B. subtilis* (2 µg/mL), *M. luteus* (4 µg/mL),	
		Β-mangostin (**9**)	*B. cereus* (0.25 µg/mL), *B. subtilis* (4 µg/mL)	
		Cowagarcinone E (**10**)	*B. cereus* (4 µg/mL), *B. subtilis* (4 µg/mL), *M. luteus* (8 µg/mL)	
		Rubraxanthone (**11**)	*B. cereus* (2 µg/mL), *B. subtilis* (1 µg/mL), *M. luteus* (2 µg/mL)	
	Garcinia smeathmannii (stem barks)	1,3,5,8-Tetrahydroxy-2- (3-methy but-2-enyl)-4- (3,7-dimethylocta-2,6- dienyl) xanthone (**12**)	*E. faecalis* (8 µg/mL)	[33]
		Cheffouxanthone (**13**)	*E. faecalis* (8 µg/mL)	
		Ananixanthone (**14**)	*E. faecalis* (2 µg/mL)	

Family	Species (part)	Compound name	Test microorganisms (MIC in μg/mL)	Refs.
Clusiaceae	*Allanblackia gabonensis (fruits)*	Morelloflavone (**15**)	ATCC8739 (8 μg/mL)	[34]
Myristicaceae	*Pycnanthus angolensis (roots)*	Pycnanthulignene A (**16**)	MRSA (9.8 μg/mL)	[35]
		3,4-Dimethoxy-3′,4′-methylenedioxy-7,7′-epoxylignan (**17**)	*M. smegmatis* (9.8 μg/mL)	
		4,5-Dimethoxy-3′,4′-methylenedioxy-2,7′-cycloligna-7,7′-diene (**18**)	*M. tuberculosis* (9.8 μg/mL)	
Dioscoreaceae	*Dioscorea bulbifera* (Bulbil)	Bafoudiosbulbins C (**19**)	*M. smegmatis* ATCC700084, *M. tuberculosis* ATCC27294 and *M. tuberculosis* MTCS2 (8 μg/mL)	[36]
Clusiaceae	*Garcinia nobilis* (stem bark)	4-Prenyl-2-(3,7-dimethyl-2,6-octadienyl)-1,3,5,8tetrahydroxyxanthone (**20**)	*M. tuberculosis* ATCC27294 and *M. tuberculosis* MTCS2 (8 μg/mL)	[37]
Moraceae	*Dorstenia manii* (roots)	Dorsmanin C (**21**)	*P. aeruginosa* PA 01 and *E. coli* ATCC 10536 (4 μg/mL)	[38]
		Dorsmanin F (**22**)	*P. aeruginosa* PA 01 and *E. coli* ATCC 10536 (4 μg/mL), *K. pneumonia* PA01 (8 μg/mL)	
		Dorsmanin E (**23**)	*Candida albicans* TCC9002 (8 μg/mL)	
	Ficus exasperata (stem bark)	(S)-(−) Oxypeucedanin hydrate (**24**)	*B. cereus* (9.76 μg/mL)	[39]
		(R)-(+) Oxypeucedanin hydrate (25)		
	Trilepisium madagascariense (stem bark)	Dihydrokaempferol (**26**)	*E. coli* ATCC8739 (8 μg/mL)	[40]
Rutaceae	*Fagara tessmannii* (roots)	Bergenin (**27**)	E.AG102 and *K. pneumoniae* ATCC 11296 (4 μg/mL) *E. coli* ATCC 8739, *K. pneumoniae* ATCC 11296, *K. pneumoniae* KP 55, *P. stuartii* PS 299645 and *P. aeruginosa* PA01 (8 μg/mL)	[41]
Hypericaceae	*Harungana madagascariensis* (bark)	Ferruginin (**28**)	*E. coli* ATCC 10536, *K. pneumoniae* K2 and *E. cloacae* BM 67 (4 μg/mL) *E. aerogenes* ATCC 13048, *E. aerogenes* EA 294, *P. aeruginosa* PA01 and *K. pneumoniae* KP 55 (8 μg/mL) *K. pneumonia* ATCC 11296, *E. cloacae* BM 47 and *E. coli* ATCC 8739, *E. aerogenes* ATCC 13048, *K. pneumoniae* KP 55, *P. stuartii* (8 μg/mL)	[42]
Fabaceae	*Erythrina sigmoidea* (leaves)	Neobavaisoflavone (29)	*E. coli* ATCC 8739, *E. cloacae* ECC 169, *K. pneumoniae* KP 55, *P. stuartii* NAE16 and	[43]

Family	Species (part)	Compound name	Test microorganisms (MIC in μg/mL)	Refs.
			P. aeruginosa PA01 (8 μg/mL) *P. stuartii* ATCC 29916, *E. cloacae* BM 47 (4 μg/mL)	
Moraceae	*Milicia excels* (roots and leaves)	2-(3,5-Dihydroxyphenyl) benzofuran-5,6-diol (**30**)	*E. coli* ATCC 8739, *K. pneumonia* ATCC 11296, *E. cloacae* BM 47 (4 μg/mL) *E. coli* AG 102 and *K. pneumoniae* KP 55 (8 μg/mL)	[44–46]
		Candidone (**31**)		
Myristicaceae	*Myristica fragrans* (seeds)	3′,4′,7-Trihydroxyflavone (**32**)	*E. coli* ATCC 8739 (8 μg/mL) *P. stuartii* ATCC (199645) (4 μg/mL)	[47, 48]
Hypericaceae	*Hypericum roeperianum* (stem Bark)	1,4,6,7-Tetrahydroxyxanthone (**33**)	*P. aeruginosa* PA01 (2 μg/mL)	[49]
Fabaceae	*Entada abyssinica* (leaves)	Entadanin (**34**)	*S. typhimurium* (1.56 μg/mL)	[50]
		Quercitrin (**35**)	*S. typhimurium* (3.12 μg/mL)	
Meliaceae	*Pseudocedrela kotschyi* (stem bark)	3,4-Secotirucalla- 4 (28),7,24-trien-3,21- dioic acid (**36**)	*S. aureus* (4 μg/mL)	[51]
		3,4-Secotirucalla- 4 (28),7,24-trien-3,21- dioic acid (**36**) and 3- methyl ester 3,4- secotirucalla-4 (28),7,24- trien-3,21-dioic (**37**) (1:1)	*S. aureus* (8 μg/mL)	
Rubiaceae	*Crossopteryx febrifuga* (stem bark)	18-epi-3β-D-Glucopyranosylurs- 12,20 (30)diene-27,28- dioic acid (**38**)	*K. pneumoniae* ATCC11296 (8 μg/mL)	[52]
Lamiaceae	*Leucosceptrum canum* (aerial part)	4-En-3-keto-stigmasterol (**39**)	*M. luteus* (9.5 μg/mL)	[53]
		Stigmast-5-en-3-acetate (**40**)	*M. luteus* (4.2 μg/mL)	
		4′,5,7-Trihydroxy-6 methoxyflavone (**41**)	*M. luteus* (6.5 μg/mL)	
		Leucoperoxyterpene (**42**)	*M. luteus* (5.4 μg/mL) and *S. minor* (5.4 μg/mL)	
Caryophyllaceae	*Silene rubella* (aerial part)	Oleanolic acid (**43**)	VRE (6.36 μg/mL)	[54]
Fabaceae	*Entada abyssinica* (leaves)	Ursolic acid (**44**)	*B. cereus* (6.25 μg/mL)	[55]

Methicillin-resistant Staphylococcus aureus (MRSA), vancomycin resistant Enterococcus (VRE), methicillin-sensitive Staphylococcus aureus (MSSA), vancomycin-sensitive Enterococcus (VSE), Staphylococcus aureus (S. aureus), Bacillus subtilis (B. subtilis), Bacillus cereus (B. cereus), Micrococcus luteus (M. luteus), Mycobacteria smegmatis (M. smegmatis), Mycobacteria tuberculosis (M. tuberculosis), Pseudomonas aeruginosa (P. aeruginosa), Escherichia coli (E. coli), Klebsiella pneumonia (K. pneumonia), Candida albicans (C. albicans), Bacillus cereus (B. cereus), Escherichia coli (E. coli), Klebsiella pneumoniae (K. pneumoniae), Providencia stuartii (P. stuartii), Pseudomonas aeruginosa (P. aeruginosa), Enterobacter cloacae (E. cloacae), Enterobacter aerogenes (E. aerogenes), Salmonella typhimurium (S. typhimurium), Staphylococcus aureus (S. aureus), Micrococcus luteus (M. luteus), Streptococcus minor (S. minor), Vancomycin resistant Enterococcus (VRE).

Table 1.
Examples of plant-based natural products with significant antiinfective properties.

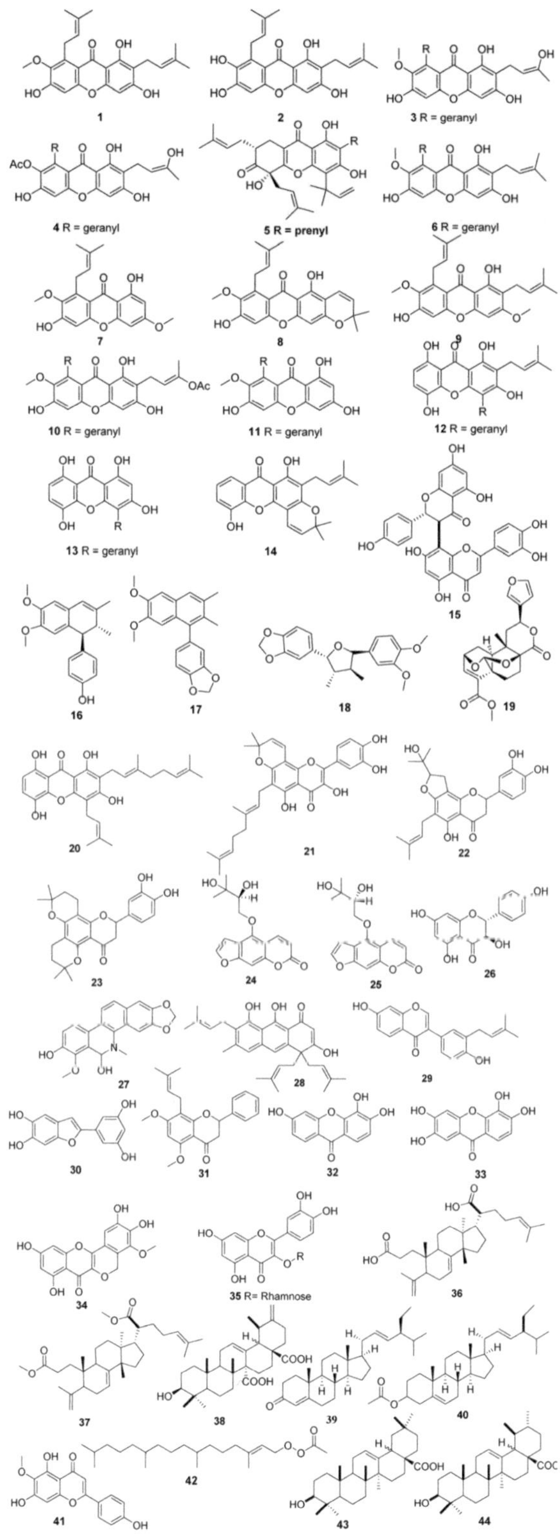

Figure 1.
Bioactive compounds against infective bacteria and fungi.

main bottlenecks in drug development because essential oil present a low bioavailability. But when isolated, some of their constituents are water soluble e.g. 1.25 mg/mL at 25°C for carvacrol (https://pubchem.ncbi.nlm.nih.gov/compound/carvacrol#section=Solubility) or 1.59 mg/mL at 25°C (https://pubchem.ncbi.nlm.nih.gov/compound/6549) for linalool and are being studied and used as excipient in drug formulation. Essential oil are however reputed in therapies which promoted local application, inhalation or bath modes of treatment like aromatherapy [27, 56]. They are associated to numerous of ailments including depression, indigestion, headache, insomnia, muscular pain, respiratory problems, skin ailments, swollen joints or urine complications [56]. Unlike essential oil, secondary metabolites mainly found in the solid part of a plant extract can present significant activities with considerable bioavailability and hence, constitute the main research object in natural product domain.

Since 2010 at least 44 metabolites have been reported with MIC values below 10 μg/mL. Phenolic compounds (**1–18, 20–26, 28–35, 41**) were the group of compounds mostly active among the metabolites found. Besides, benzophenanthridine (**27**), steroids (**36–37, 40**), pentacyclic triterpenoids (**38–39, 43–44**) and diterpenoids (**19, 42**) have been found active against various infective strains (**Table 1** and **Figure 1**). Some of the strains studied are among the microbes listed by WHO as highly harmful and needed new drugs.

3. Plant-based compounds with anticancer features

The chemistry and biology of plants to fight against malignant cells are wide and diverse. Various classes of metabolites have been reported to possess valuable anticancer properties. As a recall, taxol, one of the mostly used anticancer drug in chemotherapy, is a complex diterpene-based metabolite; vinblastine, used in the therapy of various cancer as well, is made up of terpenic indol-type alkaloids and artemisinin or parthenolide actually in active clinical trials for cancer drugs possess a sesquiterpenoid lactone backbone. Since 2010, more than 72 compounds have

Family	Species (part)	Compound name	Cancer cell lines (IC_{50})	Refs.
Asparagaceae	*Bellevalia eigii* (bulbs)	5,7,3′-triihydroxy-4′- methoxy Homoisoflavanone (**45**)	MDA-MB-435 (1.0 μM)	[57]
		5,3′-dihydroxy-4′,7,8-trimethoxy Homoisoflavanone (**46**)	MDA-MB-435 (1.1 μM)	
		7-*O*-methyl-3′-hydroxypunctatin (**47**)	MDA-MB-435 (4–6 μM)	
	Bellevalia flexuosa (bulbs)	3′-hydroxy-3,9-dihydroeucomin (**48**)	MDA-MB-435 (1–6 μM); MDA-MB- 231 (9–5 μM)	[58]
Asparagaceae	*Urginea depressa* (whole plant)	Urgineanin A (**49**)	H522-T1 (0–071 μM); A2780 (0.32 μM); A2058 (0.068 μM)	[59]
		Urgineanin B (**50**)	H522-T1 (6.78 μM); A2780 (3.4 μM)	
		Urgineanin C (**51**)	H522-T1 (0.74 μM); A2780 (1.35 μM), A2058 (0.69 μM)	
		Urgineanin D (**52**)	H522-T1 (0.43 μM); A2780 (0.35 μM), A2058 (0.38 μM)	
		Urgineanin E (**53**)	A2780 (1.44 μM)	
		Urgineanin F (**54**)	A2780 (2.3 μM)	

Family	Species (part)	Compound name	Cancer cell lines (IC_{50})	Refs.
Convallariaceae	*Ophiopogon japonicus* (tubers)	Homoisopogon A **(55)**	KB (0.51 μM); LU-1 (0.66 μM); SK- Mel-2 (0.66 μM)	[60]
	Ophiopogon japonicus (roots)	5,7,4′-Trihydroxy-3′-methoxy-6,8-dimethylhomoisoflavanone **(56)**	A549 (6.40 μM)	[61]
		Methylophiopogonanone B **(57)**	A549 (0.84 μM)	
		Methylophiopogonanone A **(58)**	A549 (1.66 μM)	
Liliaceae	*Scilla persica* (bulbs)	Scillapersicene **(59)**	AGS (8.4 μM)	[62]
Amaryllidaceae	*Crinum zeylanicum* (whole plant)	Ungeremine **(60)**	CCRF-CEM (4.89 μM); MDA-MB - *231-pcDNA* (5.47 μM); MDA-MB- 231-*BCRP* (3.67 μM); HCT116 (*p53*$^{+}$/$^{+}$) (6.45 μM) HCT116 (*p53*$^{-}$/$^{-}$) (7.06 μM); U87MG (5.38 μM)	[63]
Euphorbiaceae	*Macaranga balansae* (fruits)	6,8-Diprenyl-4-methyl-naringenin **(61)**	Pan C1 (7.89 μM)	[64]
		(2*S*)-6-Farnesylnaringenin **(62)**	P388 (3.27 μg/mL)	
		6-Farnesyl-3′,4′,5,7-tetrahydroxy flavanone **(63)**	P388 (2.61 μg/mL)	
	Macaranga triloba (inflorescences)		HeLa (1.3 μg/mL), HL-60 (3.3 μg/mL)	[65]
Euphorbiaceae	*Macaranga tanarius* (fruits)	Vedelianin **(64)**	KB (0.050 μM), MCF-7 (0.050 μM)	[66]
		Schweinfurthin E **(65)**	KB (0.050 μM), MCF-7 (0.030 μM)	
		Schweinfurthin F **(66)**	KB (0.10 μM), MCF-7 (0.12 μM)	
		Schweinfurthin H **(67)**	KB (0.26 μM)	
Thelypteridaceae	*Cyclosorus parasiticus* (leaves)	Parasiticin C **(68)**	SW1990 (2.33 μM), MDA-MB-231 (4.88 μM), MCF-7 (4.16 μM), HepG2 (1.6 μM), A.549 (5.50 μM), ALLSIL (6.06 μM)	[67]
		2′,4′-Dihydroxy-6′-methoxy-3′,5′-Dimethylchalcone **(69)**	SW1990 (6.64 μM), MDA-MB-231 (9.67 μM), MCF-7 (8.49 μM), HepG2 (2.82 μM), A549 (7.89 μM), ALLSIL (9.50 μM)	
Moraceae	*Artocarpus obtusus* (stem bark)	Pyranocycloartobiloxanthone A **(70)**	HL60 (0.5 μg/mL), K562 (2.0 μg/mL)	[68]
Guttiferae	*Calophyllum soulattri* (stem bark)	Soulattrin **(71)**	Raji (1.01 μM/mL), LS174T (1.25 μM/mL), IMR-32 (0.27 μg/mL), SK-MEL-28 (0.57 μM/mL).	[69]
Guttiferae	*Garcinia xanthochymus* (stem bark)	1,3,5,6-T etrahydroxy-4,7,8-tri (3-methylbut-2-enyl) xanthone **(72)**	PC-3 (6.8 μM)	[70]

Family	Species (part)	Compound name	Cancer cell lines (IC_{50})	Refs.
Papaveraceae	*Macleaya microcarpa* (roots)	Maclekarpine A (**73**)	BGC-823 (0.7 μM)	[71]
		Maclekarpine C (**74**)	HCT-8 (1.9 μM), Bel-7402 (2.1 μM), A2780 (1.6 μM), A549 (3.4 μM)	
		Maclekarpine D (**75**)	HCT-8 (1.9 μM), BGC-823 (0.2 μM), A2780 (2.0 μM)	
		Maclekarpine E (**76**)	BGC-823 (0.1 μM)	
		6-Methoxydihydrochelerythrine (**77**)	HCT-8 (1.1 μM), Bel-7402 (0.9 μM) BGC-823 (0.8 μM), A2780 (1.8 μM)	
		Dihydrosanguinarine (**78**)	HCT-8 (1.3 μM), Bel-7402 (2.3 μM) BGC-823 (0.1 μM), A2780 (2.1 μM)	
		Dihydrochelerythrine (**79**)	HCT-8 (1.4 μM), BGC-823 (0.4 μM), A2780 (3.5 μM)	
		6-Butoxydihydrochelerythrine (**80**)	HCT-8 (1.7 μM), Bel-7402 (1.3 μM) BGC-823 (0.7 μM) A2780 (1.8 μM),	
Papaveraceae		Bis[6-(5,6-dihydrochelerythrinyl)] ether (**81**)	HCT-8 (1.6 μM), Bel-7402 (2.1 μM) BGC-823 (0.1 μM), A2780 (1.6 μM),	
		6-Methoxydihydrosanguinarine (**82**)	HCT-8 (0.5 μM), Bel-7402 (0.5 μM) BGC-823 (0.6 μM), A2780 (0.5 μM), A549 (0.6 μM)	
Amaryllidaceae	*Zephyranthes candida* (whole plant)	*N*-methylhemeanthidine Chloride (**83**)	HL-60 (0.91 μM), K562 (1.0 μM), A549 (1.1 μM), HepG2 (1.5 μM), HT-29 (1.2 μM)	[72]
		Hemeanthamin (**84**)	HL-60 (1.4 μM), K562 (2.5 μM), A549 (2.5 μM), HepG2 (4.8 μM), HT-29 (2.1 μM)	
		Lycorine (**85**)	HL-60 (1.6 μM), K562 (2.3 μM), A549 (1.9 μM), HepG2 (3.7 μM), HT-29 (3.2 μM)	
		N-phenethylcrinasiadine (**86**)	HL-60 (1.6 μM), K562 (2.3 μM), A549 (1.9 μM), HepG2 (3.7 μM), HT-29 (3.2 μM)	
Asparagaceae	*Bellevalia flexuosa* (bulbs)	Urginin B (**87**)	A2780 (0.011 μM), A2058 (0.060 μM), H522-T1 (0.044 μM)	[59]
		Urginin C (**88**)	A2780 (0.041 μM), A2058 (0.076 μM), H522-T1 (0.051 μM)	
		14*β*-Bydroxy-19*β*-oxobufa-4,20, 22-trienolide-3*β*-O-*β*-D-gluco pyranoside (**89**)	A2780 (0.024 μM), A2058 (0.048 μM), H522-T1 (0.034 μM)	
		14*β*-Hydroxybufa-4,20,22-trien olide-3*β*-*O*-{α-L-rhamnopyranosyl-[(1 → 4)-*β*-D-glucopyranosyl]-(1 → 3)-*α*-L-rhamnopyranoside} (**90**)	A2780 (0.111 μM), A2058 (0.18 μM), H522-T1 (0.11 μM)	
Asteraceae	*Leptocarpha rivularis*	Leptocarpin (**91**)	DU-145 (2.0 μM), PC-3 (4.5 μM), HT-29 (3.8 μM), MCF7 (3.1 μM), MDA-MB-231	[73]

Family	Species (part)	Compound name	Cancer cell lines (IC_{50})	Refs.
			(6.4 µM), CCD 841 CoN (5.2 µM)	
	Smallanthus sonchifolius (leaves)	Enhydrin (**92**)	CCRF-CEM (3.6 µM)	[74]
		Uvedalin (**93**)	CCRF-CEM (9.2 µM)	
		Polymatin B (**94**)	CCRF-CEM (0.8 µM), CEM-ADR5000 (1.3 µM), MIA-PaCa-2 (3.7 µM)	
		Sonchifolin (**95**)	CCRF-CEM (3.1 µM), CEM-ADR5000 (3.1 µM), MIA-PaCa-2 (7.4 µM)	
		8β-Angeloyloxy-9α-hydroxy-14-oxo-acanthospermolide (**96**)	CCRF-CEM (2.2 µM), CEM-ADR5000 (6.7 µM), MIA-PaCa-2 (8.9 µM)	
		Fluctuanin (**97**)	CCRF-CEM (0.6 µM), CEM-ADR5000 (1.4 µM), MIA PaCa-2 (4.4 µM)	
	Ambrosia cumanensis (aerial parts)	2,3-Dehydropsilostachyn C (**98**)	Jurkat (6.0 µM), U937 (8.0 µM)	[75]
	Sonchus palustris (roots)	15-*p*-Hydroxyphenylacetyllactucin (**99**)	CEM (5.1 µM), BJ (9.8 µM)	[76]
		15-*p*-Methoxyphenylacetyllactucin (**100**)	CEM (3.9 µM), BJ (8.4 µM)	
Compositae	*Carpesium abrotanoides* (whole plant)	Caroguaianolide A (**101**)	MDA-MB-231 (7.96 µM)	[77]
		Caroguaianolide B (**102**)	MDA-MB-231 (4.25 µM), HGC-2 (6.47 µM)	
	Carpesium abrotanoides (whole plant)	Caroguaianolide C (**103**)	MDA-MB-231 (2.67 µM), HGC-2 (4.83 µM)	
		Akihalin (**104**)	MDA-MB-231 (4.83 µM), HGC-2 (7.35 µM)	
		4β-Hydroxy,l0β-hydroperoxyl,5αh,7αh,8βh-guaia-1,11(13)- dien-8α,12-olide (**105**)	MDA-MB-231 (5.79 µM)	
		4α-Hydroxy-1βh-guaia-9,11 (**13**)-dien-12,8α-olide (106)	MDA-MB-231 (4.07 µM), HGC-2 (8.95 µM)	
		(3ar,4as,5S,7as,8S,9ar)-5-Hydroxy-4a,8-dimethyl-3-methylen-decahy droazuleno[6,5-b]furan- 2 (3H)-on (**107**)	MDA-MB-231 (5.32 µM)	
	Carpesium faberi (whole plant)	Guaianodilactones A (**108**)	CCRF-CEM (9.13 µM)	[78]
		Guaianodilactones C (**109**)	CCRF-CEM (4.74 µM)	
		Guaianodilactones B (**110**)	CCRF-CEM (2.03 µM)	
Asteraceae	*Inula japonica* (aerial part)	Neojaponicone B (**111**)	Jurkat (5.9 µM), 6 T-CEM (4.4 µM)	[79]
		Inulanolide E (**112**)	Jurkat (5.5 µM), 6 T-CEM (4.6 µM)	
		Inulanolide A (**113**)	Jurkat (5.8 µM), 6 T-CEM (4.3 µM)	

Family	Species (part)	Compound name	Cancer cell lines (IC_{50})	Refs.
		Japonicone Q (**114**)	Jurkat (3.3 μM), 6 T-CEM (2.7 μM)	
		Japonicone N (**115**)	Jurkat (2.5 μM), 6 T-CEM (2.4 μM)	
		Japonicone S (**116**)	Jurkat (4.5 μM), 6 T-CEM (3.3 μM)	
		Japonicone A (**117**)	Jurkat (3.1 μM) 6 T-CEM (2.2 μM)	

Melanoma (MDA-MB-435, A2058 and SK-Mel-2); human non-small-cell lung (H522-T1), ovarian cancer (A2780), human epidermoid carcinoma (KB), human lung adenocarcinoma (LU-1), breast (MDA-MB-231), gastric (AGS), human myeloid leukemia (K562), human gastric (SGC-7901), human-lung-tumor (A549). Murine leukemia (P-388), human pancreatic (Pan C1 and SW1990), human cervical carcinoma (Hela), mouse leukemia (P388), human leukemia (HL-60 and ALL-SIL), mouth epidermal carcinoma cells (KB), breast cancer (MCF-7 and MDA-MB-231) lung cancer (A549), hepatocellular carcinoma (HepG2), human promyelocytic leukemia (HL60), human chronic myeloid leukemia (K562), B-lymphocyte (Raji), colon carcinoma (LS174T), human neuroblastoma (IMR-32), skin carcinoma (SK-MEL-28), colon (HCT-8), liver (Bel-7402), stomach (BGC-823), ovarian (A2780), lung (A549). Myeloid leukemia (HL-60 and K562), lung (A549), hepatocellular carcinoma (HepG2), colon (HT-29), ovarien (A2780), melanoma (A2058), non-small-cell lung cancer (H522-T1), lymphoblastic leukemia cell line (CCRF-CEM), resistant T-cell leukemia cell line (CEM-ADR5000), the pancreatic, carcinoma cell line (MIA PaCa-2)–, and on peripheral blood mononuclear cells (PBMC) from healthy human subjects, HeLa (cervical carcinoma), Jurkat (T-cell leukemia), and U937 (monocytic leukemia) cell lines, human acute lymphoblastic leukemia (CEM), and normal human skin fibroblasts (BJ), human breast (MDA-MB-231), human gastric (HGC-27), human leukemia (CCRF-CEM).

Table 2.
Examples of significant secondary metabolites with antiproliferative properties.

been reported with considerable antiproliferative activity against different cancer cell lines with IC_{50} ranging from 0.001–10 μM.

These compounds are distributed in homoisoflavonoids (**45–60**), isoprenyl-flavonoids (**61–63, 68–70**), stilbenoids (**64–67**), xanthones (**71–72**), benzophenanthridines (**73–82**), *Amaryllidaceae*-type alkaloids (**83–86**), cardenolides (**87–90**) and sesquiterpenoid lactones (**91–117**). Their respective sensibility toward tumor cell lines are depicted in **Table 2** and their respective structures in **Figure 2**. As expected, sesquiterpenoid lactones were the most exploited metabolites. They are reputed for their ability to induce apoptosis in cancer cell lines with good selectivity. Homoisoflavonoids were the second most important group of compounds found to exhibit high cytotoxicity herein. The interest in this class of metabolites for anti-cancer solution is most likely related to their potency as inhibitor of angiogenesis both *in vitro* and *in vivo*, without showing any toxicity [80]. On the other hand, benzophenanthridines are reputed for their bioavailability because they contained more often ionic bond besides their bioactivity. Their mode of action in cancer therapy includes either the inhibition of mitosis via a reaction of the imine bond with the sulfhydryl nucleophile in protein and enzyme or the enzymatic activities of DNA Topoisomerase I and Topoisomerase II by implantation into DNA molecules to retard the fast proliferation of tumor cells [81].

4. Natural products in active development for drug discovery

Knowing the categories of compounds which has been screened and choose for clinical trials is quite important. It can help redefine our objectives and outlines in research. However, such information is not accessible easily. Almost all pharmaceutical makers keep this information for private uses. Nevertheless, available reports before 2010 on valuable compounds in development for cancer therapy for instance can continue to be used and analyze. There are privilege structures with

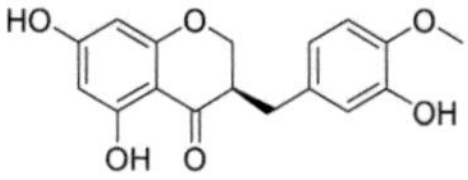

45

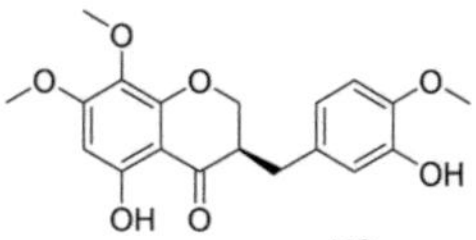

46

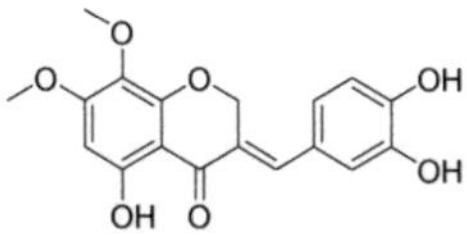

47

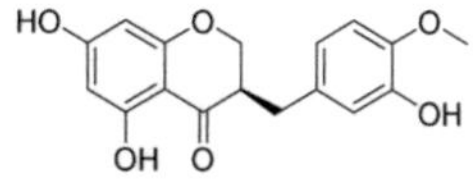

48

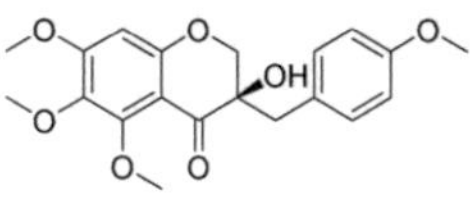

49

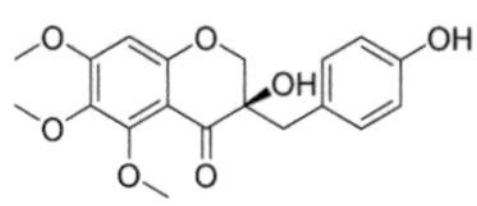

50

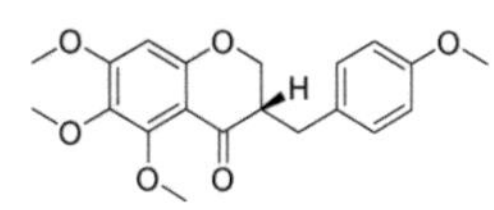

51

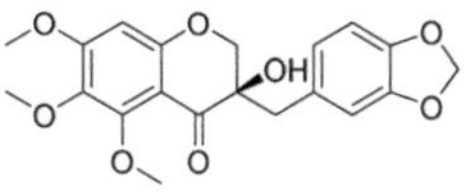

52

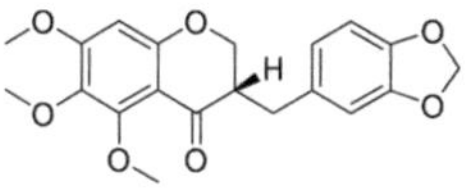

53

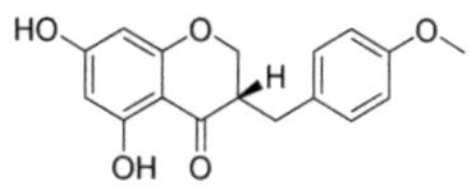

54

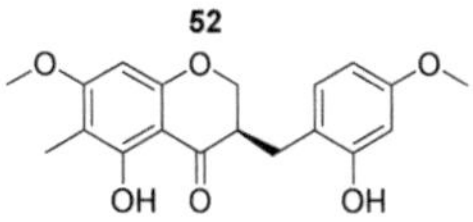

55

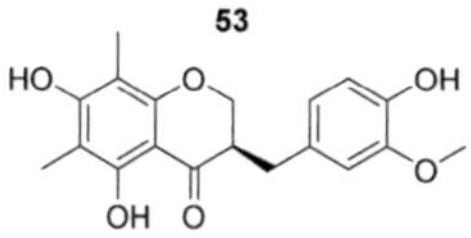

56

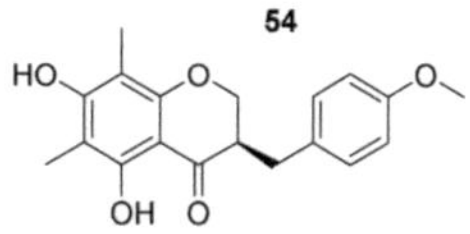

57

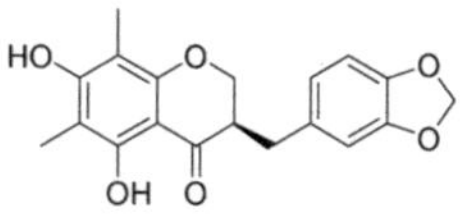

58

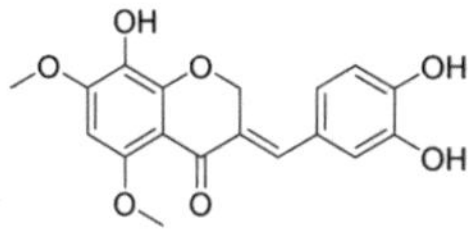

59

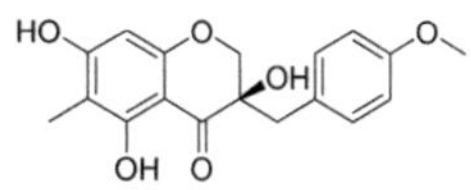

60

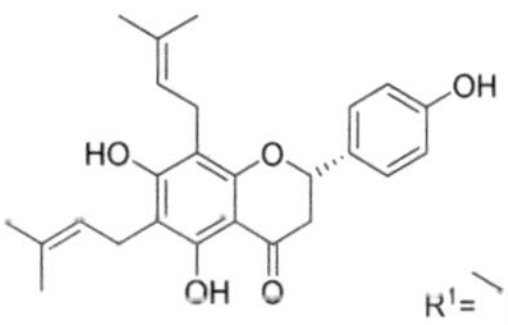

61

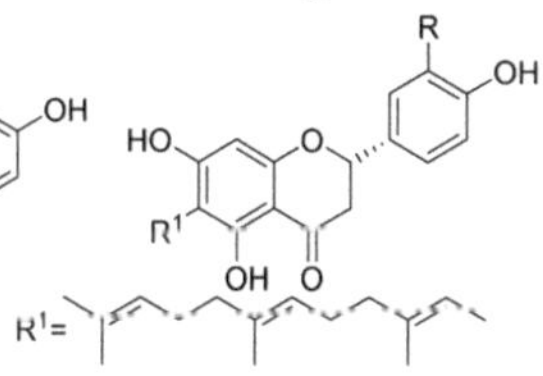

62 R= H, R^1
63 R= OH, R^1

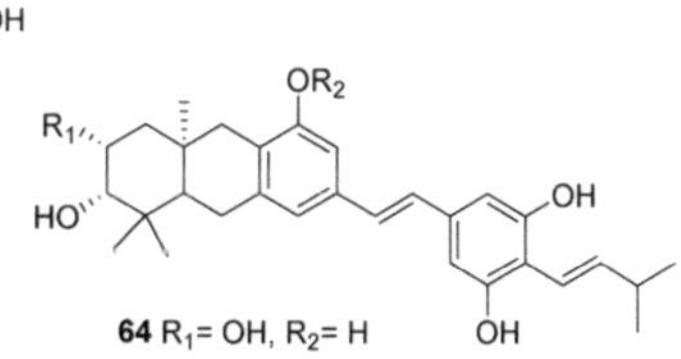

64 R_1= OH, R_2= H
65 R_1= OH, R_2= CH_3
66 R_1= H, R_2= CH_3

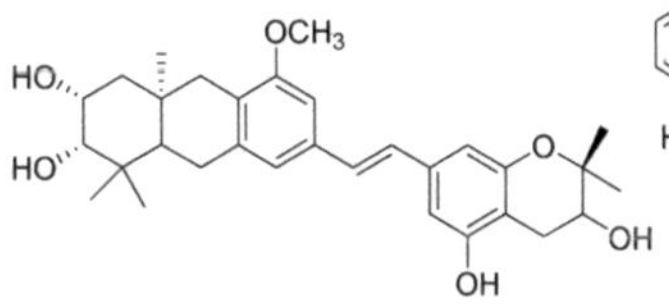

67

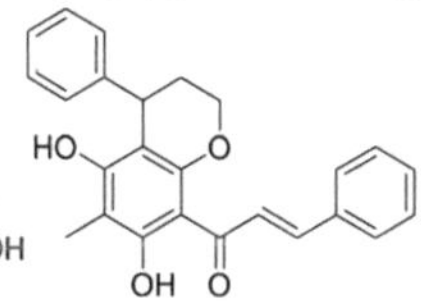

68

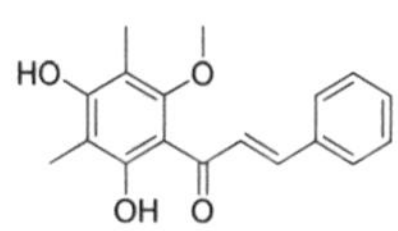

69

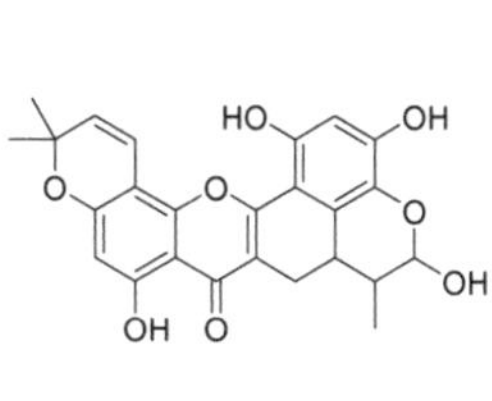

70

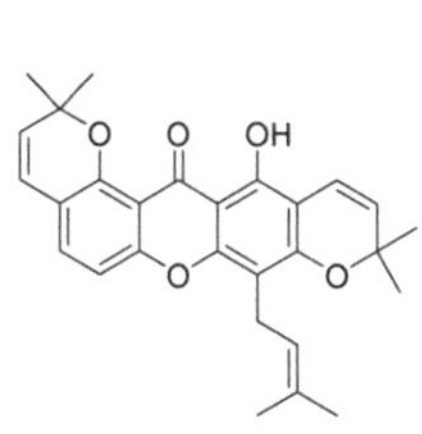

71

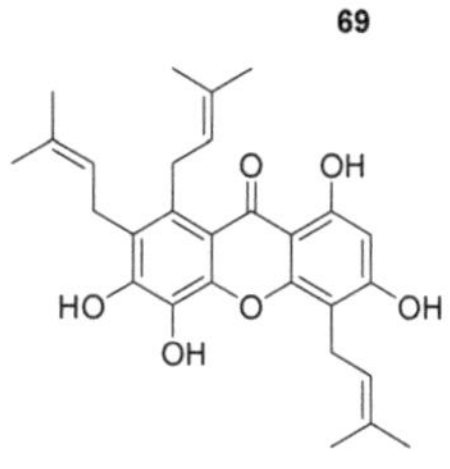

72

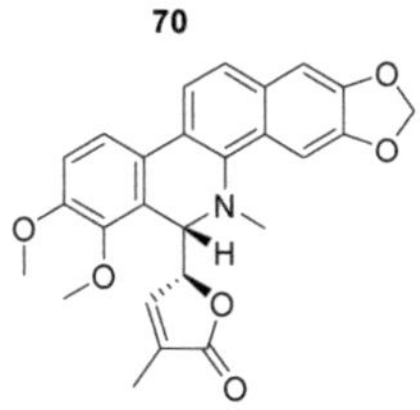

73

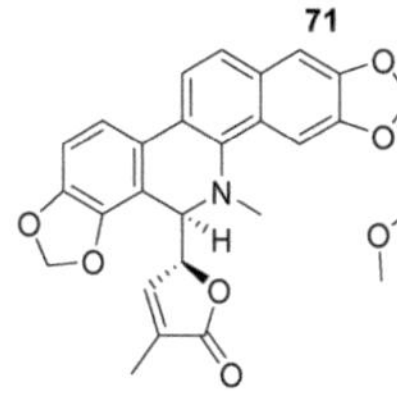

74

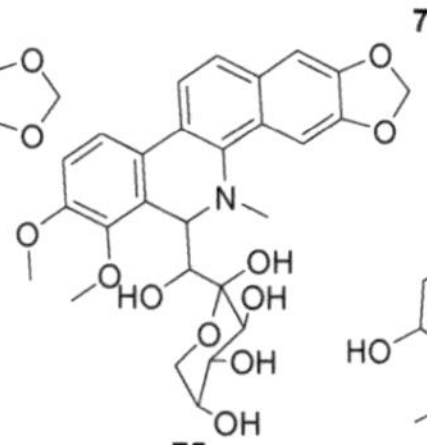

75

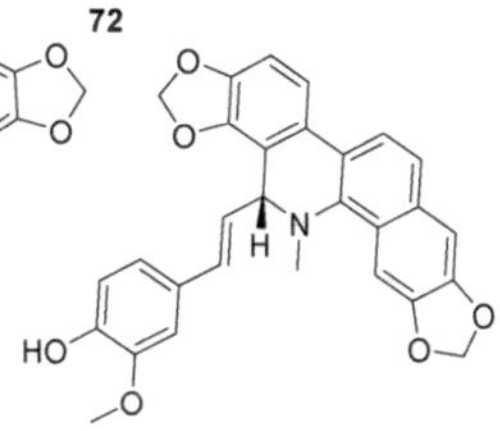

76

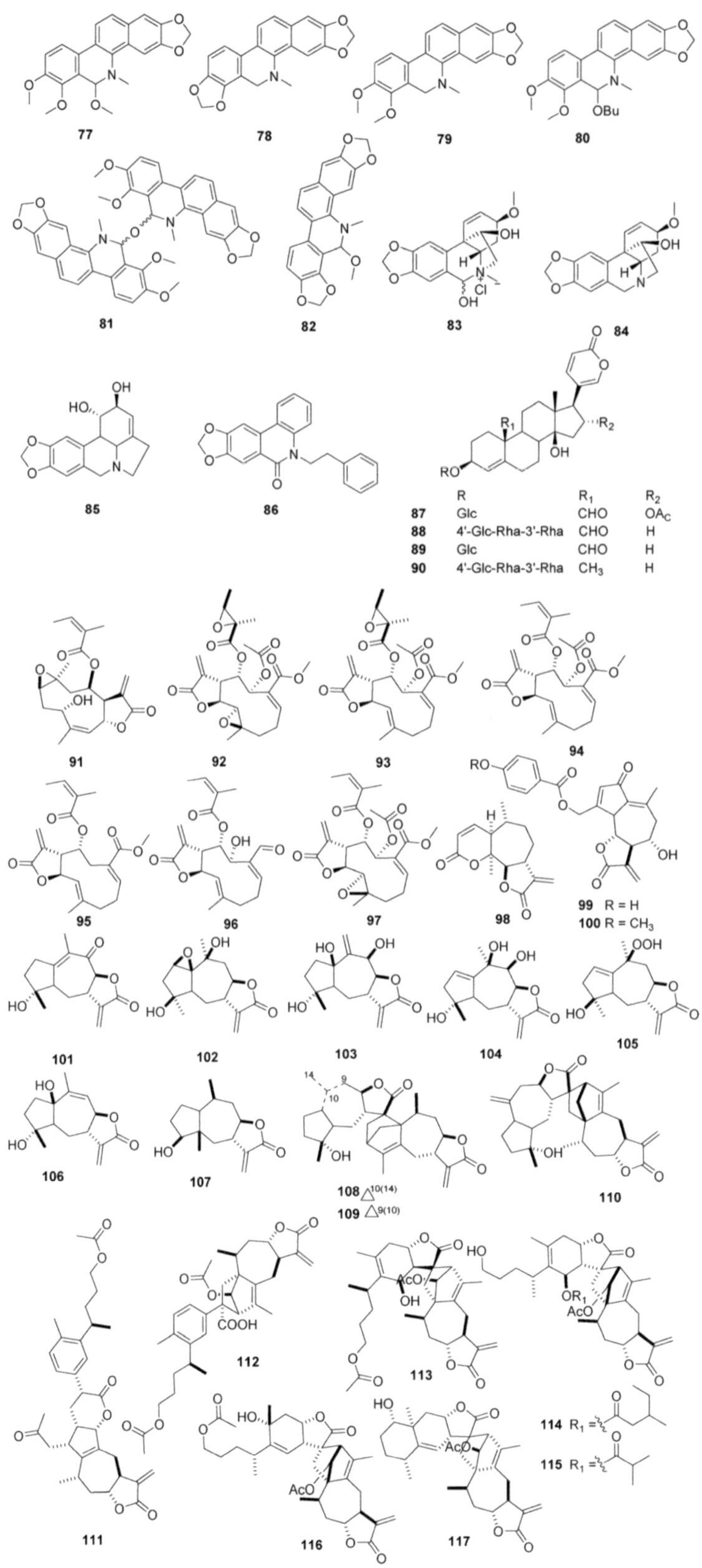

Figure 2.
Bioactive compounds against cancer cell lines.

unique structurally subunits which confer to drugs distinctive therapeutic affinities to a biological system. These core molecules include β-lactam unit like in penicillin; cyclopentanoperhydrophenanthrene fragment like in testosterone; pyrone, coumarins, isoflavone, or chalcone moieties and alkaloids like quinoline, isoquinoline or indole units.

As an example, since the large-scale screening for anticancer agents launched in the USA in 1960, more than 3000 sesquiterpene lactones have been reported. Most of them are with cytotoxic properties. Sesquiterpene lactones are well-known for their ability to bind sulfhydryl-containing peptides, mainly in proteins, presented as important route in well- programed death of a cell [82–85]. This property and other have raised up interests in this class of compounds. Many members of this class are currently in clinical trials for drug development including parthenolide, artemisinin or thapsigargin among others.

Another most important class of phytocompounds in cancer therapy is phenolic compounds. Members of these classes of metabolites are reputed in caspase activation causing apoptosis in tumor cell lines. Research found that furanocoumarins for instance in grapefruit showed significant effects towards breast cancer, the second World leading cause of cancer-related death among Women [86]. In the same line, coumarine-type of compounds known as calanolides, isolated from *Calophyllum* species have been found to be active against lymphoblastic cells infected with HIV-1 [87]. They are currently in clinical trials Phase II to drug development. Likewise, all other phenolic compounds listed above can also undergo similar interactions with cancer cells. Anthraquinones, and quinones, in general form the basic core of many anticancer drugs known as anthracyclines. Resveratrol, a stilbene-like metabolite, is being continuously checked to explain issues encountered during laboratory trials against cancer in animal model. However, association of resveratrol with established anticancer drugs like clofarabine has been proved against mesothelioma cell lines [88].

5. Conclusion

The World is facing an unprecedented drastic climate change that impacts negatively not only on human beings but also plants. New metabolism routes have surely emerged leading to compounds with unprecedented structures for some and with relevant bioactivities for others. However, nothing is being done to take advantages of this wealth for our health care always relying on "modern drugs." We should start exploring ways to use natural products with anticancer effects along with standard chemotherapy treatments to increase potency while reducing side effects of actual drugs. This strategy is currently being used in the USA. We highlighted relevant bio-sensibility of some compounds and they should now be investigated as main constituents to a standardization process of their respective plant extracts. The present survey can also help researchers in developing countries working on plants, to re-focus their research works.

Acknowledgements

The authors would like to acknowledge the Yaounde-Bielefeld Graduate School of Natural Products with Antiparasite and Antibacterial Activities (YaBiNaPA, www.yabinapa.de) for the research stay granted at the University of Bielefeld in Germany under the project No. 57316173.

Conflict of interest

The authors declare no conflict.

Author details

Gabin Thierry M. Bitchagno[1*], Vaderament-A. Nchiozem-Ngnitedem[1,2], Nadine Tseme Wandji[3], Guy Cedric T. Noulala[3], Serge Alain T. Fobofou[4] and Bruno Ndjakou Lenta[3]

1 Department of Chemistry, University of Dschang, Dschang, Cameroon

2 Department of Chemistry, University of Nairobi, Nairobi, Kenya

3 Department of Chemistry, Higher Teacher's Training College, University of Yaounde I, Yaounde, Cameroon

4 Department of Biological Chemistry and Molecular Pharmacology, Harvard Medical School, USA

*Address all correspondence to: gabin1256@gmail.com

References

[1] Bitchagno GTM, Schüffler A, Simo IK, Krumb M, Tane P, Opatz T. Neo-clerodane diterpenoids from *Conyza pyrrhopappa* Sch.Bip. ex A. Natural Product Research. 2019. DOI: 10.1080/14786419.2019.1690490

[2] Bitchagno MGT, Tankeo BS, Tsopmo A, Mpetga SDJ, Tchinda TA, Fobofou TSA, et al. Lemairones A and B: Two new antibacterial tetraflavonoids from the leaves of *Zanthoxylum lemairei* (Rutaceae). Phytochemistry Letters. 2015;**14**:1-7. DOI: 10.1016/j.phytol.2015.08.012

[3] Nganou BK, Mbaveng AT, Fobofou SAT, Fankam AG, Bitchagno GTM, Mpetga JDS, et al. Furoquinolines and dihydrooxazole alkaloids with cytotoxic activity from the stem bark of *Araliopsis soyauxii*. Fitoterapia. 2019;**133**:193-199. DOI: 10.1016/j.fitote.2019.01.003

[4] Mbaveng AT, Bitchagno GTM, Kuete V, Tane P, Efferth T. Cytotoxicity of ungeremine towards multi-factorial drug resistant cancer cells and induction of apoptosis, ferroptosis, necroptosis and autophagy. Phytomedecine. 2019; **60**:152832. DOI: 10.1016/j.phymed

[5] Sonfack G, Tchinda CF, Simo IK, Bitchagno GTM, Nganou BK, Çelik I, et al. Saponin with antibacterial activity from the roots of *Albizia adianthifolia*. Natural Product Research. 2019. DOI: 10.1080/14786419.2019.1672689

[6] Barreiro EJ. Privileged scaffolds in medicinal chemistry: An introduction. In: Bräse S, editor. Privileged Scaffolds in Medicinal Chemistry: Design, Synthesis, Evaluation. RSC Drug Discovery Series No. 50: Cambridge; 2015. p. 476. DOI: 10.1039/9781782622246-00001

[7] Butler MS, Robertson AAB, Cooper MA. Natural product and natural product derived drugs in clinical trials. Natural Product Reports. 2014;**31**: 1612-1661. DOI: 10.1039/C4NP00064A

[8] Butler MS. Natural products to drugs: Natural product derived compounds in clinical trials. Natural Product Reports. 2005;**22**:162-195. DOI: 10.1039/B514294F

[9] Dias DA, Urban S, Roessner U. A historical overview of natural products in drug discovery. Metabolites. 2012;**2**: 303-336. DOI: 10.3390/metabo2020303

[10] Der Marderosian A, Beutler JA, editors. The Review of Natural Products. 2nd ed. Seattle, WA, USA: Facts and Comparisons; 2002. p. 794

[11] Newman DJ, Cragg GM. Natural products as sources of new drugs over the last 25 years. Journal of Natural Products. 2007;**70**:461-477. DOI: 10.1021/np068054v

[12] Kedei N, Lundberg DJ, Toth A, Welburn P, Garfield SH, Blumberg PM. Characterization of the interaction of ingenol 3-angelate with protein kinase C. Cancer Research. 2004;**64**:3243-3255. DOI: 10.1158/0008-5472.CAN-03-3403

[13] Ogbourne SM, Suhrbier A, Jones B. Antitumour activity of ingenol 3-angelate: Plasma membrane and mitochondrial disruption and necrotic cell death. Cancer Research. 2004;**64**: 2833-2839. DOI: 10.1158/0008-5472.CAN-03-2837

[14] Aniszewski T. Alkaloids—Secrets of life. In: Aniszewski T, editor. Alkaloid Chemistry, Biological Significance, Applications and Ecological Role. Amsterdam: The Netherlands: Elsevier Science; 2007. p. 334

[15] Hartman H, Matsuno K. The Origin and Evolution of the Cell. Singapore: World Scientific Publishing Co Pte Ltd;

1993. p. 446. DOI: 10.1142/9789814536219

[16] American Academy of Microbiology FAQ: Human Microbiome. Archived 31 December 2016 at the Wayback Machine

[17] Rosner JL. Ten times more microbial cells than body cells in humans? Microbe. 2014;**9**:47. DOI: 10.1128/microbe.9.47.2

[18] Abbott A. Scientists bust myth that our bodies have more bacteria than human cells. Nature. 2016. DOI: 10.1038/nature.2016.19136

[19] Abreu AC, McBain AJ, Simões M. Plants as sources of new antimicrobials and resistance-modifying agents. Natural Product Reports. 2012;**29**: 1007-1021. DOI: 10.1039/c2np20035

[20] Kuete V. Potential of Cameroonian plants and derived products against microbial infections: A review. Planta Medica. 2010;**76**:1479-1491. DOI: 10.1055/s-0030-1250027

[21] Kuete V, Efferth T. African flora has the potential to fight multidrug resistance of cancer. BioMed Research International. 2015;**2015**:914813. DOI: 10.1155/2015/914813

[22] Cowan MM. Plant products as antimicrobial agents. Clinical Microbiology Reviews. 1999;**12**:564-582. DOI: 10.1128/CMR.12.4.564

[23] Watnick P, Kolter R. Biofilm, city of microbes. Journal of Bacteriology. 2000; **182**:2675-2679. DOI: 10.1128/jb.182.10.2675-2679.2000

[24] O'Toole GA, Kolter R. Initiation of biofilm formation in Pseudomonas fluorescens WCS365 proceeds via multiple, convergent signalling pathways: A genetic analysis. Molecular Microbiology. 1998;**28**:449-461. DOI: 10.1046/j.1365-2958.1998.00797.x

[25] Briandet R, Herry J, Bellon-Fontaine M. Determination of the van der Waals, electron donor and electron acceptor surface tension components of static Gram-positive microbial biofilms. Colloids and Surfaces. B, Biointerfaces. 2001;**21**:299-310. DOI: 10.1016/S0927-7765(00)00213-7

[26] Takahashi H, Suda T, Tanaka Y, Kimura B. Cellular hydrophobicity of listeria monocytogenes involves initial attachment and biofilm formation on the surface of polyvinyl chloride. Letters in Applied Microbiology. 2010;**50**: 618-625. DOI: 10.1111/j.1472-765X.2010.02842.x

[27] Man A, Santacroce L, Jacob R, Mare A, Man L. Antimicrobial activity of six essential oils against a group of human pathogens: A comparative study. Pathogen. 2019;**8**:15. DOI: 10.3390/pathogens8010015

[28] Lambert RJW, Skandamis PN, Coote PJ, Nychas G-JE. A study of the minimum inhibitory concentration and mode of action of oregano essential oil, thymol and carvacrol. Journal of Applied Microbiology. 2001;**91**:453-462. DOI: 10.1046/j.1365-2672.2001.01428.x

[29] Saad IK, Hassan L, Ghizlane Z, Hind M, Adnane R. Carvacrol and thymol components inhibiting Pseudomonas aeruginosa adherence and biofilm formation. African Journal of Microbiology Research. 2011;**5**: 3229-3232. DOI: 10.5897/AJMR11.275

[30] Sakagami Y, Thevanesam V, Programme NP, Lanka S, Lanka S, Sakagami Y. Antibacterial activity of xanthones from *Garcinia mangostana* (L.) and their structure – Activity relationship studies. Natural Product Research. 2013;**27**:938-941. DOI: 10.1080/14786419.2012.678348

[31] Auranwiwat C, Trisuwan K, Saiai A, Pyne SG, Ritthiwigrom T. Antibacterial

tetraoxygenated xanthones from the immature fruits of *Garcinia cowa*. Fitoterapia. 2014;**98**:179-183. DOI: 10.1016/j.fitote.2014.08.003

[32] Siridechakorn I, Phakhodee W, Ritthiwigrom T, Promgool T. Antibacterial dihydrobenzopyran and xanthone derivatives from *Garcinia cowa* stem barks. Fitoterapia. 2012;**83**: 1430-1434. DOI: 10.1016/j.fitote.2012.08.006

[33] Fouotsa H, Lannang AM, Dzoyem JP, Tatsimo SJN, Neumann B, Eloff JN, et al. Antibacterial and antioxidant Xanthones and Benzophenone from *Garcinia smeathmannii*. Planta Medica. 2015;**81**: 594-599. DOI: 10.1055/s-0035-1545841

[34] Nganou BK, Konga IS, Fankam AG, Bitchagno GTM, Sonfack G, Nayim P, et al. Guttiferone BL with antibacterial activity from the fruits of *Allanblackia gabonensis*. Natural Product Research. 2018;**6419**:1-9. DOI: 10.1080/14786419.2018.1465424

[35] Nono ECN, Mkounga P, Kuete V, Marat K, Hultin PG, Nkengfack AE. Pycnanthulignenes A-D, antimicrobial cyclolignene derivatives from the roots of *Pycnanthus angolensis*. Journal of Natural Products. 2010;**73**:213-216. DOI: 10.1021/np9007393

[36] Kuete V, Nono ECN, Mkounga P, Marat K, Hultin PG, Nkengfack AE. Antimicrobial activities of the CH2Cl2–CH3OH (1:1) extracts and compounds from the roots and fruits of *Pycnanthus angolensis* (Myristicaceae). Natural Product Research. 2010;**25**:432-443. DOI: 10.1080/14786419.2010.522577

[37] Kuete V, Teponno RB, Mbaveng AT, Tapondjou AL, Meyer JJM, Barboni L, et al. Antibacterial activities of the extracts, fractions and compounds from *Dioscorea bulbifera*. BMC Complementary and Alternative Medicine. 2012;**12**:228. DOI: 10.1186/1472-6882-12-228

[38] Fouotsa H, Mbaveng AT, Mbazoa CD, Nkengfack AE, Farzana S, Iqbal C, et al. Antibacterial constituents of three Cameroonian medicinal plants: *Garcinia nobilis*, *Oricia suaveolens* and *Balsamocitrus camerunensis*. BMC Complementary and Alternative Medicine. 2013;**13**:81. DOI: 10.1186/1472-6882-13-81

[39] Mbaveng AT, Kuete V, Ngameni B, Beng VP, Ngadjui BT, Meyer JJM, et al. Antimicrobial activities of the methanol extract and compounds from the twigs of *Dorstenia mannii* (Moraceae). BMC Complementary and Alternative Medicine. 2012;**12**:83. DOI: 10.1186/1472-6882-12-83

[40] Dongfack MDJ, Lallemand M-C, Kuete V, Mbazoa CD, Wansi JD, Dufat H, et al. A new Sphingolipid and Furocoumarins with antimicrobial activity from *Ficus exasperate*. Chemical & Pharmaceutical Bulletin. 2012;**60**: 1072-1075. DOI: 10.1248/cpb.c12-00279

[41] Ango YP, Kapche DWFG, Kuete V, Ngadjui BT, Bezabih M, Abegaz BM. Chemical constituents of *Trilepisium madagascariense* (Moraceae) and their antimicrobial activity. Phytochemistry Letters. 2012;5:524-528. DOI: 10.1016/j.phytol.2012.05.006

[42] Tankeo SB, Damen F, Awoufack MD, Mpetga J, Tane P, Eloff JN, et al. Antibacterial activities of the methanol extracts, fractions and compounds from *Fagara tessmanii*. Journal of Ethnopharmacology. 2015;**169**: 275-279. DOI: 10.1016/j.jep.2015.04.041

[43] Tankeo SB, Damen F, Sandjo LP, Tane P, Kuete V. Antibacterial activities of the methanol extracts, fractions and compounds from *Harungana madagascariensis*. Journal of Ethnopharmacology. 2016;**190**:100-105. DOI: 10.1016/j.jep.2016.06.005

[44] Djeussi DE, Sandjo LP, Noumedem JAK, Omosa LK, Ngadjui BT, Kuete V. Antibacterial

activities of the methanol extracts and compounds from *Erythrina sigmoidea* against Gram-negative multi-drug resistant phenotypes. BMC Complementary and Alternative Medicine. 2015;**15**:453. DOI: 10.1186/s12906-015-0978-8

[45] Mbaveng AT, Sandjo LP, Tankeo SB, Ndifor AR, Pantaleon AA, Nagdjui BT, et al. Antibacterial activity of nineteen selected natural products against multi-drug resistant Gram-negative phenotypes. Springerplus. 2015;**4**:823. DOI: 10.1186/s40064-015-1645-8

[46] Ouete JLN, Sandjo LP, Kapche DDW, Yeboah SO, Mapitse P, Abegaz BM, et al. Excelsoside: A new benzylic diglycoside from the leaves of *Milicia excelsa*. Zeitschrift für Naturforschung. 2014;**69**:271-275. DOI: 10.5560/ZNC.2014-0087

[47] Ouete JLN, Sandjo LP, Kapche DDW, Liermann JC, Opatz T, Simo IK, et al. A new flavone from the roots of *Milicia excelsa* (Moraceae). Zeitschrift für Naturforschung. Section C. 2013;**68**:259-263. DOI: 10.5560/ZNC.2013.68c0259

[48] Nganou BK, Konga IS, Fankam AG, Bitchagno GTM, Sonfack G, Nayim P, et al. Guttiferone BL with antibacterial activity from the fruits of *Allanblackia gabonensis*. Natural Product Research. 2018;**33**:2638-2646. DOI: 10.1080/14786419.2018.1465424

[49] Dzotam JK, Simo IS, Bitchagno GTM, Celik I, Sandjo LP, Tane P, et al. In vitro antibacterial and antibiotic modifying activity of crude extract, fractions and 3′,4′,7-trihydroxyflavone from *Myristica fragrans* Houtt against MDR Gram-negative enteric bacteria. BMC Complementary and Alternative Medicine. 2018;**18**:15. DOI: 10.1186/s12906-018-2084-1

[50] Damen F, Demgne OMF, Bitchagno GTM, Celik I, Mpetga JDS, Tankeo SB, et al. A new polyketide from the bark of *Hypericum roeperianum* Schimp. (Hypericaceae). Natural Product Research. 2019. DOI: 10.1080/14786419.2019.1677655

[51] Mambou CS, Nono RN, Chouna JR, Tamokou JDD, Nkeng-Efouet-Alango P, Sewald N. Antibacterial secotirucallane triterpenes from the stem bark of *Pseudocedrela kotschyi*. Zeitschrift für Naturforschung. 2018;**73**:241-246. DOI: 10.1515/znc-2017-0207

[52] Chouna JR, Tamokou JDD, Nkeng-Efouet-Alango P, Lenta BN, Sewald N. Antimicrobial triterpenes from the stem bark of *Crossopteryx febrifuga*. Zeitschrift für Naturforschung. 2015;**70**:169-173. DOI: 10.1515/znc-2014-4168

[53] Devkota KP, Lenta BN, Wansi JD, Sewald N. Antibacterial constituents from *Leucosceptrum canum*. Phytochemistry Letters. 2009;**3**:24-28. DOI: 10.1016/j.phytol.2009.10.007

[54] Hussein IA, Srivedavyasasri R, El-Hela AA, Mohammad AI, Ross SA. Antimicrobial secondary metabolites from *Silene rubella* growing in Egypt. Journal of Biomedical and Pharmaceutical Research. 2019;**8**. DOI: 10.32553/jbpr.v8i6.694

[55] Dzoyem JP, Melong R, Tsamo AT, Tchinda AT, Kapche DGWF, Ngadjui BT, et al. Cytotoxicity, antimicrobial and antioxidant activity of eight compounds isolated from *Entada abyssinica* (Fabaceae). BMC Research Notes. 2017;**10**:118. DOI: 10.1186/s13104-017-2441-z

[56] Ali B, Al-Wabel NA, Shams S, Ahamad A, Khan SA, Anwar F. Essential oils used in aromatherapy: A systemic review. Asian Pacific Journal of Tropical Biomedicine. 2015;**5**:601-611. DOI: 10.1016/j.apjtb.2015.05.007

[57] Alali F, El-elimat T, Albataineh H, Al-balas Q, Al-gharaibeh M,

Falkinham JO, et al. Cytotoxic Homoisoflavones from the bulbs of *Bellevalia eigii*. Journal of Natural Products. 2015;**78**:1708-1715. DOI: 10.1021/acs.jnatprod.5b00357

[58] El-elimat T, Rivera-chávez J, Burdette JE, Czarnecki A, Alhawarri MB, Al-gharaibeh M, et al. Cytotoxic homoisoflavonoids from the bulbs of *Bellevalia flexuosa*. Fitoterapia. 2018;**127**:201-206. DOI: 10.1016/j.fitote.2018.02.022

[59] Dai Y, Harinantenaina L, Brodie PJ, Goetz M, Shen Y, Tendyke K, et al. Antiproliferative Homoisoflavonoids and Bufatrienolides from *Urginea depressa*. Journal of Natural Products. 2013;**76**:865-872. DOI: 10.1021/np300900a

[60] Dang NH, Chung ND, Tuan HM, Hiep NT, Dat NT. Cytotoxic Homoisoflavonoids from *Ophiopogon japonicus* tubers. Biological & Pharmaceutical Bulletin. 2016;**65**: 204-207. DOI: 10.1248/cpb.c16-00743

[61] Zhou C, Zou L, Mo J, Wang X, Yang B, He Q. Homoisoflavonoids from *Ophiopogon japonicus*. Helvetica Chimica Acta. 2013;**96**:1397-1405. DOI: 10.1002/hlca.201200493

[62] Ghoran SH, Saeidnia S, Babaei E, Kiuchi F, Hussain H, Plants M, et al. Scillapersicene: A new homoisoflavonoid with cytotoxic activity from the bulbs of *Scilla persica*. Natural Product Research. 2015;**30**: 1309-1314. DOI: 10.1080/14786419.2015.1054286

[63] Mbaveng AT, Bitchagno GTM, Kuete V, Tane P, Thomas E. Phytomedicine cytotoxicity of ungeremine towards multi-factorial drug resistant cancer cells and induction of apoptosis , ferroptosis, necroptosis and autophagy. Phytomedicine. 2019; **60**:152832. DOI: 10.1016/j.phymed.2019.152832

[64] Doan H, Mai T, Toan TP, Huu GT, Le TN, Thi V, et al. New flavonoid and stilbene derivatives from the fruits of *Macaranga balansae*. Natural Product Research. 2019. DOI: 10.1080/14786419.2019.1587425

[65] Zakaria I, Ahmat N, Jaafar FM, Widyawaruyanti A. Flavonoids with antiplasmodial and cytotoxic activities of *Macaranga triloba*. Fitoterapia. 2012;**83**: 968-972. DOI: 10.1016/j.fitote.2012.04.020

[66] Mai T, Doan H, Nguyen TL, Van TT, Vu VN, Phi TD, et al. Cytotoxic phenolic compounds from fruit glandular trichomes of *Macaranga tanarius*. Journal of Analytical Methods in Chemistry. 2019;**2019**:2917032. DOI: 10.1155/2019/2917032

[67] Wei H, Zhang X, Wu G, Yang X, Pan S, Wang Y, et al. Chalcone derivatives from the fern *Cyclosorus parasiticus* and their anti-proliferative activity. Food and Chemical Toxicology. 2013;**60**:147-152. DOI: 10.1016/j.fct.2013.07.045

[68] Hashim NM, Rahmani M, Cheng G, Ee L, Sukari MA, Yahayu M, et al. Antiproliferative activity of xanthones isolated from *Artocarpus obtusus*. BioMed Research International. 2012; **2012**:130627. DOI: 10.1155/2012/130627

[69] Mah SH, Cheng G, Ee L, Teh SS, Sukari MA. Antiproliferative xanthone derivatives from *Calophyllum inophyllum* and *Calophyllum soulattri*. Pakistan Journal of Pharmaceutical Sciences. 2012;**28**:425-429

[70] Ji F, Li Z, Liu G, Niu S, Zhao N, Liu X, et al. Xanthones with Antiproliferative effects on prostate cancer cells from the stem bark of *Garcinia xanthochymus*. Natural Product Communications. 2012;7:53-56. DOI: 10.1177/1934578X1200700119

[71] Deng A-J, Qin H-L. Cytotoxic dihydrobenzophenanthridine alkaloids

from the roots of *Macleaya microcarpa*. Phytochemistry. 2010;**71**:816-822. DOI: 10.1016/j.phytochem.2010.02.007

[72] Luo Z, Wang F, Zhang J, Li X, Zhang M, Hao X. Cytotoxic alkaloids from the whole plants of *Zephyranthes candida*. Journal of Natural Products. 2012;**75**:2113-2120. DOI: 10.1021/np3005425

[73] Bosio C, Tomasoni G, Martínez R, Olea AF, Villena J. Cytotoxic and apoptotic effects of leptocarpin, a plant-derived sesquiterpene lactone, on human cancer cell lines. Chemico-Biological Interactions. 2015;**242**: 415-421. DOI: 10.1016/j.cbi.2015.11.006

[74] Ford CD, Ulloa JL, Catalán CAN, Grau A, Martino VS, Muschietti LV, et al. The sesquiterpene lactone polymatin B from *Smallanthus sonchifolius* induces different cell death mechanisms in three cancer cell lines. Phytochemistry. 2015;**117**:332-339. DOI: 10.1016/j.phytochem.2015.06.020

[75] Jimenez-usuga S, Malafronte N, Cotugno R, Leo MD, Osorio E, Tommasi ND. New sesquiterpene lactones from *Ambrosia cumanensis* Kunth. Fitoterapia. 2016;**113**:170-174. DOI: 10.1016/j.fitote.2016.07.019

[76] Shulha O, Çiçek SS, Piccolella S, Rárová L, Strnad M, Sönnichsen F, et al. Sesquiterpene lactones from *Sonchus palustris* L. (Asteraceae, Cichorieae). Phytochemistry. 2020;**170**:112196. DOI: 10.1016/j.phytochem.2019.112196

[77] Carpesium L, Wang L, Qin W, Tian L, Zhang X, Lin F, et al. Caroguaianolide A-E, five new cytotoxic sesquiterpene lactones from *Carpesium abrotanoides* L. Fitoterapia. 2016;**127**: 349-355. DOI: 10.1016/j.fitote.2018.03.015

[78] Xu X-K, Ye J, Chen L, Zhang W, Yang Y, Li H. Four new isomeric sesquiterpene lactone dimers from *Carpesium faberi*. Tetrahedron Letters. 2015;**56**:6381-6384. DOI: 10.1016/j.tetlet.2015.09.127

[79] Xu X-Y, Sun P, Guo D, Liu X, Liu J, Hu L. Cytotoxic sesquiterpene lactone dimers isolated from *Inula japonica*. Fitoterapia. 2015;**101**:218-223. DOI: 10.1016/j.fitote.2015.01.011

[80] Shim JS, Kim JH, Lee JY, Kim SN, Kwon HJ. Anti-angiogenic activity of a homoisoflavanone from *Cremastra appendiculata*. Planta Medica. 2004;**70**: 171-173. DOI: 10.1055/s-2004-815496

[81] Han N, Yang Z, Liu Z, Liu H, Yin J. Research progress on natural Benzophenanthridine alkaloids and their pharmacological functions: A review. Natural Product Communications. 2016;**11**:1181-1188. DOI: 10.1177/1934578X1601100838

[82] Garcia-Pineres AJ, Castro V, Mora G, Schmidt TJ, Strunck E, Pahl HL, et al. Cysteine 38 in p65/NF-kappaB plays a crucial role in DNA binding inhibition by sesquiterpene lactones. The Journal of Biological Chemistry. 2001;**276**:39713. DOI: 10.1074/jbc.M101985200

[83] Ghantous A, Gali-Muhtasib H, Vuorela H, Saliba NA, Darwiche N. What made sesquiterpene lactones reach cancer clinical trials? Drug Discovery Today. 2010;**15**:668-678. DOI: 10.1016/j.drudis.2010.06.002

[84] Kwok BH, Koh B, Ndubuisi MI, Elofsson M, Crews CM. The anti-inflammatory natural product parthenolide from the medicinal herb feverfew directly binds to and inhibits IkappaB kinase. Chemistry & Biology. 2001;**8**:759. DOI: 10.1016/s1074-5521(01)00049-7

[85] Zunino SJ, Ducore JM, Storms DH. Parthenolide induces significant apoptosis and production of reactive oxygen species in high-risk pre-B

leukemia cells. Cancer Letters. 2007; **254**:119-127. DOI: 10.1016/j. canlet.2007.03.002

[86] Hung W-L, Suh JH, Wang Y. Chemistry and health effects of furanocoumarins in grapefruit. Journal of Food and Drug Analysis. 2017;**25**: 71-83. DOI: 10.1016/j.jfda.2016.11.008

[87] Kashman Y, Gustafson KR, Fuller RW, Cardellin JH, McMahon JB, Currens MJ, et al. The calanolides, a novel HIV inhibitory class of coumarin derivatives from the tropical rainforest tree, *Calophyllum lanigerum*. Journal of Medicinal Chemistry. 1992;**35**: 2735-2743. DOI: 10.1021/jm00093a004

[88] Lee YJ, Lee YJ, Im JH, Won SY, Kim YB, Cho MK, et al. Synergistic anti-cancer effects of resveratrol and chemotherapeutic agent clofarabine against human malignant mesothelioma MSTO-211H cells. Food and Chemical Toxicology. 2012;**52**:61-68. DOI: 10.1016/j.fct.2012.10.06

Chapter 5

Role of Tea Polyphenols in Metabolic Syndrome

Telma Angelina Faraldo Corrêa,
Adriana Campos Rozenbaum and Marcelo Macedo Rogero

Abstract

Metabolic syndrome (MetS) increases the risk of type 2 diabetes and cardiovascular diseases (CVD). Tea (*Camellia sinensis*), one of the most consumed beverages in the world, is rich in polyphenols, mainly catechins. Tea polyphenols may ameliorate obesity by reducing body weight, increasing energy expenditure and fat oxidation, stimulating lipolysis, and improving thermogenesis. Tea polyphenols also reduce the risks of type 2 diabetes (T2D), hypertension, hyperlipidemia, and inflammation. Results of clinical trials on the effects of the consumption of tea beverage, tea extracts, or isolated tea polyphenols on biomarkers of metabolic syndrome will be reviewed in this study. The effects of tea polyphenols on antioxidant status and low-grade chronic inflammation and the molecular mechanisms involved will also be discussed.

Keywords: *Camellia sinensis*, catechins, inflammation, insulin resistance, dyslipidemia, hypertension, obesity

1. Introduction

Metabolic syndrome (MetS) is a cluster of interrelated prejudicial conditions that leads to type 2 diabetes (T2D) and cardiovascular disease (CVD). These conditions include elevated fasting plasma glucose level (hyperglycemia), abdominal/visceral obesity, dyslipidemia, and hypertension [1, 2]. The International Diabetes Federation (IDF) estimates that around 20–25% of the global adult population suffer from MetS and are more likely to die from a heart attack or stroke compared with people without MetS [1].

Since there is no specific treatment for MetS, individual characteristics must be taken into consideration. There is a need for long-term studies to determine whether existing and new therapeutic agents benefit patients with MetS, reducing the effects of MetS and preventing the appearance of associated diseases and to evaluate the potential of novel candidates as effective treatment options [3]. Several clinical studies demonstrate that lifestyle modification, especially dietary changes, is an effective strategy to reduce several factors responsible for the development of MetS. Introducing foods rich in dietary phytochemicals, such as polyphenols, into the diet of an individual is an effective lifestyle modification for the prevention of several diseases, including MetS [4, 5].

Polyphenols (phenolic compounds), one of the most relevant families of phytochemicals with health benefits, are biomolecules found in natural products. Several preclinical studies report that some polyphenols exert protective effects in

 IntechOpen

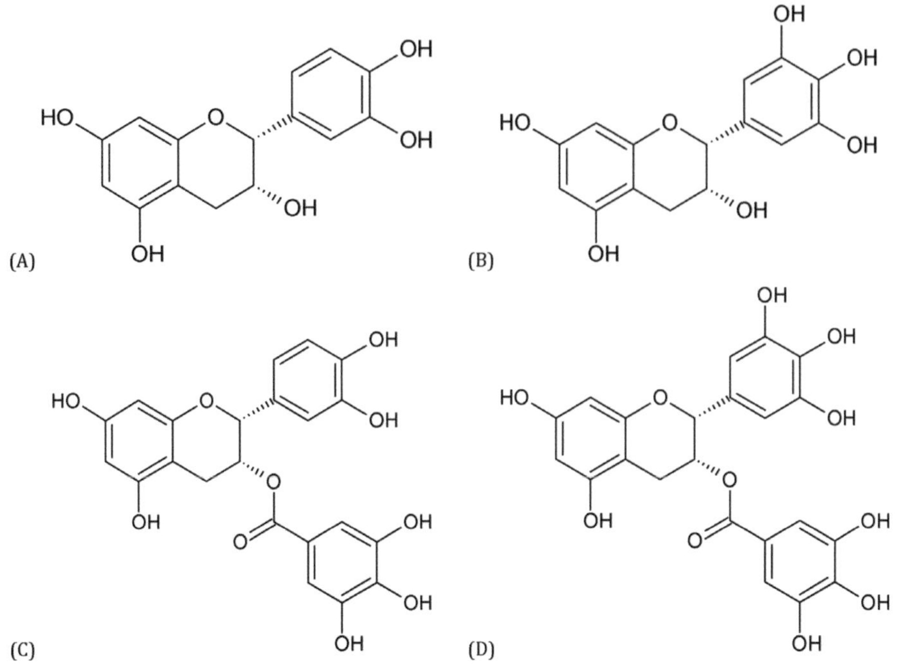

Figure 1.
Chemical structure of green tea catechins: epicatechin (A), epigallocatechin (B), epicatechin-3-gallate (C), epigallocatechin-3-gallate (D).

many diseases, including CVD and MetS, both triggered by oxidative stress [6, 7]. These compounds present antioxidant and anti-inflammatory properties and may be able to delay or prevent MetS by decreasing blood pressure, blood glucose levels, and body weight, as well as by improving lipid metabolism [7, 8]. One of the main sources of polyphenols is tea prepared from the processed leaves of *Camellia sinensis*, an herbal plant belonging to the Theaceae family. The chemical composition of tea is characterized by the presence of polyphenols (especially catechins), phenolic acids, amino acids, proteins, and fats. The catechins most commonly found in tea include epigallocatechin-3-gallate (EGCG), epigallocatechin (EGC), epicatechin-3-gallate (ECG), and epicatechin (EC) (**Figure 1**). These compounds constitute up to 30% of the dry leaf weight. A typical green tea beverage, with 2.5 g of tea leaves and 250 mL of hot water, contains 240–320 mg catechins and 60–65% EGCG [7, 9, 10]. Although the bioavailability of tea polyphenols is not clearly known, it depends on the molecular size and the number of phenolic groups. A review demonstrated that the consumption of 2 or 3 cups of tea daily resulted in 0.2–0.3 μM peak plasma levels of tea catechins [9, 11]. Notably, the health-promoting properties of green tea are due to the presence of the catechins mentioned earlier, mainly because of their antioxidant (scavenging of reactive oxygen species, inhibition of the formation of free radicals, and lipid peroxidation) and anti-inflammatory effects [10].

2. Tea, obesity, and inflammation

Obesity is a major health concern in the developed and developing world. Obesity leads to an inflammatory condition that is directly involved in the etiology of CVD, T2D, and certain types of cancer. Furthermore, the accumulation of adipose tissue in the abdominal region is a significant risk factor for the development of MetS and associated morbidities. It should be noted that inflammation is a common feature implicated in the pathophysiology of many obesity-associated disorders. The inflammatory response in obese and MetS individuals manifests systemically and is characterized by a chronic low-intensity reaction, unlike classical inflammation [12–16].

The anti-inflammatory and anti-obesity effects of *Camellia sinensis* have been associated with its catechin content, and EGCG is the most abundant and pharmacologically active catechin. Green tea, which is more effective than black tea, has been shown to significantly alleviate MetS symptoms, such as abdominal adiposity indicated by waist circumference in obese subjects. The anti-obesity mechanisms of tea polyphenols are associated with two major mechanisms: (i) decreasing the absorption of lipids and proteins in the intestine by tea constituents, thus reducing calorie intake, and (ii) activating adenosine monophosphate-activated protein kinase (AMPK) by tea polyphenols that are bioavailable in the liver, skeletal muscle, and adipose tissues. The relative importance of these two mechanisms depends on the types of tea and diet consumed by individuals. It should be noted that AMPK activation can reduce gluconeogenesis and fatty acid synthesis, leading to bodyweight reduction and MetS alleviation [17, 18].

Clinical trials [19–21] verified that green tea reduced body weight and other biomarkers linked to MetS (**Table 1**).

Cellular, animal, and human experiments demonstrated that green tea and its major component, EGCG, have anti-inflammatory effects. Moreover, EGCG

Participants	Study type	Intervention	Outcomes	Ref.
35 subjects with obesity and MetS	Randomized, controlled prospective trial	Green tea (4 cups/day), green tea extract (2 capsules and 4 cups of water/day), or placebo for 8 weeks	Both interventions ↓ body weight and BMI. Green tea beverage also ↓ lipid peroxidation	[19]
70 moderately overweight subjects	Intervention	Green tea extract (4 capsules/day—375 mg of catechins) for 12 weeks	↓ Body weight	[20]
23 overweight subjects	Double-blind study	Green tea beverage containing 588 mg or 126 mg catechins for 12 weeks	↓ Body fat parameters	[21]
56 obese, hypertensive subjects	Double-blind, placebo-controlled trial	1 capsule (379 mg green tea extract) or placebo for 3 months	↓ TNF-α and CRP serum levels ↑ antioxidant status and HDL-C ↓ TC, LDL-C, and TG	[22]
1.704 overweight or obese subjects	Meta-analysis	Green tea/green tea extract (126–800 mg catechins) for 12–24 weeks	↓ LDL-C and TC	[23]
1356 subjects	Meta-analysis	EGCG (107 to 857 mg/day—2 to 8 cups of green tea per day) for 4 to 14 weeks	↓ LDL-C	[24]

BMI, body mass index; CRP, C-reactive protein; EGCG, epigallocatechin gallate; HDL-C, high-density lipoprotein cholesterol; LDL-C, low-density lipoprotein cholesterol; MetS, metabolic syndrome; TC, total cholesterol; TG, triglycerides; TNF-α, tumor necrosis factor-alpha; ↑, Increase; ↓, reduction.

Table 1.
Clinical trials showing the effects of tea or tea catechins on inflammation, obesity, and lipid profile.

inhibits the in vitro activation of the transcription factor NF-κB and attenuates the IκB-α degradation induced by tumor necrosis factor-alpha (TNF-α) activation. The anti-inflammatory mechanism of EGCG seems to be associated with a decrease in the activity of the IKK-β protein, involved in the phosphorylation of IκB-α. Because of this effect on the NF-κB signaling pathway, catechins can reduce the gene expression of COX-2. In addition, EGCG demonstrates anti-inflammatory activities in the MAPK pathway by inhibiting the phosphorylation of p38. Catechins also reduce the gene expression of c-Jun N-terminal kinase (JNK) protein and the transcription factor AP-1 [25, 26].

It should be noted that only a limited number of studies on humans provided strong evidence related to the anti-inflammatory activity of green tea. One example is a double-blind, placebo-controlled trial, in which 56 obese, hypertensive subjects received green tea extract or placebo for 3 months [22]. Green tea extract reduced diabetes and inflammation risk, increased total antioxidant status, and improved the lipid profile.

3. Tea and lipid profile

Hyperlipidemia, characterized by increased levels of total cholesterol (TC) and low-density lipoprotein (LDL-C), is a major risk factor for CVD. Several clinical trials demonstrated that the ingestion of polyphenols such as flavonoids and phenolic acids can improve the concentrations of TC, LDL-C, and high-density lipoprotein (HDL-C) [8].

Green tea beverage consumption and green tea extract supplementation can also improve lipid profile, reducing blood TC and LDL-C concentrations, especially when used for a long time. These changes are due to the presence of major tea polyphenols, namely, the catechins [27, 28]. A study conducted on rats fed with atherogenic diet demonstrated that the supplementation with green tea preparation consisting of 66.5% EGCG and other catechins could decrease plasma TC and LDL-C levels and increase plasma HDL-C levels [29]. Another study on rats and atherogenic diet indicated that EGCG can significantly reduce TC, LDL-C, very low-density lipoprotein cholesterol (VLDL-C), triacylglycerols (TG), and cardiac risk ratio values while increasing the concentration of HDL-C [30].

Studies on humans also reported that EGCG can improve lipid profile. Its mechanisms may be associated with decreasing the absorption of lipids, inhibiting the lipogenesis pathway, and attenuating inflammation [23, 24, 31] (**Table 1**).

Other green tea catechins may have a beneficial effect on plasma TC and LDL-C levels in humans. Kim et al. [32] reviewed 20 trials and verified that the intake of green tea catechins, at doses of 145 to 3.000 mg per day, reduced TC by 5.5 mg/dL and LDL-C by 5.3 mg/dL, while there were no changes in plasma HDL-C levels.

More importantly, green tea can decrease plasma TC and LDL-C levels in overweight or obese people with no side effects, especially with long-term consumption [23].

Consumption of green tea extract catechin complex (843 mg of EGCG, 202 mg of ECG, 107 mg of EGC, and 107 mg of EC), for 12 months, significantly reduced (compared with the placebo group) plasma TC, LDL-C, and non-HDL-C levels in postmenopausal women. In hypercholesterolemic participants, green tea extract supplementation resulted in a reduction of 8.5% in TC and 12.4% in LDL-C concentrations. This study suggests that green tea extract, with high concentrations of catechins, may be recommended for lowering cholesterol, especially in those with high cholesterol concentrations [33].

4. Tea and blood pressure

Hypertension is a multifactorial clinical condition characterized by constant elevation of systolic blood pressure (SBP) levels ≥140 and/or diastolic blood pressure (DBP) ≥ 90 mmHg. It is often associated with metabolic disorders and functional and/or structural changes in target organs, aggravated by the presence of other risk factors, such as dyslipidemia, abdominal obesity, glucose intolerance, and T2D [34]. Hypertension is one of the leading risk factors for CVD, and it is a major cause of premature death worldwide; it affects about 1 billion people worldwide [34].

Tea flavonoids can reduce the risk of hypertension and consequently the risk of CVD [9, 35]. Catechins act as antioxidants and vasodilators and inhibit endothelial dysfunction and thrombogenesis [9, 36]. Catechins might reduce blood pressure by enhancing nitric oxide signaling [9]. The health benefits of tea for blood pressure were demonstrated in healthy subjects, diabetic subjects, and obese and/or hypertensive subjects [37]. Clinical trials showing the effect of tea on blood pressure are summarized in **Table 2**. It should be noted that several factors may influence the effect of tea consumption on blood pressure such as the duration and frequency of consumption, dosage, tea bioactive compounds, the evaluated population, and the degree of hypertension [37].

4.1 Molecular mechanisms of tea regulating blood pressure

Evidence indicates that vascular superoxide anion inactivates nitric oxide (NO) and plays a critical role in the development of hypertension. NO reacts with superoxide anion to form peroxynitrite. Peroxynitrite can cause protein tyrosine nitration, which modifies protein structure and function and affects cell homeostasis, oxidizes LDL-C, and leads to reduced activity of endothelial nitric oxide synthase (eNOS) [44, 45]. Angiotensin II generates vascular superoxide anion by activating nicotinamide adenine dinucleotide phosphate (NADPH) oxidase. Superoxide anion contributes to increased blood pressure, endothelial dysfunction, vascular remodeling, and sodium retention, consequently contributing to the development of hypertension (**Figure 2**). Possibly, green tea extract reduces the risk of hypertension by reducing vascular reactive oxygen species (ROS) formation and NADPH oxidase activity [37]. In rats, decaffeinated green tea extract stimulated the activation of the eNOS via the phosphatidylinositol 3-kinase (PI3-kinase)/Akt pathway [46].

Caveolin-1, the major negative regulator of eNOS activity, has its gene expression attenuated by green tea polyphenols via the activation of extracellular signal-regulated kinase 1/extracellular signal-regulated kinase 2 (ERK1/ERK2) and inhibition of p38 mitogen-activated protein kinase (MAPK) signaling pathways [47].

Another mechanism by which tea consumption can reduce the risk of hypertension is by inhibiting renin activity. The study conducted by Li et al. [48] showed that oolong and black tea extracts inhibited renin activity. The beneficial effect was attributed to thearubigins. However, monomeric catechins did not contribute to the inhibitory effect promoted by the tea extracts.

Endothelin-1 may contribute to hypertension by enhancing vascular superoxide anion production via ETA/NADPH oxidase. Evidence indicates that epigallocatechin gallate reduces endothelin-1 expression and secretion from endothelial cells, partly via Akt- and AMPK-stimulated forkhead box *transcription factor* class O1 (FOXO1) regulation of the endothelin-1 promoter [37].

Participants	Study type	Intervention	Outcomes	Ref.
4579 older Chinese (>60 years) without hypertension or antihypertensive treatment	Cross-sectional	Green tea (~90% of subjects) nonhabitual drinkers, 1–5x/week; ≥6x/week	Reduction of 16 and 22% with tea consumption 1–5 times per week and ≥6 times per week, respectively	[36]
20 subjects with T2D (mean 53 years old; BMI, 30 kg/m^2)	Parallel, double-blind, placebo-controlled	Decaffeinated green tea extract (400 mg/ day) or placebo for 12 weeks	No difference in blood pressure, anthropometric, and metabolic parameters when compared to placebo	[38]
100 mildly hypertensive patients with diabetes	Randomized clinical trial	3 g/150 mL of sour tea or 3 g/150 mL of green tea (3x/day) for 4 weeks	↓ Systolic and diastolic blood pressure in both groups	[39]
56 obese, hypertensive subjects	Double-blind, placebo-controlled trial	1 capsule (379 mg green tea extract) or placebo for 3 months	↓ Systolic and diastolic blood pressure compared with placebo	[22]
19 hypertensive subjects	Randomized, double-blind, controlled, cross-over study	Black tea (129 mg flavonoids) or placebo (2x/day) for 8 days	↓ Systolic (3.2 mmHg) and diastolic (2.6 mmHg) blood pressure and prevented blood pressure increase after a fat load	[40]
123 prediabetic subjects	Randomized controlled clinical trial	600 mL/day of green tea or control (warm water) for 14 weeks	↓ Mean arterial pressure ↓ Waist/hip ratio compared to control Did not affect fasting plasma glucose nor HbA1C level	[41]
1697 subjects (22–74 years)	Meta-analysis	Green tea (208 to 1344 mg/day of catechin) for 3–16 weeks	↓ Systolic (1.17 mmHg) and diastolic (1.24 mmHg) blood pressure, no effect of caffeine, effect of low catechin dose was greater than high catechin	[42]
971 overweight or obese subjects (29–54 years)	Meta-analysis	Green tea/ reen tea extract (320–1207 mg of catechin) for 3–16 weeks	↓ Systolic and diastolic blood pressure in normotensive and hypertensive subjects. Significant reduction only for low catechin dose (250 mg of EGCG/ day) and intervention ≥3 months Caffeine did not interfere with the results	[43]

BMI, body mass index; EGCG, epigallocatechin gallate; HbA1C, glycated hemoglobin A1C; T2D, type 2 diabetes; ↓, reduction.

Table 2.
Clinical trials showing the tea effect on blood pressure.

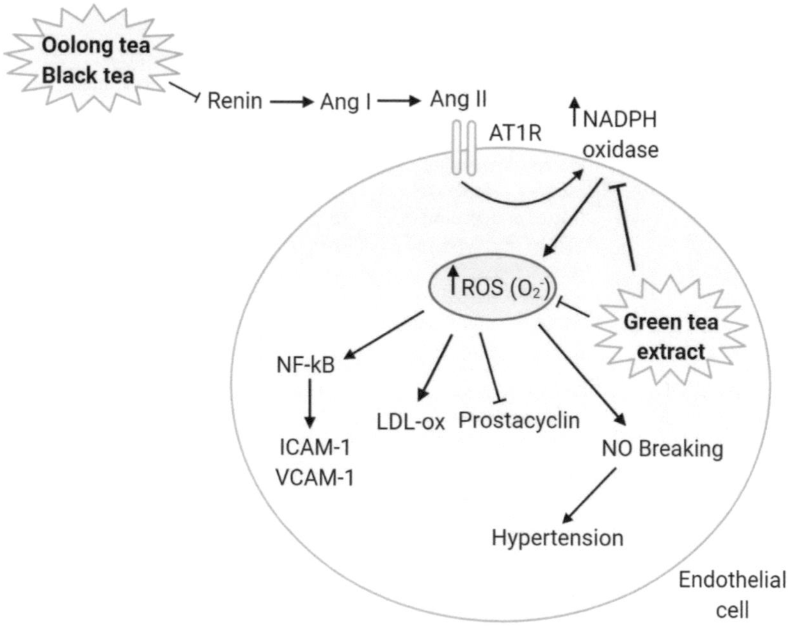

Figure 2.
In hypertension, there is excessive ROS generation in endothelial cells induced by angiotensin II. ROS excess can stimulate the NF-kB pathway and increase endothelial inflammation. ROS excess may also increase LDL-C oxidation and inhibit prostacyclin; superoxide anion reacts with NO to form peroxynitrite, which is cytotoxic. The NO loss may reduce vasorelaxation and contribute to endothelial dysfunction and hypertension. Green tea extract may inhibit the ROS production as well as Oolong tea and black tea may inhibit renin, and consequently angiotensin II. Ang II: angiotensin II; AT1R: angiotensin II type 1 receptor; ICAM-1: intercellular adhesion molecule 1; LDL-ox: oxidized LDL-C; NF-kB: nuclear factor kappa B; NO: nitric oxide; ROS: reactive oxygen species; VCAM-1: vascular cell adhesion molecule 1. →: stimulation; ⊥: inhibition.

5. Tea and insulin resistance/diabetes

Insulin resistance is a key feature of MetS and an important risk factor for CVD and T2D. Diabetes is a global health issue with high morbidity and mortality. The global prevalence of diabetes was 8.5% in 2014. In 2016, about 3.7 million deaths were caused by high blood glucose levels and diabetes. Almost half of the deaths caused by high blood glucose levels occur before the age of 70. T2D is linked to insulin resistance, altered lipid profile, hypertension, and endothelial dysfunction [49].

Recent evidence indicates that tea consumption improves insulin sensitivity and reduces the risk of T2D [9, 50]. Possibly, tea polyphenols act on gut microbiota, increase the probiotic species in the intestine, and attenuate the gene expression of enzymes involved in gluconeogenesis (phosphoenolpyruvate carboxykinase (PEPCK) and glucose-6-phosphatase) and glucose production in the liver, mediated by AMPK activation [9].

In vitro, in vivo, and clinical studies have shown that green tea catechins, mainly EGCG, have various antidiabetic activities [50–53]. However, some studies have shown that there are no beneficial effects of tea on T2D [51, 52, 54, 55] (**Table 3**).

5.1 Molecular mechanisms of tea regulating insulin resistance/T2D

Green tea enhances glucose-stimulated insulin secretion through the cyclic adenosine monophosphate (cAMP)/Akt pathway. Moreover, EGCG could activate

Participants	Study design	Outcome	Ref.
13 normal subjects and 11 prediabetic subjects	Randomized, double-blind, placebo-controlled crossover study Placebo, low dose of black tea polyphenol (110 g) or high dose (220 g) of polyphenol + sucrose (50 g) Samples collected at 0, 30, 60, 90, and 120 min from tea intake	Low dose and high dose of polyphenol reduced incremental blood glucose area under the curve (AUC) compared with placebo in both normal and prediabetic subjects. No significant difference between low dose and high dose of polyphenol	[56]
92 subjects with T2D and lipid abnormalities	Double-blinded, randomized and placebo-controlled clinical trial 500 mg green tea extract (3x/day) or control (cellulose) for 16 weeks	↓ Triacylglycerols ↓ HOMA-IR ↑ HDL-C ↑ GLP-1	[57]
49 subjects with T2D (average age of 65 years; median duration of diabetes, 6 years) 80% of them using hypoglycemic medication	A double-blind, placebo-controlled, randomized trial Placebo, 375 mg or 750 mg/day for 3 months	Extract of green and black tea did not show a hypoglycemic effect	[54]
Overweight or obese male subjects (40–65 years)	Placebo-controlled, randomized trial 400 mg capsules of EGCG or placebo (lactose)—2x/day for 8 weeks	No effect on insulin sensitivity, insulin secretion, or glucose tolerance	[55]
1584 subjects	Meta-analysis Green tea catechins with or without caffeine ≥12 weeks <12 weeks	↓ Fasting blood glucose Glucose-lowering effect was observed when follow-up ≥12 weeks	[58]
30 subjects with T2D	Randomized controlled trial 600 mL/day of black tea or 200 mL/day of black tea for 12 weeks	↓ HbA1C with 600 mL/day ↓ Pro-inflammatory $CD3^+$ $CD4^+$ $IL\text{-}17^+$ cells	[59]
56 obese, hypertensive subjects	Double-blind, placebo-controlled clinical trial 1 capsule/day (379 mg of green tea extract) or placebo for 3 months	↓ Systolic and diastolic blood pressure ↓ Fasting serum glucose ↓ Insulin resistance (HOMA-IR) ↓ TNF-α and CRP ↓ Total antioxidant status ↓ Total cholesterol, LDL-C, and triacylglycerols ↑ HDL-C	[22]
68 overweight subjects with T2D (20–65 years)	Randomized, double-blind, placebo-controlled clinical trial 1500 mg of decaffeinated green tea extract/day or placebo (cellulose) for 16 weeks	↓ HbA1C, ↓ HOMA-IR, ↑ ghrelin, ↓ waist circumference	[60]

Participants	Study design	Outcome	Ref.
66 subjects with T2D (32–73 years)	Randomized controlled trial Green tea extract (456 mg catechin)/ day or control (just followed) for 2 months	Blood glucose, HbA1c, insulin levels, and HOMA-IR did not differ from the control group	[61]

CRP, C-reactive protein; HbA1C, glycated hemoglobin A1C; HDL-C, high-density lipoprotein cholesterol; HOMA-IR, homeostasis model assessment of insulin resistance index; GLP-1, glucagon-like peptide 1; LDL-C, low-density lipoprotein cholesterol; TNF, tumor necrosis factor; T2D, type 2 diabetes; ↓, Reduction; ↑, increase.

Table 3.
Studies that showed the effect of tea on insulin resistance and diabetes.

AMPK to improve the shutdown of the insulin stress signal pathway caused by serine phosphorylation of insulin receptor substrate-1 (IRS-1), improving insulin resistance [50]. High plasma glucose level increases ROS production, while EGCG improved insulin resistance by scavenging ROS. ROS plays a key role in increasing JNK and IRS-1 serine phosphorylation and reducing the transduction of insulin signal [62]. Green tea catechin increases insulin sensitivity by directly activating peroxisome proliferator-activated receptor (PPAR) γ [50]. In addition to insulin sensitivity, EGCG can also inhibit glucose absorption by competitively binding with the sodium-glucose transporter-1 (SGLT-1) in intestinal epithelial cells and enhance glucose uptake in muscles and adipocytes via enhancement of the GLUT4 expression [44, 53].

Most studies showing the beneficial effects of EGCG on glucose homeostasis were performed in vitro. EGCG showed an insulin-like activity through the reduction of gluconeogenic enzymes (glucose-6-phosphatase and PEPCK) in hepatocytes by suppressing their gene expression [51, 52]. In myocytes, green tea or EGCG stimulates GLUT4 translocation and glucose uptake via the PI3-kinase/ Akt signaling pathway; alternatively, muscle glucose uptake occurs via AMPK [52]. In laboratory animals, EGCG improved insulin sensitivity in peripheral organs and inhibited gluconeogenesis [51]. In humans, EGCG can protect pancreatic β cell from cytokines, inhibiting the NF-kB activation [63]. Tea also reduces carbohydrate absorption by inhibiting α-amylase, β-glucosidase, and sodium-glucose transporters [51, 56].

Black tea rich in theaflavins decreases the risk of T2D by inhibiting obesity through AMPK phosphorylation and promoting the browning of white adipose tissue [50]. The pro-inflammatory cytokines TNF-α and interleukin (IL)-1 are involved in obesity-associated insulin resistance and T2D. Black tea consumption has a potential role in downregulating serum TNF-α and IL-1 levels and upregulating IL-10, an anti-inflammatory cytokine [63].

6. Conclusion

Tea contains polyphenols that may provide an important source of dietary antioxidants in humans. Moderate consumption of tea seems to reduce the risk of MetS and/or MetS-related diseases. EGCG, the main catechin in tea, presents major health benefits. Green tea seems to have the best potential antioxidant effects when compared to other teas. It is worth mentioning that most of the studies have not demonstrated toxicity due to tea consumption or supplementation; however, further research, especially in humans, should be conducted to confirm this property and evaluate the underlying mechanisms of action.

Conflict of interest

The authors have no conflict of interest to disclose.

Author details

Telma Angelina Faraldo Corrêa[1,2]*, Adriana Campos Rozenbaum[1,2]
and Marcelo Macedo Rogero[2,3]

1 Department of Food and Experimental Nutrition, School of Pharmaceutical Sciences, University of São Paulo, São Paulo, Brazil

2 Food Research Center (FoRC), CEPID-FAPESP, Research Innovation and Dissemination Centers, São Paulo Research Foundation, São Paulo, Brazil

3 Nutritional Genomics and Inflammation Laboratory, Department of Nutrition, School of Public Health, University of São Paulo, São Paulo, Brazil

*Address all correspondence to: telma.nutri@yahoo.com.br

References

[1] IDF. The IDF consensus worldwide definition of the metabolic syndrome. IDF Communications. 2006:1-23. Available from: https://www.idf.org/e-library/consensus-statements/60-idfconsensus-worldwide-definitionof-the-metabolic-syndrome.html

[2] Eisvand F, Razavi BM, Hosseinzadeh H. The effects of *Ginkgo biloba* on metabolic syndrome: A review. Phytotherapy Research. 2020. DOI: 10.1002/ptr.6646

[3] Alberti KGMM, Zimmet P, Shaw J. The metabolic syndrome—A new worldwide definition. Lancet. 2005;**366**(9491):1059-1062. DOI: 10.1016/S0140-6736(05)67402-8

[4] Islam M, Alam F, Solayman M, Khalil M, Kamal MA, Gan SH. Dietary phytochemicals: Natural swords combating inflammation and oxidation-mediated degenerative diseases. Oxidative Medicine and Cellular Longevity. 2016;**2016**. DOI: 10.1155/2016/5137431

[5] Finicelli M, Squillaro T, Di Cristo F, Di Salle A, Melone MAB, Galderisi U, et al. Metabolic syndrome, Mediterranean diet, and polyphenols: Evidence and perspectives. Journal of Cellular Physiology. 2019;**234**(5):5807-5826. DOI: 10.1002/jcp.27506

[6] González-Castejón M, Rodriguez-Casado A. Dietary phytochemicals and their potential effects on obesity: A review. Pharmacological Research. 2011;**64**(5):438-455. DOI: 10.1016/j.phrs.2011.07.004

[7] Chiva-Blanch G, Badimon L. Effects of polyphenol intake on metabolic syndrome: Current evidences from human trials. Oxidative Medicine and Cellular Longevity. 2017;**2017**. DOI: 10.1155/2017/5812401

[8] Liu K, Luo M, Wei S. The bioprotective effects of polyphenols on metabolic syndrome against oxidative stress: Evidences and perspectives. Oxidative Medicine and Cellular Longevity. 2019;**2019**. DOI: 10.1155/2019/6713194

[9] Yang CS, Wang H, Sheridan ZP. Studies on prevention of obesity, metabolic syndrome, diabetes, cardiovascular diseases and cancer by tea. Journal of Food and Drug Analysis. 2018;**26**(1):1-13. DOI: 10.1016/j.jfda.2017.10.010

[10] Musial C, Kuban-Jankowska A, Gorska-Ponikowska M. Beneficial properties of green tea catechins. International Journal of Molecular Sciences. 2020;**21**(5):1744. DOI: 10.3390/ijms21051744

[11] Yang CS, Sang S, Lambert JD, Lee MJ. Bioavailability issues in studying the health effects of plant polyphenolic compounds. Molecular Nutrition & Food Research. 2008;**52**(S1):S139-SS51. DOI: 10.1002/mnfr.200700234

[12] Innes JK, Calder PC. Omega-6 fatty acids and inflammation. Prostaglandins, Leukotrienes, and Essential Fatty Acids. 2018;**132**:41-48. DOI: 10.1016/j.plefa.2018.03.004

[13] Rogero MM, Calder PC. Obesity, inflammation, toll-like receptor 4 and fatty acids. Nutrients. 2018;**10**(4):432. DOI: 10.3390/nu10040432. PubMed PMID: 29601492. PubMed Central PMCID: 5946217

[14] Calder PC, Bosco N, Bourdet-Sicard R, Capuron L, Delzenne N, Doré J, et al. Health relevance of the modification of low grade inflammation in ageing (inflammageing) and the role of nutrition. Ageing Research Reviews. 2017;**40**:95-119. DOI: 10.1016/j.arr.2017.09.001

[15] Minihane AM, Vinoy S, Russell WR, Baka A, Roche HM, Tuohy KM, et al. Low-grade inflammation, diet composition and health: Current research evidence and its translation. The British Journal of Nutrition. 2015;**114**(7):999-1012. DOI: 10.1017/S0007114515002093

[16] Hotamisligil GS. Inflammation, metaflammation and immunometabolic disorders. Nature. 2017;**542**(7640):177. DOI: 10.1038/nature21363

[17] Cabrera C, Artacho R, Giménez R. Beneficial effects of green tea—A review. Journal of the American College of Nutrition. 2006;**25**(2):79-99. DOI: 10.1080/07315724.2006.10719518

[18] Westerterp-Plantenga MS, Lejeune MPGM, Kovacs EMR. Body weight loss and weight maintenance in relation to habitual caffeine intake and green tea supplementation. Obesity Research. 2005;**13**(7):1195-1204. DOI: 10.1038/oby.2005.142

[19] Basu A, Sanchez K, Leyva MJ, Wu M, Betts NM, Aston CE, et al. Green tea supplementation affects body weight, lipids, and lipid peroxidation in obese subjects with metabolic syndrome. Journal of the American College of Nutrition. 2010;**29**(1):31-40. DOI: 10.1080/07315724.2010.10719814

[20] Chantre P, Lairon D. Recent findings of green tea extract AR25 (Exolise) and its activity for the treatment of obesity. Phytomedicine. 2002;**9**(1):3-8. DOI: 10.1078/0944-7113-00078

[21] Hase T, Komine Y, Meguro S, Takeda Y, Takahashi H, Matsui Y, et al. Anti-obesity effects of tea catechins in humans. Journal of Oleo Science. 2001;**50**(7):599-605. DOI: 10.5650/jos.50.599

[22] Bogdanski P, Suliburska J, Szulinska M, Stepien M, Pupek-Musialik D, Jablecka A. Green tea extract reduces blood pressure, inflammatory biomarkers, and oxidative stress and improves parameters associated with insulin resistance in obese, hypertensive patients. Nutrition Research. 2012;**32**(6):421-427. DOI: 10.1016/j.nutres.2012.05.007

[23] Yuan F, Dong H, Fang K, Gong J, Lu F. Effects of green tea on lipid metabolism in overweight or obese people: A meta-analysis of randomized controlled trials. Molecular Nutrition & Food Research. 2018;**62**(1):1601122. DOI: 10.1002/mnfr.201601122

[24] Momose Y, Maeda-Yamamoto M, Nabetani H. Systematic review of green tea epigallocatechin gallate in reducing low-density lipoprotein cholesterol levels of humans. International Journal of Food Sciences and Nutrition. 2016;**67**(6):606-613. DOI: 10.1080/09637486.2016.1196655

[25] Tsuchida T, Itakura H, Nakamura H. Reduction of body fat in humans by long-term ingestion of catechins. Progress in Medicine. 2002;**22**(9):2189-2220

[26] Rahman I, Biswas SK, Kirkham PA. Regulation of inflammation and redox signaling by dietary polyphenols. Biochemical Pharmacology. 2006;**72**(11):1439-1452. DOI: 10.1016/j.bcp.2006.07.004

[27] Zheng X-X, Xu Y-L, Li S-H, Liu X-X, Hui R, Huang X-H. Green tea intake lowers fasting serum total and LDL cholesterol in adults: A meta-analysis of 14 randomized controlled trials. The American Journal of Clinical Nutrition. 2011;**94**(2):601-610. DOI: 10.3945/ajcn.110.010926

[28] Onakpoya I, Spencer E, Heneghan C, Thompson M. The effect of green tea on blood pressure and lipid profile: A systematic review and meta-analysis of randomized clinical trials. Nutrition, Metabolism,

and Cardiovascular Diseases. 2014;**24**(8):823-836. DOI: 10.1016/j.numecd.2014.01.016

[29] Bornhoeft J, Castaneda D, Nemoseck T, Wang P, Henning SM, Hong MY. The protective effects of green tea polyphenols: Lipid profile, inflammation, and antioxidant capacity in rats fed an atherogenic diet and dextran sodium sulfate. Journal of Medicinal Food. 2012;**15**(8):726-732. DOI: 10.1089/jmf.2011.0258

[30] Xu X, Pan J, Zhou X. Amelioration of lipid profile and level of antioxidant activities by epigallocatechin-gallate in a rat model of atherogenesis. Heart, Lung & Circulation. 2014;**23**(12):1194-1201. DOI: 10.1016/j.hlc.2014.05.013

[31] Eng QY, Thanikachalam PV, Ramamurthy S. Molecular understanding of Epigallocatechin gallate (EGCG) in cardiovascular and metabolic diseases. Journal of Ethnopharmacology. 2018;**210**:296-310. DOI: 10.1016/j.jep.2017.08.035

[32] Kim A, Chiu A, Barone MK, Avino D, Wang F, Coleman CI, et al. Green tea catechins decrease total and low-density lipoprotein cholesterol: A systematic review and meta-analysis. Journal of the American Dietetic Association. 2011;**111**(11):1720-1729. DOI: 10.1016/j.jada.2011.08.009

[33] Samavat H, Newman AR, Wang R, Yuan J-M, Wu AH, Kurzer MS. Effects of green tea catechin extract on serum lipids in postmenopausal women: A randomized, placebo-controlled clinical trial. The American Journal of Clinical Nutrition. 2016;**104**(6):1671-1682. DOI: 10.3945/ajcn.116.137075

[34] WHO. Hypertension. World Health Organization. 2019. Available from: https://www.who.int/news-room/fact-sheets/detail/hypertension

[35] Yarmolinsky J, Gon G, Edwards P. Effect of tea on blood pressure for secondary prevention of cardiovascular disease: A systematic review and meta-analysis of randomized controlled trials. Nutrition Reviews. 2015;**73**(4):236-246. DOI: 10.1093/nutrit/nuv001

[36] Yin J, Duan S, Liu FC, Yao QK, Tu S, Xu Y, et al. Blood pressure is associated with tea consumption: A cross-sectional study in a rural, elderly population of Jiangsu China. The Journal of Nutrition, Health & Aging. 2017;**21**(10):1151-1159. DOI: 10.1007/s12603-016-0829-4

[37] Li D, Wang R, Huang J, Cai Q, Yang CS, Wan X, et al. Effects and mechanisms of tea regulating blood pressure: Evidences and promises. Nutrients. 2019;**11**(5):1115. DOI: 10.3390/nu11051115

[38] Quezada-Fernández P, Trujillo-Quiros J, Pascoe-González S, Trujillo-Rangel WA, Cardona-Müller D, Ramos-Becerra CG, et al. Effect of green tea extract on arterial stiffness, lipid profile and sRAGE in patients with type 2 diabetes mellitus: A randomised, double-blind, placebo-controlled trial. International Journal of Food Sciences and Nutrition. 2019;**70**(8):977-985. DOI: 10.1080/09637486.2019.1589430

[39] Mozaffari-Khosravi H, Ahadi Z, Barzegar K. The effect of green tea and sour tea on blood pressure of patients with Type 2 diabetes: A randomized clinical trial. Journal of Dietary Supplements. 2013;**10**(2):105-115. DOI: 10.3109/19390211.2013.790333

[40] Grassi D, Draijer R, Desideri G, Mulder T, Ferri C. Black tea lowers blood pressure and wave reflections in fasted and postprandial conditions in hypertensive patients: A randomised study. Nutrients. 2015;7(2):1037-1051. DOI: 10.3390/nu7021037

[41] Toolsee NA, Aruoma OI, Gunness TK, Kowlessur S, Dambala V, Murad F, et al. Effectiveness of green

tea in a randomized human cohort: Relevance to diabetes and its complications. BioMed Research International. 2013;**2013**. DOI: 10.1155/2013/412379

[42] Xu R, Yang K, Ding J, Chen G. Effect of green tea supplementation on blood pressure: A systematic review and meta-analysis of randomized controlled trials. Medicine (Baltimore). 2020;**99**(6). DOI: 10.1097/MD.0000000000019047

[43] Li G, Zhang Y, Thabane L, Mbuagbaw L, Liu A, Levine MAH, et al. Effect of green tea supplementation on blood pressure among overweight and obese adults: A systematic review and meta-analysis. Journal of Hypertension. 2015;**33**(2):243-254. DOI: 10.1097/HJH.0000000000000426

[44] Legeay S, Rodier M, Fillon L, Faure S, Clere N. Epigallocatechin gallate: A review of its beneficial properties to prevent metabolic syndrome. Nutrients. 2015;**7**(7):5443-5468. DOI: 10.3390/nu7075230

[45] Bartesaghi S, Radi R. Fundamentals on the biochemistry of peroxynitrite and protein tyrosine nitration. Redox Biology. 2018;**14**:618-625. DOI: 10.1016/j.redox.2017.09.009. PubMed PMID: 29154193. Epub: 09/19.eng

[46] Ihm S-H, Jang S-W, Kim O-R, Chang K, Oak M-H, Lee J-O, et al. Decaffeinated green tea extract improves hypertension and insulin resistance in a rat model of metabolic syndrome. Atherosclerosis. 2012;**224**(2):377-383. DOI: 10.1016/j.atherosclerosis.2012.07.006

[47] Li Y, Ying C, Zuo X, Yi H, Yi W, Meng Y, et al. Green tea polyphenols down-regulate caveolin-1 expression via ERK1/2 and p38MAPK in endothelial cells. The Journal of Nutritional Biochemistry. 2009;**20**(12):1021-1027. DOI: 10.1016/j.jnutbio.2008.12.001

[48] Li F, Ohnishi-Kameyama M, Takahashi Y, Yamaki K. Tea polyphenols as novel and potent inhibitory substances against renin activity. Journal of Agricultural and Food Chemistry. 2013;**61**(40):9697-9704. DOI: 10.1021/jf403710b

[49] WHO. Diabetes. World Health Organization. 2018. Available from: https://www.who.int/news-room/fact-sheets/detail/diabetes [cited 09 October 2019]

[50] Meng J-M, Cao S-Y, Wei X-L, Gan R-Y, Wang Y-F, Cai S-X, et al. Effects and mechanisms of tea for the prevention and management of diabetes mellitus and diabetic complications: An updated review. Antioxidants. 2019;**8**(6):170. DOI: 10.3390/antiox8060170

[51] Miyoshi N. Anti-Diabetic Effects of Green Tea. Efficacy of Tea in Human Health: Overview [Internet]. 2020. pp. 118-125 Available from: http://shizuoka-cha.com/files/6314/4374/8053/Scientific_Evidence_for_the_Health_Benefits_of_Green_Tea2.pdf#page=119 [cited 28 February 2020]

[52] Keske MA, Ng HLH, Premilovac D, Rattigan S, Jeong K, Munir K, et al. Vascular and metabolic actions of the green tea polyphenol epigallocatechin gallate. Current Medicinal Chemistry. 2015;**22**(1):59-69

[53] InterAct C, van Woudenbergh GJ, Kuijsten A, Drogan D, Van der ADL, Romaguera D, et al. Tea consumption and incidence of type 2 diabetes in Europe: The EPIC-InterAct case-cohort study. PLoS One. 2012;7(5):e36910-e. DOI: 10.1371/journal.pone.0036910. PubMed PMID: 22666334. Epub: 05/30.eng

[54] MacKenzie T, Leary L, Brooks WB. The effect of an extract of green and black tea on glucose control in adults with type 2 diabetes mellitus:

Double-blind randomized study. Metabolism. 2007;**56**(10):1340-1344. DOI: 10.1016/j.metabol.2007.05.018

[55] Brown AL, Lane J, Coverly J, Stocks J, Jackson S, Stephen A, et al. Effects of dietary supplementation with the green tea polyphenol epigallocatechin-3-gallate on insulin resistance and associated metabolic risk factors: Randomized controlled trial. The British Journal of Nutrition. 2008;**101**(6):886-894

[56] Butacnum A, Chongsuwat R, Bumrungpert A. Black tea consumption improves postprandial glycemic control in normal and pre-diabetic subjects: A randomized, double-blind, placebo-controlled crossover study. Asia Pacific Journal of Clinical Nutrition. 2017;**26**(1):59. DOI: 10.6133/apjcn.112015.08

[57] Liu C-Y, Huang C-J, Huang L-H, Chen IJ, Chiu J-P, Hsu C-H. Effects of green tea extract on insulin resistance and glucagon-like peptide 1 in patients with type 2 diabetes and lipid abnormalities: A randomized, double-blinded, and placebo-controlled trial. PLoS One. 2014;**9**(3). DOI: 10.1371/journal.pone.0091163

[58] Zheng X-X, Xu Y-L, Li S-H, Hui R, Wu Y-J, Huang X-H. Effects of green tea catechins with or without caffeine on glycemic control in adults: A meta-analysis of randomized controlled trials. The American Journal of Clinical Nutrition. 2013;**97**(4):750-762. DOI: 10.3945/ajcn.111.032573

[59] Mahmoud F, Haines D, Al-Ozairi E, Dashti A. Effect of black tea consumption on intracellular cytokines, regulatory T cells and metabolic biomarkers in type 2 diabetes patients. Phytotherapy Research. 2016;**30**(3):454-462. DOI: 10.1002/ptr.5548

[60] Hua C, Liao Y, Lin S, Tsai T, Huang C, Chou P. Does supplementation with green tea extract improve insulin resistance in obese type 2 diabetics? A randomized, double-blind, and placebo controlled clinical trial. Alternative Medicine Review. 2011;**16**(2):157-163

[61] Fukino Y, Shimbo M, Aoki N, Okubo T, Iso H. Randomized controlled trial for an effect of green tea consumption on insulin resistance and inflammation markers. Journal of Nutritional Science and Vitaminology. 2005;**51**(5):335-342. DOI: 10.3177/jnsv.51.335

[62] Ma SB, Zhang R, Miao S, Gao B, Lu Y, Hui S, et al. Epigallocatechin-3-gallate ameliorates insulin resistance in hepatocytes. Molecular Medicine Reports. 2017;**15**(6):3803-3809. DOI: 10.3892/mmr.2017.6450

[63] Roy N, Bhattacharjee K, Bandyopadhyay SK, Chatterjee S, Saha AK, Chatterjee A. Role of black tea in type 2 diabetes mellitus and metabolic syndrome. American Journal of Phytomedicine and Clinical Therapeutics. 2016;**53**:354-359

Chapter 6

Nutraceuticals of Tea *(Camellia sinensis)* for Human Health

Kula Kamal Senapati

Abstract

This book chapter describes about the phytochemicals in tea which are significantly useful in preparing nutraceutical products. The polyphenols along with other bioactive compounds present in tea have many pharmacological properties which attribute to the development of various food products where tea constitutes as an active ingredient. This chapter also discusses the potential uses of tea and their bioactive constituents in treatment and prevention of diseases in human which infer the potentiality of developing and popularizing nutraceuticals of tea.

Keywords: nutraceuticals, phytochemicals, catechins, polyphenols, antioxidant, therapeutic properties

1. Introduction

Nutraceuticals are promising class of natural products that encompasses the combined terms "nutrition" and pharmaceuticals" which incorporates the characteristics of both nutritional and pharmaceutical and thus exhibit several health benefits. The food or food products having nutraceuticals properties can be used as medicine in addition to nutritional values and have been used to support proper functioning of the body, treat and prevention of diseases, and increase the life expectancy of human beings [1–4]. Currently nutraceuticals have been explored in several diseases in prevention and cure such cancer, diabetics, cardiovascular, diseases etc.

Tea (*Camellia sinensis* L.) is one of the most popular and widely consumed beverage world-wide. Commercially tea is mostly available in three varieties viz. black (red tea), green and oolong (yellow tea) tea which differ in their physical and chemical characteristics arised from their different manufacturing process. Black tea consumption is highest in western countries which accounts for around 78% of worldwide consumption. The green tea is mostly consumed in Japan and china and accounts for 20% whereas oolong tea is consumed 2% only. Black tea is widely consumed in India and India is one of the largest tea producers in the world and it occupies about 70% of domestic consumption of the total tea production in the country.

The black tea is fully fermented and known for its characteristics brown liquor whereas green tea is unfermented and known for its light greenish yellow liquor. The oolong is semi-fermented and liquor characteristic lies in between the black and green tea. All the three types differ in their taste and flavor and accordingly their chemical profiling also varies.

Tea contains a variety of bioactive compounds such as polyphenols, polysaccharides, vitamins, amino acids etc. having medicinal properties which can be used as food additives in preparation of nutraceuticals [5–8]. Tea is mainly chemically

IntechOpen

characterized by their polyphenolic compounds and their polymeriszed products along other bioactive compounds in minor quantities.

The polyphenols which are produced by the plant as secondary metabolites are the major constituents in tea. Other secondary metabolites present in tea are phenolic acids, purine alkaloids, tannins, flavonols and their glycosides. The polyphenols in tea are catechins (C) and their conjugated products viz. epigallocatechingallate (EGCG), epigallocatechins (EGC), epicatechingallate (ECG) and epicatechins (EC) and present in higher quantities in green tea. The other two class of polyphenols found in black tea are theaflavins (TFs) and thearubigins (TRs) which are present in significant quantities in black tea and less quantity in oolong tea. The catechin polyphenols in tea are health-promising due to their antioxidant properties and are very useful in preparation of nutraceuticals from tea [9–14].

In addition, tea contains other beneficiary substances to our health such as fluoride, caffeine, minerals, trace elements e.g. manganese, chromium etc. which also added nutraceutical value of tea and its food additive which also add nutraceutical value of tea or its food additives [15]. Looking to the therapeutic properties of these phytochemicals in tea, different food products have been prepared with tea or its extract which are rich in these active ingredients. These products include ready to drink beverages, instant tea, confections, cereal bars, pet foods, candy etc. So, there is a huge possibility of using bioactive constituents in tea as food additive which provide medical or health benefit in prevention and treatment of disease and thus tea can be potentially used for the development of nutraceuticals.

In this chapter, various bioactive compounds in tea (green, black, oolong and white teas) and their properties which can be employed as nutraceuticals in the form of food or part of food products has been discussed. Further, we have addressed the role of these tea nutraceuticals and their application in prevention and treatment of various diseases.

2. Phytochemicals in tea

There are a variety of chemicals in tea which are therapeutically important and they also vary in different types of commercially available teas viz. green, black and oolong tea. The chemical characteristics of tea are also varying with variation of plant species, geographic locations, climatic condition, horticultural practices as well as plucking parameters of the tea leaves [16, 17]. Polyphenols comprises of a large group of phytochemicals, the major polyphenolic constituents in tea which account for maximum 30% total dry weight of tea leaves are the flavanols, also known as catechins. Of these polyphenolic catechins, eight catechins are appeared in significant quantities which include (+)-catechin, (-)- epicatechin, (-)-gallocatechin, (-)-epigallocatechin, (-)-catechin gallate, (-)-epicatechin gallate, (-)-gallocatechin gallate, (-)-epigallocatechin gallate.

Tea beverage is chemically characterized mainly by polyphenolic constituents (mostly flavonoids) and their polymerized compounds. Four polyphenolic catechins which are in large quantities in green tea are (–)–epigallocatechin- 3-gallate (EGCG), (–)–epigallocatechin (EGC), (–)–epicatechin-3- gallate (ECG), and (–)–epicatechin (EC) (**Table 1**). These catechins are highest in green tea (30-40% in dry weight) and least in black tea (10% in dry weight) in which catechins undergo condensation and polymerization into other phenolic compounds theaflavins and thearubigins during enzymatic oxidation of catechins during fermentation process of black tea manufacturing [18, 19]. The oolong tea has substantial amount of catechins with other oligomeric polyphenols [20].

Apart from polyphenols, there are a number of phytochemicals which are significantly important in medicinal use. These phytochemicals are oxyaromatic acids (gallic, caffeic, quinine, chlorogenic and n-coumaric acids, flavonols (Quercetin, kaempferol, myricetin), pigments (carotenoids, chlorophyll), alkaloids (caffeine, theophylline, theobromine), amino acids, lipids, polysaccharides, vitamins, lignans and saponins which also attribute to the widely explored medicinal value of tea [21–23].

The major polyphenolic compounds in different types of tea which differ due to their manufacturing processes are highlighted in **Table 2**. The white tea is least processed of all the tea types and thus it intakes maximum polyphenols whereas the black tea is fully fermented through enzyme mediated oxidation of polyphenols into oligomeric and polymeric flavanols (theaflavins, thearubigins and other oligomers) with characteristics flavor and color. In black tea processing, tea leaves (standard plucking two and a bud) are crushed to undergo enzymatic oxidation (polyphenol oxidase) and subsequent condensation of tea catechins in to the formation of theaflavins (TFs) (oligomeric) and thearubigins (TRs) (polymeric) products. This fermentation process is being limited in case of oolong tea. In green tea this enzymatic oxidation is prevented by steaming (or pan-frying) of fresh tea leaves. Therefore, the green tea infusion contains major quantities of polyphenol catechins (30-40% dry weight of four major catechins viz. (–)-Epigallocatechin-3-gallate (EGCG), (–)-epigallocatechin (EGC), (–)-epicatechin-3-gallate (ECG), and (–)-epicatechin (EC)). Since all major polyphenols are oxidized in black tea, it contains less amount of catechins (3-10%) along with theaflavins (2-6%), thearubigins (>20%) and gallic acids in significant quantities. The oolong contains green tea catechins and less amount of black tea theaflavins and thearubigins due to its partial fermentation process.

In water extract, all types of tea also contain caffeine in quantity of 2-5%.

3. Nutraceutical properties of tea phytochemicals

The therapeutic properties of tea extract are associated with the polyphenolic contents which have the highest antioxidant capacity amongst other bio-active compounds. The green tea extract has higher antioxidant capacity than that of black tea or oolong tea due the higher quantities of catechin polyphenols viz. EGCG, EGC, ECG, EC [24]. The EGCG has the highest anti-oxidant activity followed by ECG and EC whereas EGC has the lowest activity [25]. The EGCG can inhibit the production of hydrogen peroxide and superoxide radicals by tumor promoter-activated neutrophils in our body. Among the three major types of tea (green, black, oolong), green tea has the highest antioxidant activity and black tea has the lowest one. The antioxidant activity of green tea and its polyphenols have been studied in a number of *in vitro* and *in vivo* experiments and in most of these researches EGCG (major constituent in green tea) is well documented and significantly important. These nonnutritive phytochemicals are potential nutraceuticals and their easy bioavailability makes their useful consumption in prevention of diseases. The higher antioxidant properties of black tea are associated with minor quantities of catechins and other oligomeric polyphenols. In addition to different types of tea which differ in chemical contents, the health-promising factors of tea also varies with cultivars to cultivars and their geographic locations. Therefore, it is very important to determine the bioactivity of tea of different varieties in obtaining nutraceutical values.

Recent researches have gained attention in therapeutic use of tea polyphenols of green and black tea in diseases associated to metabolic syndromes. Green tea

Tea polyphenols in green and black tea

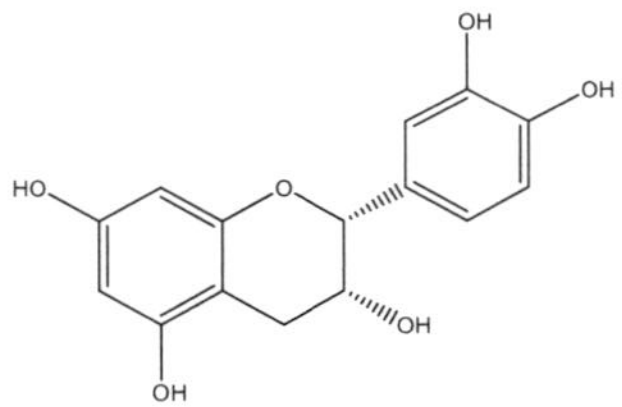

(-)-Epicatechin

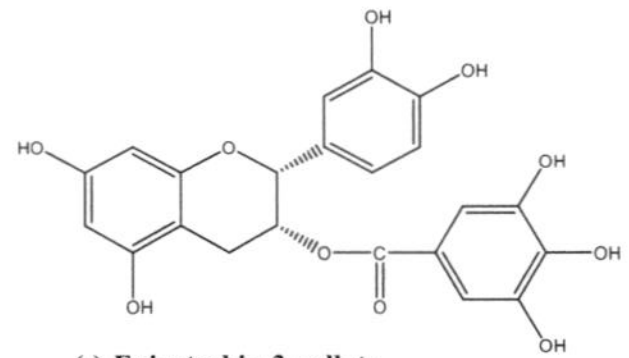

(-)-Epicatechin-3-gallate

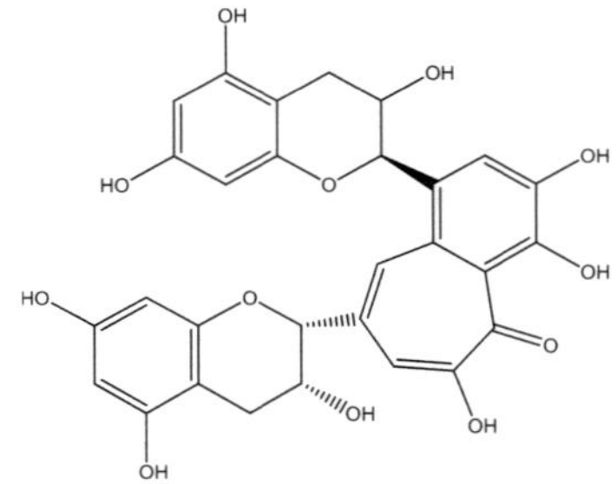

Theaflavin

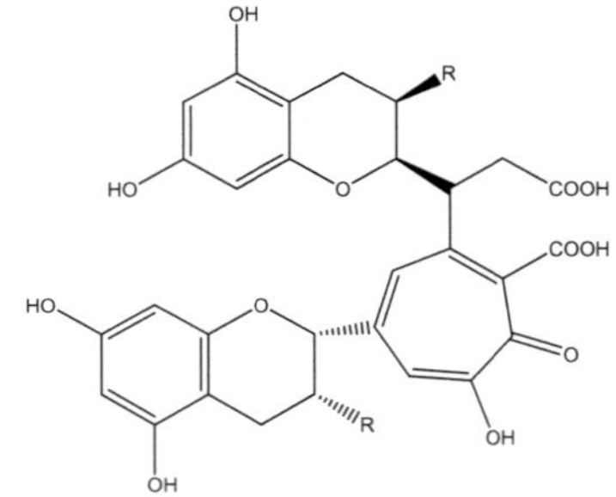

Thearubigins
(R= Gallate or other group)

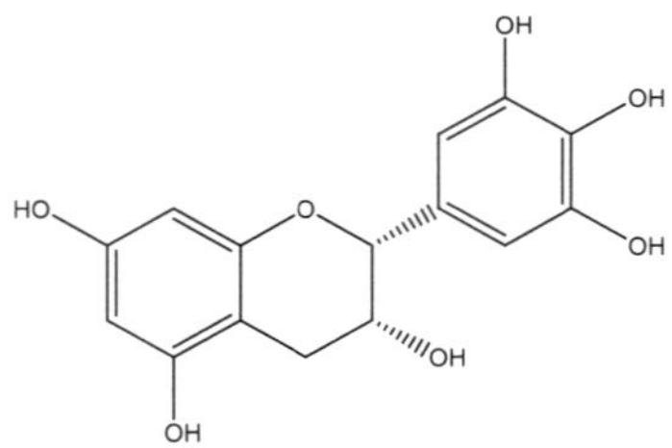

(-)-Epigallocatechin

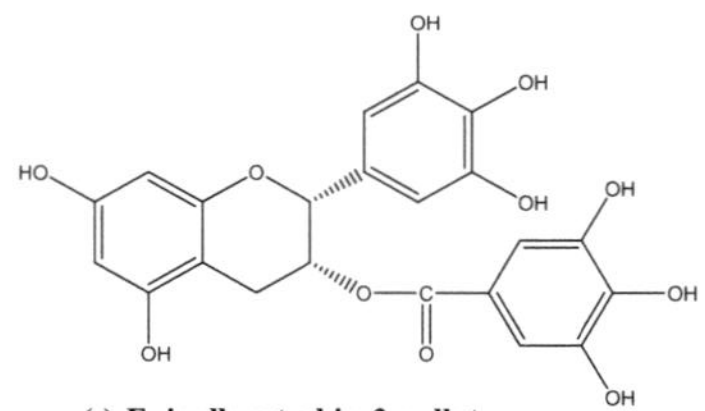

(-)-Epigallocatechin-3-gallate

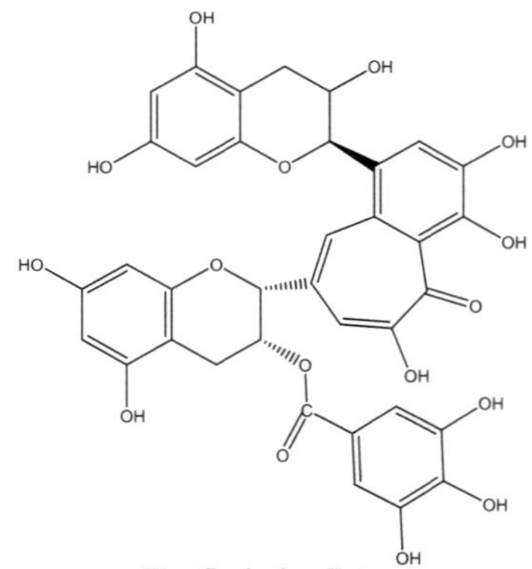

Theaflavin-3-gallate

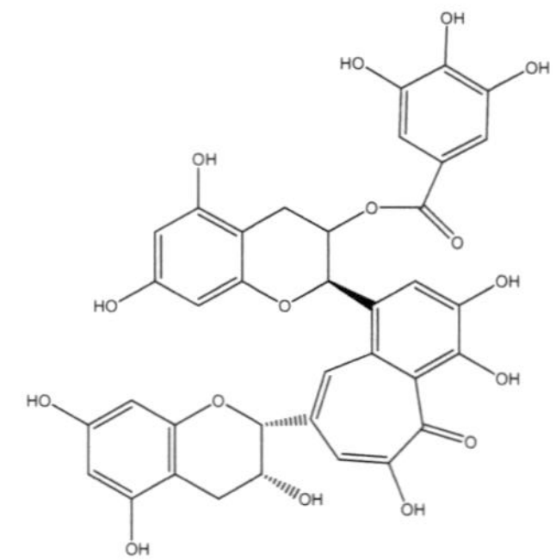

Theaflavin-3'-gallate

DOI: http://dx.doi.org/10.5772/intechopen.96506

Tea polyphenols in green and black tea			
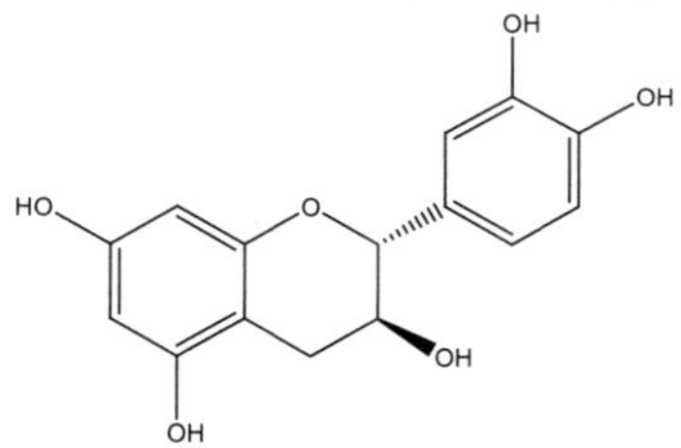 (+)-Catechin	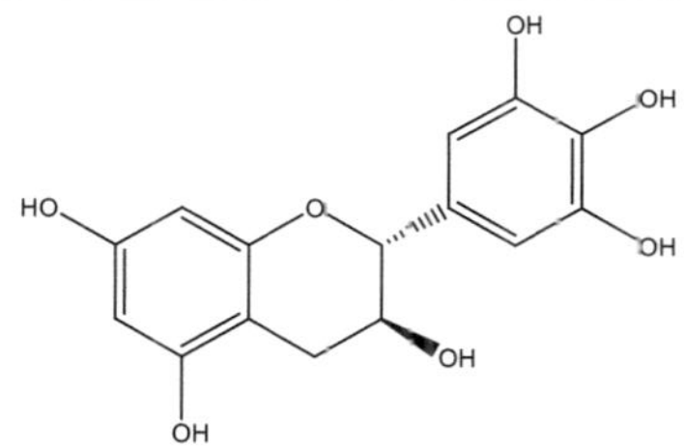 (+)-Gallocatechin	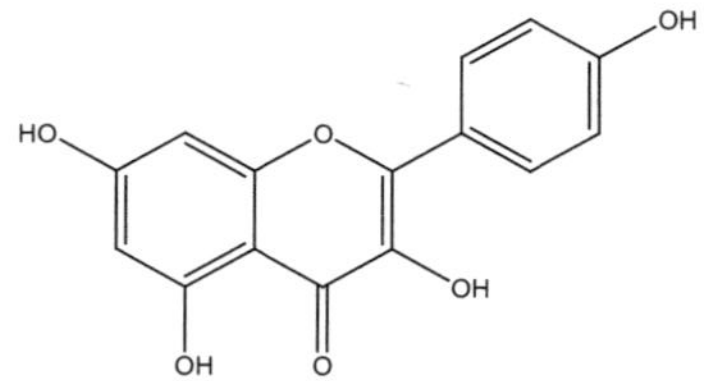 Kaempferol	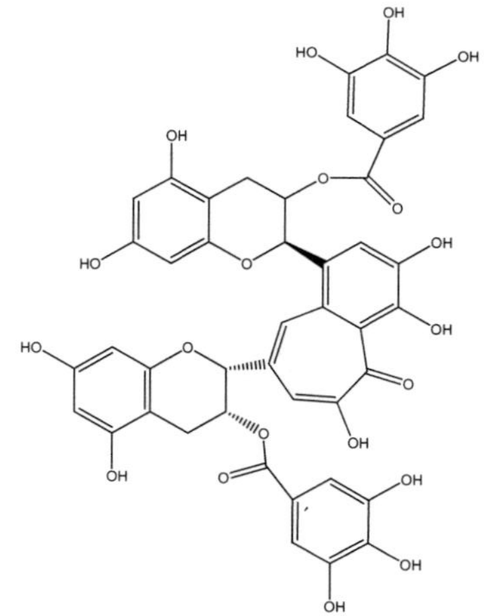 Theaflavin-3, 3'-gallate
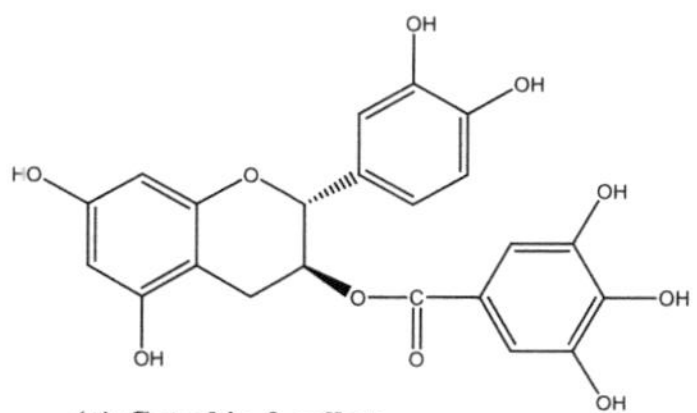 (+)-Catechin-3-gallate	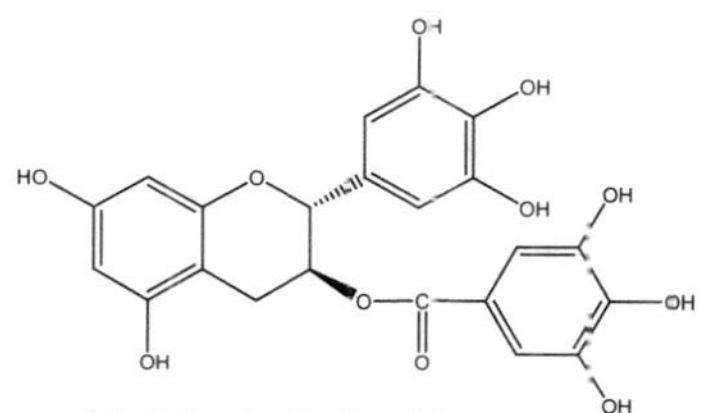 (+)-Gallocatechin-3-gallate	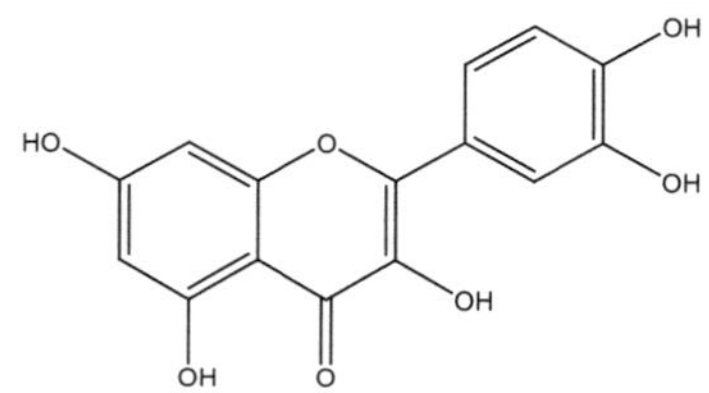 Quercetin	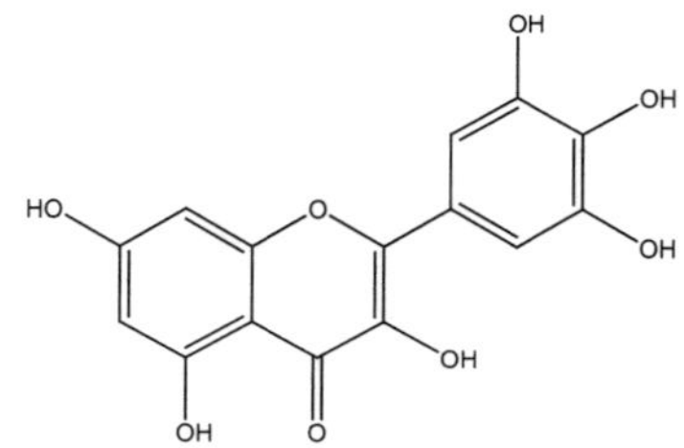 Myricetin

Table 1.
Chemical structures of major polyphenols (flavonoids: flavanols and flavonols) in Green tea and Black tea.

Type of tea	Processing technique	Physio-chemical changes	Major polyphenols
Green	Withering, steaming (Japan) or roasting (China), rolling and drying	Deactivation of enzymes to prevent fermentation (oxidation of polyphenol)	Catechins (Flavan-3-ols)
Black (red tea)	Withering, crushing, rolling, drying	Fermented (in presence of enzymes (polyphenol oxidase and peroxidase enzymes)	Catechins, theaflavins, thearubigins
Oolong	Withering, rolling, drying	Partially fermented (partial oxidation)	Catechins, theaflavins, procyanidins
White	Withering, drying	Unfermented	Catechins

Table 2.
Major polyphenols and processing conditions of different varieties of tea.

catechins have been explored to antioxidant, anticarcinogenic, anti-inflammatory, probiotic, anti-inflammatory, thermogenic, and antimicrobial activities. These have been found to inhibit carcinogenesis of the skin, lung, esophagus, stomach, liver, small intestine, colon, bladder, prostate, and mammary gland in animal studies [26]. Black and green tea can protect oxidative damage of red blood cells [27].

Looking in to the various health benefits of tea phytochemicals, tea extracts in liquid or powder form can be effectively used in food products such as sweets, biscuits, bread, cake, candies, ready-to-drink beverages as well as other polyphenol rich food supplements. The tea nutraceuticals are also available in various form of tablets, capsules and health drinks. In addition, tea also contains other nutritional substances such as dietary fiber and proteins after polyphenols extraction in tea for use in nutraceuticals which also give health benefits [28]. Consumption of tea with lemon juice (ascorbic acid) has manifold benefits in making more accessible of tea antioxidants (catechins) to get absorbed as well as enhance the iron absorption in our body [15]. Further, compounds in tea can inhibit iron absorption from foods; drinking green tea with lemon reduces that effect. The combination of honey with green tea has been used since very past as they are very effective in healing wounds. The health-promoting activities of tea also vary cultivar to cultivar and thus it is important to identify the cultivars having specific bioactivity so as to use the nutraceutical properties properly. The phytochemicals in tea and their nutraceutical properties are depicted in **Figure 1**.

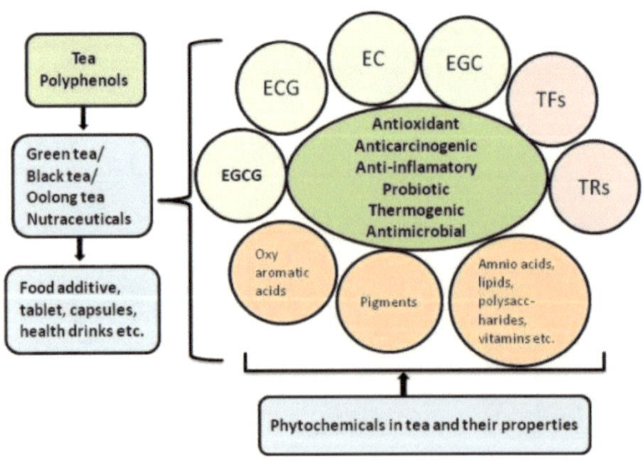

Figure 1.
Role and mechanism of nutraceutical properties of tea phytochemicals.

4. Tea phytochemicals as Nutraceuticals in different diseases

Tea is beneficial in protection and prevention of numerous critical diseases such as different cancers, diabetes, neurodegenerative and cardiovascular diseases [29–31]. The polyphenols, the major phytochemicals present in tea are known for their antioxidant properties due to their abilities to scavenge free radical species such as hydroxyl radical (OH), hydrogen peroxide (H_2O_2) and superoxide anion ($O2.^-$) and thus their consumption confer them as potential cancer chemopreventive agents as well as other free radical induced diseases. The polyphenol extract of tea has been effectively used in cancer prevention, protection cardiovascular diseases, reduce weight-loss, anti-allergic, prebiotics, osteoarthritis protection etc. [32, 33]. Green tea extract comprises of nearly 40% of polyphenols of which nearly 25% accounts for different catechins and their gallates of which EGCG is the highest in quantity (about 11%), the most studied polyphenols in therapeutic properties. Black tea polyphenols theaflavins, theaflavin gallates (which are dimers of two different catechins and their gallates) and thearubigins with 2-6% in quantity in addition to 3-10% of catechins gain attention in different diseases [34].

Herein, we wish to discuss the health benefits of tea in common and major diseases such as diabetes, cardiovascular diseases, obesity, cancers, as well as others minor diseases related to microbial and inflammatory types. The role of major phtochemicals of tea in different diseases is outlined in **Table 3**.

4.1 Tea phytochemicals in diabetics

Diabetes mellitus, a highly alarming disease worldwide is associated with high levels of blood glucose other than normal which is associated with insufficient insulin production or its ineffectiveness.

Polyphenols in green and black tea are very effective in reducing hyperglycemia and insulin resistance which can manage the blood glucose in diabetes [31, 35, 36].

Thus, tea polyphenols present potential nutraceuticals for various facets of type 2 diabetes mellitus. In type 2 diabetes, insulin tolerance is developed following major metabolic disorder which led to numerous health complications reducing the quality of life and increasing the mortality rate [37]. It has become evident from various *in vivo* experiments that polyphenol phytochemicals and their additive food products modulate metabolism of carbohydrates and lipids, reduce hyperglycemia, insulin resistance, and improve many other metabolic process related to prevention and suppression of diabetics [35]. Consumption of black or green tea on daily can suppress the diabetic cataract and lower the glucose level in blood [36]. Similarly

Phytochemicals of tea	Health benefits
Polyphenol extract of green tea	Cancer prevention, protection cardiovascular diseases, reduce weight-loss, anti-allergic, prebiotics, osteoarthritis protection etc.
Epigallocatechin-3-Gallate (EGCG), Epigallocatechin (EGC), epicatechingallate (ECG)	Anticancer, cardioprotection, Neuroprotection, Obesity management, osteoprotection, Anti-inflammation, Diabetes control and renal protection, Antimicrobial and skin care etc.
Theaflavins, Thearubigins, bisflavonols	Antioxidative, antithrombogenic, and anti-inflammatory, anti-carcinogenic, anti-mutagenic, neuroprotective etc.

Table 3.
Role of major phytochemicals of tea in different diseases.

oolong tea has profound effect in treatment of type 2 diabetes [35]. However, further clinical study is required to develop nutraceutical products to apply in diabetic patients for prevention and cure. It was reported in an investigation in Japan that daily intake of about 6 cups of tea could reduce the diabetes affected people [38]. Recently, significant contribution of black tea polyphenol viz. theaflavins and thearubigins were found very effective in oxidative stress related diseases such as diabetes [31]. Therefore, tea nutraceuticals are very much promising in control and treatment of diabetes.

It has been reported that polyphenols have the hypoglycemic effects which are associated in reducing intestinal absorption of dietary carbohydrates affecting glucose metabolism, improvement of β-cell function and insulin action, inducing insulin secretion, and anti-inflammatory and anti-oxidative activity [39].

4.2 Tea photochemicals in cardiovascular diseases

The cardiovascular diseases such as hypertension (high blood pressure), coronary heart diseases (heart attack), cerebrovascular disease (stroke), heart failure, peripheral vascular diseases, etc. have been increasing day by day. The major causes of these are foods habits, chemicals used in food and food products, chemicals used in cultivation as well as lack of awareness.

Green tea consumption has been found to be effective in reduction of cardiovascular related disorders through decreased serum cholesterol and triglyceride. It has been reported in a study carried out in Japan that regular green tea consumption (≥500 mL) reduced the mortality rate in women in cardio vascular diseases by 31% as compared to those who consume non-regularly [40].

The green consumption has the effect of reducing the risk of coronary heart diseases by lowering the hyperlipidemia and total body cholesterol in body [41]. Moreover, EGCG the major catechin in green tea has been shown its usefulness in cardioprotective effects (inhibiting the formation of cardiac hypertrophy), antithrombotic and antiplatelet activities which are associated with cardiovascular diseases [42]. The theaflavin and thearubigins in black tea functional drinks are valuable against lipid and glucose related abnormalities especially high cholesterol and LDL levels [31].

4.3 Tea phytochemicals in obesity

Obesity has been considered as a serious concern globally effecting large number of peoples. It is associated with accumulation of excess body fats which stimulate various disorders in body such as hypertension, osteoarthritis, hyperlipidemia, heart attack, etc. [43] It is identified as a major factor in many metabolic syndrome such as stroke, heart failure, hypertension, diabetics, arthritis, etc.. The oolong tea food products also were also found to be very effective in decrease of in weight of obese people [44]. The green tea catechin, EGCG has been found to be beneficial on obesity [45]. Catechins in tea helps in weight management and consumption of green tea catechins results significant reduction in body weight in human [46].

4.4 Tea phytochemicals in oral diseases

Tea phytochemicals are useful in treatment and prevention of various types oral diseases which include tooth decay and oral cancers [47, 48], dental plaque and dental caries [49]. The tea extract and fluoride in tea can prevent the carcinogenic activity of oral bacteria such as *Escherichia coli, Streptococcus salivarius*, and *Streptococcus mutans* [50].

4.5 Tea phytochemicals in cancers

Tea nutraceutical acts as powerful antioxidants are very much promising in prevention of cancers. Experimental studies have demonstrated the use of tea bioactive compounds in prevention of cancers in different organs [51]. The tea catechins has been found to act as anticarcinogenic and chemopreventive agents. Several animal model studies in mice showed the inhibiting of chemical induced 4-(methylnitrosamino)-1-(3-pyridyl)-1-butanone (NNK) tumorigenesis, tumor cell proliferation, progression of adenoma to carcinoma, as well as lung carcinogenesis. [52]. It has been conferred from several experimental evidences that tea can be effectively used in inhibition of carcinogenesis in different parts of our body such as skin, lung, stomach, esophagus, liver, small intestine, pancreas, colon, bladder, etc. [53]. In these experimental studies it was found that polyphenols in green tea acted on cell apoptosis, cell proliferation, cell cycle in tumour growth and migration. All these activities resulted in reduced the risk of certain types cancers such as skin, colorectal, prostate, breast etc. through their activity in cell apoptosis, cell proliferation, cell cycle in tumour growth and cell migration. Among four major catechins in green tea, EGCG has been extensively studied and found to be most effective polyphenols [54]. Theaflavins in black tea inhibit the DNA damage which is a major cause of induction of cancer [55] The anti-carcinogenic activity of tea polyphenols is associated with their capacity to bind with carcinogens and thereby enabled the metabolism. Cancer preventive activity of black tea is known from the report of induction of apoptosis in human leukemia cells by black tea and its polyphenols [56]. Black tea is also effective in protecting immunocytes from tumor-induced apoptosis [57].

4.6 Other applications of tea phytochemicals

Consumption of tea has other several promising-health benefits such as it decrease the risk of hypertension [58], lowers the risk of osteoporosis, protects

Green tea supplements	Observed effects	*In vivo* study
Polyphenol extract	Increase in hydroxyproline content and catalase activity, Decrease in protein carbonyl content, Inhibit protein oxidation induced by UV radiations.	Mice
Aqueous extracts	Increase the level of collagen and elastin fibers. Reduced expression of MMP-3 enzymes.	Mice
Green tea extract	Reduce muscle atrophy and mediate insulin resistance, Reduce fat accumulation and lipid droplets.	Mice, *C. elegans*
Epigallocatechin gallate (EGCG)	Reduce the number of cells affected by sunburn, Maintain equilibrium during redox reaction.	
EGCG	Extend lifespan.	Drosophila, *C. elegans*
Cholinergic acid	Delay age-related decline in body movements	*C. elegans*
Theanine	Stress resistance and lifespan extension.	C. elegans
Crude green tea extract	Reduction in total body iron.	Fruit flies

Table 4.
Effect of green tea on photoaging, stress resistance, neuroprotection and associated health complications: in vivo studies [67].

Green tea supplements	Observed effects	*Clinical* study (human)
Polyphenol extract	Conjugate metabolites in plasma, blister fluid, and skin biopsy samples	Consume capsules of green tea polyphenols
Green tea extract	Reduction in the level of cells with sunburn; Lesser DNA damage when compared to vehicle control; Reduced the prevalence of Alzheimer's disease and cognitive impairment; Free radical scavenging and anti-wrinkle effects	Topical application
EGCG, EC, and EGC	Decreased sunburn cells (Lesser activity when compared to the crude extracts).	Topical application
Polyphenol extract	Improvement in facial skin and in controlling erythema	Oral supplements of green tea polyphenols

Table 5.
Effect of green tea on photoaging, stress resistance, neuroprotection and associated health complications: Clinical trials [67].

against the risk of hip fractures [59, 60], improves the metabolic activity of the bacteria in intestinal tract [61].

The bioactive compounds in oolong tea such as polyphenol, caffeine have the antistress and antioxidant activities which can prevent the diseases related to stress [62]. The green tea catechin, EGCG is useful in inhibition of HIV infection and HIV-1 replication and multi-drug resistant *Staphylococcus aureus* infections [63]. Tea also shows antifungal activity which is higher in black tea followed by green tea and white tea [64]. Black is also effective in asthma disease which is ascribed by the antihistaminic and anti-inflammatory activities of the flavonol glycosides contained in black tea [65]. Tea extract also shows antiallergic activity by suppressing histamine activity controlling the allergic response including inflammation, urticaria, mastocytosis, asthma, and dermatitis [66].

Some other beneficial effects of the constituents are antiaging, stress control, ant-bacterial, anti-viral, and useful in Parkinson's disease, Alzheimer's disease etc. Recently, Tencomnao et al. [67] discussed the results of some *in vivo* studies on the effects of green tea supplements on various health benefits such photoaging, stress resistance, neuroprotection, and associated health complications and some of these are outlined in **Table 4**. Some of the major clinical trials on these various health benefits are depicted in **Table 5**.

5. Conclusions

This chapter discusses the nutraceutical properties of bioactive compounds in tea in particular polyphenols and their role in health benefits. The catechins in tea viz. EGCG, EGC, ECG and EC have therapeutic properties such as antioxidant, anti-diabetic, anti-microbial, anti-inflammatory as well as anti-cancer properties. The black tea polyphenols the flavins and thearubigins also show many profound therapeutic values. In addition, oolong tea and white tea are also pharmacologically important. Therefore, consumption of tea in its various form viz. green tea, black tea, oolong tea and other food products with tea as active ingredient have been increased now a days. There are many more food products with tea and or its extracted components as additive have been developed which includes instant tea beverage, confections, cereal bars, pet foods etc. Considering the growing interest

DOI: http://dx.doi.org/10.5772/intechopen.96506

in nutraceuticals due to their several health values, the aim of the present chapter is to outline the nutraceutical potential of phytochemicals of tea with their known therapeutic properties. Due to their ready bioavailability, the possibilities of using nutraceuticals from tea as natural healers in treatment and prevention of diseases have been explored.

Acknowledgements

Author specially thanks to the Tocklai Tea Research Institute, India for giving support during the tenure of working on tea research as scientist.

Author details

Kula Kamal Senapati
CIF, Indian Institute of Technology Guwahati, Guwahati, India

*Address all correspondence to: kulasenapati@gmail.com

References

[1] Eskin M, Tamir S. Dictionary of Nutraceuticals and Functional Foods. Boca Raton, USA: CRC Press. 2019. ISBN:9780367391508

[2] Jemimah G, Keerthana M, Reddy TS, Bonagiri Y, Gupta AVSSS, Sharma JVC. Nutraceuticals for prevention of various disease conditions. Indo Am. J. P. Sci, 2019; **06**(11). DOI:10.5281/zenodo.3533809

[3] Brower V. Nutraceuticals: poised for a healthy slice of the healthcare market? Nat Biotechnol, 1998; **16**(8):728-731. DOI:10.1038/nbt0898-728

[4] Kalra EK. Nutraceutical – definition and introduction. AAPS Pharmsci, 2003; 5:12081212. DOI:10.1208/ps050325

[5] Stagg GV, Millin DJ. The nutritional and therapeutic value of tea: A Review. Journal of Science of Food and Agriculture.1975;**26**:1439-1459. DOI: 10.1002/jsfa.2740261002

[6] Xing L, Zhang H, Qi R, Tsao R, Mine Y. Recent Advances in the Understanding of the Health Benefits and Molecular Mechanisms Associated with Green Tea Polyphenols. Journal of Agricultural and Food Chemistry. 2019; **67**(4):1029-1043. doi:10.1021/acs.jafc.8b06146

[7] Harbowy ME, Balestine DA, Davies PA, Cai Y. Tea Chemistry. Critical Reviews in Plant Sciences.1997; **16**(5):415-480. DOI: 10.1080/07352689709701956

[8] Senanayake SN. Green tea extract: Chemistry, antioxidant properties and food applications—A review. Journal of Functional Foods.2013, **5** (4): 1529–1541. DOI: 10.1016/j.jff.2013.08.011

[9] Wang Y, Ho CT. Polyphenolic chemistry of tea and coffee: a century of progress. Journal of Agricultural Food Chemistry. 2009;**57**(18):8109-8114. DOI:10.1021/jf804025c

[10] Lee KW, Lee HJ, Lee CY. Antioxidant activity of black tea vs. green tea. The Journal of Nutrition. 2002;**132**(4):785-786. DOI:10.1093/jn/132.4.785

[11] Higdon JV, Frei B. Tea catechins and polyphenols: health effects, metabolism, and antioxidant functions. Critical Reviews in Food Science and Nutrition.2003; **43** (1), 89–143. DOI: 10.1080/10408690390826464

[12] Xing L, Zhang H, Qi R, Tsao R, Mine Y. Recent Advances in the Understanding of the Health Benefits and Molecular Mechanisms Associated with Green Tea Polyphenols. Journal of Agricultural and Food Chemistry. 2019; **67** (4): 1029-1043. DOI:10.1021/acs.jafc.8b06146

[13] Dufresne CJ, Farnworth ER. A review of latest research findings on the health promotion properties of tea. Journal of Nutritional Biochemistry. 2001;**12**(7):404-421. DOI:10.1016/s0955-2863(01)00155-3

[14] Sajilata MG, Bajaj PR, Singhal RS. Tea Polyphenols as Nutraceuticals. Comprehensive Reviews in Food Science and Food Safety.2008:7:229-254. DOI: 10.1111/j.1541-4337.2008.00043.x

[15] Ahmad M, Baba WN, Shah U, Gani A, Gani A, et al. Nutraceutical Properties of the Green Tea Polyphenols. Journal of Food Processing & Technology.2014; **5**: 390. doi:10.4172/2157-7110.1000390

[16] Robertson A. The chemistry and biochemistry of black tea production—the non-volatiles. In: Willson KC, Clifford MN, editors. *Tea: Cultivation to*

Consumption. London, UK: Chapman & Hall. 1992:pp. 555–601

[17] Sharma V, Gulati A, Ravindranath SD. Extractability of tea catechins as a function of manufacture procedure and temperature of infusion. Food Chemistry. 2005; **93** (1): 141-148. DOI:10.1016/j.foodchem.2004.10.016

[18] Yanishlieva-Maslarowa, NV, Heinonen, IM. Sources of natural antioxidants: vegetables, fruits, herbs, spices and teas. In: Antioxidants in food - practical applications. Eds. Pokorny J, Yanishlieva N, Gordon M. CRC Press, Woodhead Publishing, Cambridge, England. 2001:210- 249.

[19] Balentine DA, Wiseman SA, Bouwens LCM. The chemistry of tea flavonoids. Critical Reviews in Food Science and Nutrition. 1997; **3**:693-704. DOI: 10.1080/10408399709527797

[20] Wang K, Liu F, Liu Z, Huang J, Xu Z, Li Y, Chen J, Gong Y &Yang X. Analysis of chemical components in oolong tea in relation to perceived quality. International Journal of Food Science and Technology 2010;**45**: 913–920. DOI:10.1111/j.1365-2621.2010.02224.x

[21] Lakenbrink C, Lapczynski S, Maiwald B, Engelhardt UH. Flavonoids and otherpolyphenols in consumer brews of tea and other caffeinated beverages. Journal of Agricultural and Food Chemistry. 2000, **48**, 2848–2852. DOI: 10.1021/jf9908042

[22] Lin JK, Lin CL, Liang, YC, Lin-Shiau SY, Juan IM. Survey of catechins, gallicacid, and methylxanthines in green, oolong, pu'erh, and black teas. Journal of Agricultural and Food Chemistry. 1998; **46**, 3635-3642. DOI: 10.1021/jf980223x

[23] Weisburger JH. Tea antioxidants and health. Handbook of antioxidants. New York. 1996.

[24] Cabrera C, Giménez R, López MC. Determination of tea components with antioxidant activity. Journal of Agricultural and Food Chemistry. **2003;** 51: 4427-4435. DOI:10.1021/jf0300801

[25] Namal Senanayake SPJ. Green tea extract: Chemistry, antioxidant properties, and food applications- A Review. Journal of Functional Foods.2013;**5**(4):1529-1541. DOI: 10.1016/j.jff.2013.08.011

[26] Yang CS, Maliakal P, Meng X. Inhibition of carcinogenesis by tea. Annual Review of Pharmacology and Toxicology.2002;**42**:25-54. DOI:10.1146/annurev.pharmtox.42.082101.154309

[27] Grinberg LN, Newmark H, Kitrossky N, Rahamim E, Chevion M, Rachmilewitz EA. Protective effects of tea polyphenols against oxidative damage to red blood cells. Biochemical Pharmacology.1997;**54**(9):973-978. DOI: 10.1016/s0006-2952(97)00155-x

[28] Lu TM, Lee CC, Maud JL, Lin SD. Quality and antioxidant property of green tea sponge cake. Food Chemistry.2010;**119**: 1090-1095. DOI: 10.1016/j.foodchem.2009.08.015.

[29] Butt M, Ahmad R, Sultan M, Nasir QM, Naz A. Green tea and anticancer perspectives: Updates from last decade. Critical Reviews in Food Science and Nutrition.2015; **55**:6, 792-805. DOI: 10.1080/10408398.2012.680205

[30] Mandel S, Yodim M. (2004). Catechin polyphenols: Neurodegeneration and neuroprotection in neurodegenerative diseases. Free Radical Biology and Medicine.2004;**37**: 304-307. DOI: 10.1016/j.freeradbiomed.2004.04.012

[31] Yang T, Koo M. Inhibitory effect of Chinese green tea on endothelial cell-induced LDL oxidation.

Atherosclerosis.2000; **148**: 67-73. DOI: 10.1016/S0021-9150(99)00239-7

[32] Imran A, Butt MS, Arshad MS, et al. Exploring the potential of black tea based flavonoids against hyperlipidemia related disorders. Lipids in Health and Disease. 2018;**17**(1):57. DOI:10.1186/s12944-018-0688-6

[33] Alipoor B, Rad AH. A review on the therapeutical effects of tea. Asian J Clin Nutri. 2012;**4**(1):1–15.DOI: 10.3923/ajcn.2012.1.15

[34] Butt MS, Imran A, Sharif MK, et al. Black tea polyphenols: a mechanistic treatise. Critical Reviews in Food Science and Nutrition. 2014;**54**(8): 1002-1011. DOI:10.1080/10408398.2011.623198

[35] Tachebele B, Abebe M, Birhan W, Addis Z. Health Benefits of Green Tea with Emphasis on Diabetes Mellitus. American Journal of Food and Nutrition. 2014; **2**(5):85-91. DOI: 10.12691/ajfn-2-5-2.

[36] Vinson JA, Zhang J. 2005. Black and green teas equally inhibit diabetic cataracts in a streptozotocin-induced rat model of diabetes. Journal of Agricultural and Food Chemistry. 2005; **53**(9):3710–3713.DOI:10.1021/jf0480521

[37] Santaguida PL, Balion C, Hunt D, et al. Diagnosis, prognosis, and treatment of impaired glucose tolerance and impaired fasting glucose. Evidence Report/Technology Assessment (Summary).2005;**128**:1-11.PMCID: PMC4780988

[38] Thielecke F, Boschmann M. The potential role of green tea catechins in the prevention of the metabolic syndrome - a review. Phytochemistry. 2009;**70**(1):11-24. DOI:10.1016/j.phytochem.2008.11.011

[39] Bahadoran Z, Mirmiran P, Azizi F. Dietary polyphenols as potential nutraceuticals in management of diabetes: a review. Journal of Diabetes and Metabolic Disorders. 2013;**12**(1):43. DOI:10.1186/2251-6581-12-43

[40] Kuriyama S. The relation between green tea consumption and cardiovascular disease as evidenced by epidemiological studies. The Journal of Nutrition.2008;**138**(8):1548S-1553S. DOI:10.1093/jn/138.8.1548S

[41] Pang J, Zhang Z, Zheng T, Yang YJ, Li N, Bai M, Peng Y, Zhang J, Li Q, Zhang B. Association of green tea consumption with risk of coronary heart disease in Chinese population. International Journal of Cardiology. 2015;**179**:275-278. DOI: 10.1016/j.ijcard.2014.11.093. Epub 2014 Nov 11. PMID: 25464464; PMCID: PMC4843789.

[42] Zheng J, Lee HC, Bin Sattar MM, Huang Y, Bian JS. Cardioprotective effects of epigallocatechin-3-gallate against doxorubicin-induced cardiomyocyte injury. European Journal of Pharmacology. 2011;**652**(1-3):82-88. DOI:10.1016/j.ejphar.2010.10.082

[43] Rossner S. Obesity: the disease of the twenty-first century. International journal of obesity and related metabolic disorders. 2002;26(Suppl. 4), S2–S4. DOI: 10.1038/sj.ijo.0802209]10.1038/sj.ijo.0802209].

[44] Sae-tan S., Grove K. A., Lambert J. D. (2011). Weight control and prevention of metabolic syndrome by green tea. Pharmacological Research.*2011;* **64**:146–154. DOI: 10.1016/j.phrs.2010.12.013

[45] Suzuki T, Pervin M, Goto S, Isemura M, Nakamura Y. Beneficial Effects of Tea and the Green Tea Catechin Epigallocatechin-3-gallate on Obesity.Molecules. 2016;21(10):1305. DOI:10.3390/molecules21101305

[46] Yang CS, Zhang J, Zhang L, Huang J, Wang Y. Mechanisms of body

weight reduction and metabolic syndrome alleviation by tea. Molecular Nutrition and Food Research. 2016;60 (1):160-174. DOI:10.1002/ mnfr.201500428

[47] Lee MJ, Lambert JD, Prabhu S, Meng X, Lu H, Maliakal P, Ho CT, Yang CS. **2004**. Delivery of tea polyphenols to the oral cavity by green tea leaves and black tea extract. Cancer Epidemiology, Biomarkers & Prevention. 2004;**13**:132–137: DOI: 10.1158/1055-9965.epi-03-0040

[48] Hamilton-Miller JMT. Anti-cariogenic properties of tea (Camellia sinensis). Journal of Medical Microbiology.2001;**50**(4):299-302. DOI: 10.1099/0022-1317-50-4-299

[49] Rasheed A, Haider M. Antibacterial activity of Camellia sinensis extracts against dental caries. Archives of Pharmacal Research. 1998;**21**(3): 348-352. DOI:10.1007/BF02975300

[50] McKay DL, Blumberg JB. The role of tea in human health: an update. Journal of the American College of Nutrition. 2002;**21**(1):1-13. DOI: 10.1080/07315724.2002.10719187.

[51] Yang CS, Wang X, Lu G, Picinich SC. (2009) Cancer prevention by tea: animal studies, molecular mechanisms and human relevance. Nature Reviews Cancer. 2009; **9(6)**: 429-439. DOI: 10.1038/nrc2641

[52] Lambert JD, Hong J, Yang GY, Liao J, Yang CS. Inhibition of carcinogenesis by polyphenols: evidence from laboratory investigations. American Journal of Clinical Nutrition. 2005;**81**(1 Suppl):284S-291S. DOI: 10.1093/ajcn/81.1.284S

[53] Das L, Bhaumik E, Raychaudhuri U, Chakraborty R. Role of nutraceuticals in human health. Journal of Food Science and Technology. 2012;**49**(2):173-183. DOI:10.1007/s13197-011-0269-4

[54] Du GJ, Zhang Z, Wen XD, et al. Epigallocatechin Gallate (EGCG) is the most effective cancer chemopreventive polyphenol in green tea. Nutrients. 2012;**4**(11):1679-1691. DOI:10.3390/ nu4111679

[55] Feng Q, Torii Y, Uchida K, Nakamura Y, Hara Y, Osawa T. Black tea polyphenols, theaflavins, prevent cellular DNA damage by inhibiting oxidative stress and suppressing cytochrome P450 1A1 in cell cultures. Journal of Agricultural and Food Chemistry. 2002;**50**(1):213-220. DOI: 10.1021/jf010875c

[56] Kundu T, Dey S, Roy M, Siddiqi M, Bhattacharya RK. Induction of apoptosis in human leukemia cells by black tea and its polyphenol theaflavin. Cancer Letters. 2005;**230**(1):111-121. DOI: 10.1016/j.canlet.2004.12.035

[57] Bhattacharyya A, Mandal D, Lahiry L, Sa G, Das T. Black tea protects immunocytes from tumor-induced apoptosis by changing Bcl-2/Bax ratio. Cancer Letters. 2004;**209**(2):147-154. DOI: 10.1016/j.canlet.2003.12.025.

[58] Hodgson JM, Puddey IB, Burke V, Beilin LJ, Jordan N. Effects on blood pressure of drinking green and black tea. Journal of Hypertension. 1999; **17** (4): 457–463. DOI:10.1097/ 00004872-199917040-00002

[59] Gardner EJ, Ruxton CHS, Leeds AR. Black tea – helpful or harmful? A review of the evidence. European Journal of Clinical Nutrition.2007; **61** (1) 3-18. DOI:10.1038/sj.ejcn.1602489

[60] Kanis J, Johnell O, Gullberg B, et al. Risk factors for hip fracture in men from southern Europe: the MEDOS study. Mediterranean Osteoporosis Study. Osteoporosis International. 1999; **9**(1):45-54. DOI:10.1007/ s001980050115

[61] Weisburger JH. **2003**. Prevention of coronary heart disease and cancer by

tea, a review. Environmental Health and Preventive Medicine. 2003; 7:283–288. DOI: 10.1007/BF02908887

[62] Kurihara H, Chen L, Zhu B, He Z, Shibata H, Kiso Y, Tanaka T, Yao X. Anti-stress effect of oolong tea in women loaded with vigil. Journal of Health Science.2003; **49**(6):436–43. DOI:

[63] Williamson MP, McCormick TG, Nance CL, Shearer WT. Epigallocatechin gallate, the main polyphenol in green tea, binds to the T-cell receptor, CD4: potential for HIV-1 therapy. Journal of Allergy and Clinical Immunology. 2006; **118(6)**:1369–1374. DOI:10.1016/j.jaci.2006.08.016

[64] Camargo LEA, Pedroso LS, Vendrame SC, Mainardes RM, & Khalil NM. Antioxidant and antifungal activities of Camellia sinensis (L.) Kuntze leaves obtained by different forms of production. Brazilian Journal of Biology, 2016;76(2), 428–434. DOI: 10.1590/1519-6984. 18814.

[65] Fortunato LR, Alves CDF, Teixeira MM, Rogerio AP, Quercetin: a flavonoid with the potential to treat asthma, Brazilian Journal of Pharmaceutical Science.2012;**48** (4): 589–599.

[66] Toyoda M, Tanaka K, Hoshino K, Akiyama H, Tanimura A, Saito Y, Profiles of potentially antiallergic flavonoids in 27 kinds of health tea and green infusions, Journal of Agricultural and Food Chemistry **1997** 45 (7), 2561-2564. DOI:10.1021/jf970024y

[67] Prasanth MI, Sivamaruthi BS, Chaiyasut C,Tencomnao T, A Review of the Role of Green Tea (*Camellia sinensis*) in Antiphotoaging, Stress Resistance, Neuroprotection, and Autophagy, *Nutrients* **2019**11(2), 474. DOI:10.3390/nu11020474

Chapter 7

Alkaloids and Their Pharmacology Effects from *Zanthoxylum* Genus

Nguyen Xuan Nhiem, Pham Minh Quan and Nguyen Thi Hong Van

Abstract

Zanthoxylum genus (Rutaceae) comprises about 212 species distributed in warm temperature and subtropical areas in the worldwide. *Zanthoxylum* species have been used in traditional for the treatment of tooth decay, snakebites, blood circulation problems, stomach problems, inflammation, rheumatic, and parasitic diseases. The chemical investigations of *Zanthoxylum* have been studied by many scientists over the world. Several classes of compounds have been isolated from this genus such as alkaloids, coumarins, and monoterpenes. Of these, alkaloids are the main components and play an important role in *Zanthoxylum* species. Alkaloids have been shown the potential promise about biological activities: cytotoxic, antimalarial, leishmanicidal, anti-inflammatory, analgesic, antiviral, and antibacterial activities. This chapter will focus on the structure elucidation and pharmacological activities of alkaloids from *Zanthoxylum* species. In addition, the absolute configuration of some alkaloids from *Zanthoxylum* genus will be also discussed.

Keywords: *Zanthoxylum*, Rutaceae, alkaloids, ^{13}C-NMR, circular dichroism

1. Introduction

Zanthoxylum genus is one of the biggest genera belonging to the Rutaceae family, including 212 species in the world and widely distributed in the warm or tropic temperate zones. Research findings showed that *Zanthoxylum* genus have many interesting biological activities such as antifungal, antibacterial, antiviral, antimalarial, anti-inflammatory, antioxidant, tuberculosis, cardiovascular, and liver protective activities, especially cytotoxic activities. From the *Zanthoxylum* species, many compounds have been isolated, including alkaloids, lignans, coumarins, flavonoids, terpenoids, steroids, etc.; they are the specific classes of compounds in *Zanthoxylum* genus. The main components presented in this genus are alkaloids and coumarins, with significant biological activities, especially anticancer activities. In particular, this genus contains high levels of benzophenanthridine alkaloids that not only shown their potential cytotoxic *in vitro* but also their ability to inhibit tumor *in vivo* through many mechanisms, resistant against many pathogenics including MRSA strain (methicillin-resistant *Staphylococcus aureus*)—a bacterium caused dangerous infections in hospital [1] and also shown anti-inflammatory activity [2] (**Figure 1**).

Zanthoxylum nitidum

Zanthoxylum setulosum

Zanthoxylum ovalifolium

Zanthoxylum rhoifolium

Zanthoxylum sprucei

Zanthoxylum monogynum

Zanthoxylum panamense

Zanthoxylum ekmanii

Zanthoxylum zanthoxyloides

Zanthoxylum caribaeum

Figure 1.
Photographs of the Zanthoxylum *species. The images were obtained from http://tropical.theferns.info.*

2. Alkaloids constituents from *Zanthoxylum* genus

A total of 35 *Zanthoxylum* species have been studied and showed the presence of alkaloids: *Z. acanthopodium*, *Z. ailanthoides*, *Z. americanum*, *Z. arborescens*, *Z. atchoum*, *Z. austrosinense*, *Z. avicennae*, *Z. bouetense*, *Z. budrunga*, *Z. bungeanum*, *Z. caribaeum*, *Z. chiloperone*, *Z. clava-herculis*, *Z. colantrillo*, *Z. coriaceum*, *Z. culantrillo*, *Z. cuspidatum*, *Z. dimoncillo*, *Z. fagara*, *Z. integrifoliolum*, *Z. lemairei*, *Z. monophyllum*, *Z. myriacanthum*, *Z. nitidum*, *Z. ovalfolium*, *Z. paracanthum*, *Z. procerom*, *Z. rhoifolium*, *Z. riedelianum*, *Z. rubescens*, *Z. schinifolium*, *Z. simulans*, *Z. tingoassuiba*, *Z. usambarense*, and *Z. williamsii*.

2.1 Benzophenanthridine

Benzophenanthridine alkaloids (**1–51**) were isolated from *Zanthoxylum* species. Of these, nitidine (**1**), chelerythrine (**2**), and arnottianamide (**48**) were found in almost *Zanthoxylum* species (**Figure 2** and **Table 1**).

2.2 Aporphines and benzylisoquinolines, and furoquinolines

Aporphines and benzylisoquinolines, and furoquinolines (**52–75**) were reported from *Zanthoxylum* species. Magnoflorine (**52**), lauriforine (**55**), skimmianine (**69**), γ-fagarine (**70**), and dictamnine (**71**) were found in *Zanthoxylum* species such as *Z. americanum*, *Z. bouetense*, *Z. budrunga*, *Z. caribaeum*, *Z. clava-herculis*, *Z. cuspidatum*, *Z. dimoncillo*, *Z. fagara*, *Z. monophyllum*, *Z. nitidum*, *Z. ovalifolium*, *Z. rubescens*, *Z. schinifolium*, *Z. simulans*, *Z. usambarense*, and *Z. williamsii* (**Figure 3** and **Table 2**).

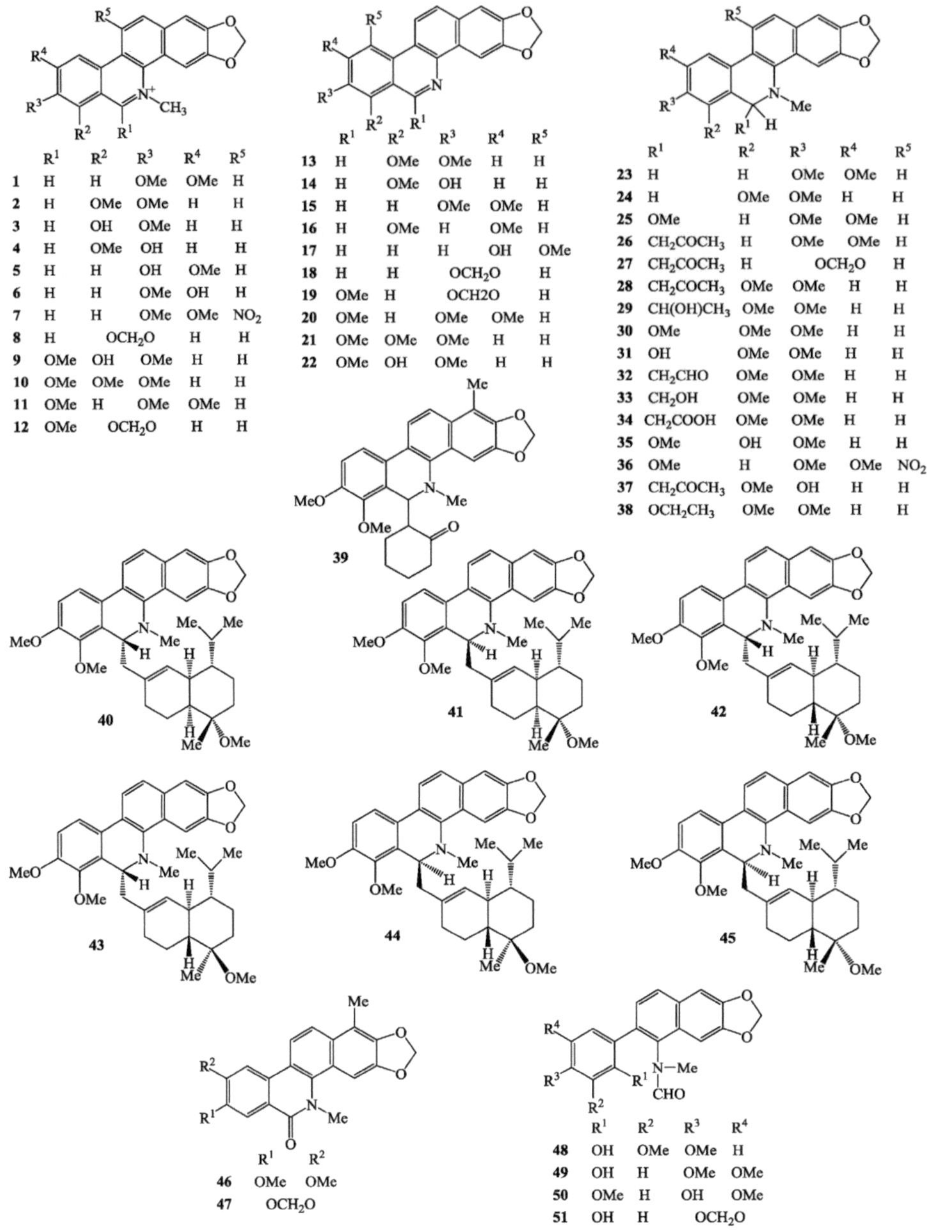

	R^1	R^2	R^3	R^4	R^5
1	H	H	OMe	OMe	H
2	H	OMe	OMe	H	H
3	H	OH	OMe	H	H
4	H	OMe	OH	H	H
5	H	H	OH	OMe	H
6	H	H	OMe	OH	H
7	H	H	OMe	OMe	NO_2
8	H	OCH_2O		H	H
9	OMe	OH	OMe	H	H
10	OMe	OMe	OMe	H	H
11	OMe	H	OMe	OMe	H
12	OMe	OCH_2O		H	H

	R^1	R^2	R^3	R^4	R^5
13	H	OMe	OMe	H	H
14	H	OMe	OH	H	H
15	H	H	OMe	OMe	H
16	H	OMe	H	OMe	H
17	H	H	H	OH	OMe
18	H	H	OCH_2O		H
19	OMe	H	OCH2O		H
20	OMe	H	OMe	OMe	H
21	OMe	OMe	OMe	H	H
22	OMe	OH	OMe	H	H

	R^1	R^2	R^3	R^4	R^5
23	H	H	OMe	OMe	H
24	H	OMe	OMe	H	H
25	OMe	H	OMe	OMe	H
26	CH_2COCH_3	H	OMe	OMe	H
27	CH_2COCH_3	H	OCH_2O		H
28	CH_2COCH_3	OMe	OMe	H	H
29	$CH(OH)CH_3$	OMe	OMe	H	H
30	OMe	OMe	OMe	H	H
31	OH	OMe	OMe	H	H
32	CH_2CHO	OMe	OMe	H	H
33	CH_2OH	OMe	OMe	H	H
34	CH_2COOH	OMe	OMe	H	H
35	OMe	OH	OMe	H	H
36	OMe	H	OMe	OMe	NO_2
37	CH_2COCH_3	OMe	OH	H	H
38	OCH_2CH_3	OMe	OMe	H	H

39 **40** **41** **42** **43** **44** **45**

	R^1	R^2
46	OMe	OMe
47	OCH_2O	

	R^1	R^2	R^3	R^4
48	OH	OMe	OMe	H
49	OH	H	OMe	OMe
50	OMe	H	OH	OMe
51	OH	H	OCH_2O	

Figure 2.
The structures of alkaloids **1–51**.

2.3 Quinolines, quinolones, and quinazolines

There were 19 quinolines, quinolones, and quinazolines (**79–97**) isolated from *Zanthoxylum* species. They are mainly found in *Z. simulans* and *Z. nitidum* (**Figure 4** and **Table 3**).

2.4 Indolopyridoquinazolines, acridones, and canthinones

There are 10 indolopyridoquinazolines, acridones, and canthinones (**98–107**) isolated from *Zanthoxylum* plants (*Z. atchoum*, *Z. simulans*, and *Z. ovalfolium*).

No.	Compound names	Sources	Ref.
1	Nitidine	*Z. myriacanthum*, *Z. williamsii*, *Z. clava-herculis*, *Z. americanum*, *Z. bouetense*, *Z. nitidum*, *Z. usambarense*, *Z. ovalifolium*, *Z. lemairei*, *Z. atchoum*	[3–18]
2	Chelerythrine	*Z. williamsii*, *Z. monophyllum*, *Z. clava-herculis*, *Z. americanum*, *Z. bouetense*, *Z. nitidum*, *Z. usambarense*, *Z. simulans*, *Z. lemairei*, *Z. atchoum*	[4–8, 11, 13, 14, 17]
3	Fagaridine	*Z. nitidum*, *Z. atchoum*	[6, 17]
4	Isofagandine	*Z. nitidum*	[6]
5	Terihanine	*Z. ovalifolium*	[10]
6	Isoterihanine	*Z. ovalifolium*	[10]
7	11-Nitronitidine	*Z. atchoum*	[17]
8	Sanguinarin	*Z. nitidum*	[11, 13]
9	Methoxyfagaridine	*Z. atchoum*	[17]
10	9-Methoxy chelerythrine chloride	*Z. rubescens*	[5]
11	8-Methoxynorchelerythrine	*Z. nitidum*	[9]
12	8-Methoxysanguinarine	*Z. nitidum*	[19]
13	Norchelerythrine	*Z. nitidum*, *Z. simulans*	[17, 20–23]
14	Decarine	*Z. nitidum*, *Z. simulans*	[13, 20–24]
15	*N*-Nortidine	*Z. myriacanthum*	[23, 25]
16	7,9-Dimethoxy-2,3-methylen dioxybenzophenantridine	*Z. myriacanthum*	[25]
17	Zanthoxyline	*Z. rhoifolium*, *Z. nitidum*	[18, 26]
18	Noravicine		[23]
19	Rhoifoline A	*Z. rhoifolium*, *Z. nitidum*	[13, 26, 27]
20	Rhoifoline B	*Z. rhoifolium*	[26]
21	6,7,8-Trimethoxy-2,3-methylen dioxybenzophenantridine	*Z. nitidum*	[11]
22	8-Methoxyisodecarine	*Z. nitidum*	[19]
23	Dihydronitidine	*Z. myriacanthum*, *Z. nitidum*	[3, 12]
24	Dihydrochelerythrine	*Z. coriaceum*, *Z. nitidum*, *Z. simulans*	[9, 11–13, 18, 20, 22, 28]
25	5,6-Dihydro-6-methoxynitidine	*Z. nitidum*	[29]
26	6-Acetonyldihydronitidine	*Z. rhoifolium*, *Z. nitidum*	[12, 26, 30]
27	6-Acetonyldihydroavicine	*Z. rhoifolium*	[26]
28	6-Acetonyldihydrochelerythrine	*Z. rhoifolium*, *Z. nitidum*	[12, 18, 22, 23, 26]
29	(*R*)-8-(1-hydroxyethyl) dihydrochelerythrine	*Z. nitidum*	[9, 23, 31]
30	8-Methoxydihydrochelerythrine	*Z. nitidum*, *Z. bungeanum*	[9, 13, 23]
31	8-Hydroxydihydrochelerythrine	*Z. nitidum*	[9, 13, 23]
32	Dihydrochelerythrinyl-8-acetaldehyde	*Z. nitidum*	[13]

No.	Compound names	Sources	Ref.
33	Bocconoline	*Z. nitidum*	[18]
34	Carboxymethyl dihydrochelerythrine	*Z. nitidum*	[18, 23]
35	6-Methoxy-7-hydroxydihydro chelerythrine	*Z. nitidum*	[23]
36	6-Nitro-8-methoxy-7,8-dihydronitidine	*Z. atchoum*	[17]
37	8-(2′-Cyclohexanone)-7,8-dihydrochelerythrine	*Z. nitidum*	[31]
38	6-Acetonyl-*N*-methyl-dihydrodecarine	*Z. lemairei*, *Z. riedelianum*, *Z. nitidum*	[14, 18]
39	Ethoxychelerythrine	*Z. nitidum*	[32]
40	Zanthomuurolanine	*Z. nitidum*	[33]
41	*epi*-Zanthomuurolanine	*Z. nitidum*	[33]
42	Zanthocadinanine A	*Z. nitidum*	[33]
43	Zanthocadinanine B	*Z. nitidum*	[33]
44	*epi*-Zanthocadinanine B	*Z. nitidum*	[33]
45	*epi*-Zanthocadinanine A	*Z. nitidum*	[22]
46	Oxynitidine	*Z. nitidum*	[17, 22]
47	Oxyavicine	*Z. nitidum*, *Z. ailanthoides*	[9, 11, 13, 22, 23]
48	Arnottianamide	*Z. nitidum*, *Z. simulans*, *Z. bungeanum*, *Z. ailanthoides*, *Z. austrosinense*	[13, 17, 20–23]
49	Isoarnottianamide	*Z. nitidum*, *Z. myriacanthum*	[13]
50	10-*O*-demethyl-17-*O*-methylisoarnottianamide	*Z. lemairei*	[14]
51	Integriamide	*Z. nitidum*	[13]

Table 1.
Benzophenanthridines from Zanthoxylum *species.*

Until now, only a small number of this class of compounds have been published (**Figure 5** and **Table 4**).

2.5 Other alkaloids

Amines were mainly found in *Z. coriaceum*. But tryptamines were only found in *Z. nitidum.* 16 amines and 6 tryptamines have been reported (**Figure 6** and **Table 5**).

3. Biological activities of alkaloids

The abundance and diversity as well as the valuable properties in terms of chemical compositions and biological activities of the *Zanthoxylum* genus have attracted the attention of many research scientists. The studies have shown that the extracts and alkaloids from *Zanthoxylum* species have many valuable biological

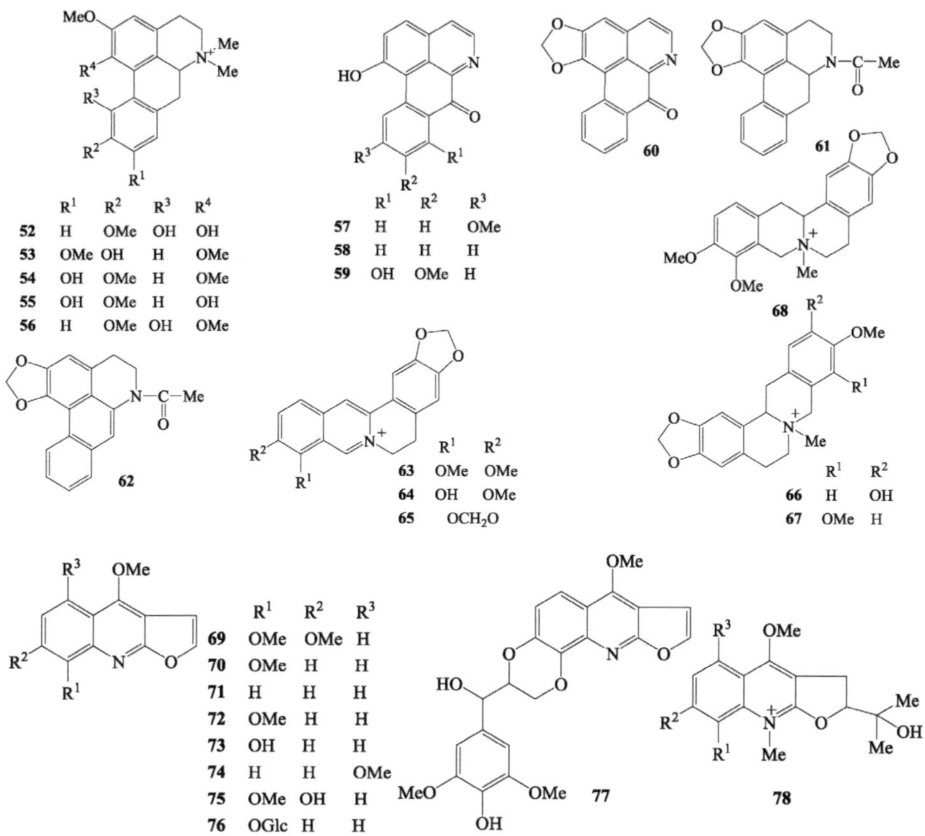

Figure 3.
The structures of alkaloids 52–78.

activities: anticancer, antibacterial, antifungal, antiviral, anti-inflammatory, and antioxidant activities. Many trials of biological properties of these species have been studied and evaluated promising applications in medicine. However, the most prominent compounds with cytotoxic activity in the genus *Zanthoxylum* are amides and alkaloids.

3.1 Cytotoxic activities

In folk medicine, many species of *Zanthoxylum* are used as drugs to treat cancer, such as: the people in Kakamega, Kenya use the leaves and roots of *Z. gilletii* to treat breast and skin cancers [48]; fruits of *Zanthoxylum* species are used in Indians and South Korea for chemopreventive effects [49, 50], while Cameron people use them to treat anemia disease sickle erythrocytes [51] and Japanese people use as one of the main components in the traditional medicine daikenchuto to treat gastrointestinal and chronic diseases [52]. The chloroform-soluble fraction of *Z. ailanthoides* showed cytotoxic activity against HL-60 and WEHI-3 cell lines with IC_{50} values of 73.06 and 42.22 μg/ml, respectively [53].

The methanol, hexane, and chloroform extracts from *Z. usambarense* were evaluated for cytotoxicity against two breast cancer cell lines, MDA-MB-231 and MCF-7 and one brain tumor cell line, U251 using MTT assay [54]. The crude extract of *Z. setulosum* collected in Monteverde, Costa Rica showed potent cytotoxic activity (100% cells killed at 100 μg/ml) on three cancer cell lines, MCF-7, MDA-MB-231,

No.	Compound names	Sources	Ref.
	Aporphines		
52	Magnoflorine	*Z. fagara, Z. williamsii, Z. monophyllum, Z. clava-herculis, Z. americanum, Z. usambarense, Z. nitidum*	[4, 7, 8, 34]
53	Cocsarmine	*Z. tingoassuiba*	[4]
54	Xanthoplanine	*Z. tingoassuiba*	[4]
55	Lauriforine	*Z. fagara, Z. williamsii, Z. clava-herculis, Z. americanum*	[4, 34]
56	*N*-methyl isocorydine	*Z. caribaeum, Z. coriaceum*	[28, 34]
57	Zanthoxoaporphine A	*Z. paracanthum*	[35]
58	Zanthoxoaporphine B	*Z. paracanthum*	[35]
59	Zanthoxaporphine C	*Z. paracanthum*	[35]
60	Liriodenine	*Z. nitidum*	[11, 13, 20, 22, 32, 36]
61	(−)-*N*-acetylanonanine	*Z. simulans, Z. nitidum*	[21, 22]
62	*N*-acetyldehydroanonaine	*Z. simulans, Z. nitidum*	[21, 22]
	Benzylisoquinolines		
63	Berberine	*Z. caribaeum, Z. monophyllum, Z. clava-herculis*	[34, 4]
64	Berberubine	*Z. nitidum*	[11, 13]
65	Coptisine	*Z. nitidum*	[11, 13]
66	(−)-Usambarine	*Z. usambarense*	[7]
67	(−)-cis-*N*-methylcanadine	*Z. usambarense, Z. nitidum*	[7, 8]
68	*N*-methylcanadine	*Z. coriaceum*	[28]
	Furoquinolines		
69	Skimmianine	*Z. dimoncillo, Z. caribaeum, Z. fagara, Z. williamsii, Z. americanum, Z. rubescens, Z. bouetense, Z. simulans, Z. nitidum, Z. atchoum*	[4, 5, 17, 21, 22, 29, 34, 37]
70	γ-Fagarine	*Z. americanum, Z. simulans, Z. nitidum, Z. cuspidatum*	[4, 21, 22, 24, 29, 37]
71	Dictamnine	*Z. budrunga, Z. ovalifolium, Z. nitidum, Z. schinifolium, Z. avicennae, Z. acanthopodium*	[10, 13, 29, 37, 38]
72	8-Methoxy dictamnine	*Z. rubescens*	[5]
73	Robustine	*Z. simulans, Z. nitidum*	[21, 24]
74	5-Methoxydictamine	*Z. ovalifolium, Z. nitidum*	[10, 29]
75	Haplopine	*Z. nitidum*	[37]
76	4-Methoxyfuro[2,3-b] quinoline-8-*O*-β-D-glucopyranoside	*Z. nitidum*	[24]
77	Zanthonitidine A	*Z. nitidum*	[24]
78	(+)-*N*-methylplatydesmine	*Z. usambarense*	[7]

Table 2.
Aporphines and benzylisoquinolines, and furoquinolines from Zanthoxylum *species.*

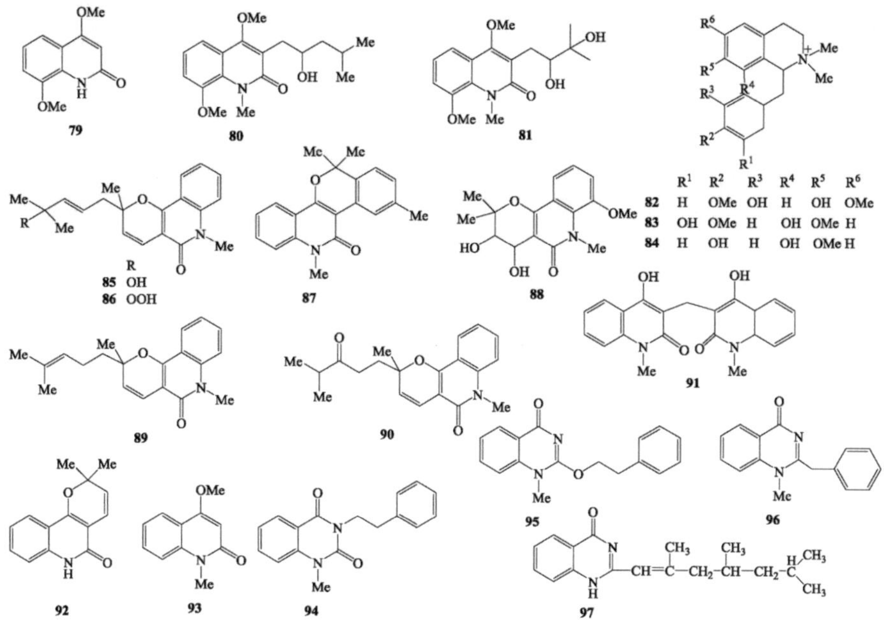

Figure 4.
The structures of alkaloids 79–97.

and MDA-MB-468 [55]. The methanol extract of *Z. avicennae* inhibited the highly metastatic HA22T liver cancer cell migration and invasion effects through PP2A activation [56]. Most recently, the methanol extract of *Z. alatum* showed the apoptotic activity on Ehrlich ascites tumor in Swiss albino mice [57].

A screening study of cytotoxic activity of the extracts from 11 species used as salad in Korea showed that the methanol extract of *Z. schinifolium* had the strongest cytotoxic against Calu-6 cell line with the IC_{50} values < 25.0 μg/ml, meanwhile the methanol extract of *Z. piperitum* exhibited antioxidant effects through ability to arrest radical DPPH. Through the results of this study, the authors suggested that these salad vegetables can be used as functional foods to support cancer treatment [58]. The linear fatty acid amides of the sandshool class are the major ingredient found in seeds of *Z. piperitum* exhibited cytotoxicity in the A-549 cell line [59]. Glycoprotein from the seeds of *Z. piperitum* prevented damage to liver tissue caused by *N*-nitrosodiethylamine in the experimental mouse model [49].

Thirteen benzophenanthridines were isolated from *Z. nitidum* by Wang et al. [23]. The research indicated that 6-methoxy-7-hydroxydihydrochelerythrine exhibited the moderate cytotoxic activity against A549, Hela, SMMC-7721 and EJ, with the IC_{50} values of 27.50, 37.50, 16.95 and 60.42 μM, respectively. 6-Methoxydihydrochelerythrin and 8-(10-hydroxyethyl)-7,8-dihydrochelerythrine also showed strong cytotoxicity when tested against the four human cancer cell lines (A549, Hela, SMMC-7721 and EJ). These results suggested that benzophenanthridines may become a valid alternative of potential basis for new anti-proliferative agents [23]. Methyl 7-(β-D-mannopyranosyloxy)-1*H*-indole-2-carboxylate (**126**), methyl 7-[(3-*O*-acetyl-β-D-mannopyranosyl)oxy]-1*H*-indole-2-carboxylate (**127**), and 2-methyl-1*H*-indol-7-yl β-D-mannopyranoside (**128**) were isolated from the ethanol extract of *Z. nitidum* roots. Biological evaluation revealed that these alkaloids possess significant cytotoxicities against all the tested tumor cell lines with the IC_{50} values of less than 30 μM [46]. Liriodenine (**60**) was the active compound against the MCF-7, NCI-H460, and SF-268 cell lines with IC_{50} values of 2.19, 2.38, and 3.19 μg/ml, respectively [22]. In addition, normelicopidine (**101**)

No.	Compound names	Sources	Ref.
	Quinolines		
79	Edulitine	*Z. simulans*, *Z. nitidum*	[21, 24, 37]
80	Lunacridine	*Z. budrunga*	[39]
81	Edulinine	*Z. williamsii*, *Z. nitidum*	[4, 37]
82	Tembetarine	*Z. fagara*, *Z. usambarense*, *Z. nitidum*	[4, 7, 8]
83	(*R*)-(+)-isotembetarine	*Z. nitidum*	[8]
84	(−)-Oblongine	*Z. usambarense*	[7]
85	Simulenoline	*Z. simulans*	[21]
86	Peroxysimulenolin	*Z. simulans*	[21]
87	Benzosimulin	*Z. simulans*	[21]
88	Zanthodioline	*Z. simulans*, *Z. nitidum*	[21, 24, 37]
89	Zanthosimuline	*Z. simulans*	[21]
90	Huajiaosimuline	*Z. simulans*	[21]
91	Zanthobisquinolone	*Z. simulans*	[21]
	Quinolones		
92	Flindersine	*Z. nitidum*	[22]
93	4-Methoxy-1-methyl-2-quinolone	*Z. nitidum*	[22, 24]
	Quinazolines		
94	1-Methyl-3-(2′-phenylethyl)-lH,3Hquinazoline-2,4-dione	*Z. arborescens*	[34]
95	1-Methyl-3-[2′-(4″-methoxyphenyl) ethyl]-lH,3H quinazoline-2,4-dione	*Z. arborescens*	[34]
96	Arborine	*Z. budrunga*	[38]
97	2-(2′,4′,6′-Trimethyl-heptenyl)-4-quinozolone	*Z. budrunga*	[38]

Table 3.
Quinolines, quinolones, and quinazolines from Zanthoxylum *species.*

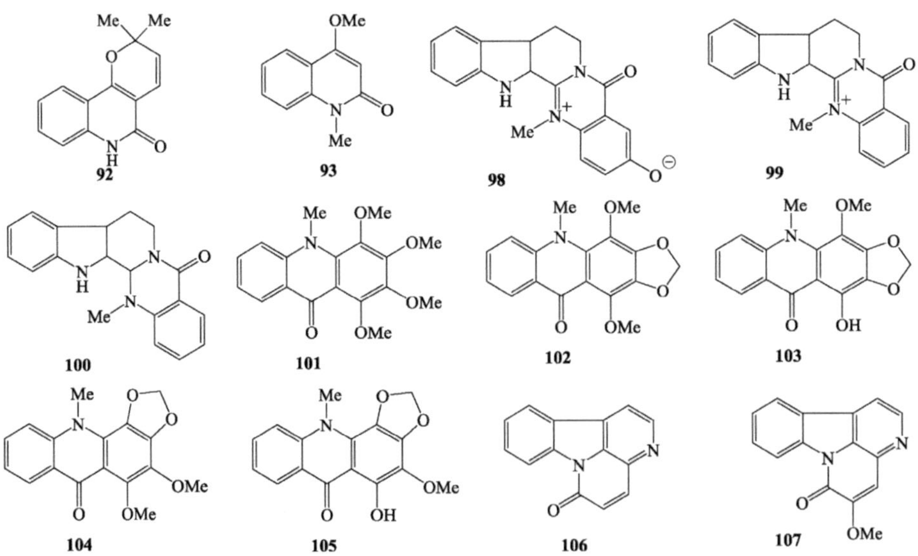

Figure 5.
The structures of alkaloids 98–107.

No.	Compound names	Sources	Ref.
	Indolopyridoquinazolines		
98	3-Hydroxydehydroevodiamine	*Z. atchoum*	[17]
99	Dehydroevodiamine	*Z. atchoum*	[17]
100	Evodiamine	*Z. atchoum*	[17]
	Acridones		
101	Normelicopidine	*Z. simulans*	[40]
102	Normelicopine	*Z. simulans*	[40]
103	Melicopine	*Z. simulans*	[40]
104	Melicopidine	*Z. simulans*	[40]
105	Melicopicine	*Z. simulans*	[40]
	Canthinones		
106	6-Canthinone	*Z. ovalfolium*	[10, 41]
107	5-Methoxycanthin-6-one	*Z. chiloperone*	[42]

Table 4.
Indolopyridoquinazolines, acridones, and canthinones from Zanthoxylum *species.*

from *Z. simulans* showed the cytotoxic activities against PC-3M, LNCaP, and Dd2 with the IC_{50} values of 12.5, 21.1, and 18.9 μg/ml respectively.

Acridone alkaloid derivatives isolated from the roots and fruits of *Z. leprieurii* showed the selective moderately active against two cancer cell lines, A549 and DLD-1 in comparison to normal cell line, WS1 [60]. Liriodenine (**60**) was also isolated from *Z. nitidum* and showed significant cytotoxic activity against three human cancer cell lines, MCF-7, NCI-H460, and SF-268 with IC_{50} values of 2.19, 2.38, and 3.19 μg/ml, respectively. A series of benzo[c]phenanthridine alkaloids isolated from *Zanthoxylum* species showed significant cytotoxic activities: huajiaosimuline (**90**) and zanthosimuline (**89**) isolated from *Z. simulans* showed significant antiplatelet aggregation activity and induced terminal differentiation with cultured HL-60 cells [61], 7,8-dehydro-1-methoxyrutaecarpine, norchelerythrine (**13**), ethoxychelerythrine (**39**), 6-acetonyldihydrochelerythrine (**29**), γ-fagarine (**70**), skimmianine (**69**), (−)-matairesinol, and canthin-6-one (**106**) isolated from the roots of *Z. integrifoliolum* exhibited cytotoxic activities on two human cancer cell lines, P-388 and HT-29 (IC_{50} values < 4 μg/ml) [62]. A new benzophenanthridine-*type* alkaloid, rutaceline isolated from the stem bark powder of *Z. madagascariense* and induced cell cycle arrest in the GO/G1 phase, decreased of cells in S phase as well as induced DNA fragmentation in both cancer cell lines (human colorectal adenocarcinoma (Caco-2) and the African green monkey kidney (Vero) cell lines) [63]. Three others alkaloids isolated from the rhizome of *Z. capense* exhibited strong anticancer activity in HCT-116 colon carcinoma cell line [64].

Nitidine (**1**), a specific compound in *Zanthoxylum* species: *Z. myriacanthum*, *Z. williamsii*, *Z. clava-herculis*, *Z. americanum*, *Z. bouetense*, *Z. nitidum*, *Z. usambarense*, *Z. ovalifolium*, *Z. lemairei*, *Z. atchoum* inhibited gastric tumor cell growth, induced tumor cell apoptosis *in vitro* and effectively suppressed the volume, weight, and microvessel density of human SGC-7901 gastric solid tumors at a dosage of 7 mg/kg/d (intraperitoneal injection) [15], suppressed the growth and pro-apoptotic effects on renal cancer cells both *in vitro* and *in vivo* [16]. Nitidine could inhibit breast cancer cell migration and invasion both *in vitro* and *in vivo* [65]. Chelerythrine (**2**) was found in *Z. williamsii*, *Z. monophyllum*, *Z. clava-herculis*,

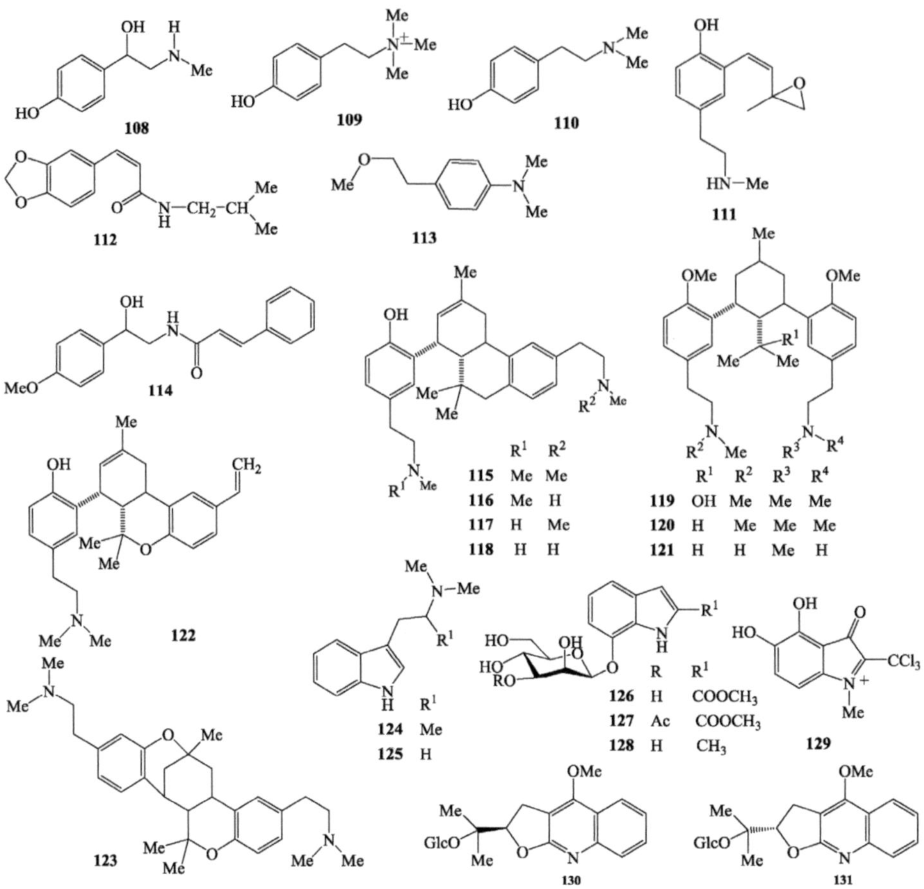

Figure 6.
The structures of alkaloids ***108–131****.*

No.	Compound names	Sources	Ref.
108	Synephrine	*Z. fagara*, *Z. culantrillo*	[4]
109	Candicine	*Z. clava-herculis*, *Z. americanum*	[4]
110	Hordenine	*Z. coriaceum*	[28]
111	4-(2-*N*-methyltyraminyl)-(*Z*)-1,2-epoxy-2-ethylbut-3-ene	*Z. coriaceum*	[28]
112	Fagaramide	*Z. rubescens*	[5]
113	-(2-methoxyethyl)-*N*,*N*-dimethyl benzenamine	*Z. nitidum*	[43]
114	(+)-Aegiline	*Z. coriaceum*	[28]
115	Alfileramine	*Z. coriaceum*, *Z. integrifoliolum*	[28, 44]
116	*N′*-demethylalfileramine	*Z. coriaceum*	[28]
117	*N*-demethylalfileramine	*Z. coriaceum*	[28]
118	*N*,*N′*-demethylalfileramine	*Z. coriaceum*	[28]
119	Culantraraminol	*Z. procerom*, *Z. colantrillo*	[45]
120	Culantraramine	*Z. coriaceum*	[28]
121	*N*,*N′*-demethylculantraramine	*Z. coriaceum*	[28]

No.	Compound names	Sources	Ref.
122	Integramine	*Z. integrifoliolum*	[44]
123	Isoalfileramine	*Z. coriaceum*	[28]
124	*N,N,N*-trimethyltryptamine	*Z. nitidum*	[8]
125	*N*-trimethyltryptamine	*Z. nitidum*	[8]
126	Methyl 7-(β-D-mannopyranosyloxy)-1H-indole-2-carboxylate	*Z. nitidum*	[46]
127	Methyl 7-[(3-*O*-acetyl-β-D-mannopyranosyl)oxy]-1H-indole-2-carboxylate	*Z. nitidum*	[46]
128	2-Methyl-1H-indol-7-yl *β*-D mannopyranoside	*Z. nitidum*	[46]
129	4,5-Dihydroxy-1-methyl-3-oxo-2-(trichloromethyl)-3H-indolium chloride	*Z. nitidum*	[43]
130	Zanthonitiside A	*Z. nitidum*	[47]
131	Zanthonitiside B	*Z. nitidum*	[47]

Table 5.
Other alkaloids from Zanthoxylum *species.*

Z. americanum, *Z. bouetense*, *Z. nitidum*, *Z. usambarense*, *Z. simulans*, *Z. lemairei*, and *Z. atchoum.* Chelerythrine increased cellular ROS level, leading to endoplasmic reticulum stress, inactivating STAT3 activities and inducing apoptosis in RCC cells which were suppressed by NAC, a special ROS inhibitor [66]. Chelerythrine significantly reduced the gastric ulcer index, myeloperoxidase activities, macroscopic and histological score in a dose-dependent manner [67].

Magnoflorine (**52**) could inhibit the apoptosis of the cells stimulated with TNF-α/IFN-γ. Further animal experiments confirmed that magnoflorine significantly attenuated the AD-like symptom and inhibited the AD-induced increases in IgE/IL-4, as compared with positive control [68]. Doxorubicin effects on the inhibition of migration and invasion of breast cancer cells was significantly promoted by magnoflorine. Doxorubicin-induced cell distribution in G2/M phase was markedly elevated when co-treated with magnoflorine. It is observed that apoptosis process were enhanced through doxorubicin/magnoflorine combinatory treatment rather than using doxorubicin alone through inducing Caspase-3 cleavage. In addition, magnoflorine markedly promoted the role of doxorubicin in autophagy induction by elevating light chain 3 (LC3)-II expression [69].

Liriodenine (**60**) was commonly found in *Zanthoxylum* genus. The effect of liriodenine induced significant apoptosis and suppression of cell growth of the MCF-7 cell line. The results indicated that the anticancer effects of liriodenine suppress cell growth and induce the apoptosis of human breast cancer MCF-7 cells through inhibition of Bcl-2, cyclin D1 and VEGF expression, and upregulation of p53 expression [70].

Skimmianine (**69**) significantly inhibit the growth of non-small cell lung cancer cells and markedly induce apoptosis in non-small cell lung cancer cells [71].

3.2 Inflammatory effects

Inflammation defines as the immune system responses to injury or infection with foreign organisms such as bacteria and viruses. However, excessive chronic inflammation represents the basis of inflammatory diseases including rheumatoid arthritis, diabetes, and chronic hepatitis. Several research groups have reported the inflammatory activity of *Zanthoxylum* genus. In LPS-induced endotoxemic mice, nitidine (**1**) increased IL-10 production, suppressed inflammatory responses, and reduced mortality remarkably. In LPS-stimulated RAW264.7 cells and in peritoneal macrophages from endotoxemic mice, nitidine significantly enhanced the activation of Akt, a critical signal transducer for IL-10 production, and inhibition of Akt prevented nitidine from enhancing IL-10 production and ameliorating endotoxemia [72]. Chelerythrine (**2**) markedly suppressed TNF-α, IL-6, and IL-1β production and oxidative LPS-induced [73]. Chelerythrine was found to inhibit NO production, pro-inflammatory IL-6 and TNF-α level in serum and gastric mucosal in the mice exposed to ethanol induced ulceration in a dose-dependent manner [67]. Skimmianine (**69**) significanly decreased in the mRNA levels of TNF-α and IL-6, which are upstream events of the inflammatory cascade. The levels of PGE2 and NO and the activities of COX-2 and 5-LOX were also significantly reduced after skimmianine treatment [71].

3.3 Antifungal and antibacterial activities

Besides cytotoxic activities, the *Zanthoxylum* species has also showed antifungal and antibacterial activities. In traditional medicine, many *Zanthoxylum* species are used commonly to treat skin diseases, purulent dermatitis, diarrhea, hepatitis and nephritis. Aqueous-ethanol 90% extracts of leaves, roots, and stem barks of *Z. leprieurii* and *Z. xanthoxyloides* inhibited the *in vitro* growth of *Candida albicans*, *Cryptococcus neoformans* and seven filamentous fungi tested [74]. Ethanolic extracts of the *Z. fagara*, *Z. elephantiasis*, and *Z. martinicense* showed antifungal activity [75]. Antifungal activity was also found in all extracts of leaves, fruits, twigs, bark, and roots of *Z. americanum* [76, 77]. Canthin-6-one (**106**) and 5-methoxycanthin-6-one (**107**) are major components in *Z. chiloperone* showed the broad-spectrum antifungal activity [78, 79]. In addition, benzophenanthridines such as dictamnine (**71**), γ-fagarine (**70**), 5-methoxydictamnine from *Z. nitidum* [29], liriodenine from *Z. tetraspermum* showed significant antifungal activity [80].

The screening *in vitro* and *in vivo* activity against the tuberculosis bacterium of compounds isolated from *Z. capense* showed that a benzophenanthridine alkaloid, decarine (**14**) and a *N*-isobutylamide *N*-isobutyl-(2E,4E)-2,4-tetradecadienamide exhibited antibacterial activity against *Mycobacterium tuberculosis* H37Rv (MIC value of 1.6 μg/ml) [81]. 6-Acetonyldihydronitidine (**26**) and 6-acetonyldihydroavicine (**27**) isolated from the stem bark of *Z. tetraspermum* [80] and from the bark and twigs of *Z. rhoifolium* and *Z. tetraspermum* [26], showed significant antibacterial activity.

In particular, benzophenanthridine alkaloids from *Zanthoxylum* genus exhibited strong activity against methicillin-resistant *Staphylococcus aureus* (MRAS) such as: dihydrochelerythrine (**24**) from *Z. rhetsa* [82], decarine (**14**), norchelerythrine (**13**), dihydrochelerythrine (**24**), 6-acetonyldihydrochelerythrine (**28**), tridecanonchelerythrine, and 6-acetonyldihydronitidine (**26**) from Z. *capense* [83], bis-[6-(5,6-dihydro-chelerythrinyl)] ether, 6-ethoxy-chelerythrine, and 4-methoxy-*N*-methyl-2-quinolone from *Z. monophyllum* [83], chelerythrine (**2**) from *Z. clava-herculis* [31]. The polymeric proanthocyanidins from *Z. piperitum* also showed antibacterial activity against MRAS [84]. 4-Methoxy-*N*-methyl-2-quinolone from *Z. monophyllum* exhibited significant inhibitory activity against MRSA bacteria with the IC_{50} value of 1.5 μg/ml [1].

Chelerythrine showed strong antibacterial activities against Gram-(+) bacteria, *Staphylococcus aureus*, Methicillin-resistant *S. aureus*, and extended spectrum β-lactamase *S. aureus*. Chellerythrine experiments on three bacteria resulted in

MICs were all 0.156 mg/ml. It suggest the primary anti-bacterial mechanism of this compound could be originated from the destruction of the channels across the bacterial cell membranes which lead to protein leakage to the outside of the cell and its inhibition on protein biosynthesis [85].

3.4 Other biological effect

Besides above mentioned biological activities, the alkaloid from *Zanthoxylum* plants also showed antivirus, cardioprotective, liver protective, antidiabetic, and antimalarial activities. Benzophenanthridine alkaloids, 5,6-dihydro-6-methoxynitidine, skimmianine, and 5-methoxydictamnine from *Z. nitidum* showed significant antiviral activities against hepatitis B virus [29], decarine, γ-fagarine, (+)-tembamide from the root bark of *Z. ailanthoides* against HIV with EC_{50} values < 0.1 μg/ml [86]. Nitidine showed similar *in vitro* activity in CQ-sensitive and resistant strains, and also a satisfying selectivity index (>10) when compared with a non-cancerous cells line. Nitidine can be considered a potential anti-malarial lead compound [87].

4. Structure elucidation of benzophenanthridine alkaloids from *Zanthoxylum* genus

4.1 NMR methods

Benzophenanthridine alkaloids are the most popular class of compounds isolated from *Zanthoxylum* genus. Structures of benzophenanthridines were elucidated by ^{1}H-, ^{13}C-NMR, DEPT, COSY, HSQC, HMBC, NOESY, and ROESY. The absolute configurations of these compounds were also determined by XRAY, and experimental CD as well as calculated CD.

Study on the structures of benzophenanthridine from *Zanthoxylum* genus, we found some following specifics: dioxymethylene group at C-2 and C-3, unsaturated and saturated bond at N/C-6; some substitutions at C-6 such as sesquiterpenes. **Tables 6** and 7 summarized ^{13}C-NMR characteristics of benzophenanthridine as follows:

1. When dioxymethylene group at C-2/C-3, ^{13}C-NMR chemical shift was about 102.0 ppm.

2. The *N*-methyl group at N was confirmed by chemical shift about 50.1–53.0 ppm when the presence of double bond at N/C-6; chemical shift about 41.1–41.2 ppm when the presence of single bond at N/C-6.

3. When C-substitution at C-6, chemical shifts at C-6 appeared around 57.3–66.7 ppm (methine carbon).

4. The positions of methoxy groups at benzophenanthridines normally appear at C-6, C-7, C-8, and C-9 with chemical shift around 55.7–62.8 ppm. Especially when the presence of single bond at N/C-6, the chemical shift of methoxy group at C-6 as 40.9–41.2 ppm.

5. When substitution groups at C-6 appear, they will have additional signals such as sesquiterpene.

C	1	7	11	12	13	14	19	20	22	24
1	107.3	108.0	104.7	106.2	104.4	104.4	104.7	104.7	104.5	104.2
2	151.0	154.0	147.6	150.4	148.2	147.9	147.6	147.4	147.4	147.1
3	150.5	153.5	147.1	150.3	148.2	148.1	147.0	147.0	145.9	147.7
4	103.9	106.0	102.6	104.7	100.7	100.8	100.6	102.6	101.4	100.6
4a	132.6	123.0	121.1	121.2	126.9	128.3	120.8	121.0	120.0	126.2
4b	152.2	138.0	135.7	132.7	136.9	138.7	152.4	135.8	128.0	142.6
6	134.6	155.0	162.7	163.4	145.5	145.7	164.0	164.3	162.7	48.6
6a	134.5	122.0	119.8	129.2	120.6	121.4	135.9	119.0	126.4	126.1
7	109.7	110.0	150.3	147.0	144.1	142.1	106.6	108.6	145.7	146.0
8	154.2	155.0	152.8	151.1	149.4	147.5	148.2	149.6	148.1	152.2
9	161.4	160.0	118.0	127.0	120.5	123.5	131.1	153.5	126.4	110.9
10	105.6	105.0	117.9	119.1	118.6	118.5	102.6	102.7	118.6	118.6
10a	121.6	133.0	129.0	120.2	127.3	126.4	132.0	128.9	118.1	126.2
10b	128.3	119.0	117.3	126.4	120.0	120.0	116.8	116.7	123.7	123.7
11	120.0	144.0	118.5	117.9	118.4	118.5	118.5	118.3	118.7	120.0
12	131.9	128.0	123.4	132.1	127.6	127.0	123.2	123.2	127.1	123.6
12a	122.3	132.0	131.8	133.9	129.5	129.2	120.9	131.8	129.1	130.8
2,3-OCH_2O	104.4	103.0	101.6	103.4	101.4	101.4	101.5	101.5	101.8	100.9
7,8-OCH_2O				102.3						
8,9-OCH_2O							101.9			
NCH_3	52.2	53.0		50.1						41.2
6-OCH_3			40.9	49.7			41.1	41.2	41.2	
7-OCH_3			61.8		61.5	61.1				60.9
8-OCH_3	57.2	58.0	56.7		56.7			56.2	59.9	55.7
9-OCH_3	57.9	58.0						56.1		
Solv.	m	m	m	m	d	d	c	c	m	c
Ref.	[71]	[17]	[9]	[19]	[72]	[72]	[26]	[26]	[19]	[72]

c, recorded in chloroform-d_1; d, DMSO-d_6; m, methanol-d_4.

Table 6.
^{13}C-NMR data of benzophenanthridine alkaloids.

4.2 Circular dichlorism

Circular dichroism (CD), a spectroscopic technique based on differential absorption of left- and right-handed circularly polarized light, is ideally disposed to analyze molecular structure, composition and interactions of chiral systems. Quantum mechanical calculations based on density functional theory (DFT) and its time-dependent formulation theory (TD-DFT) could be used to determine the theoretical chiroptical response of all the possible conformations of complexed-structures;

C	26	29	36	37	38	40	41	42	43	44
1	123.3	106.9	106.0	101.2	104.0	105.2	105.1	105.0	104.9	105.0
2	148.7	149.9	152.5	147.6	147.5	148.4	148.4	148.3	148.3	148.4
3	149.0	150.9	152.0	147.9	146.5	148.9	148.8	148.5	148.5	148.8
4	104.3	101.6	101.0	104.2	99.3	101.9	102.0	102.6	102.7	102.4
4a	123.8	128.8	130.5	131.0	123.1	132.1	132.1	132.0	132.0	132.1
4b	130.9	140.0	141.0	140.0	137.8	141.3	141.1	141.1	141.1	141.1
6	60.0	66.7	92.0	56.2	54.3	58.5	57.4	56.6	56.3	57.4
6a	123.5	126.0	121.0	126.2	121.3	131.1	131.1	131.5	131.4	131.3
7	100.4	149.3	112.0	146.7	149.5	147.0	147.0	147.0	147.0	147.1
8	147.5	154.4	150.0	151.9	143.8	153.0	153.0	153.0	153.0	153.0
9	148.2	114.4	150.0	111.3	116.0	112.1	111.8	112.0	112.0	111.9
10	106.4	121.4	110.0	119.1	118.7	119.3	119.3	119.3	119.3	119.2
10a	139.0	127.2	127.0	125.3	130.1	125.8	125.8	125.9	125.9	125.8
10b	127.0	126.7	119.0	123.2	127.4	124.8	124.8	124.8	124.8	124.9
11	119.6	121.9	144.0	119.6	119.5	120.7	120.7	120.8	120.7	120.7
12	110.4	126.7	121.0	123.5	123.6	124.4	124.5	124.4	124.3	124.4
12a	127.3	133.4	130.0	127.4	126.4	128.7	128.7	128.7	128.7	128.6
2,3-OCH_2O	101.3	103.4	104.0	101.0	101.0	101.2	101.3	101.3	101.3	101.3
NCH_3	42.4	44.2	40.0	42.3	42.4	43.3	43.2	43.2	43.1	43.2
6-OCH_3			55.0							
7-OCH_3		62.8		60.8	60.0	60.9	60.9	61.0	61.0	61.0
8-OCH_3	56.1	57.9	57.0	55.7		55.8	55.6	55.7	55.7	55.7
9-OCH_3	56.0		57.0							
1′	148.4	69.3		53.3	47.2	48.2	42.6	50.7	47.9	47.3
2′	207.9	20.4		211.9	206.1	21.7	21.6	23.6	23.4	23.6
3′	31.5			41.8	30.0	30.8	30.8	29.9	29.5	29.4
4′				28.9		135.7	134.6	135.5	136.4	136.3
5′				23.8		128.2	128.2	126.0	125.2	125.9
6′				30.4		35.2	35.0	38.0	40.3	40.3
7′						44.6	44.6	46.7	46.5	47.1
8′						20.0	20.0	20.4	17.7	22.3
9′						32.4	32.8	34.5	36.3	36.2
10′						76.1	76.1	74.3	76.3	76.3
11′						43.1	43.0	44.1	44.2	43.4
12′						27.4	27.3	26.6	26.2	26.4
13′						15.9	15.8	15.6	15.4	15.4
14′						22.4	22.2	21.8	21.8	22.0
15′						23.0	23.1	23.2	22.3	18.2
10′-OCH_3						48.1	48.3	48.7	48.4	48.3
Solv.	c	m	m	c	d	p	p	p	p	p
Ref.	[30]	[9]	[17]	[31]	[14]	[33]	[33]	[33]	[33]	[33]

c, recorded in chloroform-d_1; d, DMSO-d_6; m, methanol-d_4; p, pyridine-d_5.

Table 7.
^{13}C-NMR data of benzophenanthridine alkaloids (continued).

by comparison with the experimental CD spectra. This approach can lead to the elucidation of possible absolute structure in the absence of X-ray crystallography or NMR data.

Van et al. isolated four new compounds from *Z. nitidum*. Of these compounds **130** and **131** have the same constitution. This suggested the aglycone could be enantiomer. Thus, the absolute configuration at C-11 of **130** and **131** were elucidated by the comparison of its experimental ECD spectra with those calculated spectra. The TD-DFT calculated ECD spectra [47] of a pair of epimers (**130a** and **131a**) are shown in **Figure 7**. The CD spectra of **130** and **131** were found to be similar to **130a** and **131a** indicating the absolute configuration at C-11 as *R* and S, respectively.

Yang et al., isolated five novel dihydrobenzo[*c*]phenanthridine alkaloids, zanthomuurolanine (**40**), *epi*-zanthomuurolanine (**41**), zanthocadinanine A (**42**), zanthocadinanine B (**43**), and *epi*-zanthocadinanine B (**44**) from *Z. nitidum* [33]. The absolute configurations of these compounds were determined by XRAY and also CD spectra.

Zhao et al. isolated a pair of new enantiomeric furoquinoline alkaloids, zanthonitidine A (77) from *Z. nitidum.* There is no obvious absorption of electronic circular dichroism indicated that zanthonitidine A was proposed to be a racemate

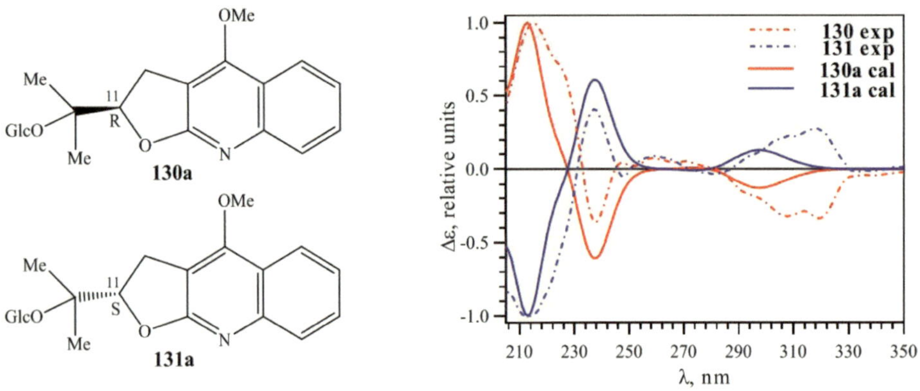

Figure 7.
Experimental CD and calculated ECD spectra of **130** *and* **131** *(calculated spectra are shifted by −8 nm). The figure was cited from Van et al [47].*

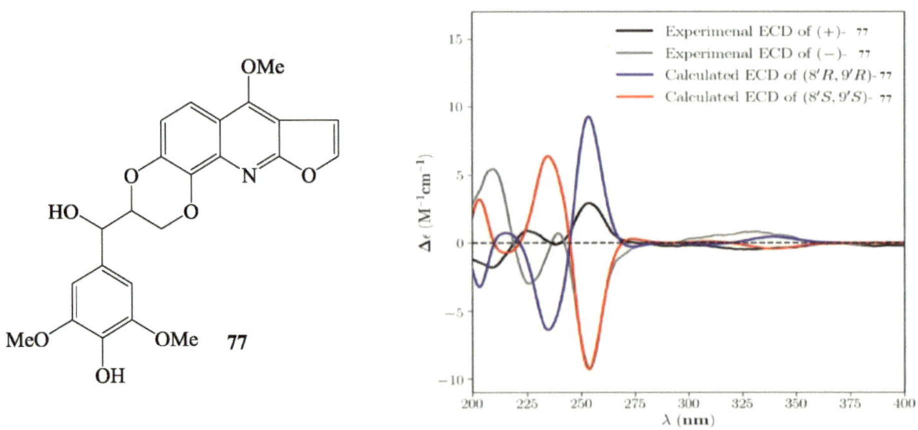

Figure 8.
Two possible stereochemical structures of 77; experimental ECD spectra of (+)-77/(−)-77 and calculated ECD spectra of (8′R, 9′R)/(8′S, 9′S) of 77. The figure was cited from Zhao et al [24].

mixture. Thus, they used Chiralpak ID column chromatography to separate the mixtures to obtain the enantiomers, (+) and (-)-zanthonitidine A. The absolute configurations of the enantiomers were then determined by comparing the experimental CD to the calculated ECD using TD-DFT of the Gaussian 9.0. By analyzing ECD spectra at the same theory level, the absolute configurations of (+) and (−)-zanthonitidine A were evaluated as (8′*R*,9′*R*)-zanthonitidine A and (8′*S*,9′*S*)-zanthonitidine A [24] (**Figure 8**).

Overall, experimental and calculated ECD spectra could play an important role for determine absolute configurations of alkaloids from *Zanthoxylum* species.

5. Conclusions

Alkaloids are the main constituents of *Zanthoxylum* species, present in the fruits, leaves, bark and root of plants. There are different types of skeletons of these alkaloids, including benzophenanthridines, aporphines, benzylisoquinolines, furoquinolines, quinolines, quinolones, quinazolines, indolopyridoquinazolines, acridones, canthinones, amines and tryptamines; in which benzophenanthridines are the main ingredient. Alkaloids from *Zanthoxylum* species have been displayed a variety of valuable biological activities, such as antibacterial, antifungal, antiviral, anti-inflammatory, antioxidant, cardiovascular protect and especially anti-cancer effects. Some alkaloids of which shown their potential to become natural healing agents, this has increasingly attracted scientists' interest in the genus *Zanthoxylum*. The data collected in this chapter has clearly shown that *Zanthoxylum* alkaloids with abundance of chemical structures and a wide range of cytotoxic activities on many the cancer cell lines. These could be good sources of potential cancer chemo-preventive agents. Further studies should be carried out to know more clearly the anticancer mechanisms of these alkaloids.

Conflict of interest

The authors declared no potential conflicts of interest with respect to the research, authorship, and/or publication of this book chapter.

Author details

Nguyen Xuan Nhiem[1], Pham Minh Quan[2] and Nguyen Thi Hong Van[2*]

1 Institute of Marine Biochemistry, Vietnam Academy of Science and Technology, Hanoi, Vietnam

2 Institute of Natural Products Chemistry, Vietnam Academy of Science and Technology, Hanoi, Vietnam

*Address all correspondence to: van762004@yahoo.com

References

[1] Rodriguez-Guzman R, Radwan MM, Burandt CL, Williamson JS, Ross SA. Xenobiotic biotransformation of 4-methoxy-N-methyl-2-quinolone, isolated from *Zanthoxylum monophyllum*. Natural Product Communications. 2010;**5**:1463-1464

[2] Hu J, Zhang WD, Liu RH, Zhang C, Shen YH, Li H-L, et al. Benzophenanthridine alkaloids from *Zanthoxylum nitidum* (ROXB.) DC, and their analgesic and anti-inflammatory activities. Chemistry & Biodiversity. 2006;**3**:990-995

[3] Waterman PG. Alkaloids from the root bark of *Zanthoxylum myriacanthum*. Phytochemistry. 1975;**14**:2530

[4] Stermitz FR, Caolo MA, Swinehart JA. Chemistry of *Zanthoxylum*. Part 5. Alkaloids and other constituents of *Zanthoxylum williamsii*, *Z. monophyllum*, and *Z. fagara*. Phytochemistry. 1980;**19**:1469-1472

[5] Moody JO, Sofowora A. Leaf alkaloids of *Zanthoxylum rubescens*. Planta Medica. 1984;**50**:101-103

[6] Fang SD, Wang LK, Hecht SM. Inhibitors of DNA topoisomerase I isolated from the roots of *Zanthoxylum nitidum*. Journal of Organic Chemistry. 1993;**58**:5025-5027

[7] Kato A, Moriyasu M, Ichimaru M, Nishiyama Y, Juma FD, Nganga JN, et al. Examination of alkaloidal constituents of *Zanthoxylum usambarense* by a combination of ion-pair extraction and ion-pair chromatography using sodium perchlorate. Phytochemical Analysis. 1995;**6**:89-95

[8] Moriyasu M, Ichimaru M, Nishiyama Y, Kato A, Wang J, Zhang H, et al. (R)-(+)-Isotembetarine, a quaternary alkaloid from *Zanthoxylum nitidium*. Journal of Natural Products. 1997;**60**:299-301

[9] Patino-Ladino OJ, Cuca-Suarez LE. Benzophenanthridine alkaloids from *Zanthoxylum quinduensis*. Actualidades Biologicas. 2005;**27**:53-58

[10] Halstead CW, Forster PI, Waterman PG. Alkaloids from the stem bark of an Australian population of *Zanthoxylum ovalifolium*. Natural Product Research. 2006;**20**:940-945

[11] Liang M, Zhang W, Hu J, Liu R, Zhang C. Simultaneous analysis of alkaloids from *Zanthoxylum nitidum* by high performance liquid chromatography-diode array detector-electrospray tandem mass spectrometry. Journal of Pharmaceutical and Biomedical Analysis. 2006;**42**:178-183

[12] Cai M, Zhou Y, Wang X, Li R, Liao X, Ding L. Rapid structural characterization of isomeric benzo[c] phenanthridine alkaloids from the roots of *Zanthoxylum nitidium* by liquid chromatography combined with electrospray ionization tandem mass spectrometry. Rapid Communications in Mass Spectrometry. 2007;**21**: 1931-1936

[13] Hu J, Zhang W-D, Shen Y-H, Zhang C, Xu L, Liu R-H, et al. Alkaloids from *Zanthoxylum nitidum* (Roxb.) DC. Biochemical Systematics and Ecology. 2007;**35**:114-117

[14] Talontsi FM, Matasyoh JC, Ngoumfo RM, Chepkorir R. Mosquito larvicidal activity of alkaloids from *Zanthoxylum lemairei* against the malaria vector Anopheles gambiae. Pesticide Biochemistry and Physiology. 2011;**99**:82-85

[15] Chen J, Wang J, Lin L, He L, Wu Y, Zhang L, et al. Inhibition of STAT3 signaling pathway by nitidine chloride

suppressed the angiogenesis and growth of human gastric cancer. Molecular Cancer Therapeutics. 2012;**11**:277-287

[16] Fang Z, Tang Y, Jiao W, Xing Z, Guo Z, Wang W, et al. Nitidine chloride induces apoptosis and inhibits tumor cell proliferation via suppressing ERK signaling pathway in renal cancer. Food and Chemical Toxicology. 2014;**66**: 210-216

[17] Akoua Yao-Kouassi P, Caron C, Ramiarantsoa H, Prost E, Harakat D, Le Magrex-Debar E, et al. New nitro-benzo[c]phenanthridine and indolopyridoquinazoline alkaloids from *Zanthoxylum atchoum*. Comptes Rendus Chimie. 2015;**18**:891-897

[18] Tuyen TT, Van NTH, Quan PM, Inh CT, Minh PTH, Thuy TTT, et al. Non-alkaloid compounds isolated from *Zanthoxylum nitidum* plant. Vietnam Journal of Chemistry. 2018;**56**:396-400

[19] Cui XG, Zhao QJ, Chen QL, Xu L, Song Y, Jin YS, et al. Two new benzophenanthridine alkaloids from *Zanthoxylum nitidum*. Helvetica Chimica Acta. 2008;**91**:155-158

[20] Deyun K, Gray AI, Hartley TG, Waterman PG. Alkaloids from an Australian accession of *Zanthoxylum nitidum* (Rutaceae). Biochemical Systematics and Ecology. 1996;**24**:87-88

[21] Chen I-S, Tsai I-W, Teng C-M, Chen J-J, Chang Y-L, Ko F-N, et al. Pyranoquinoline alkaloids from *Zanthoxylum simulans*. Phytochemistry. 1997;**46**:525-529

[22] Yang CH, Cheng MJ, Lee SJ, Yang CW, Chang HS, Chen IS. Secondary metabolites and cytotoxic activities from the stem bark of *Zanthoxylum nitidum*. Chemistry & Biodiversity. 2009;**6**:846-857

[23] Wang CF, Fan L, Tian M, Du SS, Deng ZW, Feng JB, et al. Cytotoxicity of benzophenanthridine alkaloids from the roots of *Zanthoxylum nitidum* (Roxb.) DC. var. *fastuosum* How ex Huang. Natural Product Research. 2015;**29**: 1380-1383

[24] Zhao L-N, Guo X-X, Liu S, Feng L, Bi Q-R, Wang Z, et al. (±)-Zanthonitidine A, a pair of enantiomeric furoquinoline alkaloids from *Zanthoxylum nitidum* with antibacterial activity. Natural Products and Bioprospecting. 2018;**8**:361-367

[25] Sukari MA, Salim WSW, Ibrahim NH, Rahmani M, Aimi N, Kitajima M. Phenantridine alkaloids from *Zanthoxylum myriacanthum*. Fitoterapia. 1999;**70**:197-199

[26] Gonzaga WA, Weber AD, Giacomelli SR, Dalcol II, Hoelzel SCS, Morel AF. Antibacterial alkaloids from *Zanthoxylum rhoifolium*. Planta Medica. 2003;**69**:371-374

[27] Hu J, Shi X, Mao X, Chen J, Zhu L, Zhao Q. Antinociceptive activity of Rhoifoline A from the ethanol extract of *Zanthoxylum nitidum* in mice. Journal of Ethnopharmacology. 2013;**150**:828-834

[28] Marcos M, Villaverde MC, Riguera R, Castedo L, Stermitz FR. *Zanthoxylum coriaceum* alkaloids related to bishordeninyl terpenes. Phytochemistry. 1990;**29**:2315-2319

[29] Yang G, Chen D. Alkaloids from the roots of *Zanthoxylum nitidum* and their antiviral and antifungal effects. Chemistry & Biodiversity. 2008;**5**: 1718-1722

[30] Wang XL, Yu KB, Li QF, Peng SL, Ding LS. 8-Acetonyldihydronitidine. Archive of Acta Crystallographica Section E. 2006;**62**:o2247-o2248

[31] Geng D, Li D-X, Shi Y, Liang J-Y, Min Z-D. A new benzophenanthridine alkaloid from *Zanthoxylum nitidum*. Zhongguo Tianran Yaowu. 2009;7: 274-277

[32] Liu H, Feng J, Feng K, Lai M. Optimization of the extraction conditions and quantification by RP-LC analysis of three alkaloids in *Zanthoxylum nitidum* roots. Pharmaceutical Biology. 2014;**52**:255-261

[33] Yang CH, Cheng MJ, Chiang MY, Kuo YH, Wang CJ, Chen IS. Dihydrobenzo[*c*]phenanthridine alkaloids from stem bark of *Zanthoxylum nitidum*. Journal of Natural Products. 2008;**71**:669-673

[34] Dreyer DL, Brenner RC. Chemotaxonomy of the Rutaceae. Part XIII. Alkaloids of some Mexican *Zanthoxylum* species. Phytochemistry. 1980;**19**:935-939

[35] Samita FN, Sandjo LP, Ndiege IO, Hassanali A, Lwande W. Zanthoxoaporphines A-C: Three new larvicidal dibenzo[de,g]quinolin-7-one alkaloids from *Zanthoxylum paracanthum* (Rutaceae). Beilstein Journal of Organic Chemistry. 2013;**9**: 447-452, No. 447

[36] Liu Y-C, Chen Z-F, Liu L-M, Peng Y, Hong X, Yang B, et al. Divalent later transition metal complexes of the traditional chinese medicine (TCM) liriodenine: Coordination chemistry, cytotoxicity and DNA binding studies. Dalton Transactions. 2009:10813-10823

[37] Yang Z-D, Zhang D-B, Ren J, Yang M-J. Skimmianine, a furoquinoline alkaloid from *Zanthoxylum nitidum* as a potential acetylcholinesterase inhibitor. Medicinal Chemistry Research. 2012;**21**: 722-725

[38] Ruangrungsi N, Tantivatana P, Borris RP, Cordell GA. Traditional medicinal plants of Thailand. III. Constituents of *Zanthoxylum budrunga* (Rutaceae). Journal of the Science Society of Thailand. 1981;**7**:123-127

[39] Ahmad MU, Rahman MA, Huq E, Chowdhury R. Alkaloids of *Zanthoxylum budrunga*. Fitoterapia. 2003;**74**:191-193

[40] Wang C, Wan J, Mei Z, Yang X. Acridone alkaloids with cytotoxic and antimalarial activities from *Zanthoxylum simulans* Hance. Pharmacognosy Magazine. 2014;**10**:73-76, 76a

[41] Talapatra SK, Dutta S, Talapatra B. Alkaloids and terpenoids of *Zanthoxylum ovalifolium*. Phytochemistry. 1973;**12**:729-730

[42] Ferreira ME, Rojas de Arias A, Torres de Ortiz S, Inchausti A, Nakayama H, Thouvenel C, et al. Leishmanicidal activity of two canthin-6-one alkaloids, two major constituents of *Zanthoxylum chiloperone* var. *angustifolium*. Journal of Ethnopharmacology. 2002;**80**:199-202

[43] Hu J, Zhang W-D, Shen Y-H, Zhang C, Liu R-H, Xu X-K, et al. Two novel alkaloids from *Zanthoxylum nitidum*. Helvetica Chimica Acta. 2007; **90**:720-722

[44] Liu S-L, Tsai I-L, Ishikawa T, Harayama T, Chen I-S. Bishordeninyl terpene alkaloids from *Zanthoxylum integrifoliolum*. Journal of the Chinese Chemical Society. 2000;**47**:571-574

[45] Schroeder DR. 1985

[46] Hu J, Shi X, Mao X, Chen J, Li H. Cytotoxic mannopyranosides of indole alkaloids from *Zanthoxylum nitidum*. Chemistry & Biodiversity. 2014;**11**: 970-974

[47] Van NTH, Tuyen TT, Quan PM, Long PQ, Nhiem NX, Tai BH, et al. Alkaloid glycosides and their cytotoxic constituents from *Zanthoxylum nitidum*. Phytochemistry Letters. 2019;**32**:47-51

[48] Ochwang'i DO, Kimwele CN, Oduma JA, Gathumbi PK, Mbaria JM, Kiama SG. Medicinal plants used in treatment and management of cancer in Kakamega County, Kenya. Journal of Ethnopharmacology. 2014;**151**: 1040-1055

[49] Lee J, Lim KT. Inhibitory effect of phytoglycoprotein (24 kDa) on hepatocarcinogenesis in N-nitrosodiethylamine-treated ICR mice. Journal of Pharmacy and Pharmacology. 2011;**63**:840-848

[50] Rajamani P, Banerjeet S, Rao AR. Chemoprotective influence of *Zanthoxylum* sps. on hepatic carcinogen metabolizing and antioxidant enzymes and skin papillomagenesis in murine model. Indian Journal of Experimental Biology. 2011;**49**:857-863

[51] Misra LN, Wouatsa NAV, Kumar S, Venkatesh Kumar R, Tchoumbougnang F. Antibacterial, cytotoxic activities and chemical composition of fruits of two Cameroonian *Zanthoxylum* species. Journal of Ethnopharmacology. 2013; **148**:74-80

[52] Tokita Y, Yuzurihara M, Sakaguchi M, Satoh K, Kase Y. The pharmacological effects of daikenchuto, a traditional herbal medicine, on delayed gastrointestinal transit in rat postoperative Ileus. Journal of Pharmacological Sciences. 2007;**104**: 303-310

[53] Chou ST, Chan HH, Peng HY, Liou MJ, Wu TS. Isolation of substances with antiproliferative and apoptosis-inducing activities against leukemia cells from the leaves of *Zanthoxylum ailanthoides* Sieb. & Zucc. Phytomedicine. 2011;**18**:344-348

[54] Ozkan M, Mutiso PB, Nahar L, Liu P, Brown S, Wang W, et al. *Zanthoxylum usambarense* (Engl.) Kokwaro (Rutaceae) extracts inhibit the growth of the breast cancer cell lines MDA-MB-231 and MCF-7, but not the brain tumour cell line U251 in vitro. Phytotherapy Research. 2013;**27**:787-790

[55] Walker TM, Vogler B, Moriarity DM, Haber WA, Setzer WN. A phytochemical investigation of *Zanthoxylum setulosum*. Natural Product Communications. 2011;**6**:1807-1808

[56] Dung TD, Feng CC, Kuo WW, Pai P, Chung LC, Chang SH, et al. Suppression of plasminogen activators and the MMP-2/-9 pathway by a *Zanthoxylum avicennae* extract to inhibit the HA22T human hepatocellular carcinoma cell migration and invasion effects in vitro and in vivo via phosphatase 2A activation. Bioscience, Biotechnology, and Biochemistry. 2013; **77**:1814-1821

[57] Karmakar I, Haldar S, Chakraborty M, Chaudhury K, Dewanjee S, Haldar PK. Regulation of apoptosis through bcl-2/bax proteins expression and DNA damage by *Zanthoxylum alatum*. Pharmaceutical Biology. 2016;**54**:503-508

[58] Chon SU, Heo BG, Park YS, Kim DK, Gorinstein S. Total phenolics level, antioxidant activities and cytotoxicity of young sprouts of some traditional Korean salad plants. Plant Foods for Human Nutrition. 2009;**64**: 25-31

[59] Jang KH, Chang YH, Kim DD, Oh KB, Oh U, Shin J. New polyunsaturated fatty acid amides isolated from the seeds of *Zanthoxylum piperitum*. Archives of Pharmacal Research. 2008;**31**:569-572

[60] Ngoumfo RM, Jouda JB, Mouafo FT, Komguem J, Mbazoa CD, Shiao TC, et al. In vitro cytotoxic activity of isolated acridones alkaloids from Zanthoxylum leprieurii Guill. et Perr. Bioorganic & Medicinal Chemistry. 2010;**18**:3601-3605

[61] Chen I-S, Wu S-J, Tsai I-L, Wu T-S, Pezzuto JM, Lu MC, et al. Chemical and bioactive constituents from *Zanthoxylum simulans*. Journal of Natural Products. 1994;**57**:1206-1211

[62] Chen JJ, Fang HY, Duh CY, Chen IS. New indolopyridoquinazoline, benzo[c]

phenanthridines and cytotoxic constituents from *Zanthoxylum integrifoliolum*. Planta Medica. 2005;**71**: 470-475

[63] Pachon G, Rasoanaivo H, Azqueta A, Rakotozafy JC, Raharisololalao A, De Cerain AL, et al. Anticancer effect of a new benzophenanthridine isolated from *Zanthoxylum madagascariense* (Rutaceline). In Vivo. 2007;**21**:417-422

[64] Mansoor TA, Borralho PM, Luo X, Mulhovo S, Rodrigues CM, Ferreira MJ. Apoptosis inducing activity of benzophenanthridine-type alkaloids and 2-arylbenzofuran neolignans in HCT116 colon carcinoma cells. Phytomedicine. 2013;**20**:923-929

[65] Pan X, Han H, Wang L, Yang L, Li R, Li Z, et al. Nitidine chloride inhibits breast cancer cells migration and invasion by suppressing c-Src/FAK associated signaling pathway. Cancer Letters. 2011;**313**:181-191

[66] He H, Zhuo R, Dai J, Wang X, Huang X, Wang H, et al. Chelerythrine induces apoptosis via ROS-mediated endoplasmic reticulum stress and STAT3 pathways in human renal cell carcinoma. Journal of Cellular and Molecular Medicine. 2020;**24**:50-60

[67] Li WF, Hao DJ, Fan T, Huang HM, Yao H, Niu XF. Protective effect of chelerythrine against ethanol-induced gastric ulcer in mice. Chemico-Biological Interactions. 2014;**208**:18-27

[68] Wu S, Yu D, Liu W, Zhang J, Liu X, Wang J, et al. Magnoflorine from Coptis chinese has the potential to treat DNCB-induced Atopic dermatits by inhibiting apoptosis of keratinocyte. Bioorganic & Medicinal Chemistry. 2019;**28**:115093

[69] Wei T, Xiaojun X, Peilong C. Magnoflorine improves sensitivity to doxorubicin (DOX) of breast cancer cells via inducing apoptosis and autophagy through AKT/mTOR and p38 signaling pathways. Biomedicine & Pharmacotherapy. 2020;**121**:109139

[70] Li ZH, Gao J, Hu PH, Xiong JP. Anticancer effects of liriodenine on the cell growth and apoptosis of human breast cancer MCF-7 cells through the upregulation of p53 expression. Oncology Letters. 2017;**14**:1979-1984

[71] Ratheesh M, Sindhu G, Helen A. Anti-inflammatory effect of quinoline alkaloid skimmianine isolated from *Ruta graveolens* L. Inflammation Research. 2013;**62**:367-376

[72] Yang N, Yue R, Ma J, Li W, Zhao Z, Li H, et al. Nitidine chloride exerts anti-inflammatory action by targeting Topoisomerase I and enhancing IL-10 production. Pharmacological Research. 2019;**148**:104368

[73] Fan L, Fan Y, Liu L, Tao W, Shan X, Dong Y, et al. Chelerythrine attenuates the inflammation of lipopolysaccharide-induced acute lung inflammation through NF-kappaB signaling pathway mediated by Nrf2. Frontiers in Pharmacology. 2018;**9**:1047

[74] Ngane AN, Biyiti L, Zollo PH, Bouchet P. Evaluation of antifungal activity of extracts of two Cameroonian rutaceae: *Zanthoxylum* leprieurii Guill. et Perr. and *Zanthoxylum* xanthoxyloides Waterm. Journal of Ethnopharmacology. 2000;**70**:335-342

[75] Dieguez-Hurtado R, Garrido-Garrido G, Prieto-Gonzalez S, Iznaga Y, Gonzalez L, Molina-Torres J, et al. Antifungal activity of some Cuban *Zanthoxylum* species. Fitoterapia. 2003; **74**:384-386

[76] Smith ML, Gregory P, Bafi-Yeboa NF, Arnason JT. Inhibition of DNA polymerization and antifungal specificity of furanocoumarins present in traditional medicines. Photochemistry and Photobiology. 2004;**79**:506-509

[77] Bafi-Yeboa NF, Arnason JT, Baker J, Smith ML. Antifungal constituents of northern prickly ash, *Zanthoxylum americanum* mill. Phytomedicine. 2005; **12**:370-377

[78] Ferreira ME, Cebrian-Torrejon G, Corrales AS, Vera de Bilbao N, Rolon M, Gomez CV, et al. *Zanthoxylum chiloperone* leaves extract: First sustainable Chagas disease treatment. Journal of Ethnopharmacology. 2011; **133**:986-993

[79] Thouvenel C, Gantier JC, Duret P, Fourneau C, Hocquemiller R, Ferreira ME, et al. Antifungal compounds from *Zanthoxylum chiloperone* var. *angustifolium*. Phytotherapy Research. 2003;**17**: 678-680

[80] Nissanka AP, Karunaratne V, Bandara BM, Kumar V, Nakanishi T, Nishi M, et al. Antimicrobial alkaloids from *Zanthoxylum tetraspermum* and *caudatum*. Phytochemistry. 2001;**56**: 857-861

[81] Luo X, Pires D, Ainsa JA, Gracia B, Duarte N, Mulhovo S, et al. *Zanthoxylum capense* constituents with antimycobacterial activity against *Mycobacterium tuberculosis* in vitro and ex vivo within human macrophages. Journal of Ethnopharmacology. 2013; **146**:417-422

[82] Tantapakul C, Phakhodee W, Ritthiwigrom T, Yossathera K, Deachathai S, Laphookhieo S. Antibacterial compounds from *Zanthoxylum rhetsa*. Archives of Pharmacal Research. 2012;**35**:1139-1142

[83] Luo X, Pedro L, Milic V, Mulhovo S, Duarte A, Duarte N, et al. Antibacterial benzofuran neolignans and benzophenanthridine alkaloids from the roots of *Zanthoxylum capense*. Planta Medica. 2012;**78**:148-153

[84] Hatano T, Kusuda M, Inada K, Ogawa TO, Shiota S, Tsuchiya T, et al. Effects of tannins and related polyphenols on methicillin-resistant *Staphylococcus aureus*. Phytochemistry. 2005;**66**:2047-2055

[85] He N, Wang P, Wang P, Ma C, Kang W. Antibacterial mechanism of chelerythrine isolated from root of *Toddalia asiatica* (Linn) Lam. BMC Complementary Medicine and Therapies. 2018;**18**:261

[86] Cheng MJ, Lee KH, Tsai IL, Chen IS. Two new sesquiterpenoids and anti-HIV principles from the root bark of *Zanthoxylum ailanthoides*. Bioorganic & Medicinal Chemistry. 2005;**13**: 5915-5920

[87] Bouquet J, Rivaud M, Chevalley S, Deharo E, Jullian V, Valentin A. Biological activities of nitidine, a potential anti-malarial lead compound. Malaria Journal. 2012;**11**:67

Chapter 8

Flavonoid-Mediated Modulation of CYP3A Enzyme and P-Glycoprotein Transporter: Potential Effects on Bioavailability and Disposition of Tyrosine Kinase Inhibitors

Muzaffar Iqbal

Abstract

The consumption of herbal products and dietary supplements along with conventional medicines has raised concerns regarding herb-drug interactions. The available literature from experimental and clinical studies suggested that the consumption of herbs or dietary supplements that modulate efflux proteins, especially P-glycoprotein (P-gp) and metabolic enzyme CYP3A, may cause clinically relevant herb-drug interactions by alteration of bioavailability and disposition profiles of targeted drug. It has been also hypothesized that both CYP3A and P-gp work synergistically to limit systemic exposure of orally administered substrate drugs. Many *in vitro* and *in vivo* studies suggested that co-administration of flavonoids significantly enhances the bioavailability of orally administered drugs, which may be due to inhibition of the CYP3A enzyme and P-gp transporter. Recently, a large number of orally administered tyrosine kinase inhibitors (TKIs) have been clinically approved for cancer chemotherapy, and many are currently estimated to be under development. TKIs are all primarily metabolized by CYP3A, and most of them are also substrates of P-gp. Numerous studies have suggested that the plasma exposure of orally administered TKIs increases when co-administered with other drugs due to their dual inhibitory activities against P-gp and CYP3A. However, limited data are available regarding the interaction between flavonoids and TKIs. The objective of this article is to review the potential role of flavonoids in modulation of CYP3A enzyme and P-gp transporter and their influence on bioavailability and disposition of TKIs.

Keywords: flavonoids, tyrosine kinase inhibitors, CYP3A4, P-glycoprotein, bioavailability, disposition

1. Introduction

Due to common belief that natural medicines are much safer than synthetic one, the use of complementary and alternative medical therapies (CAMs) has become a common trend around the world. It can be used either alone or in combination with prescription medicines [1]. According to an estimate of World Health Organization (WHO), approximately 80% of the developing countries' population relies on CAMs for their primary healthcare needs [2, 3]. CAMs have been also become popular around the developed countries, and this has led to a tremendous growth in international herbal drug market for the last 15 years [4–6]. Consumption of CAMs is more pronounced in patients diagnosed with cancer or human immune virus (HIV) infection, especially with regard to various antidepressant and energy treatments to cope with their mental and physiological instability [7]. The bioavailability and the distribution characteristics are the key factor for the therapeutic effects of pharmaceuticals at their site(s) of action in the tissue [8]. Due to higher consumption of CAMs (herbal extracts and dietary supplements) with prescription medicines, there is a growing awareness that herbal remedies and/or phytoconstituents may affect the bioavailability and disposition characteristics of conventional pharmaceuticals [9]. The medical and scientific literature supported by *in vitro* and *in vivo* laboratory studies including preclinical and clinical trials suggested that the co-administration of natural products or its phytoconstituents may affect the metabolism and bioavailability of prescription drugs, which significantly increasing the risk of serious (clinical) adverse reactions or therapeutic failure. The primary mechanisms underlying the herb-drug interactions involve either the induction or the inhibition of intestinal drug efflux pumps (including efflux proteins, such as P-gp and MRPs) and the intestinal and hepatic metabolism mediated by cytochrome P450 enzymes (CYP3A) [10–12]. Hence, the consumption of herbs that can modulate efflux proteins and/or CYP3A may cause clinically relevant herb-drug interactions and alter drug bioavailability [13, 14]. Any inhibitory effect of herbal extracts/constituents on efflux proteins and/or CYP3A may result in high exposure of substrate drugs in plasma and tissue and lead to toxicity, whereas any inductive effect may cause low exposure, leading to a decrease in efficacy and treatment failure.

The biological effects produced by CAMs are due to the presence of various classes of phytochemicals present there, that is, alkaloids, flavonoids, terpenoids, carotenoids, polyketides, and phenylpropanoids. Among them, flavonoids have attracted much interest due to their numerous pharmacological activities and health benefit in the form of their antioxidant, anti-inflammatory, antimutagenic, antibacterial, antiangiogenic and enzyme modulatory, antiallergic, and anticancer activities [15, 16] and become the main components of herbal products where it presents in the form of quercetin, genistein, hypericin, kaempferol, and silibinin. There is accumulated evidence in the literature, which confirms that flavonoids modulate drug metabolism. Modulation may happen by either (1) altering the expression and/or activity of P450 enzymes, (2) affecting the P-gp-mediated cellular efflux of drugs, and/or (3) inhibiting the intestinal glucuronidation of the drug. This evidence confirms that the consumption of flavonoids or flavonoid-containing dietary supplements with conventional pharmacotherapeutic regimens should need to be examined to avoid drug-flavonoid interactions [17–21].

Among the novel classes of anticancer drug development, small molecule tyrosine kinase inhibitors (TKIs) currently represent one of the most promising and rapidly expanding groups. Almost 25 TKIs (mostly in oral dosage form) have been already approved by international drug agencies, >130 are being evaluated in different phases of clinical trials, and many more are in various stages of development [22, 23]. Most of TKIs are primarily metabolized by CYP3A4 and also interact with P-gp and/or Breast Cancer Resistance Protein (BCRP) where it serves as both substrates and inhibitors. Considering the dual roles of TKIs on both CYP3A4 and drug transporters (P-gp, BCRP) and its influence in drug disposition, the potential

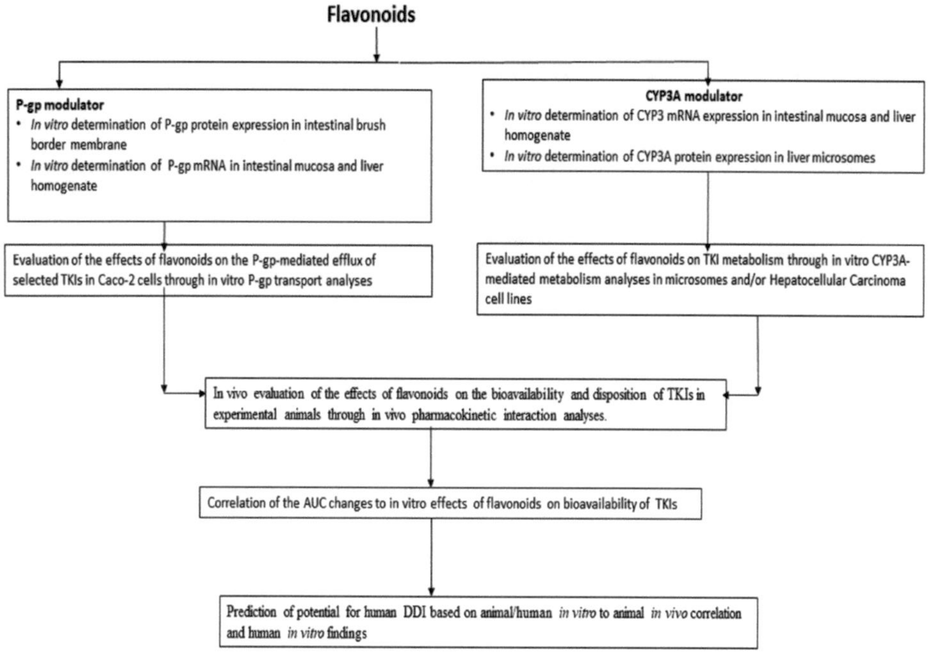

Figure 1.
Schematic layout for prediction of flavonoids in modulation of CYP3A enzyme and P-gp transporter and its influence on disposition of TKIs.

of TKI-drug interactions is an important consideration [24, 25]. In addition, most TKIs are being used orally and prescribed for long duration along with other medications, which may result in significant drug-drug interactions (DDI). This review provides a comprehensive overview of the potential role of flavonoids in modulation of CYP3A enzyme and P-gp transporter and their possible influence on bioavailability and disposition of tyrosine kinase inhibitors (**Figure 1**).

2. CYP3A

Cytochrome P450 enzymes (CYP) are the most versatile enzyme system involved in detoxification and oxidative metabolism of various endogenous substrates (steroid hormones, lipids, and bile acids) and xenobiotics (drugs, environmental pollutants, and dietary products) [26, 27]. It consists of over 400 isoforms, and their activities can be increased or decreased by many drugs either by inducing the biosynthesis or by directly inhibiting its activity, which is a major source for drug-induced toxicity via DDI. CYP3A enzymes, which constitute the predominant phase I drug-metabolizing enzymes, are estimated to metabolize between 50 and 70% of currently administered drugs [28]. Alone CYP3A4, which is the most abundant congener of the CYP3A family contributes approximately 30% of hepatic CYP activity and more than 70% of intestinal CYP activity [29]. Many drugs used in different types of therapies are substrates for CYP3A4, and it is presented at high levels in both liver and intestine [30, 31]. Reports from *in vitro* and *in vivo* studies have already established that naturally occurring dietary supplements and phytoconstituents can modulate hepatic and enterocytic CYP activity.

CYP inhibition-mediated DDI is widely recognized, and the necessity of the enzyme inhibition studies is included in the guidance from the USFDA [32]. Usually, two types of CYP inhibition occur: (1) reversible inhibition represented by competitive inhibition, which is concentration-dependent inhibition and (2) irreversible inhibition, also called mechanism-based inhibition (MBI), where during inhibition process enzyme is inactivated by stable complex formation with a metabolite. In MBI, enzyme reduction activity continues until the inactivated CYP is replaced by a newly synthesized CYP, the duration of the elevated blood concentration of a drug coadministered with a mechanism-based inhibitor is longer. Therefore, MBI requires more attention because they have been reported to cause unanticipated adverse effects [33, 34].

3. P-gp

P-gp is an adenosine triphosphatase (ATPase) energy-dependent, membrane-bound protein that belongs to the ABC efflux transporter family [35]. The ABCB1 gene, which is also known as the multidrug resistance 1 (MDR1), encodes P-gp and is responsible for cellular efflux of numerous drugs [36]. It is more prominent in various resistant human tumors, where it is believed to be the major factors responsible for multidrug resistant (MDR). P-gp-mediated transport of drugs is saturable, ATP-dependent, osmotically sensitive mechanism that generates a concentration gradient. In intestines, P-gp expressed in apical side of the epithelial cells where it pumps the drug back into the GI lumen resulting in fecal excretion. In liver, it presents in canalicular surface of hepatocytes to remove drug and metabolites from the interior of the cell. In brain, it is expressed in endothelial cells of blood brain barrier and prevented the entry of xenobiotics into brain. In kidney, it is expressed in proximal tubes of kidney to efflux drug into urine and in certain hematological cells to put drug back into circulation [37–40]. Higher expression of P-gp in excretory organs (liver and kidney) facilitates metabolism of substrate drugs via biliary excretion and renal elimination. As an efflux transporter, ABCB1 prevents intestinal absorption of orally administered drugs and limits its oral bioavailability. A broad range of clinically used drugs are substrate of P-gp, including anticancer agents (anthracyclines, vinca alkaloids, epipodophyllotoxins, methotrexate, and taxol) [41], cardiac drugs (digoxin and quinidine) [42], protease inhibitors (saquinavir, indinavir, and ritonavir) [43], immunosuppressants (cyclosporine) [44], and antibiotics (actinomycin D) [45].

4. Interplay between CYP3A and P-gp

Various preclinical and clinical studies have postulated that both CYP3A enzyme and P-gp transporter display strong effects in modulation of oral drug bioavailability and elimination of numerous drugs. Both CYP3A and P-gp act in functional collaboration during the first-pass elimination of drug [46]. Extensive overlap exists between the substrate specificities and the tissue-specific expression patterns of P-gp and CYP3A, especially in the liver (hepatocytes) and intestine (enterocytes) [47, 48]. It has also been predicted that the orally administered drugs, which are dual substrates of P-gp and CYP3A, the back-transportation mechanism by P-gp in intestinal epithelial cells, are available for further CYP3A4-mediated metabolism within these cells resulting in massive first-pass effects in intestine [49]. The combined action of these two pathways was expected to be more efficient than the sum of their individual activities, resulting in synergistic effects of P-gp and CYP3A.

Two mechanistic frameworks have been suggested to support the possible synergistic action of P-gp and CYP3A in intestine. Under the first mechanism, P-gp acts to decrease the level of intracellular concentration of a substrate drug in

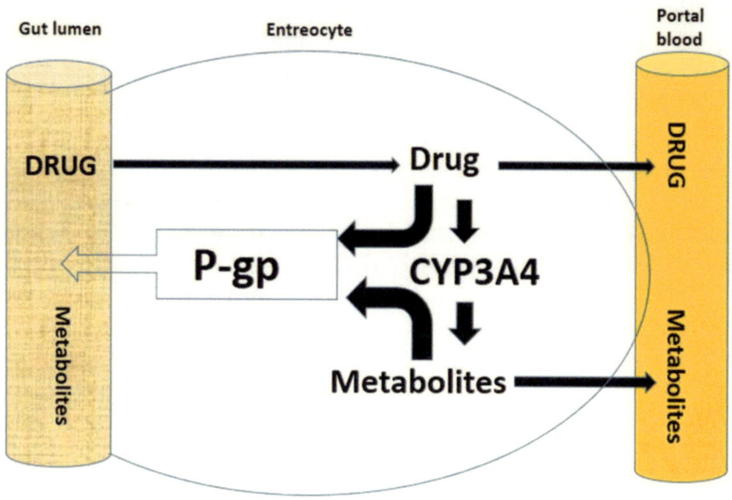

Figure 2.
Potential functional relationships between P-gp and CYP3A4 in enterocytes.

enterocytes, thus preventing possible saturation of the CYP3A enzyme by maintaining the drug within the linear range for metabolic activity. Under the second mechanism, the functional effects of P-gp, together with subsequent drug reuptake, allow the repetition of drug substrate, and therefore, probability of the drug to be metabolized increases by prolonged access to enterocyte CYP3A. The above cyclic repetition of drugs *per se* would itself increase the drug metabolism, even without considering saturating or nonsaturating conditions for CYP3A [50]. Based on above evidence, it is concluded that induction and inhibition of (intestinal) P-gp and CYP3A4 are important mechanisms underlying DDIs [51].

The hypothesis of synergistic collaboration between CYP3A and P-gp was clarified in a study that showed a dramatic increase in the systemic exposure of docetaxel and the risk of intestinal toxicity in P-gp$^{-/-}$CYP3A$^{-/-}$ double knockout mice. After oral administration of docetaxel, 3-fold and 12-fold increases in bioavailability were observed in P-gp$^{-/-}$ and CYP3A$^{-/-}$ mice, respectively, in comparison with wild-type mice. However, when both of the primary detoxification systems were missing, that is, in P-gp$^{-/-}$CYP3A$^{-/-}$ double knockout mice, bioavailability showed a disproportionate increase in >70-fold in comparison with wild-type mice [48].

Within the context described above, it is clear that drugs that are dual substrates for P-gp and CYP3A are highly susceptible to herb-drug interactions. Therefore, the consumption of herbs that modulate efflux transporters and CYP3A may cause clinically significant drug-herb interactions and alter drug bioavailability [13, 14]. The inhibitory effect of herbs on P-gp and CYP3A may result in increasing the drug concentration level in the plasma and tissues, leading to toxicity, whereas any inductive effect may result in lowering the concentration level, leading to loss of efficacy and/or treatment failure (**Figure 2**).

5. Regulation of CYP3A and P-gp expression

Pregnane X receptor (PXR) is a highly promiscuous nuclear hormone receptor, generally expressed on sites that are important to drug dispositions (e.g., small intestine, liver, and kidney). The human orthologue of PXR is also known as steroid and xenobiotics receptor (SXR) and is coded by NR 112. It has been reported that PXR regulates the expression of CYP3A and several other genes encoding protein and other enzymes, which involves in drug disposition including P-gp [52, 53]. Based on this evidence, Maglich et al. hypothesized that PXR plays a more important role in the regulation of drug metabolizing enzymes and drug transporters in the small intestines [54]. The observation that PXR regulates both CYP3A and P-gp provides further evidence for the argument that these proteins coordinately mediate detoxification of many xenobiotics during oral absorptions [46]. Synold et al. were the first to describe coregulation of drug metabolism and efflux via CYP3A and P-gp in liver and intestine by the human receptor PXR/SXR. Their results indicate that paclitaxel reduces its own oral bioavailability by activating the PXR and induces its own metabolism and biliary elimination [55].

6. Experimental models for P-gp and CYP3A-mediated DDI studies

6.1 *In vitro* model for CYP3A and P-gp studies

The *in vitro* CYP inhibition study is usually performed in human liver microsomes (HLMs) or human liver hepatocytes (HLHs), Caco-2 cell lines, or recombinant CYP (rCYP) enzymes [56]. The technology used for CYP inhibition includes luminescence, florescence, radiometric, and HPLC or LC-MS/MS assay. In florescence and luminescence assays, the metabolism of profluorescent or proluminescent substrate by CYP enzyme to give their fluorescent or luminescent product, respectively, is measured in rCYP enzyme. In radiometric method, the release of radiolabel on metabolism of substrate is measured in HLM. In HPLC or LC-MS/MS assay, the concentration of substrate/probe drugs is measured in HLM, HLH, and rCYP enzymes [57].

The *in vitro* analyses used for P-gp-mediated efflux studies include cytotoxicity assay, accumulation/efflux assays, transport assays, and ATPase assays. In cytotoxicity assays, the IC_{50} end point (concentration of P-gp substrates or inhibitors) that inhibits the growth of P-gp expressing cells is measured. In accumulation/efflux assay, the accumulation of drugs in P-gp expressing cells is measured. In transport assays, the permeation of drugs from apical-to-basolateral and basolateral-to-apical compartment in polarized epithelial cells is measured, whereas in ATPase assay, the stimulation or inhibition of P-gp ATPase enzyme activity in membranes of P-gp expressed cells is measured [58, 59].

6.2 *In vivo* models for CYP3A and P-gp-mediated DDI studies

Various strains of mice (including transgenic and knockout), rats, and nonhuman primate (monkey) have been used for CYP3A, P-gp, and dual CYP3A/P-gp-mediated DDI studies [60].

Since midazolam and triazolam are specifically extensively metabolized by CYP3A4 and ketoconazole is a potent inhibitor of CYP3A4, several groups of studies used as model for experimental CYP3A-mediated DDI studies. Moreover, midazolam is not a substrate for P-gp, whereas the ketoconazole can effectively inhibit both CYP3A and P-gp, which can confirm that the interaction between them can be sole attributed to effects on CYP3A. CYP gene knockout and humanized mice have been established for CYP3A-mediated drug interaction using midazolam as substrate [61]. Several strains of rats are most commonly used for CYP3A-mediated drug interaction using midazolam, diltiazem, nifedipine, and doxorubicin as substrates [62, 63]. In addition, cynomolgus monkey was also used as model for

Flavonoids	Dose-dependent (IC_{50}) inhibition of CYP3A4 activity	↑ Cellular accumulation of rhodamine (MCF-7/ADR cell)	References
Quercetin	1.97 μM	3–10 μM	[83]
Myricetin	7.8 μM	10–30 μM	[84]
Baicalein	9.2 μM	1–10 μM	[85]
Baicalein	9.2 mM	10–30 μM	[86]
Silibinin	1.8 μM	—	[87]

Table 1.
In vitro *effects of some common flavonoids on CYP3A4 and P-gp activity.*

CYP3A-mediated DDIs; however, its choice is limited due to its high cost, handling, and ethical concerns [64].

Valspodar, elacridar, and zosuquidar are more selective P-gp inhibitors and thought to be low affinity for CYP enzymes, and other drug transporter proteins are commonly used for *in vivo* DDI studies in mice and rats. The drugs such as paclitaxel, cyclosporine, and digoxin are most commonly used as substrates for P-gp-mediated inhibition. Moreover, Rho123 is attractive probe and has been widely used as marker to evaluate P-gp functions because it is not a substrate for CYP3A enzyme [65–67]. The cynomolgus monkey has been also used as model to evaluate effect of P-gp-mediated DDI using erythromycin and fexofenadine as substrate [64, 68].

The evaluation of flavonoid-mediated DDI by using dual substrates of CYP3A and P-gp, inhibitors, and species has been discussed in detail in Section 9 and summarized in **Table 1**.

7. Flavonoids as P-gp and CYP3A modulators

Flavonoids (a group of polyphenolic compounds) are mainly abundant in vegetables and fruits and routinely consume through our common diet and in the form of beverages (plant-derived), for example, wine and tea [37]. In addition, they are the main constituents of many herbal products/formulations. Structurally, these compounds possess a framework consisting of a chromane ring together with an aromatic ring that is attached at different positions. Based on various substitutions and the oxidation status of the ring C atoms, flavonoids can be categorized into various subclasses, including flavones, flavonols, flavonones, flavanols, isoflavones, and chalcones. Since the last decade, there has been a drastically increase in scientific work on flavonoids, with >2000 publications/year containing "flavonoids" as a keyword found in different literature sources. Flavonoids display antioxidant, anticarcinogenic, antiviral, anti-inflammatory, and antiesterogenic properties, and high intake of flavonoids has been linked with a reduced risk of cancer, cardiovascular disease, osteoporosis, and other age-related degenerative diseases [37, 69].

Due to the wide range of health benefits of flavonoids and their remarkable safety record, numerous herbal preparations containing these compounds are marketed in various formulations as dietary supplements. The total daily intake of flavonoids in the average US diet has been estimated to be more than 1 g [37]. Therefore, the concentration of flavonoid expected to be present is sufficient after the ingestion of flavonoids and/or flavonoid-containing supplements, suggesting a potential herb-drug interaction.

Thus, the consumption of higher doses of flavonoids is common in daily life, and it may increase the risk of pharmacokinetic interactions with clinically used medicine. This concern is also confirmed by increasing evidence, which showed significant or even life-threatening interactions between flavonoids or flavonoid containing products and prescription drugs [70, 71].

Initial publications cite various examples of flavonoids as P-gp transport inhibitors, thereby affecting the bioavailability and cellular uptake of anticancer drugs. These experiments include *in vitro* analyses of the effects of flavonoids on the intracellular accumulation of P-gp substrates using P-gp-overexpressing cells and a variety of clinical and animal model studies, especially involving P-gp knockout animals [37]. For example, concomitant administration of quercetin increased moxidectin oral bioavailability in lambs [72]; oral bioavailability of quinine was increased by naringin [73]; cyclosporine by baicalein and its aglycone [74]; and paclitaxel by flavones in rats [75]. Similarly, quercetin increased the oral bioavailability of paclitaxel and tamoxifen in rats [76, 77] and digoxin in pigs, which results in high toxicity [78]. All of these studies indicate that flavonoid-P-gp interactions can occur *in vivo*, resulting in pharmacokinetic interactions. In contrast to the above results, several flavonoids appear to induce P-gp transport, resulting in a decrease in the bioavailability of substrate drugs. For instance, *in vitro* studies, kaempferol and quercetin, produced inductive effects on P-gp efflux [79, 80], and therefore, the consumption of pure herbal constituents, which contains hypericin, kaempferol, quercetin, and silibinin for 10 days, may produce a significant increase in the expression of P-gp mRNA [81]. *In vivo* studies have also indicated that long-term exposure (14 days) to St. John's wort (a flavonoid-containing herbal product) leads to higher expression of MDR1 in the rat intestine [82]. Based on these findings, it has been concluded that chronic exposure to some flavonoids induces intestinal expression of P-gp, resulting in reduced intestinal drug absorption, possibly due to enhanced drug efflux; however, the inhibitory effects on P-gp-mediated efflux are based on short-term exposure. Meanwhile, subsequent *in vitro* and *in vivo* studies indicate that the pharmacokinetic interactions of drugs with flavonoids may result in the modulation not only of drug transporters (P-gp) but also of metabolizing enzymes, especially CYP3A, that is, dual inhibition of P-gp and CYP 3A. *In vitro* studies confirmed that quercetin, myricetin, baicalein, and silibinin were found to produce dose-dependent inhibition of CYP 3A4 activity in CYP inhibition assay and increased the cellular accumulation of rhodamine (MCF-7/ADR cell) in P-gp transport studies [83–87]. However, in another study, quercetin and rutin were found to induce the function of CYP 3A4 and P-gp, which may lead to increase the bioavailability of substrate drugs [88]. *In vitro* effects of some common flavonoids on CYP3A4 and P-gp activity are summarized in **Table 1**. In *in vivo* studies, coadministration of quercetin, baicalein, silibinin, epigallocatechein, and kaempeferol increased the bioavailability of oral tamoxifen dose dependently through inhibition of P-gp efflux and reduction in the first-pass metabolism through inhibition of CYP3A metabolism in the small intestine and/or liver [78, 84, 89–91]. Similarly, quercetin, silibinin, naringin, flavone (2-phenyl-4H-1-benzopyran-4-one), genistein, and morin also increased the oral bioavailability of paclitaxel mainly through inhibition of CYP3A4-mediated metabolism in the small intestine and/or liver and inhibition of the P-gp efflux in the small intestine [75–77, 92–94]. Additionally, quercetin, myricetin, and baicalein also reduced the bioavailability of doxorubicin by similar mechanism [83, 84, 95]. Similar results were also produced by morin, quercetin, and niringin with diltiazem and baicalein with nimodipine [85, 96–98]. In contrast, reduced oral bioavailability of tamoxifen by biochanin A and cyclosporine by quercetin and rutin was also reported in rats [88, 99]. *In vivo* interaction of flavonoids with dual substrates of CYP3A4 and P-gp is summarized in **Table 2**.

Flavonoids (oral dose)	Dual substrate of CYP3A and P-gp (oral dose)	Aimals/ species	Effect on bioavailability parameters			References
			% change in C_{max}	% change in AUC	Change in RBA* (fold)	
Quercetin (2.5 and 7.5 mg/kg)	Tamoxifen (10 mg/kg)	Rat	↑23–35	↑ 34–60	↑1.35–1.61	[78]
Baicalein (0.5, 3, and 10 mg/kg)	Tamoxifen (10 mg/kg)	Rat	↑ 54.8–100	↑47.6–89.1	↑1.47–1.89	[86]
Silibinin (0.5, 2.5, and 10 mg/kg)	Tamoxifen (10 mg/kg)	Rat	↑45.2–78.6	↑40.2–71.3	↑1.40–1.72	[87]
Epigallocatechein (0.5, 3, and 10 mg/kg)	Tamoxifen (10 mg/kg)	Rat	↑57.1–89.7	↑48.4–77.0	↑1.48–1.77	[90]
Kaempeferol (2.5 and 10 mg/kg)	Tamoxifen (10 mg/kg)	Rat	↑48.9–47.7	↑39.8–47.7	↑1.40–1.48	[91]
Biochanin A (100 mg/kg)	Tamoxifen (10 mg/kg)	Rat	↓23.5	↓32.3	↓ 1.32	[99]
Quercetin (2–20 mg/kg, p.o.)	Paclitaxel (40 mg/kg)	Rat	—	—	↑ 1.76–3.29	[76]
Silibinin (0.5, 2.5, and 10 mg/kg)	Paclitaxel (40 mg/kg)	Rat	↑31.0–52.9	↑65.8–101.7	↑1.15–2.02	[87]
Naringin (1, 3, 10, and 20 mg/kg)	Paclitaxel (40 mg/kg)	Rat	—	—	↑1.35–1.69	[92]
Flavone (2–20 mg/kg)**	Paclitaxel** (40 mg/kg)	Rat	—	—	↑2.4–3.1	[75]
Genistein (3.3 and 10 mg/kg)	Paclitaxel (30 mg/kg)	Rat	—	—	↑1.26–1.55	[93]
Morin (3.3–10 mg/kg)	Paclitaxel (30 mg/kg)	Rat	↑ 70–90	↑30–70	↑1.32–1.68	[95]
Quercetin (0.6, 3, and 15 mg/kg)	Doxorubicin (50 mg/kg)	Rat	↑ 35.1–125.7	↑ 31.2–136	↑1.33–2.36	[83]
Myricetin(0.4, 2, and 10 mg/kg)	Doxorubicin (40 mg/kg)	Rat	↑45–105	↑51–117	—	[84]
Baicalein (0.3, 1.5, and 6 mg/kg)	Doxorubicin (40 mg/kg)	Rat	—	—	↑1.20–1.96	[95]
Morin (1.5, 7.5, and 15 mg/kg)	Diltiazem (15 mg/kg)	Rat	↑ 30–120	—	1.38–1.80	[96]
Quercetin (2, 10, and 20 mg/kg)	Diltiazem (15 mg/kg)	Rabbit	—	—	1.75–2.76	[97]
Naringin (5 and 15 mg/kg)	Diltiazem (15 mg/kg)	Rat	—	—	↑2.07–2.20	[98]
Quercetin	Cyclosporine	Rat	↓67.8	↓43.3	—	[88]
Rutin	Cyclosporine	Rat	↓63.2	↓57.2	—	[88]
Baicalein (0.4, 2, and 8 mg/kg)	Nimodipine (12 mg/kg)	Rat	—	—	↑1.39–1.58	[85]

**RBA, relative bioavailability.*
***Flavone, 2-phenyl-4H-1-benzopyran-4-one.*

Table 2.
In vivo *interaction of some common flavonoids with dual substrates of CYP3A and P-gp.*

8. Significance of flavonoids as P-gp and CYP3A inhibitors

Various experimental and clinical studies confirmed that flavonoids produce antioxidant, anti-inflammatory, and anticarcinogenic effects. Studies also confirmed that these effects were attributed due to their inhibitions of efflux transporter enzyme (P-gp) and/or drug metabolizing enzyme (CYP3A). Therefore, flavonoids as P-gp inhibitor may use with other chemotherapeutic drugs for cancer treatment [100]. Occurrence of P-gp protein in various body tissues affects the absorption, distribution, metabolism, and excretion of drugs. Therefore, the dual effect of anticarcinogenic and P-gp modulation may synergistically act for the treatment of cancer [101]. The chemotherapeutic treatment of metastatic brain tumors is limited due to its low distribution in brain tissue by blood brain barrier and blood-cerebrospinal fluid barrier. P-gp is presented in the apical membranes of these cells, and flavonoids can improve the permeation of chemotherapeutic drugs by inhibiting the P-gp-mediated efflux [102]. Flavonoids can be used as nontoxic P-gp and/or CYP3A inhibitors and by its coadministration could improve the bioavailability of poorly unavailable drugs, especially for anticancer drugs, by interfering its clearance or inhibiting its metabolism [103]. P-gp presents in bile canaliculi and kidney suggested that it can also play a role in biliary and renal elimination of drugs. Coadministration of flavonoids (as P-gp inhibitors) can reduce the clearance of anticancer drugs, for example, vinblastine, doxorubicin, and irinotecan [104, 105]. Flavonoids can also play an important role in reversal of MDR in cancer chemotherapy. P-gp-associated MDR is a serious concern for limitation of cancer treatment. P-gp occurrence in tumor cell has been extensively characterized, and its overexpression has been confirmed during relapse. Therefore, it can be concluding that P-gp inhibitors (flavonoids) can potentially reverse the MDR during cancer chemotherapy [106].

9. Tyrosine kinase inhibitors

The evidence of protein tyrosine kinase enzyme involvement in tumor development makes it novel targets for selective chemotherapy and thus target for rational design of drug development. Now protein kinases, especially tyrosine kinases, are being used as main targets for drug development related to malignancy, resulting in the high approval rate of various TKIs by the FDA [24, 107]. Imatinib was the first of its kind, which was introduced clinically, followed by various molecules such as gefitinib, erlotinib, sorafenib, afatinib, nilotinib, bosutinib, crizotinib, ponatinib, lapatinib, sunitinib, and dasatinib, and many more are in pipeline [108]. Although mechanism of action of these compounds is same, that is, competitive ATP inhibition at the catalytic binding site of tyrosine kinases, they differ from each other in the spectrum of targeted kinase activity, pharmacokinetic profile, and compound-specific adverse effects [109]. These TKIs have been developed in oral formulations, are administered on a daily basis, and usually prescribed at a fixed dose. Although oral administration may be convenient for patients as it can reduce health care costs, improve quality of life of patients, and avoid heavy burden of day-stay infusion units, this practice also displays a disadvantage, in that the oral bioavailability of most of TKIs is highly dependent on their absorption through gastrointestinal tract and first-pass hepatic metabolism [25, 107].

Almost all TKIs are rapidly absorbed, and their maximum plasma concentration (C_{max}) was achieved in 3–6 h after oral administration except sunitinib (6–12 h). Food intake has no significant effect on the absorption of imatinib, dasatinib, gefitinib, sorafenib, or sunitinib. However, the bioavailability of lapatinib and

TKIS	CYP 3A4	P-gp	References
Imatinib	√	√	[108, 110, 111]
Gefitinib	√	√	[24, 108, 115, 118]
Erlotinib	√	√	[24, 118]
Sorafenib	√	X	[24, 108, 118]
Dasatinib	√	√	[24, 108, 118]
Sunitinib	√	√	[24, 108, 118]
Lapatinib	√	√	[24, 108, 118]
Nilotinib	√	√	[24, 108]
Crizotinib	√	√	[108, 111, 118]
Vandetanib	√	X	[108, 111]
Vemurafenib	X	√	[108, 112, 118]
Axitinib	√	√	[108, 111, 118]
Bosutinib	√	X	[108, 111, 118]
Pazopanib	√	√	[108, 111, 118]
Ponatinib	√	√	[108, 111, 118]
Dabrafenib	√	√	[108, 111, 118]
Cediranib	√	√	[114, 118]
Tandutinib	NA	√	[118]
Ibrutinib	√	X	[116, 121]
Afatinib	X	√	[108, 111]
Cabozantinib	√	X	[108, 111]
Regorafenib	√	X	[108, 111]
Ruxolitinib	√	X	[108, 117]
Osimertinib	√	√	[120]

Table 3.
Substrate potential of TKIs with CYP3A4 and/or P-gp.

nilotinib is increased pronouncedly with food intake. Almost all TKIs are high to plasma protein (>90%) and therefore extensively distribute into tissues resulting in large volume of distribution and prolong terminal half-life. Excretion of TKIs is predominantly through feces, and only a small fraction is eliminated with urine. Almost all TKIs are dual substrates of CYP3A4 (the most abundant CYP in the human liver and intestine) and P-gp efflux transporter, except sorafenib, vandetanib, bosutinib, ibrutinib, cabozantinib, regorafenib, and ruxolitinib, which are only substrate of CYP3A4, whereas vemurafenib and afatinib, which are effluxed by P-gp only [24, 108, 110–121]. **Table 3** summarizes the substrate potential of TKIs with CYP3A4 and/or P-gp.

10. Drug-drug interaction with TKI

DDIs represent a serious concern, especially for agents that influenced by efflux transporters and CYP3A4 enzyme, and can produce clinically relevant drug interactions by alteration of its bioavailability. Because the majority of TKIs are substrate of CYP3A4 and/or P-gp, DDI with CYP3A4 and/or P-gp inhibitors and inducers must be taken into account, and they must be used with caution, as advised in the package insert. Recently, it has been reported that coprescription of those medicines, which may induce or inhibit the metabolic pathways used by TKIs, is very high. Overall coprescribing rates of drugs that induce metabolism of TKI and may lead to decrease the effectiveness of TKIs ranged from 23 to 57%, whereas coprescribing rates of drugs that inhibit metabolism of TKIs and may increase its toxicity ranged from 24 to 74% [122]. For example, coadministration of imatinib with dual inhibitors CYP3A4 and P-gp increases not only the plasma concentration but also the intracellular concentration of imatinib. Dual inhibitors CYP3A4 and P-gp, such as verapamil fluconazole, itraconazole [123], erythromycin, clarithromycin [124], cyclosporine [125], and ketoconazole [126], increased the intracellular concentration of imatinib by inhibiting both its CYP3A4-mediated metabolism and its efflux through P-gp, which might result in increasing its cellular toxicity. Moreover, P-gp-mediated efflux inhibition by proton pump inhibitors, such as pantoprazole, has reported to increase the brain concentration of imatinib [127]. In contrast, coadministration of CYP3A4 inducers, such as rifampicin or certain antiepileptics, may lead to a reduction in imatinib exposure of up to 74% [24, 110]. Similarly, inhibitors of both CYP3A4 and P-gp increase both the plasma and intracellular concentrations of dasatinib as well, which are also expected to occur for verapamil, erythromycin, clarithromycin, fluconazole, itraconazole [123], cyclosporine [125], and ketoconazole [123, 128]. Concomitant administration of the CYP3A4 inducer rifampicin leads to a reduction in dasatinib exposure of 80% [24, 129, 130]. The area under curve (AUC) of nilotinib is increased 3-fold in healthy subjects receiving ketoconazole [130], whereas coadministration of CYP3A4 inducers, such as rifampicin, leads to a 4.8-fold reduction in nilotinib exposure [24, 129, 130]. Administration of gefitinib in the presence of rifampicin reduces the AUC of gefitinib by 83%, while in the presence of itraconazole, the AUC of gefitinib is increased by 78% [24, 131]. Furthermore, coadministration of ketoconazole results in a 3.6-fold increase in lapatinib plasma exposure, whereas coadministration of carbamazepine results in a decrease in the AUC of lapatinib by 72% [24, 132]. Although the result of above studies confirmed the risk of frequent DDIs among TKIs, but did not addressed the clinical consequences of these, that is, increased toxicity or therapeutic failure. Moreover, in some cases, in spite of knowing, these potential interacting combinations could have been intentionally prescribed by physicians because they considered the potential benefits to outweigh the risks or because the patient had the ability to tolerate these combinations in the past [133]. Since most of these TKIs are relatively new, their scientific evidence that supports their DDIs is limited. Therefore, it is not unexpected to observe that medical oncologists are not able to report TKI DDI pairs, which might have a high probability of causing deleterious effects in the treatment of cancer patients [134].

11. Flavonoids and TKI interaction

Compounds that are capable of inhibiting the activity of tyrosine kinase receptors (RTKs) are expected to display antiproliferative properties. Various *in vitro* and *in vivo* studies suggested that most of the flavonoids quercetin, genistein, hesperidin, and naringenin have TKI properties, which play a significant role in its anticancer effects [135–138]. Due to antiproliferative properties, flavonoids can be use along with conventional TKIs in clinical practice and therefore it definitely raise concern of pharmacokinetic interaction. So, it is understood that similar to conventional medicines, flavonoids act as dual modulators of CYP3A4 and P-gp, which may produce significant effects on the disposition kinetics of TKIs. In spite of that, limited data are available relevant to interaction of flavonoids with TKIs in the literature. St. John's wort that contains numerous flavonoids was found to increase imatinib clearance by 43% and decrease its AUC and C_{max} by 30 and 29%, respectively [139, 140] in human subjects. Similarly, St. John's wort may also decrease the plasma concentration of dasatinib, and its use is discouraged in patients receiving it [141]. Genistein (isoflavone) when administered for 15 days significantly decreased the C_{max} and AUC of imatinib, whereas its single dose did not produce any effects in rats [142]. Silybin, a constituent of silymarin, also decreased the AUC of imatinib after multiple dose administration (15 days) in rats [143]. However, apigenin in single dose increased the AUC of imatinib but decreased it in multiple dose administration (15 days) in rats [144], which suggests that apigenin may act as enzyme inhibitor in single dose and become inducer after long-term administration. Epigallocatechin-3-gallate also decreased the C_{max} and AUC of sunitinib after single dose administration in rats [145]. However, Bas 100, a novel mechanism-based CYP3A4 inhibitor, isolated from grapefruit juice increased 2.1-fold AUC of erlotinib after single dose administration [146]. Similar

Flavonoids/constituents	TKI inhibitors	Model	Pharmacokinetic effects	References
St. John's wort (300 mg, p.o)	Imatinib (400 mg, p.o.)	Healthy volunteers	↓AUC (30%) ↑ Oral clearance (43%)	[139]
St. John's wort (300 mg, p.o)	Imatinib (400 mg, p.o.)	Healthy volunteers	↓AUC (30%) ↓ C_{max} (29%)	[140]
Genistein (50, 100 mg/kg, p.o.) for 15 days	Imatinib (30 mg/kg, p.o.)	Rat	↓AUC (27–28%) ↓ C_{max} (23–25%)	[142]
Genistein (50, 100 mg/kg, p.o.) single dose	Imatinib (30 mg/kg, p.o.)	Rat	No effects	[142]
Apigenin (165 mg/kg, p.o) for 15 days	Imatinib (30 mg/kg, p.o.)	Rat	↓AUC (25%)	[144]
Apigenin (165 and 252 mg/kg, p.o) single dose	Imatinib (30 mg/kg, p.o.)	Rat	↑ AUC (25–40%)	[144]
Silybin (50 mg/kg, single and multiple for 15 days, p.o.	Imatinib (p.o.)	Rat	↓ AUC significantly	[143]
BAS 100 (10 mg /kg, p.o.)	Erlotinib 10 mg /kg, p.o.)	Mice	↑ AUC (2.1 fold)	[146]
Epigallocatechin-3-gallate (100 mg /kg, p.o.)	Sunitinib (30 mg/kg, p.o.)	Rats	↓C_{max} (47.7%) ↓AUC (51.5%)	[145]
Grapefruit juice (240 mL), p.o	Nilotinib (400 mg, p.o.)	Healthy volunteers	↑ AUC (29%) ↑ C_{max} (60%)	[147]

Table 4.
In vivo *pharmacokinetic interaction of flavonoids or flavonoids containing herbal constituents with TKIs.*

effects were produced by the intake of 240 ml of grapefruit juice (an inhibitor of CYP3A4 and P-gp), which are shown to increase the nilotinib AUC by 60%, and thus, coadministration of nilotinib with grapefruit juice is not recommended [147]. Additionally, a case study showed that patients who had already developed resistance to gefitinib treatment become responsive after the withdrawal of all CAMs [148]. **Table 4** summarizes the pharmacokinetic interaction of flavonoids or flavonoids containing herbal constituents with TKIs in both experimental animals and clinical studies.

12. Conclusions

The CYP3A enzyme accounts approximately 30% of hepatic and more than 70% of intestinal CYP activity. P-gp efflux protein is encoded by ABCB1 gene (MDR1) and is responsible for cellular efflux of numerous drugs. PXR regulates both CYP3A and P-gp and coordinately mediates detoxification of many xenobiotics during oral absorptions. The flavonoid constituent-mediated modulation of CYP3A and P-gp is the main mechanism through which the bioavailability and disposition of conventional drugs are regulated. The synergistically modulation of CYP3A and P-gp by flavonoids may increase the potency of chemotherapeutic drugs. Moreover, it increases the permeability of drugs in brain and reduces the MDR in cancer chemotherapy. Most of the TKIs are oral dosage formulations and also dual substrates of CYP3A and P-gp, so the risk of pharmacokinetic interactions with flavonoids is expected on concurrent administration. Therefore, there is a compelling need to study the cellular and molecular mechanisms involved in the flavonoid-mediated modulation of CYP3A and P-gp and their expected impact on the exposure and disposition of TKIs. Results from some experimental and clinical studies have confirmed the interaction between TKIs and flavonoids, but these unwanted clinical consequences in cancer patients have not been elaborated. However, even small changes in drug metabolism and pharmacokinetics of TKIs may lead to therapeutic failure or toxicity in cancer patients. Nevertheless, it is a challenging task to determine the clinical effects of the DDI due to the large interpatient variability in the pharmacokinetics of the TKIs.

Conflict of interest

The author declares that there is no conflict of interest.

Author details

Muzaffar Iqbal[1,2]

1 Department of Pharmaceutical Chemistry, College of Pharmacy, King Saud University, Riyadh, Saudi Arabia

2 Bioavailability Unit, Central Laboratory, College of Pharmacy, King Saud University, Riyadh, Saudi Arabia

*Address all correspondence to: muziqbal@gmail.com; muziqbal@ksu.edu.sa

References

[1] Bhattaram VA, Graefe U, Kohlert C, Veit M, Derendorf H. Pharmacokinetics and bioavailability of herbal medicinal products. Phytomedicine. 2002;**9** (Suppl 3):1-33

[2] Mukherjee PK, Wahile A. Integrated approaches towards drug development from Ayurveda and other Indian system of medicines. Journal of Ethnopharmacology. 2006;**103**:25-35

[3] Robinson MM, Zhang X. The world medicines situation. In: Traditional Medicines: Global Situation, Issues and Challenges. Geneva: World Health Organization; 2011. pp. 1-12

[4] Citarasu T. Herbal biomedicines: A new opportunity for aquaculture industry. Acquaculture International. 2010;**18**:403-414

[5] David B, Wolfender J, Dias DA. The pharmaceutical industry and natural products: Historical status and new trends. Phytochemistry Reviews. 2015;**14**:299-315

[6] Ghani A. Herbal Medicines: Present Status, Future Prospects. October 17, 2013. Available from: http://www.pharmabiz.com/NewsDetails.aspx?aid=78355&sid=21 [Accessed: 20 January 2017]

[7] Pal D, Mitra AK. MDR- and CYP3A4-mediated drug-herbal interactions. Life Sciences. 2006;**78**:2131-2145

[8] Li Y, Revalde J, Paxton JW. The effects of dietary and herbal phytochemicals on drug transporters. Advanced Drug Delivery Reviews. 2017;**116**:45-62. DOI: 10.1016/j.addr.2016.09.004

[9] Strandell J, Neil A, Carlin G. An approach to the in vitro evaluation of potential for cytochrome P450 enzyme inhibition from herbals and other natural remedies. Phytomedicine. 2004;**11**:98-104

[10] Evans AM. Influence of dietary components on the gastrointestinal metabolism and transport of drugs. Therapeutic Drug Monitoring. 2000;**22**:131-136

[11] Ioannides C. Pharmacokinetic interactions between herbal remedies and medicinal drug. Xenobiotica. 2002;**32**:451-478

[12] Wilkinson GR. The effects of diet, aging and disease-states on presystemic elimination and oral drug bioavailability in human. Advanced Drug Delivery Reviews. 1997;**27**:129-159

[13] Fugh-Berman A. Herb–drug interactions. Lancet. 2000;**355**:134-138

[14] Fugh-Berman A, Ernst E. Herb–drug interactions: Review and assessment of report reliability. British Journal of Clinical Pharmacology. 2001;**52**:587-595

[15] Cushnie TPT, Lamb AJ. Recent advances in understanding the antibacterial properties of flavonoids. International Journal of Antimicrobial Agents. 2011;**38**:99-107

[16] García-Lafuente A, Guillamón E, Villares A, Rostagno MA, Martínez JA. Flavonoidsas anti-inflammatory agents: Implications in cancer and cardiovascular disease. Inflammation Research. 2009;**58**:537-552

[17] Brantley SJ, Argikar AA, Lin YS, Nagar S, Paine MF. Herb-drug interactions: Challenges and opportunitiesfor improved predictions. Drug Metabolism and Disposition. 2014;**42**:301-317

[18] Sahu K, Siddiqui AA, Shaharyar M, Malik S. Pharmacokinetic interaction between febuxostat and morin in rats. Expert Opinion on Drug Metabolism & Toxicology. 2014;**10**:307-312

[19] Kumar KK, Priyanka L, Gnananath K, Babu PR, Sujatha S. Pharmacokinetic drug interactions between apigenin, rutin and paclitaxel mediated by P-glycoprotein in rats. European Journal of Drug Metabolism and Pharmacokinetics. 2015;**40**:267-276

[20] Ferreira A, Pousinho S, Fortuna A, Falcao A, Alves G, et al. Phytochemistry Reviews. 2014;**14**:233-272

[21] Gufford BT, Chen G, Lazarus PTN, Graf NH, Paine MF. Identification of diet-derived constituents as potent inhibitors of intestinal glucuronidation. Drug Metabolism and Disposition. 2014;**42**:1675-1683

[22] Neul C, Schaeffeler E, Sparreboom A, Laufer S, Schwab M, Nies AT. Impact of membrane drug transporters on resistance to small-molecule tyrosine kinase inhibitors. Trends in Pharmacological Sciences. 2016;**37**:904-932

[23] Shao J, Markowitz JS, Bei D, An G. Enzyme- and transporter-mediated drug interactions with small molecule tyrosine kinase inhibitors. Journal of Pharmaceutical Sciences. 2014;**103**:3810-3833

[24] van Erp NP, Gelderblom H, Guchelaar HJ. Clinical pharmacokinetics of tyrosine kinase inhibitors. Cancer Treatment Reviews. 2009;**35**:692-706

[25] Pajares B, Torres E, Trigo JM, Sáez MI, Ribelles N, Jiménez B, et al. Tyrosine kinase inhibitors and drug interactions: A review with practical recommendations. Clinical & Translational Oncology. 2012;**14**:94-101

[26] Kim SK, Ovak RF. The role of intracellular signaling in insulinmediated regulation of drug metabolizing enzyme gene and protein expression. Pharmacology & Therapeutics. 2007;**113**:88-120

[27] Nebert DW, Russell DW. Clinical importance of the cytochromes P450. The Lancet. 2002;**360**:1155-1162

[28] WatkinsPB,WrightonSA,SchuetzEG, Molowa DT, Guzelian PS. Identification of glucocorticoid-inducible cytochromes P-450 in the intestinal mucosa of rats and man. The Journal of Clinical Investigation. 1987;**80**:1029-1036

[29] Kolars JC, Schmiedlin-Ren P, Schuetz JD, Fang C, Watkins PB. Identification of rifampin-inducible P450IIIA4 (CYP3A4) in human small bowel enterocytes. The Journal of Clinical Investigation. 1992;**90**:1871-1878

[30] Shimada T, Yamazaki H, Mimura M, Inui Y, Guengerich FP. Inter individual variations in human liver cytochrome P-450 enzymes involved in the oxidation of drugs, carcinogens and toxic chemicals studies with liver microsomes of 30 Japanese and 30 Caucasians. The Journal of Pharmacology and Experimental Therapeutics. 1994;**270**:414-423

[31] de Waziers I, Cugnenc PH, Yang CS, Leroux JP, Beaune PH. Cytochrome P 450 isoenzymes, epoxide hydrolase and glutathione transferases in rat and human hepatic and extrahepatic tissues. The Journal of Pharmacology and Experimental Therapeutics. 1990;**253**:387-394

[32] Centre for Drug Evaluation and Research (CDER). Draft Guideline for In Vitro Metabolism- and Transporter Mediated Drug-Drug Interaction. 2017. Available from: https://www.fda.gov/media/108130/download

[33] Ito K, Iwatsubo T, Kanamitsu S, Ueda K, Suzuki H, Sugiyama Y. Prediction of pharmacokinetic alterations caused by drug-drug interactions: Metabolic interaction in the liver. Pharmacological Reviews. 1998;**50**(3):387-412

[34] Venkatakrishnan K, von Moltke LL, Obach RS, Greenblatt DJ. Drug metabolism and drug interactions: Application and clinical value of in vitro models. Current Drug Metabolism. 2003;**4**(5):423-459. DOI: 10.2174/1389200033489361

[35] Loo TW, Clarke DM. Recent progress in understanding the mechanism of P-glycoprotein-mediated drug efflux. The Journal of Membrane Biology. 2005;**206**:173-185

[36] Gottesman MM, Fojo T, Bates SE. Multidrug resistance in cancer: Role of ATP-dependent transporters. Nature Reviews. Cancer. 2002;**2**(1):48-58

[37] Morris ME, Zhang S. Flavonoid-drug interactions: Effects of flavonoids on ABC transporters. Life Sciences. 2006;**78**:2116-2130

[38] Cordon-Cardo C, O'Brien JP, Casals D, Rittman-Grauer L, Biedler JL, Melamed MR, et al. Multidrug-resistance gene (P-glycoprotein) is expressed by endothelial cells at blood-brain barrier sites. Proceedings of the National Academy of Sciences of the United States of America. 1989;**86**(2):695-698

[39] Thiebaut F, Tsuruo T, Hamada H, Gottesman MM, Pastan I, Willingham MC. Cellular localization of the multidrug-resistance gene product P- glycoprotein in normal human tissues. Proceedings of the National Academy of Sciences of the United States of America. 1987;**84**:7735-7738

[40] Thiebaut F, Tsuruo T, Hamada H, Gottesman MM, Pastan I, Willingham MC. Immunohistochemical localization in normal tissues of different epitopes in the multidrug transport protein P170: Evidence for localization in brain capillaries and crossreactivity of one antibody with a muscle protein. The Journal of Histochemistry and Cytochemistry. 1989;**37**:159-164

[41] Gottesman MM, Pastan I. Biochemistry of multidrug resistance mediated by the multidrug transporter. Annual Review of Biochemistry. 1993;**62**:385-427

[42] Schinkel AH, Wagenaar E, van Deemter L, Mol CA, Borst P. Absence of the mdr1a P-Glycoprotein in mice affects tissue distribution andpharmacokinetics of dexamethasone, digoxin, and cyclosporin A. The Journal of Clinical Investigation. 1995;**96**:1698-1705

[43] Lee CG, Gottesman MM, Cardarelli CO, Ramachandra M, Jeang KT, Ambudkar SV, et al. HIV-1 protease inhibitors are substrates for the MDR1 multidrug transporter. Biochemistry. 1998;**37**:3594-3601

[44] Meador J, Sweet P, Stupecky M, Wetzel M, Murray S, Gupta S, et al. Enhancement by cyclosporin A of daunorubicin efficacy in Ehrlich ascites carcinoma and murine hepatoma. Cancer Research. 1987;**47**:6216-6219

[45] Horio M, Chin KV, Currier SJ, Goldenberg S, Williams C, Pastan I, et al. Transepithelial transport of drugs by the multidrug transporter in cultured Madin–Darby canine kidney cell epithelia. The Journal of Biological Chemistry. 1989;**264**:14880-14884

[46] Christians U, Schmitz V, Haschke M. Functional interactions between P-glycoprotein and CYP3A in drug metabolism. Expert Opinion on Drug Metabolism & Toxicology. 2005;**1**(4):641-654

[47] van Waterschoot RA, Lagas JS, Wagenaar E, van der Kruijssen CM, van Herwaarden AE, Song JY, et al. Absence of both cytochrome P450 3A and P-glycoprotein dramatically increases docetaxel oral bioavailability and risk of intestinal toxicity. Cancer Research. 2009;**69**:8996-9002

[48] Wacher VJ, Wu CY, Benet LZ. Overlapping substrate specificities and tissue distribution of cytochrome P450 3A and P-glycoprotein: Implications for drug delivery and activity in cancer chemotherapy. Molecular Carcinogenesis. 1995;**13**:129-134

[49] van Waterschoot RA, Schinkel AH. A critical analysis of the interplay between cytochrome P450 3A and P-glycoprotein: Recent insights from knockout and transgenic mice. Pharmacological Reviews. 2011;**63**:390-410

[50] Benet LZ. The drug transporter-metabolism alliance: Uncovering and defining the interplay. Molecular Pharmaceutics. 2009;**6**:1631-1643

[51] Bhardwaj RK, Glaeser H, Becquemont L, Klotz U, Gupta SK, Fromm MF. Piperine, a major constituent of black pepper, inhibits human P-glycoprotein and CYP3A4. The Journal of Pharmacology and Experimental Therapeutics. 2002;**302**(2):645-650

[52] Lutz JD, Kirby BJ, Wang L, Song Q, Ling J, Massetto B, et al. Cytochrome P450 3A induction predicts P-glycoprotein induction. Part 1: Establishing induction relationships using ascending dose rifampin. Clinical Pharmacology and Therapeutics. 2018;**104**(6):1182-1190

[53] Rosenfeld JM, Vargas R Jr, Xie W, Evans RM. Genetic profiling defines the xenobiotic gene network controlled by the nuclear receptor pregnane X receptor. Molecular Endocrinology. 2003;**17**(7):1268-1282

[54] Maglich JM, Stoltz CM, Goodwin B, Hawkins-Brown D, Moore JT, Kliewer SA. Nuclear pregnane x receptor and constitutive androstane receptor regulate overlapping but distinct sets of genes involved in xenobiotic detoxification. Molecular Pharmacology. 2002;**62**(3):638-646

[55] Synold TW, Dussault I, Forman BM. The orphan nuclear receptor SXR coordinately regulates drug metabolism and efflux. Nature Medicine. 2001;**7**(5):584-590

[56] Riley RJ, Kenna JG. Cellular models for ADMET predictions and evaluation of drug-drug interactions. Current Opinion in Drug Discovery & Development. 2004;**7**(1):86-99

[57] Fowler S, Zhang H. In vitro evaluation of reversible and irreversible cytochrome P450 inhibition: Current status on methodologies and their utility for predicting drug-drug interactions. The AAPS Journal. 2008;**10**(2):410-424. DOI: 10.1208/s12248-008-9042-7

[58] Muenster U, Grieshop B, Ickenroth K, Gnoth MJ. Characterization of substrates and inhibitors for the in vitro assessment of Bcrp mediated drug-drug interactions. Pharmaceutical Research. 2008;**25**(10):2320-2326. DOI: 10.1007/s11095-008-9632-1

[59] Hochman JH, Yamazaki M, Ohe T, Lin JH. Evaluation of drug interactions with P-glycoprotein in drug discovery: In vitro assessment of the potential for drug-drug interactions with P-glycoprotein. Current Drug Metabolism. 2002;**3**(3):257-273. DOI: 10.2174/1389200023337559

[60] Marathe PH, Rodrigues AD. Multi-drug resistance in cancer. In: Zhou J, editor. Methods in Molecular Biology. Vol. 56. Humana Press; 2010

[61] Granvil CP, Yu AM, Elizondo G, et al. Expression of the human CYP3A4 gene in the small intestine of transgenic mice: In vitro metabolism and pharmacokinetics of midazolam. Drug Metabolism and Disposition. 2003;**31**(5):548-558. DOI: 10.1124/dmd.31.5.548

[62] Ogasawara A, Kume T, Kazama E. Effect of oral ketoconazole on intestinal first-pass effect of midazolam and fexofenadine in cynomolgus monkeys. Drug Metabolism and Disposition. 2007;**35**(3):410-418. DOI: 10.1124/dmd.106.011288

[63] Sutton D, Butler AM, Nadin L, Murray M. Role of CYP3A4 in human hepatic diltiazem N-demethylation: Inhibition of CYP3A4 activity by oxidized diltiazem metabolites. The Journal of Pharmacology and Experimental Therapeutics. 1997;**282**(1):294-300

[64] Yamano K, Yamamoto K, Kotaki H, Sawada Y, Iga T. Quantitative prediction of metabolic inhibition of midazolam by itraconazole and ketoconazole in rats: Implication of concentrative uptake of inhibitors into liver. Drug Metabolism and Disposition. 1999;**27**(3):395-402

[65] Schinkel AH, Mol CA, Wagenaar E, van Deemter L, Smit JJ, Borst P. Multidrug resistance and the role of P-glycoprotein knockout mice. European Journal of Cancer. 1995;**31A**(7-8):1295-1298. DOI: 10.1016/0959-8049(95)00130-b

[66] Kemper EM, Verheij M, Boogerd W, Beijnen JH, van Tellingen O. Improved penetration of docetaxel into the brain by co-administration of inhibitors of P-glycoprotein. European Journal of Cancer. 2004;**40**(8):1269-1274. DOI: 10.1016/j.ejca.2004.01.024

[67] Yumoto R, Murakami T, Sanemasa M, Nasu R, Nagai J, Takano M. Pharmacokinetic interaction of cytochrome P450 3A-related compounds with rhodamine 123, a P-glycoprotein substrate, in rats pretreated with dexamethasone. Drug Metabolism and Disposition. 2001;**29**(2):145-151

[68] Ward KW, Azzarano LM. Preclinical pharmacokinetic properties of the P-glycoprotein inhibitor GF120918A (HCl salt of GF120918, 9,10-dihydro-5-methoxy-9-oxo-N-[4-[2-(1,2,3,4-tetrahydro-6,7-dimethoxy-2-isoquinolinyl)ethyl] phenyl]-4-acridine-carboxamide) in the mouse, rat, dog, and monkey. The Journal of Pharmacology and Experimental Therapeutics. 2004;**310**(2):703-709. DOI: 10.1124/jpet.104.068288

[69] Havsteen BH. The biochemistry and medical significance of the flavonoids. Pharmacology & Therapeutics. 2002;**96**:67-202

[70] Bailey DG, Arnold JM, Munoz C, Spence JD. Grapefruit juice--felodipine interaction: Mechanism, predictability, and effect of naringin. Clinical Pharmacology and Therapeutics. 1993;**53**:637-642

[71] Ducharme MP, Warbasse LH, Edwards DJ. Disposition of intravenous and oral cyclosporine after administration with grapefruit juice. Clinical Pharmacology and Therapeutics. 1995;**57**:485-491

[72] Dupuy J, Lespine A, Sutra JF, Alvinerie M. Fumagillin a new P-glycoprotein-interfering agent able to modulate moxidectin efflux in rat hepatocytes. Journal of Veterinary Pharmacology and Therapeutics. 2006;**29**:489-494

[73] Zhang H, Wong CW, Coville PF, Wanwimolruk S. Effect of the grapefruit flavonoid naringin on pharmacokinetics of quinine in rats. Drug Metabolism and Drug Interactions. 2000;**17**:351-363

[74] Lai MY, Hsiu SL, Hou YC, Tsai SY, Chao PD. Significant decrease of cyclosporine bioavailability in rats caused by a decoction of the roots of *Scutellaria baicalensis*. Planta Medica. 2004;**70**:132-137

[75] Choi JS, Choi HK, Shin SC. Enhanced bioavailability of paclitaxel after oral

coadministration with flavone in rats. International Journal of Pharmaceutics. 2004a;**275**:165-170

[76] Choi JS, Jo BW, Kim YC. Enhanced paclitaxel bioavailability after oral administration of paclitaxel or prodrug to rats pretreated with quercetin. European Journal of Pharmaceutics and Biopharmaceutics. 2004b;**57**:313-318

[77] Wang YH, Chao PD, Hsiu SL, Wen KC, Hou YC. Lethal quercetin-digoxin interaction in pigs. Life Sciences. 2004;**74**:1191-1197

[78] Shin SC, Choi JS, Li X. Enhanced bioavailability of tamoxifen after oral administration of tamoxifen with quercetin in rats. International Journal of Pharmaceutics. 2006;**313**:144-149

[79] Chieli E, Romiti N, Cervelli F, Tongiani R. Effects of flavonols on P-glycoprotein activity in cultured rat hepatocytes. Life Sciences. 1995;**57**:1741-1751

[80] Shapiro AB, Ling V. Effect of quercetin on Hoechst 33342 transport by purified and reconstituted P-glycoprotein. Biochemical Pharmacology. 1997;**53**:587-596

[81] Patel J, Buddha B, Dey S, Pal D, Mitra AK. In vitro interaction of the HIV protease inhibitor ritonavir with herbal constituents: Changes in P-gp and CYP3A4 activity. American Journal of Therapeutics. 2004;**11**:262-277

[82] Dürr D, Stieger B, Kullak-Ublick GA, Rentsch KM, Steinert HC, Meier PJ, et al. St John's Wort induces intestinal P-glycoprotein/MDR1 and intestinal and hepatic CYP3A4. Clinical Pharmacology and Therapeutics. 2000;**68**:598-604

[83] Choi JS, Piao YJ, Kang KW. Effects of quercetin on the bioavailability of doxorubicin in rats: Role of CYP3A4 and P-gp inhibition by quercetin. Archives of Pharmacal Research. 2011a;**34**:607-613

[84] Choi JS, Shin SC, Choi JS. Effects of myricetin on the bioavailability of doxorubicin for oral drug delivery in rats: Possible role of CYP3A4 andP-glycoprotein inhibition by myricetin. Archives of Pharmacal Research. 2011b;**34**:309-315

[85] Cho YA, Choi JS, Burm JP. Effects of the antioxidant baicalein on the pharmacokinetics of nimodipine in rats: A possible role of P-glycoprotein and CYP3A4 inhibition by baicalein. Pharmacological Reports. 2011;**63**:1066-1073

[86] Li C, Kim M, Choi H, Choi J. Effects of baicalein on the pharmacokinetics of tamoxifen and its main metabolite, 4-hydroxytamoxifen, in rats: Possible role of cytochrome P450 3A4 and P-glycoprotein inhibition by baicalein. Archives of Pharmacal Research. 2009;**64**:579-583

[87] Lee CK, Choi JS. Effects of silibinin, inhibitor of CYP3A4 and P-glycoprotein in vitro, on the pharmacokinetics of paclitaxel after oral and intravenous administration in rats. Pharmacology. 2010;**85**:350-356

[88] Yu CP, Wu PP, Hou YC, Lin SP, Tsai SY, Chen CT, et al. Quercetin and rutin reduced the bioavailability of cyclosporine from Neoral, an immunosuppressant, through activating P-glycoprotein and CYP 3A4. Journal of Agricultural and Food Chemistry. 2011;**59**:4644-4648

[89] Kim CS, Choi SJ, Park CY, Li C, Choi JS. Effects of silybinin on the pharmacokinetics of tamoxifen and its active metabolite, 4-hydroxytamoxifen in rats. Anticancer Research. 2010;**30**:79-85

[90] Shin SC, Choi JS. Effects of epigallocatechin gallate on the oral

bioavailability and pharmacokinetics of tamoxifen and its main metabolite, 4-hydroxytamoxifen, in rats. Anti-Cancer Drugs. 2009;**20**:584-588

[91] Piao Y, Shin SC, Choi JS. Effects of oral kaempferol on the pharmacokinetics of tamoxifen and one of its metabolites, 4-hydroxytamoxifen, after oral administration of tamoxifen to rats. Biopharmaceutics & Drug Disposition. 2008;**29**:245-249

[92] Choi JS, Shin SC. Enhanced paclitaxel bioavailability after oral coadministration of paclitaxel prodrug with naringin to rats. International Journal of Pharmaceutics. 2005;**292**:149-156

[93] Li X, Choi JS. Effect of genistein on the pharmacokinetics of paclitaxel administered orally or intravenously in rats. International Journal of Pharmaceutics. 2007;**337**:188-193

[94] Choi BC, Choi JS, Han HK. Altered pharmacokinetics of paclitaxel by the concomitant use of morin in rats. International Journal of Pharmaceutics. 2006;**323**:81-85

[95] Shin SC, Li C, Choi JS. Effects of baicalein, an antioxidant, on the bioavailability of doxorubicin in rats: Possible role of P-glycoprotein inhibition by baicalein. Die Pharmazie. 2009;**64**:579-583

[96] Choi JS, Han HK. Pharmacokinetic interaction between diltiazem and morin, a flavonoid, in rats. Pharmacological Research. 2005;**52**:386-391

[97] Choi JS, Li X. Enhanced diltiazem bioavailability after oral administration of diltiazem with quercetin to rabbits. International Journal of Pharmaceutics. 2005;**297**:1-8

[98] Choi JS, Han HK. Enhanced oral exposure of diltiazem by the concomitant use of naringin in rats. International Journal of Pharmaceutics. 2005;**305**:122-128

[99] Singh SP, Wahajuddin M, Raju KS, Ali MM, Kohli K, Jain GK. Reduced bioavailability of tamoxifen and its metabolite 4-hydroxytamoxifen after oral administration with biochanin A (an isoflavone) in rats. Phytotherapy Research. 2012;**26**:303-307

[100] Bansal T, Jaggi M, Khar RK, Talegaonkar S. Emerging significance of flavonoids as P-glycoprotein inhibitors in cancer chemotherapy. Journal of Pharmacy & Pharmaceutical Sciences. 2009;**12**(1):46-78. DOI: 10.18433/j3rc77

[101] Bansal T, Jaggi M, Khar RK, Talegaonkar S. Status of flavonols as P-glycoprotein inhibitors in cancer chemotherapy. Current Cancer Therapy Reviews. 2009;**5**(2):46-78. DOI: 10.2174/157339409788166742

[102] Dai H, Marbach P, Lemaire M, Hayes M, Elmquist WF. Distribution of STI-571 to the brain is limited by P-glycoprotein-mediated efflux. The Journal of Pharmacology and Experimental Therapeutics. 2003;**304**(3):1085-1092. DOI: 10.1124/jpet.102.045260\

[103] Marchetti S, Mazzanti R, Beijnen JH, Schellens JH. Concise review: Clinical relevance of drug drug and herb drug interactions mediated by the ABC transporter ABCB1 (MDR1, P-glycoprotein). The Oncologist. 2007;**12**(8):927-941. DOI: 10.1634/theoncologist.12-8-927

[104] Iyer L, Ramírez J, Shepard DR, et al. Biliary transport of irinotecan and metabolites in normal and P-glycoprotein-deficient mice. Cancer Chemotherapy and Pharmacology. 2002;**49**(4):336-341. DOI: 10.1007/s00280-001-0420-4

[105] Tobin PJ, Hong Y, Seale JP, Rivory LP, McLachlan AJ. Loperamide

inhibits the biliary excretion of irinotecan (CPT-11) in the rat isolated perfused liver. The Journal of Pharmacy and Pharmacology. 2005;**57**(1):39-45. DOI: 10.1211/0022357055100

[106] Goldstein LJ, Galski H, Fojo A, et al. Expression of a multidrug resistance gene in human cancers. Journal of the National Cancer Institute. 1989;**81**(2):116-124. DOI: 10.1093/jnci/81.2.116

[107] Klümpen HJ, Samer CF, Mathijssen RH, Schellens JH, Gurney H. Moving towards dose individualization of tyrosine kinase inhibitors. Cancer Treatment Reviews. 2011;**37**:251-260

[108] Herbrink M, Nuijen B, Schellens JH, Beijnen JH. Variability in bioavailability of small molecular tyrosine kinase inhibitors. Cancer Treatment Reviews. 2015;**41**:412-422

[109] Hartmann JT, Haap M, Kopp HG, Lipp HP. Tyrosine kinase inhibitors—A review on pharmacology, metabolism and side effects. Current Drug Metabolism. 2009;**10**:470-481

[110] Peng B, Lloyd P, Schran H. Clinical pharmacokinctics of imatinib. Clinical Pharmacokinetics. 2005;**44**:879-894

[111] Shukla S, Sauna ZE, Ambudkar SV. Evidence for the interaction of imatinib at the transport-substrate site(s) of the multidrug-resistance-linked ABC drug transporters ABCB1 (P-glycoprotein) and ABCG2. Leukemia. 2008;**22**:445-447

[112] MacLeod AK, McLaughlin LA, Henderson CJ, Wolf CR. Activation status of the pregnane X receptor influences vemurafenib availability in humanized mouse models. Cancer Research. 2015;**75**:4573-4581

[113] Suttle AB, Grossmann KF, Ouellet D, Richards-Peterson LE, Aktan G, Gordon MS, et al. Assessment of the drug interaction potential and single- and repeat-dose pharmacokinetics of the BRAF inhibitor dabrafenib. Journal of Clinical Pharmacology. 2015;**55**:392-400

[114] Lassen U, Miller WH, Hotte S, Evans TR, Kollmansberger C, Adamson D, et al. Phase I evaluation of the effects of ketoconazole and rifampicin on cediranib pharmacokinetics in patients with solid tumours. Cancer Chemotherapy and Pharmacology. 2013;**71**:543-549

[115] Teo YL, Ho HK, Chan A. Metabolism-related pharmacokinetic drug-drug interactions with tyrosine kinase inhibitors: Current understanding, challenges and recommendations. British Journal of Clinical Pharmacology. 2015;**79**:241-253

[116] Marostica E, Sukbuntherng J, Loury D, de Jong J, de Trixhe XW, Vermeulen A, et al. Population pharmacokinetic model of ibrutinib, a Bruton tyrosine kinase inhibitor, in patients with B cell malignancies. Cancer Chemotherapy and Pharmacology. 2015;**75**:111-121

[117] Shi JG, Chen X, Emm T, Scherle PA, McGee RF, Lo Y, et al. The effect of CYP3A4 inhibition or induction on the pharmacokinetics and pharmacodynamics of orally administered ruxolitinib (INCB018424 phosphate) in healthy volunteers. Journal of Clinical Pharmacology. 2012;**52**:809-818

[118] Deng J, Shao J, Markowitz JS, An G. ABC transporters in multi-drug resistance and ADME-Tox of small molecule tyrosine kinase inhibitors. Pharmaceutical Research. 2014;**31**:2237-2255

[119] Scheffler M, Di Gion P, Doroshyenko O, Wolf J, Fuhr U. Clinical pharmacokinetics of tyrosine kinase inhibitors: Focus on

4-anilinoquinazolines. Clinical Pharmacokinetics. 2011;**50**:371-403

[120] US Prescribing Information of TAGRISSO™ (Osimertinib) Tablets. 2015. Available from: http://www.accessdata.fda.gov/drugsatfda_docs/label/2015/208065s000lbl.pdf [Accessed: 18 January 2017]

[121] US Prescribing Information of IMBRUVICA® (Ibrutinib) Capsules, for Oral Use. 2016. Available from: https://www.janssenmd.com/pdf/imbruvica/PI-Imbruvica.pdf [Accessed: 18 January 2017]

[122] Bowlin SJ, Xia F, Wang W, Robinson KD, Stanek EJ. Twelve-month frequency of drug-metabolizing enzyme and transporter-based drug-drug interaction potential in patients receiving oral enzyme-targeted kinase inhibitor antineoplastic agents. Mayo Clinic Proceedings. 2013;**88**:139-148

[123] Haouala A, Widmer N, Duchosal MA, Montemurro M, Buclin T, Decosterd LA. Drug interactions with the tyrosine kinase inhibitors imatinib, dasatinib, and nilotinib. Blood. 2011;**117**:e75-e87

[124] Azuma M, Nishioka Y, Aono Y, Inayama M, Makino H, Kishi J, et al. Role of alpha1-acid glycoprotein in therapeutic antifibrotic effects of imatinib with macrolides in mice. American Journal of Respiratory and Critical Care Medicine. 2007;**176**:1243-1250

[125] Yokota A, Kimura S, Masuda S, et al. INNO-406, a novel BCR-ABL/Lyn dual tyrosine kinase inhibitor, suppresses the growth of Ph leukemia cells in the central nervous system, and cyclosporine A augments its in vivo activity. Blood. 2007;**109**:306-314

[126] Dutreix C, Peng B, Mehring G, et al. Pharmacokinetic interaction between ketoconazole and imatinib mesylate (Glivec) in healthy subjects. Cancer Chemotherapy and Pharmacology. 2004;**54**:290-294

[127] Breedveld P, Pluim D, Cipriani G, et al. The effect of Bcrp1 (Abcg2) on the in vivo pharmacokinetics and brain penetration of imatinib mesylate (Gleevec): Implications for the use of breast cancer resistance protein and P-glycoprotein inhibitors to enable the brain penetration of imatinib in patients. Cancer Research. 2005;**65**:2577-2582

[128] Johnson FM, Agrawal S, Burris H, et al. Phase 1 pharmacokinetic and drug-interaction study of dasatinib in patients with advanced solid tumors. Cancer. 2010;**116**:1582-1591

[129] Tanaka C, Yin OQ, Sethuraman V, et al. Clinical pharmacokinetics of the BCR-ABL tyrosine kinase inhibitor nilotinib. Clinical Pharmacology and Therapeutics. 2010;**87**:197-203

[130] Brave M, Goodman V, Kaminskas E, et al. Sprycel for chronic myeloid leukemia and Philadelphia chromosome-positive acute lymphoblastic leukemia resistant to or intolerant of imatinib mesylate. Clinical Cancer Research. 2008;**14**:352-359

[131] Swaisland HC, Ranson M, Smith RP, Leadbetter J, Laight A, McKillop D, et al. Pharmacokinetic drug interactions of gefitinib with rifampicin, itraconazole and metoprolol. Clinical Pharmacokinetics. 2005;**44**:1067-1081

[132] Smith DA, Koch KM, Arya N, Bowen CJ, Herendeen JM, Beelen A. Effects of ketoconazole and carbamazepine on lapatinib pharmacokinetics in healthy subjects. British Journal of Clinical Pharmacology. 2009;**67**:421-426

[133] Ko Y, Tan SLD, Chan A, Wong YP, Yong WP, Ng RCH, et al. Prevalence of the coprescription of clinically important interacting drug combinations involving oral anticancer

agents in Singapore: A retrospective database study. Clinical Therapeutics. 2012;**34**:1696-1704

[134] Chan A, Tan SH, Wong CM, Yap KYL, Ko Y. Clinically significant drug–drug interactions between oral anticancer agents and nonanticancer agents: A delphi survey of oncology pharmacists. Clinical Therapeutics. 2009;**31**:2379-2386

[135] Teillet F, Boumendjel A, Boutonnat J, Ronot X. Flavonoids as RTK inhibitors and potential anticancer agents. Medicinal Research Reviews. 2008;**28**(5):715-745

[136] Li C, Wang T, Zhang C, Xuan J, Su C, Wang Y. Quercetin attenuates cardiomyocyte apoptosis via inhibition of JNK and p38 mitogen-activated protein kinase signaling pathways. Gene. 2016;**577**:275-280

[137] Larsen CA, Dashwood RH, Bisson WH. Tea catechins as inhibitors of receptor tyrosine kinases: Mechanistic insights and human relevance. Pharmacological Research. 2010;**62**:457-464

[138] Chandrika BB, Steephan M, Kumar TR, Sabu A, Haridas M. Hesperetin and Naringenin sensitize HER2 positive cancer cells to death by serving as HER2 Tyrosine Kinase inhibitors. Life Sciences. 2016;**160**:47-56

[139] Frye RF, Fitzgerald SM, Lagattuta TF, Hruska MW, Egorin MJ. Effect of St John's wort on imatinib mesylate pharmacokinetics. Clinical Pharmacology and Therapeutics. 2004;**76**:323-329

[140] Smith PF, Bullock JM, Booker BM, Haas CE, Berenson CS, Jusko WJ. Induction of imatinib metabolism by hypericum perforatum. Blood. 2004;**104**:1229-1230

[141] Wong SF. New dosing schedules of dasatinib for CML and adverse event management. Journal of Hematology & Oncology. 2009;**2**:10

[142] Wang Z, Wang L, Xia MM, Sun W, Huang CK, Cui X, et al. Pharmacokinetics interaction between imatinib and genistein in rats. BioMed Research International. 2015;**2015**:368976

[143] Wang L, Wang Z, Xia MM, Wang YY, Wang HY, Hu GX. Inhibitory effect of silybin on pharmacokinetics of imatinib in vivo and in vitro. Canadian Journal of Physiology and Pharmacology. 2014;**92**:961-964

[144] Liu XY, Xu T, Li WS, Luo J, Geng PW, Wang L, et al. The effect of apigenin on pharmacokinetics of imatinib and its metabolite N-desmethyl imatinib in rats. BioMed Research International. 2013;**2013**:789184

[145] Ge J, Tan BX, Chen Y, Yang L, Peng XC, Li HZ, et al. Interaction of green tea polyphenol epigallocatechin-3-gallate with sunitinib: Potential risk of diminished sunitinib bioavailability. Journal of Molecular Medicine (Berlin, Germany). 2011;**89**:595-602

[146] Smith NF, Baker SD, Gonzalez FJ, Harris JW, Figg WD, Sparreboom A. Modulation of erlotinib pharmacokinetics in mice by a novel cytochrome P450 3A4 inhibitor, BAS 100. British Journal of Cancer. 2008;**98**:1630-1632

[147] Yin OQ, Gallagher N, Li A, Zhou W, Harrell R, Schran H. Effect of grapefruit juice on the pharmacokinetics of nilotinib in healthy participants. Journal of Clinical Pharmacology. 2010;**50**:188-194

[148] Hwang SW, Han HS, Lim KY, Han JY. Drug interaction between complementary herbal medicines and gefitinb. Journal of Thoracic Oncology. 2008;**3**:942-943

Chapter 9

Cardioprotective Effects of Cultivated Black Chokeberries (*Aronia* spp.): Traditional Uses, Phytochemistry and Therapeutic Effects

Valentina Buda, Minodora Andor, Antal Diana, Florina Ardelean, Ioana Zinuca Pavel, Cristina Dehelean, Codruta Soica, Roxana Folescu, Felicia Andrei and Corina Danciu

Abstract

Cardiovascular diseases represent the main cause of morbidity and mortality worldwide. Obesity, sedentary life style, diet, smoking and stress are the principal inducers of hypertension, endothelium dysfunction and insulin resistance in the developed countries. The latest *in vitro* and *in vivo* studies on different type of extracts obtained from black-fruited Aronia highlight its excellent cardioprotective actions for the prevention and treatment of cardiovascular and metabolic disorders. So, this chapter aims to bring an up-to-date regarding the antioxidant, anti-inflammatory, anti-atherosclerotic, antiplatelet, blood pressure, glucose and lipid reduction properties of black-fruited Aronia, as a possible new therapeutic strategy for the primary and secondary prevention of cardiovascular pathologies.

Keywords: black chokeberries, cardioprotective, black-fruited Aronia, antioxidant, anti-atherosclerotic, antithrombotic

1. Black chokeberries: historical background, taxonomy and botanical aspects

The use of medicinal plants in the prevention and treatment of diseases is an old practice that has been maintained over time and is currently being given special attention by scientists, as well as by people aiming to preserve their health or treat a disease. In the past, the therapeutic properties of the plants were discovered by observing their effects on the animals that ate them [1–3]. People gathered knowledge about the therapeutic effects of plants and several traditional medicines were born [4]. At the beginning of the nineteenth century, the first plant compounds were isolated and later some of them served as a model for the development of synthetic drugs [4, 5]. Plant-derived products are intensively studied and they are important sources in drug discovery [1]. Studies are directed toward the identification of mechanisms of action of plant compounds [6], exploiting their capacity of reducing the side effects of synthetic drugs [7] and development of new delivery systems that can increase the efficacy of phytochemicals [8].

One of the plants whose health benefits led to numerous studies is black chokeberry, also known as black-fruited Aronia. Its fruits enjoy presently a high recognition as a health food, despite lacking the historical advantage of other well-known rosaceous berry crops (strawberries, raspberries and blackberries). The highly regarded berries (botanically: pomes), cultivated in many European countries and North America, have an unusual history, which reflects on the taxonomic confusions surrounding the plant.

Unlike other edible plants, black Aronia spread from America to Europe due to its decorative interest. The ancestor of today's cultivated black-fruited Aronia is *Aronia melanocarpa* (Michx.) Elliott, a species that grows wildly in North America [9] along with two other *Aronia* species. The small genus *Aronia* Medik. (chokeberry) includes multistemmed, deciduous shrubs that differ in the color and size of fruits (pomes), as well as in the pubescence of leaves, stems and inflorescences. Three North American and one European species are recognized [10]. Red chokeberry, *A. arbutifolia* (L.) Pers., grows up to 3 m tall and has obovate to elliptical leaves, which are shiny green on the upper side and tomentose, slightly gray on the lower side. The margins are serrated and the tip is short, acuminate. Flowers are white, in compound corymbs. In late September to early October, red fruits are produced. Black chokeberry, *A. melanocarpa* (Michx.) Elliott, has a smaller habit (up to 1.5 m) and is not pubescent [11]. Its fruits are black, shiny, with a diameter of 0.8–1.3 cm; they typically ripen in August. A third species native to North America, *A. prunifolia* (Marshall) Rehder, or purple chokeberry, is considered by certain authors to be a hybrid between the former two species, a distinct species or a variety of black or red chokeberry. It has dark purple to black fruits, and the leaf pubescence is intermediate between *A. arbutifolia* and *A. melanocarpa* [12].

A. melanocarpa is cold-hardy member of the Rosaceae family, where it is assigned to the Amygdaloideae subfamily and Maleae tribe [13]. Wild-growing black chokeberry plants were introduced to Europe in the nineteenth century as ornamental shrubs. The first record of the species in Russia (1816) mentions black Aronia under the name *Mespilus melanocarpa* in the catalog of plants of the Kremenets Botanical Garden [14]. Soon, it was grown as a cold-resistant ornament in other botanical gardens; before the twentieth century, it was however not grown for its fruits, which, though edible, are poorly palatable and astringent. The black-fruited Aronia is cultivated nowadays as a distinct morphology in comparison to its wild-growing North American counterparts: its leaves are wider, flowers are larger and more numerous, fruits have a larger diameter, and corymbs bear a higher number of fruits [15]. It has been proposed that this species is the result of breeding and selection experiments performed in the early twentieth century by Ivan Michurin, and that all *Aronia* plants cultivated in the former Soviet Union originate from the Russian pomologist's nursery [14]. Then, it spread to other European countries such as Poland, Norway, Finland, Sweden or Germany [16]. In 1976, chokeberry was introduced to Japan [17]. The distinct differences from its wild-growing progenitors and the constancy in characteristics supported the assignment of the large-fruited chokeberry to a new species, *Aronia mitschurinii* A.K.Skvortsov & Maitul [18]. Its origin is however not completely elucidated. One of the hypotheses is that the species is a hybrid obtained by backcrossing *A. melanocarpa* with an F1 x*Sorbaronia* hybrid (*Sorbus aucuparia* × *A. melanocarpa*). Leonard and co-workers using amplified fragment length polymorphism (AFLP) were able to show similarities of *A. mitschurinii* to x*Sorbaronia* hybrids [19]. Further studies with more sensitive and complex tools (like multilocus nuclear data) are proposed in order to firmly establish the origin of *A. mitschurinii* [20].

An important feature of the cultivated black-fruited Aronia is its extremely low variability due to the apomictic formation of seeds, a process occurring without fertilization of the egg. Plants resulting from apomixis are clones of the mother plant. Cultivated varieties of *A. mitschurinii*, including 'Nero,' 'Viking' and 'Galicjanka,' have in fact undistinguishable phenotypes [21] and are tetraploid (2n = 68) [14]. On the contrary, 'Hugin' and 'Elata' cultivars are considered true *A. melanocarpa* genotypes [21].

Taking into account the better understanding of *Aronia* taxonomy and genetics, in recent years, it has become accepted that the cultivated black-fruited Aronia berries, which were the subject of most biomedical and phytochemical studies, are not *A. melanocarpa* fruits as reported. Research results should rather be assigned to *A. mitschurinii*, which is the only species used for commercial fruit production [21].

Further developments on the subject of black-fruited Aronia taxonomy should certainly lead to a more stable and correct nomenclature of the chokeberry species investigated by biomedical research. A unified, proper nomenclature is essential to enable researchers to assign correctly a therapeutic activity to *A. melanocarpa*, *A. mitschurinii*, or even *X Sorbaronia mitschurinii,* all of them used in various publications to assign cultivated black-fruited chokeberry.

2. Traditional use of black chokeberries

Although wild-growing black chokeberries (*Aronia melanocarpa*) were known to the North American settlers, the taste of the fruits was dissuading for a use as foodstuff. They employed fruits and bark as an astringent. The Forest Potawatomi Native Americans used the fruits of this plant in cold treatment and the preparation of traditional pemmican [17].

In Russia and Lithuania, the cultivated black chokeberry fruits were used as adjuvant treatment for high blood pressure and as anti-atherosclerotic agent [22]. Other uses include treatment of hemorrhoids, achlorhydria, avitaminosis and convalescence [17].

Because of their astringent taste, black chokeberry fruits are not usually consumed as such, but they are used for the production of juices, wines, teas, jellies or syrups, especially in fruit blends [23, 24]. Chokeberry powders are used as a natural dye in the food industry [25].

The high content of phenolic compounds, mainly anthocyanins, which are responsible for the black color of the fruits, determined an increased interest in the medicinal properties of this plant. *Aronia* ssp. fruits have a higher antioxidant activity than other berries. Various studies emphasized antioxidant, anticancer, anti-diabetic, anti-inflammatory and hepatoprotective properties for the juice or fruit extract, which made cultivated black-fruited Aronia an important health food and dietary supplement [26].

3. Phytochemistry

Several analyses have been performed in order to evaluate the organic as well as inorganic constituents of black chokeberries. They include the research of the differences in the chemical composition of wild vs. cultivated chokeberry [21, 27], the influence of the cultivar type [28–30], the degree of fruit maturity [31], fertilization [32], the application of biosynthesis regulators [33] and geographic location [34].

Wild *Aronia* genotypes contain less water and more phenolics and have a higher antioxidant activity than cultivated black chokeberries [21]. Typical values for the dry weight of cultivated black chokeberry fruits are 17.9–26%; fresh chokeberries afford 11.1–17.4% juice and 44.6–50% pomace [35]. The majority (72%) of the dry weight is constituted by dietary fiber, higher than other fruits like bilberries and currants. The study of dietary fiber with solid-state NMR could show that its components are water-insoluble fibers (cellulose, hemicellulose, lignin, cutin and pectins), soluble fibers and other constituents, which are valuable antioxidants, anthocyanins and procyanidins [36].

The sour taste of *Aronia* ssp. berries is due to the presence of organic acids, mainly malic and citric acids. Their total content is however lower than in other berries [37]. Cultivated black chokeberries contain between 5.71 and 19.36% reducing sugars [23] and are characterized by a relatively high sorbitol amount (median content of 70 g/kg) [34] when compared to other berries. The high sorbitol content, shared as well by rowan berries (*Sorbus aucuparia*), may be related to the possible hybrid nature of cultivated chokeberry. Large-fruited Aronia cultivars have been hypothesized to contain the genomes of *Aronia melanocarpa* and *Sorbus aucuparia* in a ratio of 3:1 [19, 21]. The high sorbitol content has been proposed to serve as an analytical tool in the control of juice blends [38]. Another feature of the sugar profile in cultivated black chokeberry is the absence of sucrose; its presence in black Aronia-based products suggests the addition of sugar or other fruits [34].

The lipid content of black cultivated chokeberries is reduced (below 0.2%) [23] and is mainly owed to hydrophobic constituents of the skin and seeds. The content in proteins is also low, with values of 10.7% in the pomace. The main amino acid was glutamic acid (19.8%), followed by aspartic acid (8.9%) and arginine (7.9%) [39].

Regarding the content in minerals, chokeberries may be valuable to complement dietary potassium and zinc intake. Among vitamins, Aronia fruits contain vitamins B1, B2, B6 and C, pantothenic acid and niacin; the detailed content has been extensively reviewed [23, 38].

The analysis of the volatile constituents in Aronia berry juice afforded the detection of 74 constituents, of which the most abundant were 3-penten-2-one, 3,9-epoxy-p-menth-1-ene and benzaldehyde. Among the aroma-active compounds, ethyl-2-methyl butanoate, ethyl-3-methyl butanoate and ethyl decanoate could account for the "fruity" aroma notes, while nonanal is responsible for the "green" notes [40]. The bitter-almond scent of the fruits is due to the presence of amygdalin [34]. This compound is a cyanogenic glycoside present in many representatives of the Rosaceae family [35]. Interestingly, a recent research aimed at identifying compounds, which inhibit adipocyte differentiation, was able to isolate from the butanol fraction of a chokeberry extract amygdalin and prunasin as active compounds [41]. These two cyanogenic glycosides suppress the expressions of peroxisome proliferator-activated receptor γ, CCAAT/enhancer-binding protein α (C/EBPα), sterol regulatory element-binding protein 1c, fatty acid synthase (FAS), and adipocyte fatty-acid–binding protein (aP2) [42].

The most intensively studied compounds in cultivated black chokeberries are phenolic compounds, mainly anthocyanins, procyanidins and phenolic acids. The total phenolic content ranges from 3440 mg/100 g dry weight to 7849 mg/100 g dry weight, depending on cultivar, ripening stage at harvest, cultivation conditions or analytical methods used for the quantification [38]. Analysis of several chokeberry cultivars identified a higher polyphenolic content for the fruits of 'Hugin' cultivar compared to 'Viking,' 'Galicjanka' and 'Nero' cultivars [29]. There are differences in the polyphenols content in chokeberry products; for example, the pomace has a five-fold higher content compared to juice [16].

Polymeric proanthocyanidins represent 66% of chokeberry polyphenols, while anthocyanins represent about 25%. The fruits contain up to 5181.60 mg/100 g dried weight polymeric procyanidins [43]. The constitutive unit is mainly (−)-epicatechin and the units are connected by C4—C6 and C4—C8 bonds (B-type bonds). The degree of polymerization ranges from 2 to 23 units. The main proanthocyanidins found in the fruits and bark of chokeberry are dimeric procyanidin B2 and procyanidin B5 and trimeric procyanidin C1 (**Figure 1**) [44].

Anthocyanins are a class of flavonoids that give blue, dark red or purple color of the fruits and they are important compounds for the biological activity of chokeberries [38]. Black chokeberries have a higher content of anthocyanins than other berries, such as blackberries, strawberries or red raspberries [45]. Anthocyanins are found mainly in fruit skin [46]. They are represented by cyanidin glycosides such as

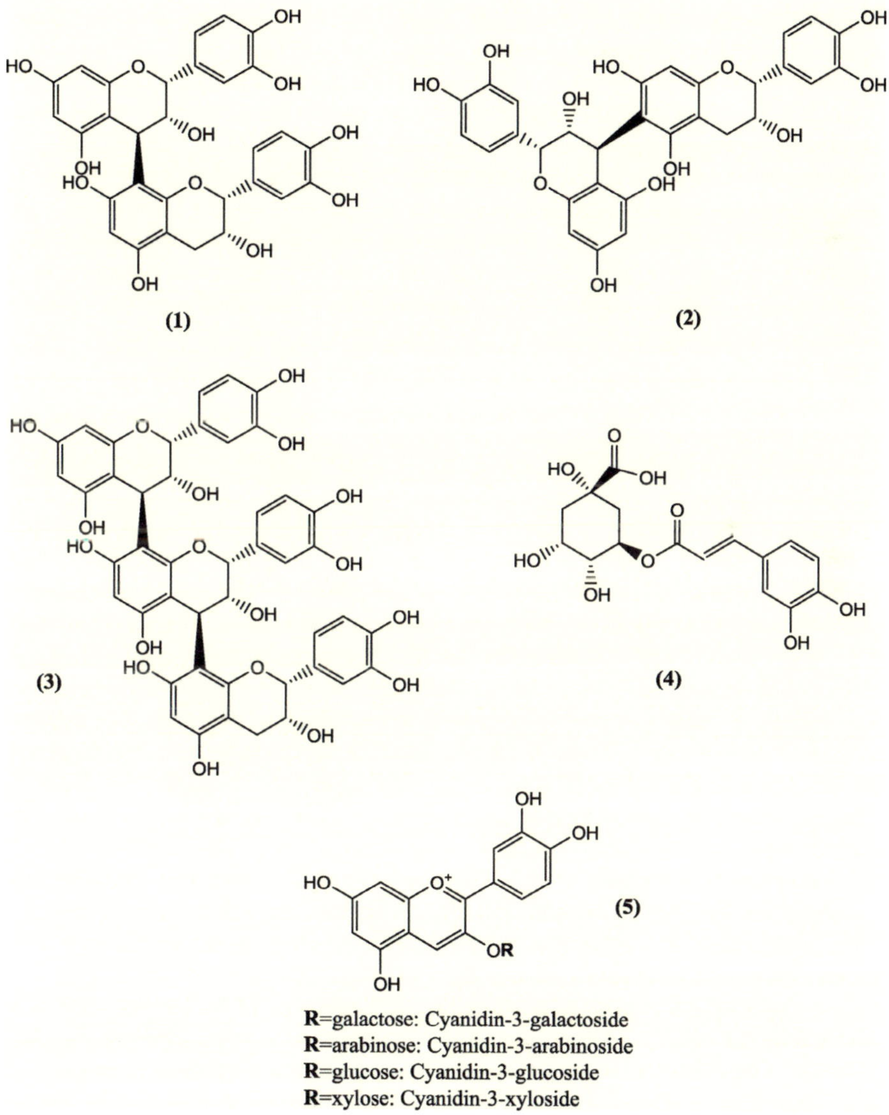

Figure 1.
Chemical structures of polyphenolic compounds from cultivated black-fruited chokeberries. (1) Procyanidin B2; (2) procyanidin B5; (3) procyanidin C1; (4) chlorogenic acid; (5) anthocyanins.

cyanidin-3-glucoside, cyanidin-3-galactoside, cyanidin-3-xyloside and cyanidin-3-arabinoside [43]. The study of the fruits of black chokeberry, 'Viking', 'Nero' and 'Galicjanka' cultivars revealed that 'Nero' and 'Viking' cultivars had higher anthocyanin content. Cyanidin-3-galactoside and cyanidin-3-arabinoside are the major anthocyanins in chokeberries, while cyanidin-3-xyloside and cyanidin-3-glucoside are found in lower amounts [47]. The fruits also contain pelargonidin-3-arabinoside and pelargonidin-3-galactoside in trace amounts [17, 38]. Other anthocyanins such as cyanidin-3-pentoside-(epi)catechin, cyanidin-3,5-hexoside-(epi)catechin, and cyanidin-3-hexoside-(epi)cat-(epi)cat were identified in black chokeberry juice and powder [46].

The main phenolic acids in black chokeberries are chlorogenic and neochlorogenic acids. They represent about 7.5% of fruit polyphenols [43]. In addition to chlorogenic and neochlorogenic acids, other phenolic acids such as cryptochlorogenic acid, 3-*O*-p-coumaroylquinic acid and di-caffeic quinic acid have been reported in chokeberries [46]. Chokeberry juice contains a greater amount of these compounds than pomace [43]. Fruits of 'Viking' cultivar and wild chokeberry have a higher content of phenolic acids compared to fruits of 'Nero' and 'Galicjanka' cultivars [47].

Chokeberry also contains flavonol glycosides such as quercetin 3-*O*-galactoside, quercetin 3-*O*-rutinoside, quercetin 3-*O*-glucoside, quercetin 3-*O*-arabinoside, isorhamnetin 3-*O*-rutinoside or kaempferol 3-*O*-glucoside [48]. Other phenolic compounds identified in chokeberry fruits and flowers are the flavanone eriodictyol-7-*O*-β-glucuronide and the flavonols quercetin-3-robinobioside and quercetin-3-vicianoside [49]. Even though the fruits were the most investigated for the polyphenols in their composition, black chokeberry leaves also contain these compounds, with a higher content in the young leaves compared to the old ones (**Figure 1**) [50].

4. Cardioprotective actions: *in vivo* and *in vitro* studies: clinical trials

Diseases of the circulatory system were reported to be the main cause of death in Europe and America [51, 52]. The affections of the circulatory system are related to high blood pressure, smoking, cholesterol and diabetes (the main pathologies of modern times), resulting in stroke or ischemic heart disease as the major pathologies causing invalidity and death in the developed countries [53].

Obesity, sedentary lifestyle, diet, smoking and stress are the principal inducers of endothelial dysfunction and insulin resistance, which will later induce hypertension and diabetes [54]. Thus, targeting endothelial dysfunction, insulin resistance and the altered metabolic state represents important strategies for decreasing the incidence and complications of cardiovascular diseases, as well as the medical costs [53, 55].

As well known, the vascular endothelium is the single layer of cells that lines the internal lumen of blood vessels [54, 56]. A healthy endothelium is essential for the cardiovascular system and is considered nowadays an organ [53, 57]. It possesses several important functions such as relaxation of vascular smooth muscle cells, inhibition of platelet aggregation, limitation of leukocyte adhesion, regulation of the vascular tone, inhibition of vascular smooth proliferation, and specialized autocrine and paracrine secretion (producing and secreting vasoactive, vasoprotective, angiogenic, inflammatory and thrombotic/antithrombotic molecules) [53, 55, 57]. It also produces growth factors and responds to physical and chemical signals. The term "endothelial dysfunction" is used not only to describe the impaired metabolism of nitric oxide (NO) or the imbalance of endothelium-derived relaxing and constrictor factors, but also to describe an aberrant endothelium activation and abnormalities between endothelium and leukocytes, platelets and other regulatory molecules [53, 58, 59]. Thus, it plays an essential role in the pathogenesis of several cardiovascular (hypertension, atherosclerosis, systemic and pulmonary hypertension, etc.), as well as other diseases (inflammatory diseases or cancer) (**Figure 2**) [60–62].

Elevated oxidative stress is another major risk factor of cardiovascular pathologies as an altered metabolic state will increase the reactive oxygen species (ROS) production, which will determine an increase in lipid peroxidation, inflammation, endothelial dysfunction, platelet aggregation and activation [63]. On endothelial cells, oxidative stress has been shown to immediately increase vessel damage. Moreover, the lesion caused by oxidative stress can affect the entire body's organs and systems (**Figure 2**) [64].

Thus, it is important to find new therapeutic strategies to improve endothelial function and decrease ROS as strategies for the primary prevention of cardiovascular diseases in the ever-aging Western societies.

In the last decade, several plant bioactive substances (called nutraceuticals) have proven to have outstanding properties, which suggest that they could play an important role in preventing the onset and development of chronic diseases. The term nutraceutical refers to a natural bioactive compound with several properties, such as health promoting, disease preventing or medicinal properties [65].

Cultivated black chokeberry fruits have been discovered to contain an enormous source of bioactive compounds such as antioxidants-especially polyphenols, flavonoids (anthocyanins, flavonols and flavanols), vitamins (C and E), minerals (potassium, magnesium and calcium), pectins, carotenoids and carbohydrates [16]. Moreover, the *in vitro* and *in vivo* studies have highlighted its antioxidant (one of the richest plants in antioxidants), anti-inflammatory, anti-proliferative,

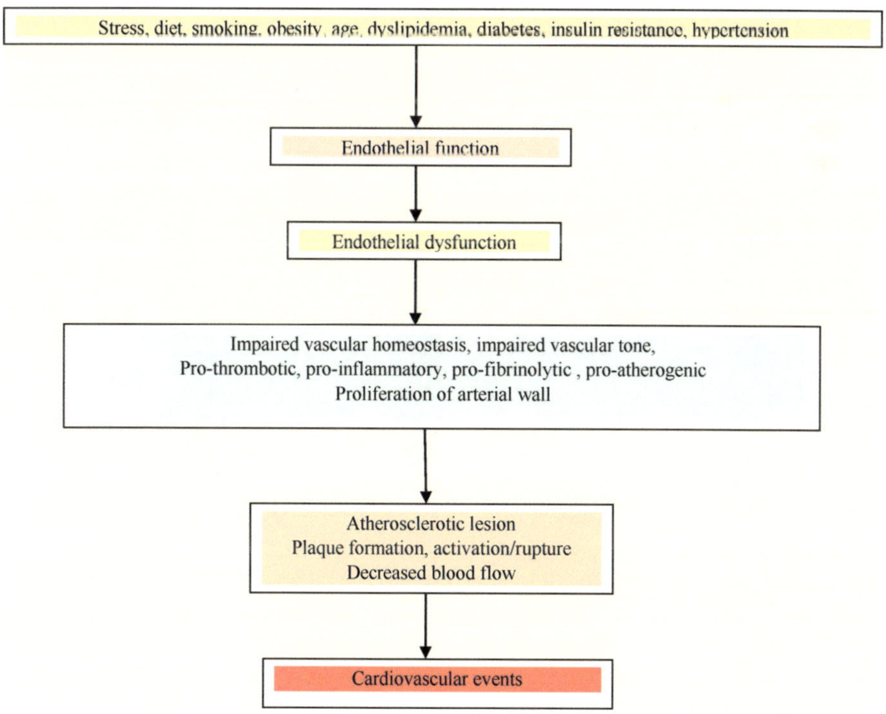

Figure 2.
Main aspects regarding the development of cardiovascular disease events.

anti-atherosclerotic, hypotensive, antiplatelet, gastroprotective, antimicrobial (antibacterial and antiviral), immunomodulatory and anti-tumor profile, making it an excellent candidate for the prevention of cardiovascular and metabolic disorders [16, 66].

Herein, the main and the latest findings regarding the cardioprotective properties of cultivated black chokeberry will be presented and discussed (**Figure 3**).

4.1 Antioxidant properties

There are currently a large number of studies that presents the antioxidant activity of black chokeberry both *in vivo* (animal studies and clinical trials) and *in vitro* (isolated cells and cell lines) explaining the capability of different types of extracts obtained from this vegetal product to protect against free radicals. Black chokeberry fruits are considered to contain a huge level of antioxidants-more precisely vitamin C and polyphenols, such as anthocyanins, phenolic acids, tannins, flavanols, flavonols and flavonoids [16, 45, 67]. The antioxidant mechanism includes radical scavenging, inhibition of reactive oxygen and nitrogen species formation, of prooxidant enzymes, stimulation of antioxidant ones and most probably cellular signaling to manage the enzymes and antioxidant substance levels [68]. The berries highlighted the highest antioxidant capacity among other berries and fruits investigated via the oxygen radical absorbing capacity (ORAC) assay [28, 38, 69, 70]. Moreover, it was noted that the pomace compared with the berries or the juice has a higher antioxidant capacity (being five times more concentrated than the juice) and the highest content of phenolic compounds (mainly polymeric proanthocyanidins and (−)-epicatechin) [43]. On the other hand, Rop et al. [71], showed the same high antioxidant property for the fruit product, pomace, fresh berry fruits and juice [16, 71]. An important amendment is the fact that the antioxidant activity levels and the polyphenols concentration are species dependent [16].

A recent *in vivo* study, published in 2019, showed the additive effect of chokeberry and walnut in increasing the expression of antioxidant enzyme genes in the

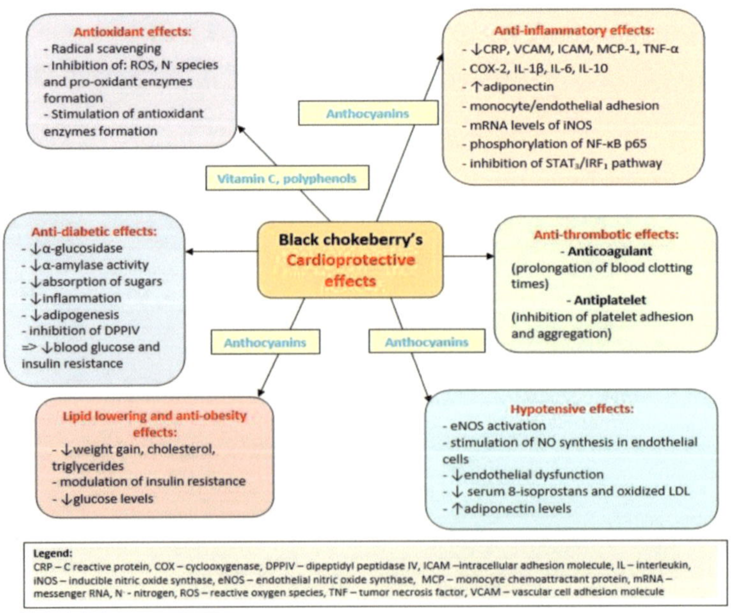

Figure 3.
The main cardioprotective effects of cultivated black chokeberry and its mechanism of action.

liver and decreasing lipid peroxidation in the liver, serum and kidneys in an aging mouse model [72]. The study performed by Jo et al. [73] described that dietary supplementation with Aronia extract increased the lifespan and improved the oxidative injuries related to the age in *Drosophila melanogaster*. Other *in vitro* studies outlined an increased antioxidant profile by decreasing the levels of ROS after incubation with *Aronia melanocarpa* bioactive compounds [74, 75]. Moreover, a decreased level of plasma lipid peroxidation was observed *in vitro*, after treatment with ziprasidone [76]. The antioxidant effect was also evidentiated in humans, as the black chokeberry juice supplementation decreased the exercise-induced oxidative damage to red blood cells in rowers [77].

The antioxidant activity is in general attributed to the polyphenolic compounds [3, 78]. It was observed that proanthocyanidin and anthocyanin content in berries, as well as neochlorogenic acid, cyaniding 3-arabinoside and (−)-epicatechin, are the main substances responsible for this action [71, 79].

4.2 Anti-inflammatoSry effect

Vascular inflammation is a primary key step process underlying endothelial dysfunction and later atherosclerosis [59]. It is responsible for endothelial activation, which will further stimulate the leukocytes' recruitment [56]. Moreover, it is a self-maintaining process, which is mediated through the expression of several molecules such as cell adhesion molecules (ICAM and VCAM), cytokines, neutrophils, fibrinogen, C-reactive protein, etc. [68]. Apart from endothelial dysfunction, it will induce smooth muscle cell migration, vascular calcification, oxidative stress, degradation of extracellular matrix and collagen, elastolysis and increased activity of metalloproteinases. Vascular inflammation was observed in patients with hypertension, metabolic syndrome, diabetes, infections, preeclampsia, coronary heart disease or peripheral arterial disease [58, 62]. Thus, the anti-inflammatory effect exhibited by several bioactive natural, synthetic or semi-synthetic compounds is important for the prevention of chronic diseases (especially the cardiovascular, metabolic and immune system disorders) and also for decreasing their complications [16]. Supplementation with black chokeberry extract for patients who suffered myocardial infarction and are under treatment with statins decreased the plasmatic concentration of several inflammatory biomarkers such as C-reactive protein (CRP), monocyte chemoattractant protein-1 (MCP-1), intercellular adhesion molecule (ICAM), vascular cell adhesion molecule (VCAM) and interleukin 6 (IL-6). Moreover, adiponectin concentration (an anti-inflammatory cytokine) was found to be increased [80]. The study performed by Zapolska-Downar et al., in 2012, on TNF-α–treated human aortic endothelial cells (HAECs) highlighted the fact that various concentrations of Aronia extract decreased the expression of ICAM-1, VCAM-1, the phosphorylation of NF-κB p65 and intracellular ROS production [26].

Other research groups reported that black chokeberry extract induces a vasorelaxation of coronary arterial rings, which is dose-dependent and endothelium-dependent, compared with other extracts (bilberry and elderberry) [81].

Recently, Iwashima et al. [82], showed that Aronia extract decreased TNF-α–induced monocyte/endothelial adhesion and decreased VCAM-1 expression, although it did not affect ICAM-1 expression on a vascular endothelial inflammation model of human umbilical vein endothelial cells (HUVECs). Moreover, it lowered the phosphorylation of signal transducer and activator of transcription 3 (STAT3), its nuclear levels and the interferon regulatory transcription factor-1 (IRF1) nuclear level. Thus, the researchers concluded that Aronia extract could exert an anti-atherosclerotic effect through the inhibition of STAT3/IRF1 pathway in vascular endothelial cells [82].

Lee et al. [83] highlighted the anti-neuroinflammatory effect of an ethanolic extract of Aronia in a mouse model of Alzheimer's disease. Aronia extract significantly reduced the generation of NO, as well as mRNA levels of iNOS (inducible nitric oxide synthase), COX-2 (cyclooxygenase 2), IL-1β (interleukin-1 beta) and TNF-α (tumor necrosis factor alpha), suggesting its neuroprotective effect against the development of Alzheimer's disease [83].

Wei et al. [84] showed that the anthocyanins from black chokeberry delayed the degenerative changes of brain, related to aging, in an aged mouse model, possibly by reducing inflammation, regulating the balance of redox system and inhibiting DNA damage [84].

In another study, it was observed that the administration of Aronia cold-pressed juice and oven-dried black chokeberry powder by patients with mildly elevated blood pressure levels reduced only the levels of TNF-α and IL-10 and did not influence other inflammatory markers [85].

The anti-adhesion effect of Aronia extract can be due to the presence of anthocyanins (alone or by a synergistic effect), more precisely cyaniding-3-glucoside, cyaniding-3-galactoside, cyaniding-3-arabinoside, peonidin-3-glucoside and delphinidin-3-glucoside [86].

Moreover, the anti-apoptotic effect of Aronia extract was also showed in an animal model of cardiomyoblasts [87]. In diabetic patients, the reversing of atherosclerosis development by Aronia polyphenols was also showed through augmenting the immune defenses and reducing inflammation, as the extract decreased the monocyte and granulocyte levels responsible for inflammation and also the number of lymphocytes, thus blocking the formation of atherosclerotic lesions [23, 88].

4.3 Hypotensive properties

Chronic inflammation can lead to cardiovascular diseases characterized by high blood pressure levels, altered metabolism of lipids, endothelial dysfunction, oxidative stress, etc. Endothelial dysfunction can be a cause, as well as a consequence of hypertension [59]. Thus, due to multiple mechanisms of action, black chokeberry can induce a cardioprotection by having positive effects on multiple risk factors for cardiovascular diseases (e.g., antioxidant effect, anti-inflammatory, hypolipidemic, antithrombotic, etc.) [16, 26, 89].

Over the years, the blood pressure lowering effects of black chokeberry were highlighted in several studies and nowadays black chokeberry preparations are recommended as a nutritional supplement in the management of essential arterial hypertension [16]. In one study, spontaneous hypertensive rats that were treated with commercial Aronia extract had decreased blood pressure levels compared with controls [90], although the effect was term-limited and maximal 3 h after intake. Other studies suggested that Aronia polyphenols could induce a positive effect on blood vessels by stimulating NO synthesis in endothelial cells (via activation of eNOS), through a mechanism related to ACE inhibitors (angiotensin I-converting enzyme inhibitors) and endothelium-dependent [68, 91]. Sihora et al. studied the effects of 2 months administration of black chokeberry preparation in patients with metabolic syndrome on the activity of ACE. They observed a 25% decrease of ACE after 1 month of administration and 30% decrease after the second month. Moreover, systolic blood pressure, diastolic blood pressure and CRP levels correlated positively with the activity of ACE [92]. The studies performed by Bell et al. [81] highlighted the vasoactive and vasoprotective potential of black chokeberry extract on the coronary arteries, due to the high concentrations of anthocyanins. The following bioactive substances were found to be responsible for the vasorelaxation and the endothelial protective effects: coumaric acid (the most potent), ferulic acid, caffeic acid and chlorogenic acid [92]. Moreover, Aronia induced a protective action in aorta and coronary arteries against atherogenic changes [93, 94].

A double-blind, placebo-controlled study performed by Naruszewicz et al. [80] showed that black chokeberry polyphenols decreased the severity of inflammation in patients after myocardial infarction and thus, Aronia can be used as a strategy treatment for the secondary prevention of ischemic heart disease [80]. Black chokeberry flavonoids significantly decreased serum 8-isoprostanes and oxidated LDL (formed as a consequence of the action of oxidative stress in the vascular wall that are contributing to the formation of foam cells, the basis of atherosclerotic lesions) [95]. Moreover, it decreased the levels of adhesion molecules (ICAM, VCAM and MCP-1) compared with the control group and increased the level of adiponectin, an anti-inflammatory molecule [80]. Anthocyanins (mostly conjugated cyanidins and chlorogenic acid) from black chokeberry juice were found to be potent stimulators of endothelial NO formation via Sirc/PI3-kinase Akt pathway, in a study performed by Kim et al. [96].

The study performed by Loo et al. [85] showed that consumption of chokeberry products can modestly lower blood pressure levels and decrease low-grade inflammation in hypertensive patients (under no regular use of anti-hypertensive drugs) with mildly elevated blood pressure levels.

4.4 Lipid-lowering and anti-obesity effects

Several studies, including large clinical trials, have showed over the time the relationship between LDL cholesterol, triglycerides and the development and progression of atherosclerosis. Moreover, it is well known that hyperlipidemia is one of the major risk factors for developing undesirable cardiovascular events [68]. Black chokeberry products had been reported to manifest lipid-lowering properties and anti-obesity effect in culture cells, in animal as well as in human studies. Qin et al. [97] showed that a dose of 10 or 20 mg/kg/body weight of anthocyanin attenuates weight gain, decreases triglycerides and cholesterol and modulates insulin resistance in Wistar rats under a fructose-rich diet of 6 weeks. Moreover, it decreased TNF-α and IL-6 [97]. The black chokeberry pomace administered in Polish Merino lambs decreased glucose and total cholesterol levels and increased HDL [98]. The juice administered *ad libitum* for 28 days decreased glucose, insulin and body weight in C57BL/6JmsSlc and KK-Ay mice [99]. Another study performed by Daskalova et al. [94] on male Wistar rats who received a daily dose of 25 ml commercial juice for 90 days showed a decrease in LDL and also retarded age-related changes in the aortic wall. Ciocoiu et al. [100] showed a decrease of total cholesterol, systolic and diastolic blood pressure levels and an increased HDL value after 8 weeks administration of an ethanol extract of black chokeberry.

Results from clinical trials have shown that administration of 100 ml juice before meal decreased the blood glucose levels of 37 healthy subjects [101]. A 12-week randomized, placebo-controlled trial showed that consummation of 500 mg Aronia extract although did not changed blood pressure levels or inflammatory and oxidative stress biomarkers, it decreased total and LDL cholesterol in healthy adults former smokers. The cholesterol-lowering capacity was closely linked with cyanidin-3-*O*-galactoside and peonidin-3-*O*-galactoside urinary levels [102]. Kardum et al. [103] showed that 100 ml of glucomannan-enriched Aronia juice-based supplement decreased the body mass index, waist circumference and systolic blood pressure, concluding that the juice had a positive effect on cellular oxidative damage, anthropometric indices of obesity and blood pressure levels.

In 2016, Shin et al. observed that adipogenesis in 3 T3-L1 preadipocytes was blocked after administration of a black chokeberry extract. Moreover, the expression of mRNA levels of some adipogenesis key genes was impacted and that the levels of PPARϒ (peroxisome proliferator-activated receptor ϒ), FABP4 (fatty acid–binding protein 4), adiponectin, MCP-1 (monocyte chemoattractant protein-1) and leptin were decreased [104].

Other studies performed concluded that anthocyanins may contribute to the prevention of obesity by lowering the sugar and lipid absorption in the digestive system [16, 105]. Although the mechanism underlying the lipid-lowering effect is not completely elucidated, it seems that the blocking of cholesterol absorption (by flavonoids), increased lipoprotein catabolism (by cyanidin), inhibition of 3-hydroxy-3-methylglutaryl-CoA reductase and thus decreased synthesis of cholesterol (by flavonoids) could explain this effect [68].

4.5 Anti-diabetic effect

Several studies describe the hypoglycemic effect of black chokeberry [16, 98, 106–108]. Moreover, its property of decreasing the insulin resistance was also highlighted. It seems that Aronia extracts can lower the risk factors associated with insulin resistance, by decreasing inflammation and adipogenesis and by modulating several pathways linked to insulin signaling [92, 97]. As a general view, metabolic disorders, more specifically plasma glucose and the lipid profile, were improved after long-term juice consumption [106–108]. It was observed that polyphenolic compounds can decrease blood glucose levels by inhibiting α-glucosidase and α-amylase activity and, thus, controlling the postprandial hyperglycemia [16]. Anthocyanins (more specifically cyaniding 3-rutinoside) might inhibit intestinal α-glucosidase and might slow down the absorption of sugars [106, 109, 110]. Moreover, anthocyanins were found to exert a positive action on normalizing the carbohydrate metabolism in diabetic patients and rats, throughout several mechanisms such as reversing the beta cells' integrity and physiology and stimulating the release of insulin [111]. Chlorogenic acid was found to be the most potent inhibitor of pancreatic α-amylase [16]. Other studies postulated that the antidiabetic effect of black chokeberry juice may be due to the cyaniding 3,5-diglucoside inhibition of DPP IV (dipeptidyl peptidase IV) [112]. It seems that in humans, black chokeberry juice can be an excellent natural alternative therapeutic strategy for the treatment of metabolic syndrome disorders, the dose varying from 100 ml to 300 ml per day, for at least 3 months [16].

A recent study, published in 2019, showed the anti-adipogenic effect of cyaniding-3-O-galactoside–enriched black chokeberry extract in C57BL/6 obese mice. The extract reduced the serum levels of insulin, leptin, triglyceride, LDL and total cholesterol and suppressed adipogenesis by decreasing the expression of several key proteins [113].

The study performed by Jakovljevic et al. [114] on a rat model with metabolic syndrome highlighted the fact that 4-week administration of Aronia extract managed to reduce blood pressure levels and to induce benefits on the heart function. It also improved glucose tolerance and oxidative stress levels and attenuated the pathological alterations of the liver, thus conferring an excellent cardioprotection, alone or in combination with other dietary regimens [114]. A fermented chokeberry extract administered for 8 weeks in obese mice decreased weight gain and increased glucose tolerance and insulin sensitivity. These results also led to the conclusion that the anti-obesity effect was not closely correlated with the cyaniding content [115].

4.6 Effects on erythrocytes

Studies showed that the anti-oxidative effect of black chokeberry may impact erythrocyte's proper functioning [23]. The juice increased the protection against oxidation in erythrocyte membranes of 25 healthy women who drank 100 ml daily for 3 months [103]. The same effect (anti-oxidative effect on erythrocytes and increased PUFA concentration) was observed in obese women who drank 100 ml juice with glucomannan for 4 weeks [103]. Moreover, in patients with hypercholesterolemia, Aronia extract decreased cholesterol level and lipid peroxidation in erythrocytes, improved the rheological properties of red cells and increased their membrane fluidity [116].

4.7 Effects on neutrophils

As neutrophils produce high concentrations of ROS, this impacts the tissue-damaging effects of inflammatory reactions. The study performed by Zielinska-Przyjemska et al. [117] showed that the oxidative metabolism of neutrophils decreased after treatment with Aronia juice, both in non-obese and in obese patients.

4.8 Antithrombotic properties

Black chokeberry exhibited *in vitro* strong anticoagulant properties by prolonging blood clotting times (APTT, prolonged PT and TT) and by decreasing the maximal velocity of fibrin polymerization in human plasma [16, 118]. Malinowska et al. showed its action on clot formation and fibrin lysis in patients with hyperhomocysteinemia [74]. Black chokeberry proved to possess properties in inhibiting also the platelet aggregation [119], as the extract decreased *in vitro* several steps of platelet activation, such as adhesion of platelets to collagen and platelet aggregation. Moreover, it decreased the production of ROS in resting blood platelets and in those activated by thrombin [119]. Sikora et al. [92] noted that 1 month of administration of Aronia extract in men did not influence the number of platelets in the blood; the extract led to prolonging the time required to reach the maximal aggregation. The study performed by Ryszawa et al. [120] assessed the effects of Aronia extract on ROS production and aggregation in the thrombocytes of smoker patients with high cardiovascular risk factors, such as hypertension, diabetes and hypercholesterolemia, who had an increased production of ROS, compared with the control group. The Aronia supplementation managed to neutralize the difference in ROS production between the studied group and the control group and induced significant anti-aggregation effects dependent on the concentration, in both groups, concluding that these effects might be independent of its capacity of modulating ROS production [120].

5. Conclusions and perspectives

From all the information presented above, one can conclude that black chokeberry is an extremely rich source of bioactive molecules that are offering an excellent cardioprotection among other fruits and that can be used in both primary and secondary prevention of cardiovascular events. Despite the low bioavailability of polyphenols and their variability, they are exceptionally important for their health-promoting properties. Based on the present findings, it can be definitely included in a healthy daily diet. However, much more research is needed to completely understand the exact mechanisms as well as the full-length actions of black chokeberry, especially in humans. Complete bioavailability studies are required concerning Aronia bioactive molecules. Moreover, the investigation of the risk of tissue accumulation, as well as appreciation of the risk/benefit ratio in humans, will be extremely beneficial. The recommended daily intake should also be established.

Conflict of interest

"The authors declare no conflict of interest."

Author details

Valentina Buda, Minodora Andor*, Antal Diana, Florina Ardelean,
Ioana Zinuca Pavel, Cristina Dehelean, Codruta Soica, Roxana Folescu,
Felicia Andrei and Corina Danciu
"Victor Babes" University of Medicine and Pharmacy, Timisoara, Romania

*Address all correspondence to: andorminodora@gmail.com

References

[1] Fürst R, Zündorf I. Evidence-based phytotherapy in Europe: Where do we stand? Planta Medica. 2015;**81**(12/13):962-967

[2] Alexa E, Danciu C, Radulov I, Obistioiu D, Sumalan RM, Morar A, et al. Phytochemical screening and biological activity of Mentha × piperita L. and *Lavandula angustifolia* Mill. extracts. Analytical Cellular Pathology (Amsterdam). 2018;**2018**:2678924. DOI: 10.1155/2018/2678924

[3] Danciu C, Pinzaru I, Coricovac D, Andrica F, Sizemore I, Dehelean C, et al. Betulin silver nanoparticles qualify as efficient antimelanoma agents in in vitro and in vivo studies. European Journal of Pharmaceutics and Biopharmaceutics. 2019;**134**:1-19. DOI: 10.1016/j.ejpb.2018.11.006

[4] Yuan H, Ma Q, Ye L, Piao G. The traditional medicine and modern medicine from natural products. Molecules. 2016;**21**(5):559

[5] Lopes CM, Lazzarini JR, Soares Júnior JM, Baracat EC. Phytotherapy: Yesterday, today, and forever? Revista da Associação Médica Brasileira. 2018;**64**(9):765-768

[6] Li D, Wang P, Luo Y, Zhao M, Chen F. Health benefits of anthocyanins and molecular mechanisms: Update from recent decade. Critical Reviews in Food Science and Nutrition. 2017;**57**(8):1729-1741

[7] Ojha S, Al Taee H, Goyal S, Mahajan UB, Patil CR, Arya DS, et al. Cardioprotective potentials of plant-derived small molecules against doxorubicin associated cardiotoxicity. Oxidative Medicine and Cellular Longevity. 2016;**2016**:5724973

[8] Xie J, Yang Z, Zhou C, Zhu J, Lee RJ, Teng L. Nanotechnology for the delivery of phytochemicals in cancer therapy. Biotechnology Advances. 2016;**34**(4):343-353

[9] Taheri R, Connolly BA, Brand MH, Bolling BW. Underutilized chokeberry (*Aronia melanocarpa*, *Aronia arbutifolia*, *Aronia prunifolia*) accessions are rich sources of anthocyanins, flavonoids, hydroxycinnamic acids, and proanthocyanidins. Journal of Agricultural and Food chemistry. 2013;**61**(36):8581-8588

[10] Brand MH. Aronia: Native shrubs with untapped potential. Arnoldia. 2010;**67**:14-25

[11] Stace C. New Flora of the British Isles. 2nd ed. Cambridge: Cambridge University Press; 1997. p. 377

[12] Hardin JW. The enigmatic chokeberries. Bulletin Torrey Botanical Club. 1973;**100**(3):178-184

[13] Liu BB, Hong DY, Zhou SL, Xu C, Dong WP, Johnson G, et al. Phylogenomic analyses of the Photinia complex support the recognition of a new genus Phippsiomeles and the resurrection of a redefined Stranvaesia in Maleae (Rosaceae). Journal of Systematics and Evolution. 2019;**57**(6):678-694

[14] Skvortsov AK, Maitulina YK, Gorbunov YN. Cultivated black-fruited Aronia: Place, time and probable mechanism of formation (in Russian). Bulletin of Moscow Society of Naturalists, Division of Biology. 1983;**88**(3):88-96

[15] Vinogradova Maitulina Y, Grygorieva O, Vergun O, Brindza J. Morphological characteristics for fruits of Aronia mitschurinii AK Skvortsov & Maitul. Potravinarstvo Slovak Journal of Food Sciences. 2017;**11**(1): 754-760

[16] Jurikova T, Mlcek J, Skrovankova S, Sumczynski D, Sochor J, Hlavacova I, et al. Fruits of black chokeberry Aronia melanocarpa in the prevention of chronic diseases. Molecules. 2017;**22**(6):944

[17] Kokotkiewicz A, Jaremicz Z, Luczkiewicz M. Aronia plants: A review of traditional use, biological activities, and perspectives for modern medicine. Journal of Medicinal Food. 2010;**13**(2):255-269

[18] Skvortsov AK, Maitulina YK. On distinctions of cultivated black-fruited Aronia from its wild ancestors (in Russian). Bull. GBS AN SSSR. 1982;**126**:35-40

[19] Leonard PJ, Brand MH, Connolly BA, Obae SG. Investigation of the origin of Aronia mitschurinii using amplified fragment length polymorphism analysis. HortScience. 2013;**48**(5):520-524

[20] Shipunov A, Gladkova S, Timoshina P, Lee HJ, Choi J, Despiegelaere S, et al. Mysterious chokeberries: New data on the diversity and phylogeny of Aronia Medik. (Rosaceae). European Journal of Taxonomy. 2019;**570**:1-14. DOI: 10.5852/ejt.2019.570

[21] Brand MH, Connolly BA, Levine LH, Richards JT, Shine SM, Spencer LE. Anthocyanins, total phenolics, ORAC and moisture content of wild and cultivated dark-fruited Aronia species. Scientia Horticulturae. 2017;**224**:332-342

[22] Hellström JK, Shikov AN, Makarova MN, Pihlanto AM, Pozharitskaya ON, Ryhänen EL, et al. Blood pressure-lowering properties of chokeberry (*Aronia mitchurinii*, var. Viking). Journal of Functional Foods. 2010;**2**(2):163-169

[23] Sidor A, Gramza-Michałowska A. Black chokeberry *Aronia melanocarpa* L.—A qualitative composition, phenolic profile and antioxidant potential. Molecules. 2019;**24**(20):3710

[24] Chrubasik C, Li G, Chrubasik S. The clinical effectiveness of chokeberry: A systematic review. Phytotherapy Research. 2010;**24**(8):1107-1114

[25] Przybył K, Gawałek J, Koszela K, Przybył J, Rudzińska M, Gierz Ł, et al. Neural image analysis and electron microscopy to detect and describe selected quality factors of fruit and vegetable spray-dried powders—Case study: Chokeberry powder. Sensors. 2019;**19**(20):4413

[26] Zapolska-Downar D, Bryk D, Małecki M, Hajdukiewicz K, Sitkiewicz D. Aronia melanocarpa fruit extract exhibits anti-inflammatory activity in human aortic endothelial cells. European Journal of Nutrition. 2012;**51**(5):563-572

[27] Sueiro L, Yousef GG, Seigler D, De Mejia EG, Grace MH, Lila MA. Chemopreventive potential of flavonoid extracts from plantation-bred and wild Aronia melanocarpa (black chokeberry) fruits. Journal of Food Science. 2006;**71**(8):C480-C488

[28] Wu X, Gu L, Prior RL, McKay S. Characterization of anthocyanins and proanthocyanidins in some cultivars of Ribes, Aronia, and Sambucus and their antioxidant capacity. Journal of Agricultural and Food Chemistry. 2004;**52**(26):7846-7856

[29] Ochmian ID, Grajkowski J, Smolik M. Comparison of some morphological features, quality and chemical content of four cultivars of chokeberry fruits (*Aronia melanocarpa*). Notulae Botanicae Horti Agrobotanici Cluj-napoca. 2012;**40**(1):253-260

[30] Wangensteen H, Bräunlich M, Nikolic V, Malterud KE, Slimestad R, Barsett H. Anthocyanins,

proanthocyanidins and total phenolics in four cultivars of aronia: Antioxidant and enzyme inhibitory effects. Journal of Functional Foods. 2014;7:746-752

[31] Lee JE, Kim GS, Park S, Kim YH, Kim MB, Lee WS, et al. Determination of chokeberry (*Aronia melanocarpa*) polyphenol components using liquid chromatography–tandem mass spectrometry: Overall contribution to antioxidant activity. Food Chemistry. 2014;**146**:1-5

[32] Skupień K, Ochmian I, Grajkowski J. Influence of mineral fertilization on selected physical features and chemical composition of aronia fruit. Acta Agrophysica. 2008;**11**(1):213-226

[33] Hudec J, Bakoš D, Mravec D, Kobida LU, Burdová M, Turianica I, et al. Content of phenolic compounds and free polyamines in black chokeberry (*Aronia melanocarpa*) after application of polyamine biosynthesis regulators. Journal of Agricultural and Food Chemistry. 2006;**54**(10):3625-3628

[34] Šnebergrová J, Čížková H, Neradova E, Kapci B, Rajchl A, Voldřich M. Variability of characteristic components of aronia. Czech Journal of Food Sciences. 2014;**32**(1):25-30

[35] Mayer-Miebach E, Adamiuk M, Behsnilian D. Stability of chokeberry bioactive polyphenols during juice processing and stabilization of a polyphenol-rich material from the by-product. Agriculture. 2012;**2**:244-258

[36] Wawer I, Wolniak M, Paradowska K. Solid state NMR study of dietary fiber powders from Aronia, bilberry, black currant and apple. Solid State Nuclear Magnetic Resonance. 2006;**30**:106-113

[37] Tanaka T, Tanaka A. Chemical components and characteristics of black chokeberry. Japanese Society for Food Science and Technology. 2001;**48**:606-610

[38] Kulling SE, Rawel HM. Chokeberry (*Aronia melanocarpa*)—A review on the characteristic components and potential health effects. Planta Medica. 2008;**74**:1625-1634

[39] Pieszka M, Gogol P, Pietras M, Pieszka M. Valuable components of dried pomaces of chokeberry, black currant, strawberry, apple and carrot as a source of natural antioxidants and nutraceuticals in the animal diet. Annals of Animal Science. 2015;**15**(2):475-491

[40] Kraujalytė V, Leitner E, Venskutonis PR. Characterization of *Aronia melanocarpa* volatiles by headspace-solid-phase microextraction (HS-SPME), simultaneous distillation/extraction (SDE), and gas chromatography-olfactometry (GC-O) methods. Journal of Agricultural and Food Chemistry. 2013;**61**(20):4728-4736

[41] Gleadow RM, Møller BL. Cyanogenic glycosides: Synthesis, physiology, and phenotypic plasticity. Annual Review of Plant Biology. 2014;**65**:155-185

[42] Kim NH, Jegal J, Kim Y, Heo JD, Rho JR, Yang M, et al. Chokeberry extract and its active polyphenols suppress adipogenesis in 3T3-L1 adipocytes and modulates fat accumulation and insulin resistance in diet-induced obese mice. Nutrients. 2018;**10**(11):1734

[43] Oszmiański J, Wojdylo A. Aronia melanocarpa phenolics and their antioxidant activity. European Food Research and Technology. 2005;**221**(6):809-813

[44] Bräunlich M, Slimestad R, Wangensteen H, Brede C, Malterud K, Barsett H. Extracts, anthocyanins and procyanidins from *Aronia melanocarpa* as radical scavengers and enzyme inhibitors. Nutrients. 2013;**5**(3):663-678

[45] Jakobek L, Šeruga M, Medvidović-Kosanović M, Novak I. Antioxidant activity and polyphenols of Aronia in comparison to other berry species. Agriculturae Conspectus Scientificus. 2007;**72**(4):301-306

[46] Oszmiański J, Lachowicz S. Effect of the production of dried fruits and juice from chokeberry (*Aronia melanocarpa* L.) on the content and antioxidative activity of bioactive compounds. Molecules. 2016;**21**(8):1098

[47] Jakobek L, Drenjančević M, Jukić V, Šeruga M. Phenolic acids, flavonols, anthocyanins and antiradical activity of "Nero","Viking","Galicianka" and wild chokeberries. Scientia Horticulturae. 2012;**147**:56-63

[48] Tian Y, Liimatainen J, Alanne AL, Lindstedt A, Liu P, Sinkkonen J, et al. Phenolic compounds extracted by acidic aqueous ethanol from berries and leaves of different berry plants. Food Chemistry. 2017;**220**:266-281

[49] Slimestad R, Torskangerpoll K, Nateland HS, Johannessen T, Giske NH. Flavonoids from black chokeberries, *Aronia melanocarpa*. Journal of Food Composition and Analysis. 2005;**18**(1):61-68

[50] Thi ND, Hwang ES. Bioactive compound contents and antioxidant activity in Aronia (*Aronia melanocarpa*) leaves collected at different growth stages. Preventive Nutrition and Food Science. 2014;**19**(3):204-212

[51] Eurostat. Cardiovascular Diseases Statistics. 2018. Available from: https://ec.europa.eu/eurostat/statistics-explained/index.php/Cardiovascular_diseases_statistics#Deaths_from_cardiovascular_diseases

[52] Heron M. Deaths: Leading causes for 2017. National Vital Statistics Reports. 2019;**68**(6):1-77

[53] Cahill PA, Redmond EM. Vascular endothelium—Gatekeeper of vessel health. Atherosclerosis. 2016;**248**:97-109. DOI: 10.1016/j.atherosclerosis.2016.03.007

[54] Buda V, Andor M, Cristescu C, Voicu M, Suciu L, Muntean C, et al. The influence of perindopril on PTX3 plasma levels in hypertensive patients with endothelial dysfunction. Farmácia. 2016;**64**(3):382-389

[55] Rajendran P, Rengarajan T, Thangavel J, Nishigaki Y, Sakthisekaran D, Sethi G, et al. The vascular endothelium and human diseases. International Journal of Biological Sciences. 2013;**9**(10):1057-1069. DOI: 10.7150/ijbs.7502

[56] Buda V, Andor M, Cristescu C, Voicu M, Cochera F, Tuduce P, et al. The effect of candesartan on pentraxin-3 plasma levels as marker of endothelial dysfunction in patients with essential arterial hypertension. Irish Journal of Medical Science. 2017;**186**(3):621-629. DOI: 10.1007/s11845-017-1580-5

[57] Gimbrone MA Jr, García-Cardeña G. Vascular endothelium, hemodynamics, and the pathobiology of atherosclerosis. Cardiovascular Pathology. 2013;**22**(1):9-15. DOI: 10.1016/j.carpath.2012.06.006

[58] Buda V, Andor M, Petrescu L, Cristescu C, Baibata DE, Voicu M, et al. Perindopril induces TSP-1 expression in hypertensive patients with endothelial dysfunction in chronic treatment. International Journal of Molecular Sciences. 2017;**18**(2). DOI: 10.3390/ijms18020348

[59] Buda V, Andor M, Baibata DE, Cozlac R, Radu G, Coricovac D, et al. Decreased sEng plasma levels in hypertensive patients with endothelial dysfunction under chronic treatment with perindopril. Drug Design Development and Therapy.

2019;**13**:1915-1924. DOI: 10.2147/DDDT.S186378

[60] Konukoglu D, Uzun H. Endothelial dysfunction and hypertension. Advances in Experimental Medicine and Biology. 2017;**956**:511-540. DOI: 10.1007/5584_2016_90

[61] Grandl G, Wolfrum C. Hemostasis, endothelial stress, inflammation, and the metabolic syndrome. Seminars in Immunopathology. 2018;**40**(2):215-224. DOI: 10.1007/s00281-017-0666-5

[62] Buda V, Andor M, Cristescu C, Tomescu MC, Muntean DM, Baibata DE, et al. Thrombospondin-1 serum levels in hypertensive patients with endothelial dysfunction after one year of treatment with perindopril. Drug Design Development and Therapy. 2019;**13**:3515-3526. DOI: 10.2147/DDDT.S218428

[63] Förstermann U, Xia N, Li H. Roles of vascular oxidative stress and nitric oxide in the pathogenesis of atherosclerosis. Circulation Research. 2017;**120**(4):713-735. DOI: 10.1161/CIRCRESAHA.116.309326

[64] Münzel T, Camici GG, Maack C, Bonetti NR, Fuster V, Kovacic JC. Impact of oxidative stress on the heart and vasculature: Part 2 of a 3-part series. Journal of the American College of Cardiology. 2017;**70**(2):212-229. DOI: 10.1016/j.jacc.2017.05.035

[65] Chanda S, Tiwari RK, Kumar A, Singh K. Nutraceuticals inspiring the current therapy for lifestyle diseases. Advances in Pharmacological Sciences. 2019;**2019**:6908716. DOI: 10.1155/2019/6908716

[66] Sidor A, Drożdżyńska A, Gramza-Michałowska A. Black chokeberry (*Aronia melanocarpa*) and its products as potential health-promoting factors—An overview. Trends in Food Science & Technology. 2019;**89**:45-60

[67] Benvenuti S, Pellati F, Melegari M, Bertelli D. Polyphenols, anthocyanins, ascorbic acid and radical scavenging activity of Rubus, Ribes, and Aronia. Journal of Food Science. 2004;**69**:164-169

[68] Denev PN, Kratchanov CG, et al. Bioavailability and antioxidant activity of black chokeberry (*Aronia melanocarpa*) polyphenols: In vitro and in vivo evidences and possible mechanisms of action: A review. Comprehensive Reviews in Food Science and Food Safety. 2012;**11**:471-489

[69] Zheng W, Wang SY. Oxygen radical absorbing capacity of phenolics in blueberries, cranberries, chokeberries, and lingonberries. Journal of Agricultural and Food Chemistry. 2003;**51**(2):502-509

[70] Moyer RA, Hummer KE, Finn CE, Frei B, Wrolstad RE. Anthocyanins, phenolics, and antioxidant capacity in diverse small fruits: Vaccinium, rubus, and ribes. Journal of Agricultural and Food Chemistry. 2002;**50**(3):519-525

[71] Rop O, Mlček J, Juríková T, Valšíková M, Sochor J, Řezníček V, et al. Phenolic content, antioxidant capacity, radical oxygen species scavenging and lipid peroxidation inhibiting activities of extracts of five black chokeberry (*Aronia melanocarpa* (Michx.) Elliot) cultivars. Journal of Medicinal Plant Research: Planta Medica. 2010;**4**:2431-2437

[72] Song EK, Park H, Kim HS. Additive effect of walnut and chokeberry on regulation of antioxidant enzyme gene expression and attenuation of lipid peroxidation in d-galactose-induced aging-mouse model. Nutrition Research. 2019;**70**:60-69. DOI: 10.1016/j.nutres.2018.09.011

[73] Jo AR, Imm JY. Effects of aronia extract on lifespan and age-related oxidative stress in

Drosophila melanogaster. Food Science and Biotechnology. 2017;**26**(5):1399-1406. DOI: 10.1007/s10068-017-0180-5

[74] Malinowska J, Babicz K, Olas B, Stochmal A, Oleszek W. Aronia melanocarpa extract suppresses the biotoxicity of homocysteine and its metabolite on the hemostatic activity of fibrinogen and plasma. Nutrition. 2012;**28**(7-8):793-798. DOI: 10.1016/j.nut.2011.10.012

[75] Kedzierska M, Olas B, Wachowicz B, et al. Effects of the commercial extract of aronia on oxidative stress in blood platelets isolated from breast cancer patients after the surgery and various phases of the chemotherapy. Fitoterapia. 2012;**83**(2):310-317. DOI: 10.1016/j.fitote.2011.11.007

[76] Dietrich-Muszalska A, Kopka J, Kontek B. Polyphenols from berries of Aronia melanocarpa reduce the plasma lipid peroxidation induced by Ziprasidone. Schizophrenia Research and Treatment. 2014;**2014**:602390. DOI: 10.1155/2014/602390

[77] Pilaczynska-Szczesniak L, Skarpanska-Steinborn A, et al. The influence of chokeberry juice supplementation on the reduction of oxidative stress resulting from an incremental rowing ergometer exercise. International Journal of Sport Nutrition and Exercise Metabolism. 2005;**15**(1):48-58

[78] Gazdik Z, Reznicek V, et al. Use of liquid chromatography with electrochemical detection for the determination of antioxidants in less common fruits. Molecules. 2008;**13**(11):2823-2836. DOI: 10.3390/molecules131102823

[79] Rugină D, Sconța Z, Leopold L, Pintea A, Bunea A, Socaciu C. Antioxidant activities of chokeberry extracts and the cytotoxic action of their anthocyanin fraction on HeLa human cervical tumor cells. Journal of Medicinal Food. 2012;**15**(8):700-706. DOI: 10.1089/jmf.2011.0246

[80] Naruszewicz M, Laniewska I, Millo B, Dłuzniewski M. Combination therapy of statin with flavonoids rich extract from chokeberry fruits enhanced reduction in cardiovascular risk markers in patients after myocardial infraction (MI). Atherosclerosis. 2007;**194**(2):e179-e184

[81] Bell DR, Gochenaur K. Direct vasoactive and vasoprotective properties of anthocyanin-rich extracts. Journal of Applied Physiology (1985). 2006;**100**(4):1164-1170

[82] Iwashima T, Kudome Y, Kishimoto Y. Aronia berry extract inhibits TNF-α-induced vascular endothelial inflammation through the regulation of STAT3. Food & Nutrition Research. 2019;**63**. DOI: 10.29219/fnr.v63.3361

[83] Lee KP, Choi NH, Kim HS, Ahn S, Park IS, Lee DW. Anti-neuroinflammatory effects of ethanolic extract of black chokeberry (*Aronia melanocapa* L.) in lipopolysaccharide-stimulated BV2 cells and ICR mice. Nutrition Research and Practice. 2018;**12**(1):13-19. DOI: 10.4162/nrp.2018.12.1.13

[84] Wei J, Zhang G, Zhang X, Xu D, Gao J, Fan J, et al. Anthocyanins from black chokeberry (*Aronia melanocarpa* Elliot) delayed aging-related degenerative changes of brain. Journal of Agricultural and Food Chemistry. 2017;**65**(29):5973-5984. DOI: 10.1021/acs.jafc.7b02136

[85] Loo BM, Erlund I, Koli R, Puukka P, Hellström J, Wähälä K, et al. Consumption of chokeberry (*Aronia mitschurinii*) products modestly lowered blood pressure and reduced low-grade inflammation in patients with mildly

elevated blood pressure. Nutrition Research. 2016;**36**(11):1222-1230. DOI: 10.1016/j.nutres.2016.09.005

[86] Krga I, Monfoulet LE, Konic-Ristic A, Mercier S, Glibetic M, Morand C, et al. Anthocyanins and their gut metabolites reduce the adhesion of monocyte to TNFα-activated endothelial cells at physiologically relevant concentrations. Archives of Biochemistry and Biophysics. 2016;**599**:51-59. DOI: 10.1016/j.abb.2016.02.006

[87] Borecki K, Zuchowski M, Siennicka A, Adler G, Jastrzebaska M. Polyphenol-rich extract of *Aronia melanocarpa* inhibits TNF-alpha induced apoptosis in H9c2 cells. Journal of Medical Science. 2016;**85**(3):185

[88] Bararu I, Bădescu L, Bădulescu O, Ciocoiu M, Bădescu M. Possibilities of limiting the inflammatory syndrome present in experimental diabetes mellitus by using natural poliphenols. Revista Annals of RSCB. 2013;**XVIII**(2):97-108

[89] Skoczyñska A, Jêdrychowska I, Porêba R, Affelska-Jercha A, Turczyn B, Wojakowska A, et al. Influence of chokeberry juice on arterial blood pressure and lipid parameters in men with mild hypercholesterolemia. Pharmacological Reports. 2007;**59**:177-182

[90] Hellström J, Shikov AN, Makarova M, Pihlanto A, Pozharitskaya O, Ryhänen E, et al. Blood pressure-lowering properties of chokeberry (*Aronia mithurinii*, var. Viking). Journal of Functional Foods. 2010;**2**:163-169. DOI: 10.1016/j.jff.2010.04.004

[91] Appeldoorn MM, Venema DP, Peters TH, et al. Some phenolic compounds increase the nitric oxide level in endothelial cells in vitro. Journal of Agricultural and Food Chemistry. 2009;**57**(17):7693-7699. DOI: 10.1021/jf901381x

[92] Sikora J, Broncel M, Mikiciuk-Olasik E. *Aronia melanocarpa* Elliot reduces the activity of angiotensin i-converting enzyme-in vitro and ex vivo studies. Oxidative Medicine and Cellular Longevity. 2014;**2014**:739721. DOI: 10.1155/2014/739721

[93] Borowska S, Brzóska MM, Gałażyn-Sidorczuk M, Rogalska J. Effect of an extract from *Aronia melanocarpa* L. berries on the body status of zinc and copper under chronic exposure to cadmium: An in vivo experimental study. Nutrients. 2017;**9**(12):E1374. DOI: 10.3390/nu9121374

[94] Daskalova E, Delchev S, Peeva Y, et al. Antiatherogenic and cardioprotective effects of black chokeberry (*Aronia melanocarpa*) juice in aging rats. Evidence-based Complementary and Alternative Medicine. 2015;**2015**:717439. DOI: 10.1155/2015/717439

[95] Han J, Nicholson AC. Lipoproteins modulate expression of the macrophage scavenger receptor. The American Journal of Pathology. 1998;**152**(6):1647-1654

[96] Kim JH, Auger C, Kurita I, et al. *Aronia melanocarpa* juice, a rich source of polyphenols, induces endothelium-dependent relaxations in porcine coronary arteries via the redox-sensitive activation of endothelial nitric oxide synthase. Nitric Oxide. 2013;**35**:54-64. DOI: 10.1016/j.niox.2013.08.002

[97] Qin B, Anderson RA. An extract of chokeberry attenuates weight gain and modulates insulin, adipogenic and inflammatory signalling pathways in epididymal adipose tissue of rats fed a fructose-rich diet. The British Journal of Nutrition. 2012;**108**(4):581-587. DOI: 10.1017/S000711451100599X

[98] Lipińska P, Atanasov AG, Palka M, Jóźwik A. Chokeberry pomace as a determinant of antioxidant parameters assayed in blood and liver tissue of polish merino and wrzosówka lambs. Molecules. 2017;**22**(11):E1461. DOI: 10.3390/molecules22111461

[99] Yamane T, Kozuka M, Konda D, et al. Improvement of blood glucose levels and obesity in mice given aronia juice by inhibition of dipeptidyl peptidase IV and α-glucosidase. The Journal of Nutritional Biochemistry. 2016;**31**:106-112. DOI: 10.1016/j.jnutbio.2016.02.004

[100] Ciocoiu M, Badescu M, Badulescu O, et al. Polyphenolic extract association with renin inhibitors in experimental arterial hypertension. Journal of Biomedical Science and Engineering. 2013;**6**(4):493-497

[101] Yamane T, Kozuka M, Yamamoto Y, et al. Protease activity of legumain is inhibited by an increase of cystatin E/M in the DJ-1-knockout mouse spleen, cerebrum and heart. Biochemistry and Biophysics Reports. 2017;**9**:187-192. DOI: 10.1016/j.bbrep.2016.12.010

[102] Xie L, Vance T, Kim B, et al. Aronia berry polyphenol consumption reduces plasma total and low-density lipoprotein cholesterol in former smokers without lowering biomarkers of inflammation and oxidative stress: A randomized controlled trial. Nutrition Research. 2017;**37**:67-77. DOI: 10.1016/j.nutres.2016.12.007

[103] Kardum N, Petrović-Oggiano G, Takic M, et al. Effects of glucomannan-enriched, aronia juice-based supplement on cellular antioxidant enzymes and membrane lipid status in subjects with abdominal obesity. ScientificWorldJournal. 2014;**2014**:869250. DOI: 10.1155/2014/869250

[104] Shin JH, Jung JH. Non-alcoholic fatty liver disease and flavonoids: Current perspectives. Clinics and Research in Hepatology and Gastroenterology. 2017;**41**(1):17-24. DOI: 10.1016/j.clinre.2016.07.001

[105] Mota KS, Dias GE, et al. Flavonoids with gastroprotective activity. Molecules. 2009;**14**(3):979-1012. DOI: 10.3390/molecules14030979

[106] Worsztynowicz P, Napierała M, Białas W, Grajek W, Olkowicz M. Pancreatic α-amylase and lipase inhibitory activity of polyphenolic compounds present in the extract of black chokeberry (*Aronia melanocarpa* l.). Process Biochemistry. 2014;**49**:1457-1463

[107] Lipińska P, Jóźwik A. Hepatoprotective, hypoglycemic, and hypolipidemic effect of chokeberry pomace on polish merino lambs. Animal Biotechnology. 2018;**29**(2):136-141. DOI: 10.1080/10495398.2017.1330209

[108] Jeon YD, Kang SH, Moon KH, et al. The effect of Aronia berry on type 1 diabetes in vivo and in vitro. Journal of Medicinal Food. 2018;**21**(3):244-253. DOI: 10.1089/jmf.2017.3939

[109] Simeonov SB, Botushanov NP, KarahanianEB, PavlovaMB, HusianitisHK, Troev DM. Effects of *Aronia melanocarpa* juice as part of the dietary regimen in patients with diabetes mellitus. Folia Medica. 2002;**44**:20-23

[110] Adisakwattana S, Yibchok-Anun S, et al. Cyanidin-3-rutinoside alleviates postprandial hyperglycemia and its synergism with acarbose by inhibition of intestinal α-glucosidase. Journal of Clinical Biochemistry and Nutrition. 2011;**49**(1):36-41. DOI: 10.3164/jcbn.10-116

[111] Valcheva-Kuzmanova S, Kuzmanov K, Tancheva S, Belcheva A.

Hypoglycemic and hypolipidemic effects of *Aronia melanocarpa* fruit juice in streptozotocin-induced diabetic rats. Methods and Findings in Experimental and Clinical Pharmacology. 2007;**29**(2):101-105

[112] Kozuka M, Yamane T, Nakano Y, et al. Identification and characterization of a dipeptidyl peptidase IV inhibitor from aronia juice. Biochemical and Biophysical Research Communications. 2015;**465**(3):433-436. DOI: 10.1016/j.bbrc.2015.08.031

[113] Lim SM, Lee HS, Jung JI, et al. Cyanidin-3-O-galactoside-enriched *Aronia melanocarpa* extract attenuates weight gain and adipogenic pathways in high-fat diet-induced obese C57BL/6 mice. Nutrients. 2019;11(5):pii: E1190. DOI: 10.3390/nu11051190

[114] Jakovljevic V, Milic P, Bradic J, et al. Standardized *Aronia melanocarpa* extract as novel supplement against metabolic syndrome: A rat model. International Journal of Molecular Sciences. 2018;20(1):pii: E6. DOI: 10.3390/ijms20010006

[115] Kim NH, Jegal J, Kim YN, et al. Antiobesity effect of fermented chokeberry extract in high-fat diet-induced obese mice. Journal of Medicinal Food. 2018;**21**(11):1113-1119. DOI: 10.1089/jmf.2017.4124

[116] Duchnowicz P, Nowicka A, et al. In vivo influence of extract from *Aronia melanocarpa* on the erythrocyte membranes in patients with hypercholesterolemia. Medical Science Monitor. 2012;**18**(9):CR569-CR574

[117] Zielinska-Przyjemska M, Ignatowicz E. Citrus fruit flavonoids influence on neutrophil apoptosis and oxidative metabolism. Phytotherapy Research. 2008;**22**(12):1557-1698

[118] Bijak M, Bobrowski M, et al. Anticoagulant effect of polyphenols-rich extracts from black chokeberry and grape seeds. Fitoterapia. 2011;**82**(6):811-817. DOI: 10.1016/j.fitote.2011.04.017

[119] Olas B, Wachowicz B, Nowak P, Kedzierska M, Tomczak A, et al. Studies on antioxidant properties of polyphenol-rich extract from berries of *Aronia melanocarpa* in blood platelets. Acta Physiologica Polonica. 2008;**59**:823-835

[120] Ryszawa N, Kawczyńska-Drózdz A, Pryjma J, et al. Effects of novel plant antioxidants on platelet superoxide production and aggregation in atherosclerosis. Journal of Physiology and Pharmacology. 2006;**57**(4):611-626

Chapter 10

Production of Medicinal Compounds from Endangered and Commercially Important Medicinal Plants through Cell and Tissue Culture Technology for Herbal Industry

Hemant Sood

Abstract

Plant cell culture technologies have made possible the production of a wide variety of pharmaceuticals such as alkaloids, terpenoids, steroids, saponins, monoterpenes, flavonoids and amino acids. The standardization of technologies for the production of plant metabolites through cell cultures helps in understanding the biology of their biosynthesis and accumulation. Various factors such as physical, chemical, nutritional, and genetic influence the production of metabolites in plant cell cultures. The controlled production of plant metabolites through cell cultures provides a suitable alternative not only in relieving pressure from natural habitats of plant species but also provides conditions suitable for year-round production of metabolites. The production of plant metabolites has been enhanced by exposing the cultured cells to biotic and abiotic elicitors. Off late, the induction of hairy roots has been found suitable in the production of metabolites synthesized in various parts of plants. The lack of proper understanding about the biology of biosynthesis of plant metabolites has been a major stumbling block, in addition to poor amenability of medicinal and aromatic plant species to *in vitro* conditions. Continuous efforts are required to be made in upscaling the production of metabolites on large scale. Least attention has been given towards working out the cost-effectiveness of metabolite production through cell cultures.

Keywords: phytochemicals, cell and tissue cultures

1. Introduction

The plant kingdom has provided a wide variety of natural products with diverse chemical structures and a vast array of biological activities, many of which found applications in health sciences. Over 80% of the approximately 30,000 known natural products are of plant origin. In 1985, 3500 new chemical structures were identified out of which 2600 were derived from higher plants and 121 clinically useful drugs were derived from plants [1]. Plants will continue to provide novel products as well as chemical models for new drugs in the near future [2].

Many of the plant species that produce medicinal herbs have been scientifically evaluated for their possible medical applications. The economic importance of phytopharmaceuticals in plants has led to their inevitable collection from their natural habitats and thus creating environmental and geopolitical instabilities posing a threat to their survival. The reckless collection of plants has put several of them under the categories of endangered or at the verge of extinction. This has prompted industries and scientists to find the alternative technologies for the production of phytopharmaceuticals so that the natural habitat of plants can be preserved.

Plant cell cultures have served as potential renewable resources for the production of valuable medicinal compounds, flavors, fragrances, pigments, dyes, cosmetics and fine chemicals. All these compounds belong to a group collectively known as secondary metabolites. The commercial importance of secondary metabolites and the possibilities of their production by means of cell culture technologies have gained great interest in the recent years. The current review is a survey and analysis of current status of various plant cell culture technologies used for the production of medicinally important metabolites. The future prospects of cell culture technologies in light of successful case studies have been reviewed and possible improvements are suggested.

2. Why plant cell cultures?

The capacity of plant cell, tissue and organ cultures to produce and accumulate many of the valuable chemical compounds has been recognized almost since the inception of in vitro technology. The strong and growing demand in today's market place for natural, renewable products has refocused attention on in vitro cell cultures as potential factories for phytochemical production. The advantage of producing plant metabolites in vitro has been in understanding the biology of their biosynthetic activity which ultimately can be enhanced by regulating physical, chemical, nutritional and genetic parameters. Medicinal compounds localized in morphologically specialized tissues or organs of native plants have been produced in culture systems not only by inducing specific organized cultures but also by undifferentiated callus/cell cultures.

The advances in plant cell culture technologies has made possible the production of a wide variety of pharmaceuticals like alkaloids, terpenoids, steroids, saponins, phenolics, flavonoids, and amino acids. The production of plant metabolites through cell cultures offer several advantages such as it makes possible to select genotypes with higher production of secondary metabolites, which can be generated on a continuous year round basis under controlled environment. Plant cell cultures eliminate potential political boundaries or geographic barriers which are otherwise to the production of a crop, such as the restriction of natural rubber production to the tropics or anthocyanin pigment production to climates with high light intensity. Many cost effective parameters have been tried for their economic production at large scale or by possible use of plant cell cultures for biotransformation of natural compounds [3].

3. Production of phytochemicals through cell culture technologies

3.1 Callus/cell suspension

Callus/cell suspension cultures have been the prime focus of various studies aimed at the production of phytochemicals of not only medicinal value but also of other industrially important metabolites. Callus is a proliferating mass of undifferentiated cells, which can be established from different explants of a plant species under in vitro conditions on suitable nutrient media. Once the callus is derived from high metabolite producing explants, their suspension cultures can be established by transferring those calli into liquid media under continuous agitation. Zenk [4] successfully established cell lines of different plants capable of producing high yields of secondary compounds in cell suspension cultures. The production of solasodine from calli of *Solanum elaeagnifolium* and cephaeline and emetine from callus cultures of *Cephaelis ipecacuanha* were successfully achieved [5]. Some of the notable cell culture methods which have been employed for large scale production of metabolites are production of taxol from cell suspension cultures of *Taxus mairei* [6]; production of paclitaxel and its related taxanes from different Taxus species; production of berberine through cell suspension culture of *Coptis japonica*; production of vincristine and vinblastine from *Catharanthus roseus* [7, 8], and production of taxoids from cell suspension cultures of *Taxus cuspidate* [9] (**Table 1**).

3.2 Hairy root cultures

Off late, the cultivation of hairy roots has been seen as a sustainable strategy for the production of medicinally important metabolites of plants not only due to the reason that harvesting roots has been destructive for the plants in nature but also

Plant species	Active ingredient	Reference
Agave amaniensis	Saponins	Andrijany et al. (1999)
Allium sativum L.	Alliin	Malpathak and David (1986)
Coptis japonica	Berberine	Suzuki et al. (1988) and Morimoto et al. (1988)
Duboisia leichhardtii	Tropane alkaloids	Yamada and Endo (1984)
Gentiana sp.	Secoiridoid glucosides	Srrzypezak et al. [10]
Panax ginseng	Saponins and sapogenins	Furuya et al. [11]
Papaver bracteatum	Thebaine	Day et al. (1986)
Rauvolfia serpentine × *Rhazya stricta* hybrid plant	3-Oxo-rhazinilam	Gerasimenko et al. (2001)
Scutellaria columnae	Phenolics	Stojakowska and Kisiel (1999)
Tecoma sambucifolium	Phenylpropanoid glycosides	Pletsch et al. (1993)
Taxus mairei	Taxol	Wu et al. [6]
Taxus spp.	Terpenes, sterols, flavonoids	Lish et al. (2002)
Catharanthus roseus	Catharanthine	Zhao et al. [7, 8]
Panax notoginseng	Gensenoside	Zhong and Zhong (1995)
Papaver somniferum	Morphine and codeine	Shia and Doran (1991)
Podophyllum hexandrum royle	Podophyllotoxin	Chattopadhay et al. [12]
Salvia fruticosa	Rosmarinic acid	Karam et al. [13]
Picrorhiza kurroa	Picroside-1	Sood et al. (2010, 2011) and [14]
Taxus cuspidate	Taxoids	Ketchum et al. [9]

Table 1.
Bioactive secondary metabolites produced through shoot/callus cultures/suspension cultures.

due to the ease of growing hairy roots in mass cultures in the absence of external hormones, absence of geotropism and high branching, etc. Furthermore, hairy roots produce secondary metabolites for larger periods of time, unlike natural roots which are not only in limited supply but are available at specific times in a year. For these reasons, switching from culturing natural plant-organs to hairy roots is considered as an attractive alternative for the production of many valuable natural secondary metabolites [15].

For establishing hairy root cultures, the plants are infected by *Agrobacterium rhizogenes* which induces hairy roots by the transfer of T-DNA from Ri plasmid into the plant genome. This ability of *A. rhizogenes* has led to studies on it as a source of root-derived pharmaceuticals [16]. Important metabolites produced through hairy roots are serpentine production from *Catharanthus roseus*, ajmalicine from *Rauvolfia micrantha* [17] and ginkgolides from hairy roots of *Gingko biloba* [18]. Large scale production of ginsenoside from *Panax ginseng* hairy roots has been achieved by optimizing organic nutrients in bioreactor for enhancing their production. Recent developments have indicated that hairy root culture technology has moved from small laboratory scale to a large scale industrial production. For example, the German Co. RooTec has been carrying out production of camptothecin and podophyllotoxin through hairy root cultures. In a cross-species co-culture system, hairy roots of *Linum flavum* have been found to increase the production of podophyllotoxin by 240% in the cell suspensions of *Podophyllum hexandrum*. It has been reported that secondary metabolites accumulating in aerial plant have also been accumulated in the hairy roots such as artemisinin which was thought to accumulate only in the aerial parts of *Artemisia annua* also accumulated in the hairy roots. Higher production of forskolin in transformed roots of *Coleus forskolli* was achieved by using various concentrations of auxins and auxin conjugates, cytokinins and GA3 [19]. The enhanced production of picroside-1 has been reported through hairy root cultures of *P. kurroa* [20].

3.3 Elicitation of phytochemicals production in callus/cell/hairy root cultures

The lower yield of phytochemicals in plant cell cultures prompted researchers to look into various other means of enhancing their production. The recognition that certain specific secondary metabolites such as phytoalexins are produced by plants in response to microorganisms has led to the concept of using such stimulators (known as elicitors) for in vitro cultures. The substances used as elicitors can be of biotic or abiotic origin [21]. The plants also elicit the same response when challenged by compounds of pathogenic origin [22]. The elicitation of cell suspension cultures or hairy root cultures with biotic or abiotic elicitors has been found to enhance the rate of production as well as the yields of plant secondary metabolites [23].

The biotic elicitors are substances of biological origin, which include fungal homogenate, chitosan, microorganisms (*Pseudomonas aeruginosa*, *Bacillus cereus*), glycoprotein or intracellular proteins whose function are coupled to receptors and act by activating or inactivating a number of enzymes or ion channels [24]. Abiotic elicitors include physical and chemical stresses such as UV radiations, temperature, antibiotics, salts of heavy metals, etc. [22].

Various fungal elicitors including cell wall fragments, polysaccharides, glycoproteins and oligosaccharides have been used for the production of secondary metabolites in many plant spp. and their cell cultures. The cell extracts and filtrates of four species of fungi were used for the production of taxol from elicited cell cultures of *Taxus* sp. [25]. The cell wall fractions of *Aspergillus niger* have been used as an elicitor in cell suspension cultures of *Taxus chinensis* thereby resulting in more than two fold increase in taxol yield and about six fold increase in total secretion.

Jasmonic acid (JA) and its methyl esters, methyl jasmonate (MJ) have been reported as key signaling compounds in the process of elicitation leading to the accumulation of various secondary metabolites. Lu et al. (2001) reported 28 fold higher saponin production in the elicited cultures of *Panax ginseng* by using yeast extract and methyl jasmonate as elicitors. Production of many valuable secondary metabolites using various elicitors have been reported successfully in various other plant species [26–29]. Enhanced production of podophyllotoxin in suspension cultures of *Linum album* was reported by using biotic (yeast extract) and abiotic (Ag^{+}, Pb^{2+} and Cd^{2+}) elicitors.

Methyl jasmonate, vanadyl sulphate and chitosan were used for enhancing the production of ginsenoside from hairy root cultures of *P. ginseng* [23].

Pitta-Alvarez and Giulietti et al. (2000) used jasmonic acid and aluminium chloride as elicitors for enhancing the production of scopolamine and hyoscyamine in hairy root cultures of *Brugmansia candida*. Bacterial elicitors like *Bacillus cereus*, *Staphylococcus aureus*, etc. have been used for enhancing scopolamine production from the adventitious hairy roots of *Scopolia parviflora* (**Table 2**).

Plant species	Secondary metabolites	Elicitor	Reference
Catharanthus roseus	Ajmalicine, vincristine, vinblastine	a. *Pythium* sp. b. *Yeast elicitor, MeJA* c. *Trichoderma viride* d. *Pythium aphanidermatum* e. *Jasmonic acid*	DiCosmo et al. (1987), Menke et al. (1999), Zhao et al. [7, 8] and Namdeo et al. (2004)
Picrorhiza kurroa	Picroside-1	Seaweed extract	[14]
Datura stramonium	Alkaloids (tropane)	*Phytophthora megasperma*	Kurosaki et al. (2001) and Dörenburg et al. (1994)
Azadirachta indica	Azadirachtin	Jasmonic acid, salicylic acid	Satdive et al. [30] and Funk et al. [31]
Papaver somniferum	Codeine, morphine	Fungal spores	Heinsterin et al. (1985)
Dioscorea deltoidea	Diosgenin	*Rhizopus arrhizus*	Rokem et al. (1984)
Hyoscyamus niger, *H. muticus*	Hyoscyamine, scopolamine	Fungal elicitor, MeJA	Singh (1995)
Rauwolfia canescens	Raucaffrincine	Yeast elicitor, MeJA	Gundlach et al. (1992) and Parchmann et al. (1997)
Panax ginseng	Saponin	Oligogalacturonic acid low energy ultra sound	Threfal and Whitehead (1988) and Hu et al. (2003a,b)
Hyoscyamus muticus	Sesquiterpenes	*Rhizoctonia solani*	Singh (1995)
Lithospermum erythrorhizon	Shikonin	Endogenous source	Fukui et al. [32]
Taxus chinensis	Taxol	Fungal elicitation	Wang et al. (2001)
Taxus brevifolia, *T. cuspidate*	Taxol, Baccatin III	Fungal elicitor	Yukimuni et al. [33], Hefner et al. (1998) and Luo et al. (2001)

Table 2.
Elicitors used for the production of secondary metabolites by cell cultures of medicinal plants.

3.4 Factors influencing the biosynthesis and accumulation of medicinal phytochemicals

Knowledge about biosynthetic pathways for secondary metabolite production open avenues for the targeted production of medicinal compounds as reported by Varun et al. [34] where he proposed the biosynthetic pathways for the production of picroside-1 and picroside-2 of *Picrorhiza kurroa* an endangered herb of North-Western Himalayas, having hepatoprotective iridoid compounds. Varun et al. [35, 36] optimized preparative RP-HPLC method for the isolation and purification of picrosides in *Picrorhiza kurroa*.

For maximizing the production and accumulation of secondary metabolites through plant cell cultures, specific physical conditions such as type and composition of nutrient media, type and source of explant for initiating cell cultures, incubation temperatures and intensity of light, etc. are of paramount importance.

3.5 Culture medium

The tissue culture media are the basic support system for the growth and development of plant cell cultured in vitro. The activities of basic primary metabolism are largely influenced by the basal media considered to be common to most of the plant species. However, the differentiation or dedifferentiation of plant tissue cultures is influenced by the combinations of growth hormones mainly auxins and cytokinins (**Table 3**). The manipulation of media components have been reported to influence the biosynthesis and accumulation of secondary metabolites in plant cell cultures. Different strategies have been employed for improving secondary metabolite production in suspension cultures. The influence of media constituents and nutrient stress affect the production of diosgenin from callus cultures of *Dioscorea deltoidea*. The production of gentipicroside and swertiamarin was enhanced on MS medium supplemented with kinetin, NAA and 3% sucrose in suspension cultures of *Gentiana davidii* [44].

The productivity of picroside-1 was increased by optimizing the concentration of nutrients in growth medium and levels of phytohormones in the shoot cultures of *Picrorhiza kurroa* [45, 46]. Elevated sucrose levels from 3 to 6% were favourable in some cultures whereas addition of fructose promoted paclitaxel production in *Taxus* cell cultures [6]. Supplementation of MS medium with seaweed extract also contributed in enhancement of picroside accumulation in shoot cultures of *Picrorhiza* species [14].

Plant species	Metabolites	Reference
Camptotheca acuminate	Camptothecin	Lorence et al. [37]
Gingko biloba	Ginkgolides	Ayadi et al. [18]
Gmelina arborea	Verbascoside	Dhakulkar et al. [38]
Linum flavum	Coniferin	Lin et al. [39]
Papaver somniferum	Morphine, sanguinarine, codeine	Le Flem et al. [40]
Panax ginseng	Ginsenoside	Palazon et al. [23]
Pueraria phaseoloides	Puerarin	Shi and Kintizos et al. [41]
Rauvolfia micrantha	Ajmalicine, ajmaline	Sudha et al. [17]
Saussurea medusa	Jaceosidin	Zhao et al. [42]
Solidago altissima	Polyacetylene (cis-dehydromatricaria ester)	Inoguchi et al. [43]

Table 3.
Pharmaceutical metabolites produced by hairy root cultures.

3.6 Type and source of explant

The type and source of explant has been of major importance in not only establishing successful tissue cultures in any plant species but also of significant importance in producing phytochemicals in vitro. The prime importance of choosing a right explant for the production of phytochemicals lies in the fact that the biosynthesis and accumulation of metabolites is very specific to tissues and organs along with their developmental stages. The tissue and developmental specific accumulation of phytochemicals thus makes it important that appropriate explant be selected for starting plant cell cultures for the production of phytochemicals.

Production of diosgenin has been carried out from cell suspension cultures of different explants of *Dioscorea doryophora* like stem-node, microtuber and intact tuber, etc., along with varying concentrations of sucrose in MS liquid media supplemented with 2 mg/L 2,4-D (0.3–3.5%). Increase in diosgenin production was obtained from tuber derived cell suspensions as compared to intact tuber explant.

Different cell lines were established on B5 medium supplemented with NAA by using stem- and needle-derived callus of *Taxus mairei* and taxol yield of upto 200 mg/L was obtained in precursor feeded cell suspensions [47].

3.7 Light and temperature

Plants tissue cultures are largely influenced by the quality and duration of light treatments. There are various case studies in the literature wherein manipulation of light parameters or the temperature regimes has resulted in the alteration in the production of secondary metabolites. Zhang et al. (2005) gave heat shocks of 35–50°C for 30–60 min in the suspension cultures of *Taxus yunnanensis* for enhancing the production of paclitaxel. Production was increased to six fold by pretreatment with abscisic acid. The production of swertiamarin and gentipicroside was enhanced in cell suspension cultures of *Gentiana davidii* by incubating at 25°C and light intensity of 2.33 Lux [44]. Increase in the concentration of glycyrrhizin was found in the root tissue of *Glycyrrhiza uralensis* grown under red light or under low and high intensity of UV-B radiations [48].

3.8 Precursor feeding

Exogenous supply of a biosynthetic precursor to culture medium also increases the yield of the desired products. The concept is based on the idea that any compound which is an intermediate, or is in the beginning of a secondary metabolite biosynthetic route, proves to be a good candidate for increasing the final yield of secondary metabolite. Varun et al. [49] has carried out exogenous feeding of immediate biosynthetic precursor, i.e., cinnamic acid and catalpol in the shoot cultures of *Picrorhiza kurroa* hence stimulated 4.2 fold production of picroside-1. The production of monoterpene alkaloids was increased in cell suspension cultures of *Catharanthus roseus* fed with precursor mevalonic acid, secologanin [50]. Callus cultures of *Dioscorea balcanica* fed with cholesterol, norflurazon as precursors were used for the production of diosgenin, phytosteroids [51]. Hallard et al. [52] used secologanin and tryptamine in cell suspension cultures of *Nicotiana tabaccum* for the production of strictosidine. Phenolics compounds were elicited from micropropagated plants of *Calligonum polygonoides* by Owis et al. [53].

Supplementation of media with amino acids has been found to enhance the production of indole alkaloids tropane alkaloid in cell suspension cultures [54, 55]. Addition of phenylalanine to cell suspension culture of *Salvia officinalis* enhanced the production of rosmarinic acid. The production of taxol from *Taxus* cultures was also increased by using the same precursor [56]. Nicotinic acid was used as a precursor in the hairy root cultures of *Nicotiana rustica* for the production of nicotine. Hakkinen et al. [57] used hyoscyamine as a foreign substrate for enhancing the production of scopolamine in the hairy roots of *N. tabaccum* and found that 85% of the converted scopolamine was released into the medium.

3.9 Genotypic variation

The biosynthesis and accumulation of secondary metabolites or phytochemicals of medicinal value is influenced by the genotype of the target plant species [58]. There are examples wherein genotypic variations have been reported for phytochemical content. However, there has been a technical problem in most of these studies because genotype collections are made from different locations, which vary in altitude, climatic conditions, etc. thus resulting in variation in accumulation of phytochemicals. It would be highly desirable and practically viable if the influence of genotypic variation on phytochemical content is investigated by collecting genotypes of a particular plant species and then growing under uniform environmental conditions. The variation for metabolite content can be done on those genotypic collections.

3.10 Metabolic engineering for the production of phytopharmaceuticals

True metabolic engineering of plant secondary metabolite pathways has been hampered due to the lack of thorough knowledge of biosynthetic pathways and their regulatory mechanisms leading to the formation of desired compounds. Methods like labeled precursor feeding, induced expression of regulatory genes and block competitive pathways and metabolism by antisense genes have been used for enhancing the production of desired metabolites. Yun et al. [59] cloned the hyoscyamine 6-beta hydroxylase gene (*h6h*) of *Hyoscyamus niger* and introduced into *Atropa belladonna* and collected scopolamine from engineered plant. In a later study, Hashimoto et al. (1993) reported fivefold higher concentration of scopolamine from *A. belladonna* hairy roots expressing the same gene than the wild-type hairy roots. Increased alkaloid production by overexpression of genes encoding key enzymes of tropane alkaloid biosynthesis pathway was reported by Palazon et al. [23] and Moyano et al. [60] in *Duboisia* hybrid, *Datura metel* and *Hyoscyamus muticis* hairy roots, respectively. Similarly, Zang et al. (2004) produced 411 mg/L scopolamine in cultivated hairy roots from the simultaneous over expression of *pmt* and *h6h* genes in *H. niger*. Elevated nicotine alkaloid production was achieved in *Nicotiana tabaccum* hairy roots carrying *h6h* gene [57]. Neha et al. [61] reported 2.6 fold increase in picroside-1 production by modulating four integrated secondary metabolic pathways, i.e., methyl erythritol phosphate, mevalonate, iridoid and phenylpropanoid pathway using seaweed extract. Moreover Sharma et al. [62] defined many strategies through metabolic engineering for stimulating the production of bio-active compounds from medicinal plants.

3.11 Upscaling the production of phytochemicals

The production of phytopharmaceuticals in cell cultures coupled with their low yield from natural sources and supply concerns of plant species has renewed interest in up scaling cell culture technology for large scale production. Bioreactors are the key step towards their commercial exploitation because it provides defined parameters for up scaling the production of phytochemicals or secondary metabolites from plant cell cultures.

Bioreactor is a large culture vessel fitted with microprocessor control unit for the control of pH, temperature, light, dissolved oxygen, gas flow rate, agitation speed, nutrient factors, cell density for optimal growth or production, handling of cultures, nutrient uptake and product harvestation, etc. The success of Mitsui Petrochemical Industry Co. Ltd. in Japan in producing shikonin on a commercial scale from *Lithospermum erythrorhizon* and that of Nitto Denko Corp. Japan in mass production of *Panax ginseng* or ginseng cells have demonstrated the practical feasibility of using cell cultures in the large scale production of secondary metabolites of pharmaceutical importance. Commercial companies like Phyton and Samyang Genex are successfully producing paclitaxel and its related taxanes on large scale [63].

Heble and Chadha (1985) reported the successful cultivation of *Catharanthus roseus* cells in 7–20 L capacity of airlift bioreactor for the production of ajmalicine and serpentine by judicious use of air lift and low agitation. Significant amounts of sanguinarine were produced in cell suspension cultures of *Papaver somniferum* using bioreactors [64]. Ginseng root tissue cultures in 20 ton bioreactor produced 500 mg/L of saponin per day [65]. Hahn et al. [66] have produced gensinoside from adventitious root culture of *Panax ginseng* through large scale bioreactor system. Chattopadhyay et al. [12] produced podophyllotoxin through cell cultures of *Podophyllum hexandrum* in a bioreactor.

Different types of culture systems have been successfully used such as airlift bioreactors were used for scaling up hairy root production of *Astragalus membranaceus* [67] and *Solanum chrysotricum* [68] and mist bioreactor for hairy root of *Tagetes patula* [69]. Flow diagram of a process for the production of picrosides from *Picrorhiza kurroa* is given below wherein callus cultures/suspension cultures have been established from different explants and accumulation of picrosides is being investigated by HPLC [46, 70] (**Figure 1**).

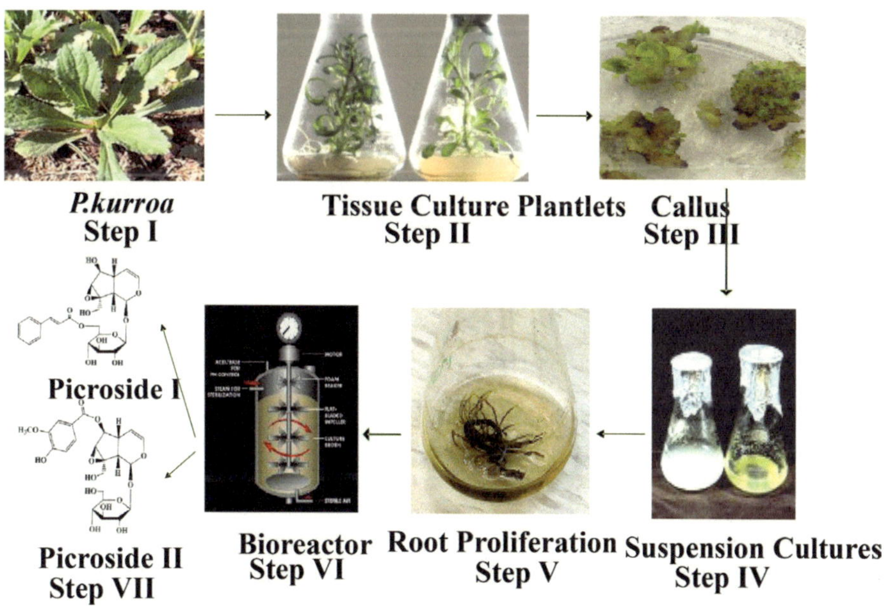

Figure 1.
Pictorial representation for picroside-1 production through plant tissue culture.

3.12 Limitations in cell culture technologies for the production of phytochemicals

The research on in vitro production of phytochemicals has been carried for the past 20 years, however, there are very few case studies where technologies have been upscaled successfully. There has been several shortcomings some of which are mentioned below:

3.12.1 General limitations

- Lack of understanding about the physical environmental and genetic factors controlling the production of pharmaceuticals
- Low yields of pharmaceuticals in tissue cultures
- Lack of information on cost effectiveness in the production of pharmaceuticals through cell cultures
- Poor amenability of most of the plant species producing pharmaceuticals to in vitro conditions
- Use of high sugar concentration (3–8%) or addition of elicitors or precursors increases the production cost considerably
- Infections due to contamination limit the progress of cell cultures
- Lack of knowledge of various molecular events that occur in secondary metabolite biosynthesis

3.12.2 Limitations pertaining to bioreactor conditions

- Cell sedimentation and death due to mass transfer of cells in large vessels limits the supply of oxygen and nutrients
- Plant cells are extremely sensitive to shear forces
- Plant cells have very low doubling time (16–24 h) therefore produce less biomass and relatively produce small amount of secondary metabolites
- For aeration of cells stirring is needed which sometimes cause damage to cells and lower the yield of products

4. Conclusions and future prospects

In spite of bottlenecks in the large scale production of phytopharmaceuticals many technological advancements and refinements have been made in the recent years right from the selection of high yielding cell lines to manipulation of basic chemical, physical and biological parameters. The identification of right explant of proper developmental stage, standardization of optimum nutrient medium resulting in maximum accumulation of pharmaceuticals, optimization of low-cost production technology are some of the areas which warrant immediate attention. Knowledge of the biosynthetic pathways of desired compounds in plants as well as in cell cultures is still rudimentary, therefore emphasis need to be made generate information based on a cellular and molecular level. Major breakthrough in the metabolomics and its integration with genomics and transcriptomics technologies will help in discovering potential genes of biosynthetic pathways so that closer understanding of the links between different levels in biological systems can lead to better understanding of the molecular biology of secondary metabolite production in plants.

Acknowledgements

The author is thankful to Prof. R.S. Chauhan, PhD scholars of Plant Biotechnology and the administration of the JUIT for providing infrastructural support in carrying out R&D on phytopharmaceuticals.

Author details

Hemant Sood
Department of Biotechnology and Bioinformatics, JUIT, Waknaghat, India

*Address all correspondence to: hemant.sood@juit.ac.in

References

[1] Payne GF, Bringi V, Prince C, Shuler ML. Plant Cell and Tissue in Liquid Systems. Munich: Hanser Publications; 1991. pp. 1-10

[2] Cox PA, Balick MJ. The ethnobotanical approach to drug discovery. Scientific American. 1994;**270**(6):82-87

[3] Ravishankar GA, Ramachandra Rao S. Biotechnological production of phyto-pharmaceuticals. Journal of Biochemistry, Molecular Biology, and Biophysics. 2000;**4**:73-102

[4] Zenk MH. The impact of plant cell culture on industry. In: Thorpe TA, editor. Frontiers of Plant Tissue Culture 1978. International Association for Plant Tissue Culture, University of Calgary; 1978. pp. 1-13

[5] Jha S, Sahu NP, Mahato SB. Production of the alkaloids emetine and cephaeline in callus cultures of *Cephaelis ipecacuanha*. Planta Medica. 1988;**54**:504-506

[6] Wu J, Wang C, Mei X. Stimulation of taxol production and excretion in *Taxus* spp cell cultures by rare earth chemical lanthanum. Journal of Biotechnology. 2001;**85**:67-73

[7] Zhao J, Hu Q, Guo Q, Zhu WH. Effects of stress factors, bioregulators, and synthetic precursor on indole alkaloid production in compact callus clusters cultures of *Catharanthus roseus*. Applied Microbiology and Biotechnology. 2001;**55**:693-698

[8] Zhao J, Zhu W, Hu Q. Enhanced catharanthine production in *Catharanthus roseus* cell cultures by combined elicitor treatment in shake flasks and bioreactors. Enzyme and Microbial Technology. 2001;**28**:673-681

[9] Ketchum REB, Rithner CD, Qiu D, Kim YS, Williams RM, Croteau RB. *Taxus* metabolites: Methyl jasmonate preferentially induces production of taxoids oxygenated at C-13 in *Taxus* media cell cultures. Phytochemistry. 2003;**62**(6):901-909

[10] Skrzypczak L, Wesolowska M, Skrzypczak E. Gentiana species XII: In vitro culture, regeneration, and production of secoirridoid glucosides. In: Bajaj YPS, editor. "Biotechnology in Agriculture and Forestry". 21, Medicinal and Aromatic Plants IV. Berlin, Heidelberg: Springer-Verlag; 1993. pp. 172-186

[11] Furuya T, Kojima H, Syono K, Ishi T, Uotani K, Nishio M. Isolation of saponin and sapogenins from callus tissue of *Panax ginseng*. Chemical and Pharmaceutical Bulletin. 1973;**21**:98-101

[12] Chattopadhyay S, Srivastava AK, Bhojwani SS, Bisaria VS. Production of podophyllotoxin by plant cell cultures of *Podophyllum hexandrum* in bioreactor. Journal of Bioscience and Bioengineering. 2002;**93**:215-220

[13] Karam NS, Jawad FM, Arikat NS, Shibli RA. Growth and rosmarinic acid accumulation in callus, cell suspension, and root cultures of wild *Salvia fruticosa*. Plant Cell, Tissue and Organ Culture. 2003;**73**:117-121

[14] Sharma N, Chauhan RS, Sood H. Seaweed extract as a novel elicitor and medium for mass propagation and picroside-1 production in an endangered medicinal herb *Picrorhiza kurroa*. Plant Cell Tissue and Organ Culture. 2015;**122**(1):57-65

[15] Shanks JV, Morgan J. Plant 'hairy root' culture. Current Opinion in Biotechnology. 1999;**10**:151-155

[16] Flores HE, Vivanco JM, Loyola-Vargas VM. Radicle biochemistry: The biology of

root-specific metabolism. Trends in Plant Science. 1999;**4**:220-226

[17] Sudha CG, Obul Reddy B, Ravishanker GA, Seeni S. Production of ajmalicine and ajmaline in hairy root cultures of *Rauvolfia micrantha* Hook f., a rare and endemic medicinal plant. Biotechnology Letters. 2003;**25**:631-636

[18] Ayadi R, Tremouillaux-Guiller J. Root formation from transgenic calli of *Ginko biloba*. Tree Physiology. 2003;**23**:713-718

[19] Mukerjee S, Ghosh B, Jha S. Higher production of forskolin in genetically transformed cultures of *Coleus forskohlii* Briq induced by growth regulators. Journal of Plant Biochemistry and Biotechnology. 2003;**12**:81-85

[20] Verma PC, Rahman L, Negi AS, Jain DR, Khanuja SPS, Baneerjee S. Agrobacterium rhizogenes-mediated transformation of *Picrorhiza kurroa* Royle ex Benth: Establishment and selection of superior hairy root clone. Plant Biotechnology Reports. 2007;**1**:169-174

[21] Eilert U. Elicitation: Methodology and aspects of application. In: Constabel F, Vasil I, editors. Cell Culture and Somatic Cell Genetics of Plants. 4th ed. San Diego: Academic Press; 1987. pp. 153-196

[22] Yeoman MM, Yeoman CL. Manipulating secondary metabolism in cultured plant cells. The New Phytologist. 1996;**134**:553-569

[23] Palazon J, Cusido RM, Bonfill M, Mallol A, Moyano E, Morales C, et al. Elicitation of different *Panax ginseng* transformed root phenotypes for an improved ginsenoside production. Plant Physiology and Biochemistry. 2003;**41**:1019-1025

[24] Veersham C. Elicitation: Medicinal Plant Biotechnology. India: CBS Publisher; 2004. pp. 270-293

[25] Ciddi V, Srinivasan V, Shuler VML. Elicitation of *Taxus* cell cultures for production of taxol. Biotechnology Letters. 1995;**17**:1343-1346

[26] Dong HD, Zhong JJ. Significant improvement of taxane production in suspension cultures of *Taxus chinensis* by combining elicitation with sucrose feed. Biochemical Engineering Journal. 2001;**8**:145-150

[27] Hu WW, Yao HU, Zhong JJ. Improvement of *Panax notoginseng* cell cultures for production of ginseng saponin and polysaccharide by high-density cultivation of pneumatically agitated bioreactors. Biotechnology Progress. 2001;**17**:838-846

[28] Wang W, Zhong JJ. Manipulation of ginsenoside heterogeneity in cell cultures of *Panax notoginseng* by addition of jasmonates. Journal of Bioscience and Bioengineering. 2002;**93**:48-53

[29] Yu KW, Gao WY, Hahn EJ, Paek KY. Jasmonic acid improves ginsenoside accumulation in adventitious root culture of *Panax ginseng* C.A. Mayer. Biochemical Engineering Journal. 2002;**11**:211-215

[30] Satdive RK, Fulzele DP, Epsan S. Enhanced production of azadirachtin by hairy root cultures of *Azadirachta indica* A. by elicitation and media optimization. Journal of Biotechnology. 2007;**128**(2):281-289

[31] Funk C, Gugler K, Brodelius P. Increased secondary product formation in plant cell suspension cultures after treatment with a yeast carbohydrate preparation (elicitor). Phytochemistry. 1987;**26**:401-405

[32] Fukui H, Tani M, Tabata M. Induction of shikonin biosynthesis by endogenous polysaccharides in *Lithospermum erythrorhizon* cell suspension cultures. Plant Cell Reports. 1990;**9**:73-76

[33] Yukimune Y, Tabata H, Higashi Y, Hara Y. Methyl jasmonate-induced overproduction of paclitaxel and baccatin III in *Taxus* cell suspension cultures. Nature Biotechnology. 1996;**14**:1129-1132

[34] Kumar V, Chauhan RS, Sood H. In vitro production and efficient quantification of major phytopharmaceuticals in an endangered medicinal herb, *Swertia chirata*. International Journal of Biotechnology and Bioengineering Research. 2013;**4**(5):495-506. ISSN 2231-1238

[35] Kumar V, Sood H, Chauhan RS. Detection of intermediates through high-resolution mass spectrometry for constructing biosynthetic pathways for major chemical constituents in a medicinally important herb, *Swertia chirayita*. Natural Product Research: Formerly Natural Product Letters. 2015;**29**(15):1449-1455

[36] Kumar V, Sharma N, Shitiz K, Singh TR, Tandon C, Sood H, et al. An insight into conflux of metabolic traffic leading to picroside-I biosynthesis by tracking molecular time course changes in a medicinal herb, *Picrorhiza kurroa*. Plant Cell, Tissue and Organ Culture. 2015;**123**(2):435-441

[37] Lorence A, Medina-Bolivar F, Nessler CL. Camptothecin and 10-hydroxycamptothecin from *Camptotheca acuminate* hairy rots. Plant Cell Reports. 2004;**22**:437-441

[38] Dhakulkar S, Ganapathi TR, Bhargav S, Bapat VA. Induction of hairy roots in *Gmelina arborea* Roxb. and production of verbascoside in hairy roots. Plant Science. 2005;**169**:812-818

[39] Lin HW, Kwok KH, Doran PM. Development of *Linum flavum* hairy root cultures for production of coniferin. Biotechnology Letters. 2003;**25**:521-525

[40] Le Flem-Bonhomme V, Laurain-Matter D, Fliniaux MA. Hairy root induction of *Papaver somniferum* var. album, a difficult-to-transform plant, by *A. rhizogenes* LBA 9402. Planta. 2004;**218**:890-893

[41] Shi HP, Kintzios S. Genetic transformation of *Pueraria phaseoloides* with *Agrobacterium rhizogenes* and puerarin production in hairy roots. Plant Cell Reports. 2003;**21**:1103-1107

[42] Zhao D, Fu C, Chen Y, Ma F. Transformation of *Saussurea medusa* for hairy roots and jaceosidin production. Plant Cell Reports. 2004;**23**:468-474

[43] Inoguchi M, Ogawa S, Furukawa S, Konda H. Production of an allelopathic polyacetylene in hairy root cultures of goldenrod (*Solidago altissima* L.). Bioscience, Biotechnology, and Biochemistry. 2003;**67**:863-868

[44] Chueh FS, Chen CC, Sagare AP, Tsay HS. Quantitative determination of secoiridoid glucoside in *in vitro* propagated plants of *Gentiana davidii* var. formosana by high performance liquid chromatography. Planta Medica. 2000;**67**:70-73

[45] Sood H, Chauhan RS. Different tissue culture parameters used for increased shoot biomass and its enrichment for a medicinal compound, picroside-1 in an endangered herb, *Picrorhiza kurroa* Royle ex Benth. Bioscientia Medicina. 2011;**2**(4):13-22

[46] Sood H, Chauhan RS. Biosynthesis and accumulation of a medicinal compound, picroside-1 in cultures of *Picrorhiza kurroa* Royle ex Benth. Plant Cell, Tissue and Organ Culture. 2010;**100**:113-117

[47] Lee CY, Lin FL, Yang CT, Wang LH, Wei HL, Tsay HS. Taxol production by cell cultures of *Taxus mairei*. In: Proceeding of the Symposium on Development and Utilization of Resources of Medicinal Plants in

Taiwan; 21 April 1995; Taiwan: Taiwan Agricultural Research Institute. TARI Special Publication 48; 1995. pp. 137-148

[48] Afreen F, Zobayed SMA, Kozai T. Spectral quality and UV-B stress stimulate glycyrrhizin concentration of *Glycyrrhiza uralensis* in hydrophonic and pot system. Plant Physiology and Biochemistry. 2005;**439**(12):1074-1081

[49] Kumar V, Sharma N, Sood H, Chauhan RS. Exogenous feeding of immediate precursors reveals synergistic effect on picroside-I biosynthesis in shoot cultures of *Picrorhiza kurroa* Royle ex Benth. Scientific Reports. 2016;**6**:29750. DOI: 10.1038/srep29750

[50] Moreno PRH, van der Heijden R, Verpoorte R. Effect of terpenoid precursor feeding and elicitation on formation of indole alkaloids in cell suspension cultures of *Catharanthus roseus*. Plant Cell Reports. 1993;**12**:702-705

[51] Savikin-Fodulovic K, Grubisic D, Gulafic L, Menkovic N, Ristic M. Diosgenin and phytosterols content in five callus lines of *Dioscorea balcanica*. Plant Science. 1998;**135**(1):63-67

[52] Hallard D, Heijden R, Verpoorte R, Lopes Cardoso MI, Pasqualij G, Memeline J, et al. Suspension cultured transgenic cells of *Nicotiana tabaccum* expressing tryptophan decarboxylase from *Catharanthus roseus* produce strictosidine upon secologanin feeding. Plant Cell Reports. 1997;**17**(1):50-54

[53] Owis AI, Abdelwaheb NS, Abul-Soad A. Elicitation of phenolics from micropropagated endangered medicinal plant *Calligonum polygonoides*. Pharmacognosy Magazine. 2016;**12**(4):465-470

[54] Silvestrini A, Pasqua G, Botta B, Monacelli B, van der Heijden R, Verpoorte R. Effect of alkaloid precursor feeding on a *Camptotheca acuminata* cell line. Plant Physiology and Biochemistry. 2002;**40**:749-753

[55] Whitmer S, Canel C, Hallard D, Goncalves C, Verpoorte R. Influence of precursor availability on alkaloid accumulation by transgenic cell lines of *Catharanthus roseus*. Plant Physiology. 1998;**116**:853-857

[56] Fett-Neto AG, Stewart JM, Nicholson SA, Pennington JJ, Di-Cosmo F. Improved taxol yield by aromatic carboxylic acid and amino acid feeding to cell cultures of *T. cuspidata*. Biotechnology and Bioengineering. 1994;**44**:967-971

[57] Hakkinen ST, Moyano E, Cusido RM, Palazon J, Pinol MT, Oksman-Caldenty KM. Enhanced secretion of tropane alkaloids in *Nicotiana tabacum* hairy roots. Journal of Experimental Botany. 2005;**56**:2611-2618

[58] Sharma TR, Singh BM, Sharma NR, Chauhan RS. Identification of high podophyllotoxin producing biotypes of *Podophyllum hexandrum* Royle from North-Western Himalaya. Journal of Plant Biochemistry and Biotechnology. 2000;**9**:49-51

[59] Yun DJ, Hasimoto T, Yamada Y. Metabolic engineering of medicinal plants: Transgenic *Atropa belladonna* an improved alkaloid composition. Proceedings of the National Academy of Sciences of the United States of America. 1992;**89**:11799-11803

[60] Moyano E, Jouhikainen K, Tammela P, Palazon J, Cusido RM, Pinol MT, et al. Effect of *pmt* gene overexpression on tropane alkaloid production in transformed root cultures of *Datura metel* and *Hyoscyamus muticus*. Journal of Experimental Botany. 2003;**54**:203-211

[61] Sharma N, Kumar V, Chauhan RS, Sood H. Modulation of picroside-I

biosynthesis in in vitro grown elicited shoots of *Picrorhiza kurroa*. Journal of Plant Growth Regulation. 2016;**35**(4):965-973

[62] Sharma M, Koul A, Sharma D, Swamy MK, Dhar MK. Metabolic engineering strategies for enhancing the production of bioactive compounds from medicinal plants. Natural Bio-active Compounds. 2019:287-316

[63] Tabata H. Paclitaxel production by plant cell culture technology. Advances in Biochemical Engineering/ Biotechnology. 2004;**87**:1-23

[64] Park JM, Yoon SY. Production of sanguinarine by suspension culture of *Papaver somniferum* in bioreactors. Journal of Fermentation and Bioengineering. 1992;**74**:292-296

[65] Charlwood BV, Charlwood KA. Terpenoid production in plant cell cultures. In: Harbourne JB, Tomas-Barberan FA, editors. Ecological Chemistry and Biochemistry of Plant Terpenoids. Oxford: Clarendon Press; 1991. pp. 95-132

[66] Hahn EJ, Kim YS, Yu KW, Jeong CS, Paek KY. Adventitious root cultures of *Panax ginseng* C.V. Meyer and ginsedoside production through large-scale bioreactor system. Journal of Plant Biotechnology. 2003;**5**:1-6

[67] Du M, Wu XJ, ding J, Hu ZB, White KN, Branford-White CJ. Astragaloside IV and polysaccharide production by hairy roots of *Astragalus membranaceus* in bioreactors. Biotechnology Letters. 2003;**25**:1853-1856

[68] Caspeta L, Quintero R, Villrreal ML. Novel airlift reactor fitting for hairy root cultures: Development and performance studies. 2005;**21**:735-740

[69] Suresh B, Bais HP, Raghavarao KSMS, Ravishankar GA, Ghildyal NP. Comparative evaluation of bioreactor design using *Tagetes patula* L. hairy roots as a model system. Process Biochemistry. 2005;**40**:1509-1515

[70] Sood H, Chauhan RS. Development of a low cost micropropagation technology for an endangered medicinal herb (*Picrorhiza kurroa*) of north-western Himalayas. Journal of Plant Sciences. 2009;**4**(2):21-31

Chapter 11

Opium Alkaloids

Mahluga Jafarova Demirkapu and Hasan Raci Yananli

Abstract

Opium alkaloids, derived from crude *Papaver somniferum L.* plant, are potent analgesic drugs, but their use is limited because of dependence and withdrawal. Opium alkaloids activate the mesocorticolimbic dopaminergic system, which project from the ventral tegmental area to the nucleus accumbens and medial prefrontal cortex, and dopamine is critically important in opioid consumption and sustaining. The reward effect resulting from the activation of the dopaminergic system leads to continued opioid consumption and occurs opioid dependence. After the development of opioid dependence, consumption continues to avoid withdrawal syndrome. Opioid dependence is accompanied with tolerance, which requires the use of high doses to achieve the same effect. When tolerance develops, the chronic consumer continues to use opioid above known toxic doses to produce the same effect, which can result in death regardless of the type of opioid used. Raw *Papaver somniferum L.* has five high-density main opium alkaloids including morphine, noscapine, codeine, thebaine, and papaverine. Some of these alkaloids bind to classical opioid receptors to produce an opioid-like effect, while the other part mediates non-opioid effects. This chapter reviews the opiod history, receptors, mechanism of action, dependence, withdrawal. In addition, general information about five main opium alkaloids, their effects, uses, routes of administration, pharmacokinetics, adverse reactions, contraindications; effects on reproduction, pregnancy, and lactation were compiled.

Keywords: morphine, noscapine, codeine, thebaine, papaverine

1. Introduction

Opium alkaloids are products obtained from the mature capsules of *Papaver somniferum L.* plant in natural way or in laboratory. Although the maintenance of *Papaver somniferum L.* is very easy, obtaining opium alkaloids naturally requires a process. Maturation of capsules occurs in May or June. First, the mature capsules are incised several times to allow the latex to flow. The exposed white latex turns brown with the effect of air. The thickened and solidified latex is collected the next day, dried under suitable conditions and made ready for use [1]. The natural way of extracting opium alkaloids for medical purposes is dried capsules [2]. The main opium alkaloids found in raw *Papaver somniferum L.* and some information are summarized in **Table 1**.

1.1 History

Papaver somniferum L. is one of the oldest medicinal plants. The first information about the production of opium alkaloids is found on Sumerian clay tablets inscribed

	Density (avg%) [3]	Molecular formula	Molecular weight (g/mol)
Morphine	11.4	$C_{17}H_{19}NO_3$	285.34
Noscapine	8.1	$C_{22}H_{23}NO_7$	413.4
Codeine	3.5	$C_{18}H_{21}NO_3$	299.4
Thebaine	3.2	$C_{19}H_{21}NO_3$	311.4
Papaverine	3.1	$C_{20}H_{21}NO_4$	339.4

Table 1.
The main opium alkaloids found in crude Papaver somniferum L. *and some information.*

in Cuneiform script in about 3000 BC. The Sumerians, whose culture developed between the Tigris and Euphrates Rivers in the south of Iraq between 4000 and 3000 BC, called the opium alkaloids "Gil" ("happiness") [4].

Opium alkaloids were first isolated in 1803 by Parisian Derosne, and named 'opium salt'. Friedrich Wilhem Adam Serturner described the 'opium salt' in detail in 1817 and named "morphine', inspired by the Morpheus (Greek god of dreams). Karl Friedrich Wilhelm Meissner first used the word 'alkaloid' in 1818, which we still use. Opioids were widely used for the first time in the Franco-Prussian War and the American Civil War for medical purposes. Tincture and pills were preferred for the purpose of analgesia in wounded soldiers. Repeated use caused opioid dependence on some soldiers, and this event was first described as "soldiers' disease" [5].

1.2 Opioid receptors

Although morphine and other opioid alkaloids are exogenous substances, they show agonistic effect by binding to the receptors of endogenous opioids. Opioid receptors were first described by Beckett and Casy in 1954 [6]. In 1965, Portoghese and colleagues shared their views on the existence of multiple opioid receptor types [7]. High-affinity and stereospecific binding sites for opioid alkaloids were also found in brain in 1973 [8]. The presence of specific opioid receptors led to the discovery of endogenous ligands. They are enkephalins [9], β-endorphin [10], and dynorphins [11]. The classic opioid receptors, were discovered in 1976–1977 and named after the prototypic drugs or tissue used in these studies: μ (mu, for morphine), δ (delta, for deferens), and κ (kappa, for ketocyclazocine) [12, 13]. These receptors show seven transmembrane domain structures specific to G protein-coupled receptors, are induced by morphine and antagonized by naloxone, and had similar analgesic effect. In 1995, the fourth opioid receptor, which is similar in structure with and closely related to classic opioid receptors, was also discovered [14, 15]. Fourth opioid receptor initially identified as ORL1 or LY132, it was later updated to N/OFQ by taking the name of its endogenous ligands (nociceptin/orphanin FQ) [16]. Although the effects of N/OFQ receptor are not fully known, they do not have a similar effect on pain as classical opioid receptors, and their sensitivity to naloxone is very low. σ (sigma), ε (epsilon), and ζ (zeta) receptors and λ (lambda) site are included in other opioid receptors [17]. The σ receptor, discovered in 1976, is not coupled to G protein, and its effects are not antagonized by naloxone [12, 18, 19]. The ε receptor is sensitive to β-endorphin [20]. The λ site regulates cell growth and is not antagonized by naloxone [21, 22]. Further information on opioid receptors is summarized in **Table 2**.

According to the studies, μ receptor was also related with addiction [38], modulation of dopaminergic system [39], learning and memory [40]. The δ receptors along with μ receptors contribute to emotional sensitivity [41].

Opioid receptors		μ receptor	δ receptor	κ receptor	N/OFQ receptor
Other names		OP3, MOP, MOPr, Mu 1	OP1, DOP, DOR, DOPr, DOR-1	OP2, KOP, KOPr, KOR-1	OP4, KOR-3, NOCIR, kappa3-related, MOR-C, nociceptin receptor ORL, XOR1, NOP-r, nociceptin/ orphanin FQ, NOPr
Regions with high distribution	CNS	Thalamus Caudate putamen Neocortex Nucleus accumbens (NAc) Amygdala İnterpeduncular complex İnferior and superior colliculi [23]	Olfactory bulb Neocortex Caudate putamen NAc Amygdala [23]	Cerebral cortex NAc Claustrum Hypothalamus [23, 24]	Cerebral cortex Anterior olfactory nucleus Lateral septum Ventral forebrain Hippocampus Hypothalamus Amygdala Substantia nigra Ventral tegmental area (VTA) Locus coeruleus Brain stem nuclei [25]
	Spinal cord	Superficial layers dorsal horn of spinal cord [26]			Dorsal horn [25]
	Non-CNS	Skin [27] Immune cells [29] Pregnant uterus [31] Gastrointestinal (GI) tract [32] Cochleae [34]	Skin [28] Immune cells [30] Pregnant uterus [31] GI tract [33] Cochleae [34]	Skin[28] Immune cells [30] Pregnant uterus [31] GI tract [33] Cochleae [34]	
Types of G-protein		Primary: Gi/Go Secondary: Gq/G11	Gi/Go	Primary: Gi/Go Secondary: G12/ G13	Primary: Gi/Go
Endogenous ligands		β-endorphin Enkephalins Endomorphin-1 and -2	β-endorphin Enkephalins	Dynorphin A Dynorphin B α-neoendorphin	Nociceptin Orphanin FQ
Agonists [from main opium alkaloids]		Morphine Codeine	Morphine	Morphine	
Antagonists		Naloxone Naltrexone	Naloxone Naltrexone	Naloxone Naltrexone	$[Nphe^1]$N/OFQ-(1-13)-NH_2 UFP-101
Effects		Analgesia Respiratory functions Cardiovascular functions GI motility Neuroendocrine functions Immune system functions Feeding Mood Thermoregulation [35]	Analgesia [36] Cardiovascular functions GI motility Mood Behaviour [15]	Analgesia Neuroendocrine functions Immune system functions Diuresis Feeding [35]	Nociception Motor and aggressive behaviours Reinforcement and reward Stress response Autonomic system functions Immune system functions [37]

Table 2.
Opioid receptors and their properties.

1.3 Mechanism of action

μ, δ, and κ opioid receptors are distributed in peripheral tissue as well as CNS. Stimulation of these receptors in the CNS results in analgesia, drowsiness, euphoria, a sense of detachment, respiratory depression, nausea and vomiting, depressed cough reflex, and hypothermia. When these receptors are stimulated in peripheral tissues, miosis, orthostatic hypotension, constipation, urinary retention etc. emerges.

After stimulation of these Gi/0-coupled opioid receptors, the adenylate cyclase enzyme is suppressed and the level of cyclic AMP decreases. In addition, the voltage-gated calcium channels in the axon ends or neuron soma are closed and intracellular calcium levels are reduced, potassium channels are opened and leading to an increase in potassium conductance. As a result, inhibition and hyperpolarization of neurons occurs when opioid receptors are stimulated [42, 43].

Analgesic or antinociceptive effects, which are indicated for use of opioids, develop at the level of the brain and spinal cord. At the brain level, attenuation of impulse spread is weakened and the perception of pain is inhibited, and at the spinal cord level, the transmission of pain impulses is suppressed [44].

1.4 Opioid dependence

Opioid dependence or addiction is a chronic, recurrent disease that changes neurotransmitter systems in the CNS and affects movement [45, 46]. Opioid dependence develops in both psychic and physical dependence. Physical opioid dependence occurs both when used for treatment and as a result of abuse. Opioid abuse is a relapsing disease with high morbidity and mortality, which is used at higher doses to produce the same effect due to tolerance. The higher the exposure time to opioids, the higher the degree of dependence and tolerance. After physical dependence develops, opioid consumption is maintained to prevent withdrawal symptoms. So the treatment is long and difficult. For this purpose, opioid agonists such as methadone, buprenorphine, an opioid antagonist naltrexone or abstinence-based treatment may be preferred. This disease, referred to as 'opioid abuse and opioid dependence' in the Diagnostic and Statistical Manual of Mental Disorders, Fourth Edition, Text Revision (DSMIV-TR), has been changed to 'opioid use disorder' in the Diagnostic and Statistical Manual of Mental Disorders, Fifth Edition (DSM-5) [47].

The estimated annual prevalence of opioids in 2010 was 0.6–0.8% of the population aged 15–64 (26×10^6 to 36×10^6). The estimated annual prevalence of opioid use is between 0.3 and 0.5% of the adult population (13×10^6 to 21×10^6 past-year users) [48].

The mesocorticolimbic dopaminergic system, which project from the VTA to the NAc and medial prefrontal cortex (mPFC) are critically important in opioid dependence [49]. Opioids act on VTA and directly or indirectly cause an increase in dopamine release in NAc region [50]. In the pathogenesis of opioid dependence, the presence of a complex mechanism including the dopaminergic system, noradrenergic, serotonergic, etc. systems should be considered [51].

1.5 Opioid withdrawal

Withdrawal is a condition that occurs when the use of an exogenous substance that is used for a long time and develops physical dependence is interrupted. In opioid dependence, the mesocorticolimbic dopaminergic system activates and induces dopamine release in the NAc region. The adaptive increase in dopaminergic activity in CNS in opioid dependence is suppressed during withdrawal and withdrawal symptoms appear [52]. In addition to dopamine, different neurotransmitters and neuromodulators, such as noradrenaline, GABA, vasopressin, substance P, neuropeptide Y, and nitric oxide, are thought to play a role in the development of opioid withdrawal [53–55]. In opioid withdrawal syndrome, symptoms such as pain, insomnia, yawning, tremor, lacrimation, rhinorrhea, sweating, dehydration, goosebumps, mydriasis, restlessness, anorexia, nausea, vomiting, diarrhea, weight loss, hyperglycemia, hypotension, decrease in respiratory rate, hyperthermia and abdominal muscle cramps are seen [45].

2. The main opium alkaloids in raw *Papaver somniferum L.* and their pharmacological properties

2.1 Morphine

International Union of Pure and Applied Chemistry (IUPAC) name: (4R,4aR,7S,7aR,12bS)-3-methyl-2,4,4a,7,7a,13-hexahydro-1H-4,12-methanobenzofuro[3,2-e]isoquinoline-7,9-diol.

ATC Code	N-Nervous system	N02-Analgesics	N02A-Opioids	N02AA-Natural opium alkaloids	N02AA01-Morphine

Morphine is one of the major opium alkaloids isolated from plant *Papaver somniferum L.*, produced synthetically and metabolized by codeine and heroin in the body. It binds to μ, δ, and κ receptors, which are widely distributed in CNS and peripheral tissue, and produce effects such as analgesia, anxiolysis, euphoria, sedation, respiratory depression, and contraction of smooth muscles of GI tract. Morphine is used in the management of severe pain, treatment of acute pulmonary edema and anesthetic procedures [56]. It can be administered orally, rectally, intramuscularly, intravenously, subcutaneously, epidurally, and intrathecally.

Absorption of morphine is variable, an almost complete absorption mainly done in the upper intestine as well as in the rectal mucosa. Morphine presents significant first-pass metabolism and oral bioavailability within 17–33% [57]. Morphine distributes to brain, skeletal muscle, liver, kidneys, lungs, intestinal tract, and spleen [58]. Hepatic metabolism occurs via glucuronic acid conjugation primarily to morphine-6-glucuronide (M6G, 10–15%) and morphine-3-glucuronide (M3G, 45–55%). Other metabolites include morphine-3,6-diglucuronide, morphine-3-ethereal sulphate, normorphine, normorphine-6-glucuronide, normorphine-3-glucuronide and codeine. M6G and normorphine show active analgesic effect by binding to opioid receptors, but M6G, which is formed more than normorphine, can contribute to analgesic effect of morphine. M3G does not contribute to the analgesic effect of morphine, because it has low affinity to opioid receptors [59]. Half-life elimination is variable according to age group: in neonates 4.5–13.3 h, in children 1–2 h, and in adults 2–4 h. Excretion occurs with urine (2–12%) and feces (7–10%).

Morphine leads to death in amounts of 0.15–0.2 g (sc) or 0.3–0.4 g (oral) in adults. Babies and young children are much more susceptible, and death has been observed at doses of 30 mg [60]. Morphine blood concentration within 10–100 μg/dL is toxically; if it is above 400 μg/dL is lethally [61].

Common adverse reactions are drowsiness, headache, constipation, nausea, vomiting, urinary retention. Although less common adverse reactions, such as depression, insomnia, paresthesia, dizziness, anxiety, abnormal dreams, confusion, seizure, myoclonus, agitation, amnesia, euphoria, pain, dyspnea, hypoventilation, respiratory depression, tremor, fever, flu-like symptoms, rhinitis, edema, hypotension, syncope, palpitations, skin rash, amblyopia, blurred vision, conjunctivitis, diplopia, miosis, nystagmus, amenorrhea, impotence, gynecomastia, urinary hesitancy, diaphoresis, anorexia, biliary colic, dyspepsia, gastroesophageal reflux disease, hiccups, xerostomia, anemia, thrombocytopenia can be seen.

Contraindications are hypersensitivity to morphine, significant respiratory depression, acute or severe bronchial asthma in the absence of resuscitative equipment, GI obstruction.

Effects on reproduction: Prolonged morphine use can cause secondary hypogonadism, which can lead to infertility in both sexes [62].

Effects on pregnancy: It is known that, morphine crosses the placenta. Morphine exposure was associated with conoventricular septal defects, atrioventricular septal defects, hypoplastic left heart syndrome, spina bifid, and gastroschisis within a 4-month period, 1 month before and 3 months after conception [63]. In addition, the use of it in the first trimester may reduce fetal heart rate from non-teratogenic effects [64]. Use of morphine late in pregnancy may result in decreased fetal breathing movements or withdrawal signs in the newborn [65, 66]. Withdrawal signs are hypothermia, hyperthermia, diarrhea, vomiting, anorexia, weight gain, high-pitched crying, hyperactivity, increased muscle tone, increased wakefulness, abnormal sleep pattern, irritability, sneezing, seizure, tremor, yawning and etc. [67, 68].

Effects on lactation: Both morphine and an active metabolite, M6G, can be detected in breast milk [69]. According to previous study, a milk:plasma ratio of morphine was 2.85, and the estimated maximum concentration in milk was 500 ng/mL [70]. Respiratory depression or drowsiness were not common in infants of breastfeeding mothers receiving morphine [71]. Although not preferred in lactation, it should be used as soon as possible and at the lowest dose if necessary [72].

2.2 Noscapine

IUPAC name: (3S)-6,7-dimethoxy-3-[(5R)-4-methoxy-6-methyl-7,8-dihydro-5H-[1,3]dioxolo[4,5-g]isoquinolin-5-yl]-3H-2-benzofuran-1-one.

ATC Code	R-Respiratory system	R05-Cough and cold preparations	R05D-Cough suppressants, excl. combinations with expectorants	R05DA-Opium alkaloids and derivatives	R05DA07-Noscapine

Noscapine, also known as narcotine, is the second opioid alkaloid according to its density in raw *Papaver somniferum L.* plant, was first isolated by Robiquet in 1817 [73]. Noscapine has no morphine-like effect, no affect on morphine withdrawal and has mild analgesic effect [74]. It is used to suppress cough frequency and intensity in bronchial asthma and pulmonary emphysema as it has central antitussive activity such as codeine [74, 75]. It is thought that noscapine exerts its antitussive effect through σ receptor which is one of the other opioid receptors. Repeated exposure to noscapine does not lead to dependence and tolerance to its antitussive effect does not develop. The potential for antineoplastic treatment is being investigated because of its antimitotic effect [76]. In the animal study, noscapine increased histamine release, leading to bronchoconstruction and hypotension, even convulsions [77]. There are also studies showing that teratogen [78]. Noscapine is used orally in the form of a combined preparation to reduce these effects.

Noscapine in terms of antitussive potency, onset, and duration of action is similar to codeine, one of the main opium alkaloids [74]. Noscapine has a relatively low bioavailability due to a first-pass metabolism [79]. Noscapine is inactivated by converting into meconin and o-demethylated metabolites. Meconin is major urinary metabolite of noscapine [80].

Adverse reactions are not expected when used in therapeutic doses [74]. When taken in high doses, drowsiness, headache, nausea, vasomotor rhinitis, conjunctivitis may be seen [81].

Effects on reproduction and pregnancy is unknown.

Effects on lactation: It is thought to have no negative effects on infant [82].

2.3 Codeine

IUPAC Name: (4R,4aR,7S,7aR,12bS)-9-methoxy-3-methyl-2,4,4a,7,7a,13-hexahydro-1H-4,12-methanobenzofuro[3,2-e]isoquinolin-7-ol

ATC Code	R-Respiratory system	R05-Cough and cold preparations	R05D-Cough suppressants, excl. combinations with expectorants	R05DA-Opium alkaloids and derivatives	R05DA04-Codeine

Codeine, a 3-methylether derivative of morphine, is the third opioid alkaloid according to its density in the raw *Papaver somniferum L.* plant, was first isolated by Robiquet in 1833. Codeine and its metabolite, morphine, act by stimulating opioid receptors. The main effects are analgesia (lighter than morphine), central antitussive and antidiarrheal effects [83]. It can also do sedation, drowsiness, and respiratory depression. Codeine is used in the management of mild to moderate pain, in the treatment of cough, persistent diarrhea and restless leg syndrome. Codeine is taken orally and intramuscularly.

Absorption is rapidly in oral use and bioavailability is higher due to less first-pass metabolism (about 53%). Codeine is distributed to a variety of tissues, with priority being to the liver, spleen and kidney [84]. Codeine-6-glucuronide, morphine, and norcodeine is formed as a result of hepatic metabolism. Morphine is then metabolized to M3G and M6G, and contributes to analgesic effects of codeine. The half-life elimination is about 3 h and excretion is mostly done through urine and with less feces.

In adults 7–14 mg/kg, in children more than 5 mg/kg intake leads to death [85]. Codeine blood concentration within 20–50 μg/dL is toxically, if it is above 60 μg/dL is lethally [86].

Abnormal dreams, insomnia, depression, paresthesia, agitation, anxiety, ataxia, dizziness, disorientation, sedation, euphoria, fatigue, hallucination, headache, bradycardia, tachycardia, circulatory depression, flushing, pruritus, skin rash, urticaria, bronchospasm, dyspnea, respiratory depression, abdominal cramps, anorexia, constipation, diarrhea, nausea, urinary hesitancy, urinary retention, blurred vision, diplopia, miosis, nystagmus, laryngospasm, muscle rigidity, tremor, hypogonadism, etc. may occur due to codeine use.

Contraindications are hypersensitivity to codeine, pediatric patients <12 years of age, postoperative management in pediatric patients <18 years of age who have undergone tonsillectomy and/or adenoidectomy, significant respiratory depression, acute or severe bronchial asthma in the absence of resuscitative equipment, GI obstruction.

Effects on reproduction: Prolonged codeine use can cause secondary hypogonadism, which can lead to infertility in both sexes [62].

Effects on pregnancy: It is known that, codeine crosses the placenta. The use in the first trimester can lead to respiratory tract malformation, pyloric stenosis, inguinal hernia, neural tube defects, cardiac and circulatory system defects, and cleft lip and palate [63, 87]. The use of codeine late in pregnancy may result in neonatal withdrawal syndrome, characterized by tremor, jitteriness, diarrhea, and poor feding [88].

Effects on lactation: Both codeine and an active metabolite, morphine, can be detected in breast milk [89]. A milk: plasma ratio of codeine is unknown. Respiratory depression, sedation and withdrawal signs can be seen in infants of breastfeeding mothers receiving codeine [90].

2.4 Thebaine

IUPAC Name: (4R,7aR,12bS)-7,9-dimethoxy-3-methyl-2,4,7a,13-tetrahydro-1H-4,12-methanobenzofuro[3,2-e]isoquinoline.

Thebaine, also known as paramorphine, which is not used for medicinal purposes and is used for the production of other opioids, is the fourth opioid alkaloid according to its density in the raw *Papaver somniferum L.* plant. Thebaine exposure can be addictive as in the use of morphine, as well as strychnine-like convulsions [91].

2.5 Papaverine

IUPAC name: 1-[(3,4-dimethoxyphenyl)methyl]-6,7-dimethoxyisoquinoline

ATC Code	A-Alimentary tract and metabolism	A03-Drugs for functional gastrointestinal disorders	A03A-Drugs for functional gastrointestinal disorders	A03AD-Papaverine and derivatives	A03AD01-Papaverine
	G-Genito urinary system and sex hormones	G04-Urologicals	G04B-Urologicals	G04BE-Drugs used in erectile dysfunction	G04BE02-Papaverine

Papaverine, which has no opioid-like effect, is the fifth opioid alkaloid based on its density in the raw *Papaver somniferum L.* plant, first isolated from opium by Merck in 1848. Papaverine affects the heart muscle and vascular smooth muscle by blocking non-selective phosphodiesterase and calcium channels [92]. Papaverine suppresses conduction and prolongs the refractory period in the heart. Vasodilatation occurs with direct effect on vascular smooth muscles including coronary and pulmonary arteries. Papaverine-mediated relaxation in smooth muscles is independent of muscle innervation and therefore does not cause paralysis in the muscles. Papaverine-mediated these effects are more pronounced especially in ischemia with arteriospasm [93]. Papaverine is used in the treatment of myocardial infarction, angina, peripheral and pulmonary embolism, peripheral vascular disease, cerebral angiospastic states. It can also be used in hypertension, urinary incontinence, prostate hyperplasia and erectile dysfunction [94]. Papaverine also has antiviral activity, which is particularly pronounced against respiratory syncytial virus, cytomegalovirus and HIV [95]. Papaverine is taken orally, intramuscularly, intravenously, and intra-arterially.

Absorption is nearly total in oral use. Oral bioavailability is higher due to less first-pass metabolism (about 54%). Papaverine is distributed to a variety of tissues, with priority being to the adipose tissue and liver. 6-Desmethylpapaverine (6-DMP, major metabolite) and 4′,6-didesmethylpapaverine (4,6-DDMP) is formed as a result of hepatic metabolism [96]. Half-life elimination is 0.5–2 h. The excretion of papaverine is through primarily urine [97]. No information on toxic blood concentrations is available. The oral median lethal dose in rats is 360 mg/kg, unknown in humans [98].

Adverse reactions are flushing, hypertension, tachycardia, headache, malaise, sedation, abdominal distress, anorexia, constipation, etc.

Use in the complete AV block is contraindicated.

Effects on reproduction and lactation is unknown.

Effects on pregnancy: It is not considered to have a adverse effect on infants.

3. Conclusions

Although information about opium alkaloids was about 3000 BC, it was first isolated in the 1800s. Opioid-like effects occur after opioid alkaloids bind to conventional opioid receptors, such as μ, δ, and κ. In particular, as a result of the agonistic effect on μ receptors, strong analgesia, physical dependence, tolerance and increased dopaminergic activity develop in the mesocorticolimbic system responsible for dependence. Increased dopaminergic activity in the mesocorticolimbic system, is a sign of developing physical dependence, decreases in the absence of opium alkaloids, leading to withdrawal syndrome. Hypogonadism in opium alkaloids users supports the idea that it has a suppressive effect on reproduction. In addition to the risk of teratogenicity, use in both pregnancy and lactation may lead to the development of abstinence syndrome in infants.

Conflict of interest

The authors declare no conflict of interest.

Author details

Mahluga Jafarova Demirkapu[1] and Hasan Raci Yananli[2]*

1 Department of Pharmacology, Faculty of Medicine, University of Tekirdag Namık Kemal, Tekirdag, Turkey

2 Department of Pharmacology, School of Medicine, University of Marmara, Istanbul, Turkey

*Address all correspondence to: hasanyananli@yahoo.com

References

[1] Mishra BK, Rastogi A, Siddiqui A, Srivastava M, Verma N, Pandey R, et al. Opium Poppy: Genetic Upgradation Through Intervention of Plant Breeding Techniques. Plant Breeding from Laboratories to Fields. Sven Bode Andersen. IntechOpen; 22 May 2013. DOI: 10.5772/53132. Available from: https://www.intechopen.com/books/plant-breeding-from-laboratories-to-fields/opium-poppy-genetic-upgradation-through-intervention-of-plant-breeding-techniques

[2] Booth M. Opium-A History [Internet]. 1998. Available from: http://. www.nytimes.com/books/first/b/booth-opium.html

[3] United Nations International Drug Control Programme. Laboratory Section. Recommended Methods for Testing Opium, Morphine and Heroin. New York: United Nations; 1998. p. 4

[4] Wink M. A short history of alkaloids. In: Roberts MF, Wink M, editors. Alkaloids. Boston, MA: Springer; 1998. pp. 35-37. DOI: 10.1007/978-1-4757-2905-4_2

[5] Mann J. Murder, Magic, and Medicine. Oxford, UK: Oxford University Press; 1992. pp. 173-186

[6] Beckett AH, Casy AF. Synthetic analgesics: Stereochemical considerations. Journal of Pharmacy and Pharmacology. 1954;**6**:986-1001. DOI: 10.1111/j.2042-7158.1954.tb11033.x

[7] Portoghese PS. A new concept on the mode of interaction of narcotic analgesics with receptors. Journal of Medicinal Chemistry. 1965;**8**:609-616. DOI: 10.1021/jm00329a013

[8] Pert CB, Snyder SH. Opiate receptor: Demonstration in nervous tissue. Science. 1973;**179**:1011-1014. DOI: 10.1126/science.179.4077.1011

[9] Hughes J, Smith TW, Kosterlitz HW, Fothergill LA, Morgan BA, Morris HR. Identification of two related pentapeptides from the brain with potent opiate agonist activity. Nature. 1975;**258**:577-580. DOI: 10.1038/258577a0

[10] Cox BM, Goldstein A, Hi CH. Opioid activity of a peptide, beta-lipotropin-(61–91), derived from beta-lipotropin. Proceedings of the National Academy of Sciences of the United States of America. 1976;**73**:1821-1823. DOI: 10.1073/pnas.73.6.1821

[11] Goldstein A, Tachibana S, Lowney LI, Hunkapiller M, Hood L. Dynorphin-(1–13), an extraordinarily potent opioid peptide. Proceedings of the National Academy of Sciences of the United States of America. 1979;**76**: 6666-6670. DOI: 10.1073/pnas.76.12.6666

[12] Martin WR, Eades CG, Thompson JA, Huppler RE, Gilbert PE. The effects of morphine- and nalorphine-like drugs in the nondependent and morphine-dependent chronic spinal dog. The Journal of Pharmacology and Experimental Therapeutics. 1976;**197**: 517-532

[13] Lord JA, Waterfield AA, Hughes J, Kosterlitz HW. Endogenous opioid peptides: Multiple agonists and receptors. Nature. 1977;**267**:495-499. DOI: 10.1038/267495a0

[14] Meunier JC, Mollereau C, Toll L, Suaudeau C, Moisand C, Alvinerie P, et al. Isolation and structure of the endogenous agonist of opioid receptor-like ORL1 receptor. Nature. 1995;**377**: 532-535. DOI: 10.1038/377532a0

[15] Reinscheid RK, Nothacker HP, Bourson A, Ardati A, Henningsen RA, Bunzow JR, et al. Orphanin FQ:

A neuropeptide that activates an opioidlike G protein-coupled receptor. Science. 1995;**270**:792-794. DOI: 10.1126/science.270.5237.792

[16] Mollereau C, Parmentier M, Mailleux P, Butour JL, Moisand C, Chalon P, et al. ORL1, a novel member of the opioid receptor family. Cloning, functional expression and localization. FEBS Letters. 1994;**341**:33-38. DOI: 10.1016/0014-5793(94)80235-1

[17] Zagon IS, Gibo DM, McLaughlin PJ. Zeta (zeta), a growth-related opioid receptor in developing rat cerebellum: Identification and characterization. Brain Research. 1991;**551**:28-35. DOI: 10.1016/0006-8993(91)90909-f

[18] Manallack DT, Beart PM, Gundlach AL. Psychotomimetic sigma-opiates and PCP. Trends in Pharmacological Sciences. 1986;7: 448-451

[19] Hanner M, Moebius FF, Flandorfer A, Knaus HG, Striessnig J, Kempner E, et al. Purification, molecular cloning, and expression of the mammalian sigma1-binding site. Proceedings of the National Academy of Sciences of the United States of America. 1996;**93**:8072-8077. DOI: 10.1073/pnas.93.15.8072

[20] Wüster M, Schulz R, Herz A. Specificity of opioids towards the mu-, delta- and epsilon-opiate receptors. Neuroscience Letters. 1979;**15**:193-198. DOI: 10.1016/0304-3940(79)96112-3

[21] Grevel J, Yu V, Sadée W. Characterization of a labile naloxone binding site (lambda site) in rat brain. Journal of Neurochemistry. 1985;**44**: 1647-1656. DOI: 10.1111/j.1471-4159.1985.tb08808.x

[22] Zagon IS, Verderame MF, Allen SS, McLaughlin PJ. Cloning, sequencing, expression and function of a cDNA encoding a receptor for the opioid growth factor, [Met(5)]enkephalin. Brain Research. 1999;**849**:147-154. DOI: 10.1016/s0006-8993(99)02046-6

[23] Mansour A, Khachaturian H, Lewis ME, Akil H, Watson SJ. Autoradiographic differentiation of mu, delta, and kappa opioid receptors in the rat forebrain and midbrain. The Journal of Neuroscience. 1987;7:2445-2464

[24] Kitchen I, Slowe SJ, Matthes HW, Kieffer B. Quantitative autoradiographic mapping of mu-, delta- and kappa-opioid receptors in knockout mice lacking the mu-opioid receptor gene. Brain Research. 1997;**778**:73-88. DOI: 10.1016/s0006-8993(97)00988-8

[25] Neal CR Jr, Mansour A, Reinscheid R, Nothacker HP, Civelli O, Akil H, et al. Opioid receptor-like (ORL1) receptor distribution in the rat central nervous system: Comparison of ORL1 receptor mRNA expression with (125)I-[(14)Tyr]-orphanin FQ binding. The Journal of Comparative Neurology. 1999;**412**:563-605

[26] Besse D, Lombard MC, Zajac JM, Roques BP, Besson JM. Pre- and postsynaptic distribution of mu, delta and kappa opioid receptors in the superficial layers of the cervical dorsal horn of the rat spinal cord. Brain Research. 1990;**521**:15-22. DOI: 10.1016/0006-8993(90)91519-m

[27] Ständer S, Gunzer M, Metze D, Luger T, Steinhoff M. Localization of mu-opioid receptor 1A on sensory nerve fibers in human skin. Regulatory Peptides. 2002;**110**:75-83. DOI: 10.1016/s0167-0115(02)00159-3

[28] Salemi S, Aeschlimann A, Reisch N, Jüngel A, Gay RE, Heppner FL, et al. Detection of kappa and delta opioid receptors in skin—Outside the nervous system. Biochemical and Biophysical Research Communications. 2005;**338**: 1012-1017. DOI: 10.1016/j.bbrc.2005.10.072

[29] Chuang TK, Killam KF Jr, Chuang LF, Kung HF, Sheng WS, Chao CC, et al. Mu opioid receptor gene expression in immune cells. Biochemical and Biophysical Research Communications. 1995;**216**:922-930. DOI: 10.1006/bbrc.1995.2709

[30] Gavériaux C, Peluso J, Simonin F, Laforet J, Kieffer B. Identification of kappa- and delta-opioid receptor transcripts in immune cells. FEBS Letters. 1995;**369**:272-276. DOI: 10.1016/0014-5793(95)00766-3

[31] Zhu Y, Pintar JE. Expression of opioid receptors and ligands in pregnant mouse uterus and placenta. Biology of Reproduction. 1998;**59**:925-932. DOI: 10.1095/biolreprod59.4.925

[32] Bagnol D, Mansour A, Akil H, Watson SJ. Cellular localization and distribution of the cloned mu and kappa opioid receptors in rat gastrointestinal tract. Neuroscience. 1997;**81**:579-591. DOI: 10.1016/s0306-4522(97)00227-3

[33] Pol O, Palacio JR, Puig MM. The expression of delta- and kappa-opioid receptor is enhanced during intestinal inflammation in mice. The Journal of Pharmacology and Experimental Therapeutics. 2003;**306**:455-462. DOI: 10.1124/jpet.103.049346

[34] Jongkamonwiwat N, Phansuwan-Pujito P, Sarapoke P, Chetsawang B, Casalotti SO, Forge A, et al. The presence of opioid receptors in rat inner ear. Hearing Research. 2003;**181**:85-93. DOI: 10.1016/s0378-5955(03)00175-8

[35] Dhawan BN, Cesselin F, Raghubir R, Reisine T, Bradley PB, Portoghese PS, et al. International Union of Pharmacology. XII. Classification of opioid receptors. Pharmacological Reviews. 1996;**48**:567-592

[36] Guan JS, Xu ZZ, Gao H, He SQ, Ma GQ, Sun T, et al. Interaction with vesicle luminal protachykinin regulates surface expression of delta-opioid receptors and opioid analgesia. Cell. 2005;**122**:619-631. DOI: 10.1016/j.cell.2005.06.010

[37] Borsodi A, Bruchas M, Caló G, Chavkin C, Christie MJ, Civelli O, et al. Opioid receptors (version 2019.4) in the IUPHAR/BPS Guide to Pharmacology Database. UPHAR/BPS Guide to Pharmacology CITE. 2019(4). DOI: 10.2218/gtopdb/F50/2019.4

[38] Hutcheson DM, Matthes HW, Valjent E, Sánchez-Blázquez P, Rodríguez-Díaz M, Garzón J, et al. Lack of dependence and rewarding effects of deltorphin II in mu-opioid receptor-deficient mice. The European Journal of Neuroscience. 2001;**13**:153-161. DOI: 10.1111/j.1460-9568.2001.01363.x

[39] Tien LT, Park Y, Fan LW, Ma T, Loh HH, Ho IK. Increased dopamine D2 receptor binding and enhanced apomorphine-induced locomotor activity in mu-opioid receptor knockout mice. Brain Research Bulletin. 2003;**61**:109-115. DOI: 10.1016/s0361-9230(03)00077-7

[40] Jamot L, Matthes HW, Simonin F, Kieffer BL, Roder JC. Differential involvement of the mu and kappa opioid receptors in spatial learning. Genes, Brain, and Behavior. 2003;**2**:80-92. DOI: 10.1034/j.1601-183x.2003.00013.x

[41] Filliol D, Ghozland S, Chluba J, Martin M, Matthes HW, Simonin F, et al. Mice deficient for delta- and mu-opioid receptors exhibit opposing alterations of emotional responses. Nature Genetics. 2000;**25**:195-200. DOI: 10.1038/76061

[42] Jordan B, Devi LA. Molecular mechanisms of opioid receptor signal transduction. British Journal of Anaesthesia. 1998;**81**:12-19. DOI: 10.1093/bja/81.1.12

[43] North RA, Williams JT, Surprenant A, Christie MJ. Mu and delta

receptors belong to a family of receptors that are coupled to potassium channels. Proceedings of the National Academy of Sciences of the United States of America. 1987;**84**:5487-5491. DOI: 10.1073/pnas.84.15.5487

[44] Malaivijitnond S, Varavudhi P. Evidence for morphine-induced galactorrhea in male cynomolgus monkeys. Journal of Medical Primatology. 1998;**27**:1-9. DOI: 10.1111/j.1600-0684.1998.tb00061.x

[45] Eddy NB, Halbach H, Isbell H, Seevers MH. Drug dependence: Its significance and characteristics. Bulletin of the World Health Organization. 1965;**32**:721-733

[46] Koob GF, Le Moal M. Opioids. In: Neurobiology of Addiction. London: Academic Press; 2006. pp. 121-171

[47] American Psychiatric Association. Diagnostic and Statistical Manual of Mental Disorders. Fifth Edition (DSM-5). Arlington: American Psychiatric Association; 2013. pp. 541-546

[48] World Drug Report. New York: United Nations Office on Drugs and Crime; 2012. p. 1

[49] Koob GF. Drugs of abuse: Anatomy, pharmacology and function of reward pathways. Trends in Pharmacological Sciences. 1992;**13**:177-184. DOI: 10.1016/0165-6147(92)90060-j

[50] Koob GF, Volkow ND. Neurocircuitry of addiction. Neuropsychopharmacology. 2010;**35**:217-238. DOI: 10.1038/npp.2009.110

[51] Gómez-Milanés I, Almela P, García-Carmona JA, García-Gutiérrez MS, Aracil-Fernández A, Manzanares J, et al. Accumbal dopamine, noradrenaline and serotonin activity after naloxone-conditioned place aversion in morphine-dependent mice. Neurochemistry International. 2012;**61**:433-440. DOI: 10.1016/j.neuint.2012.06.011

[52] Diana M, Pistis M, Muntoni A, Gessa G. Profound decrease of mesolimbic dopaminergic neuronal activity in morphine withdrawn rats. The Journal of Pharmacology and Experimental Therapeutics. 1995;**272**:781-785

[53] Yananli H, Gören MZ, Berkman K, Aricioğlu F. Effect of agmatine on brain L-citrulline production during morphine withdrawal in rats: A microdialysis study in nucleus accumbens. Brain Research. 2007;**1132**:51-58. DOI: 10.1016/j.brainres.2006. 11.028

[54] Topkara B, Yananli HR, Sakallı E, Demirkapu MJ. Effects of injection of gamma-aminobutyric acid agonists into the nucleus accumbens on naloxone-induced morphine withdrawal. Pharmacology. 2017;**100**:131-138. DOI: 10.1159/000477548

[55] Jafarova Demirkapu M, Yananli HR, Kaleli M, Sakalli HE, Gören MZ, Topkara B. The role of adenosine A1 receptors in the nucleus accumbens during morphine withdrawal. Clinical and Experimental Pharmacology & Physiology. DOI: 10.1111/1440-1681.13224

[56] Chou R, Fanciullo GJ, Fine PG, Adler JA, Ballantyne JC, Davies P, et al. Clinical guidelines for the use of chronic opioid therapy in chronic noncancer pain. The Journal of Pain. 2009;**10**:113-130. DOI: 10.1016/j.jpain.2008.10.008

[57] Masood AR, Thomas SH. Systemic absorption of nebulized morphine compared with oral morphine in healthy subjects. British Journal of Clinical Pharmacology. 1996;**41**:250-252. DOI: 10.1111/j.1365-2125.1996.tb00192.x

[58] Hunt A, Joel S, Dick G, Goldman A. Population pharmacokinetics of oral morphine and its glucuronides in children receiving morphine as immediate-release liquid or sustained-

release tablets for cancer pain. The Journal of Pediatrics. 1999;**135**:47-55. DOI: 10.1016/s0022-3476(99)70326-4

[59] Christrup LL. Morphine metabolites. Acta Anaesthesiologica Scandinavica. 1997;**41**:116-122. DOI: 10.1111/j.1399-6576.1997.tb04625.x

[60] Harvey AL. Natural and Synthetic Neurotoxins. London, England: Academic Press; 1993. p. 263

[61] Gossel TA. In: Gossel TA, Bricker JD, editors. Principles of Clinical Toxicology. 3rd ed. New York: CRC Press; 1994. p. 421

[62] Brennan MJ. The effect of opioid therapy on endocrine function. The American Journal of Medicine. 2013; **126**:S12-S18. DOI: 10.1016/j.amjmed.2012.12.001

[63] Broussard CS, Rasmussen SA, Reefhuis J, Friedman JM, Jann MW, Riehle-Colarusso T, Honein MA; National Birth Defects Prevention Study. Maternal treatment with opioid analgesics and risk for birth defects. American Journal of Obstetrics and Gynecology. 2011;**204**:314.e1-11. DOI: 10.1016/j.ajog.2010.12.039

[64] Schmid M, Kuessel L, Klein K, Metz V, Fischer G, Krampl-Bettelheim E. First-trimester fetal heart rate in mothers with opioid addiction. Addiction. 2010;**105**:1265-1268. DOI: 10.1111/j.1360-0443.2010.02982.x

[65] Kopecky EA, Ryan ML, Barrett JF, Seaward PG, Ryan G, Koren G, et al. Fetal response to maternally administered morphine. American Journal of Obstetrics and Gynecology. 2000;**183**:424-430. DOI: 10.1067/mob.2000.105746

[66] Cobrınık RW, Hood RT Jr, Chusıd E. The effect of maternal narcotic addiction on the newborn infant; review of literature and report of 22 cases. Pediatrics. 1959;**24**:288-304

[67] Dow K, Ordean A, Murphy-Oikonen J, Pereira J, Koren G, Roukema H, et al. Neonatal abstinence syndrome clinical practice guidelines for Ontario. Journal of Population Therapeutics and Clinical Pharmacology. 2012;**19**:e488-e506

[68] Hudak ML, Tan RC, Committee On Drugs, Commıttee on Fetus and Newborn, American Academy of Pediatrics. Neonatal drug withdrawal. Pediatrics. 2012;**129**:e540-e560. DOI: 10.1542/peds.2011-3212

[69] Baka NE, Bayoumeu F, Boutroy MJ, Laxenaire MC. Colostrum morphine concentrations during postcesarean intravenous patient-controlled analgesia. Anesthesia and Analgesia. 2002;**94**:184-187. DOI: 10.1097/00000539-200201000-00035

[70] Feilberg VL, Rosenborg D, Broen Christensen C, Mogensen JV. Excretion of morphine in human breast milk. Acta Anaesthesiologica Scandinavica. 1989; **33**:426-428. DOI: 10.1111/j.1399-6576.1989.tb02938.x

[71] Wittels B, Scott DT, Sinatra RS. Exogenous opioids in human breast milk and acute neonatal neurobehavior: A preliminary study. Anesthesiology. 1990;**73**:864-869. DOI: 10.1097/00000542-199011000-00012

[72] Martin E, Vickers B, Landau R, Reece-Stremtan S. ABM clinical protocol #28, peripartum analgesia and anesthesia for the breastfeeding mother. Breastfeeding Medicine. 2018;**13**:164-171. DOI: 10.1089/bfm.2018.29087.ejm

[73] Willette RE. Analgesic agents. In: Delgado JN, Remers WA, editors. Textbook of Organic Medicinal and Pharmaceutical Chemistry. 10th ed.

Philadelphia, PA: Lippincott-Raven; 1998. pp. 687-711

[74] Osol A, Hoover JE, et al. Remington's Pharmaceutical Sciences. In: Osol A, Hoover JE, editors. 15th ed. Easton, Pennsylvania: Mack Publishing Co.; 1975. p. 805

[75] Ramanathan VS, Chandra P. Recovery, separation and purification of narcotine and papaverine from Indian opium. Bulletin on Narcotics. 1981;**33**: 55-64

[76] National Cancer İnstitute. Noscapine [Internet]. 2019. Available from: https://ncit.nci.nih.gov/ncitbrowser/ConceptReport.jsp?dictionary=NCI_Thesaurus&ns=NCI_Thesaurus&code=C80589

[77] Goodman LS, Gilman A. In: Goodman LS, Gilman A, editors. The Pharmacological Basis of Therapeutics. 5th ed. New York: Macmillan Publishing Co., Inc.; 1975. p. 279

[78] Gutstein HB, Akil H. Opioid analgesics. In: Hardman JG, Limbird LE, Gilman AG, McGraw-Hill, editors. The Pharmacological Basis of Therapeutics. 10th ed. New York, NY; 2001. pp. 569-611

[79] Dahlström B, Mellstrand T, Löfdahl CG, Johansson M. Pharmacokinetic properties of noscapine. European Journal of Clinical Pharmacology. 1982;**22**:535-539. DOI: 10.1007/bf00609627

[80] Tsunoda N, Yoshimura H. Metabolic fate of noscapine. II. Isolation and identification of novel metabolites produced by C—C bond cleavage. Xenobiotica. 1979;**9**:181-187. DOI: 10.3109/00498257909038719

[81] American Medical Association, AMA Department of Drugs. AMA Drug Evaluations. 3rd ed. Littleton, Massachusetts: PSG Publishing Co., Inc.; 1977. p. 667

[82] American Academy of Pediatrics Committee on Drugs. The transfer of drugs and other chemicals into human milk. Pediatrics. 1994;**93**:137-150

[83] Troy DB. In: Troy DB, editor. Remmington: The Science and Practice of Pharmacy. 21st ed. Philadelphia, PA: Lippincott Williams & Williams; 2005. p. 1527

[84] Baselt RC, Cravey RH. Disposition of Toxic Drugs and Chemicals in Man. 4th ed. Foster City, CA, USA: Chemical Toxicology Institute; 1995. pp. 191-194

[85] Dart RC. In: Dart RC, editor. Medical Toxicology. 3rd ed. Philadelphia, PA: Lippincott Williams & Wilkins.; 2004. p. 761

[86] Gossel TA, Bricker JD. Principles of Clinical Toxicology. 3rd ed. New York, NY: Raven Press, Ltd.; 1994. p. 298

[87] Yazdy MM, Mitchell AA, Tinker SC, Parker SE, Werler MM. Periconceptional use of opioids and the risk of neural tube defects. Obstetrics and Gynecology. 2013;**122**:838-844. DOI: 10.1097/AOG.0b013e3182a6643c

[88] Nair V, Soraisham AS, Akierman A. Neonatal withdrawal syndrome due to maternal codeine use. Paediatrics & Child Health. 2012;**17**:e40-e41. DOI: 10.1093/pch/17.5.e40

[89] Meny RG, Naumburg EG, Alger LS, Brill-Miller JL, Brown S. Codeine and the breastfed neonate. Journal of Human Lactation. 1993;**9**:237-240. DOI: 10.1177/089033449300900423

[90] Lam J, Matlow JN, Ross CJ, Hayden MR, Carleton BC, Madadi P. Postpartum maternal codeine therapy and the risk of adverse neonatal

outcomes: The devil is in the details. Therapeutic Drug Monitoring. 2012;**34**: 378-380. DOI: 10.1097/FTD.0b013e31 825da19f

[91] The Merck Index. 11th ed. Title 21 Part 1308.12. 1985

[92] Kaneda T, Takeuchi Y, Matsui H, Shimizu K, Urakawa N, Nakajyo S. Inhibitory mechanism of papaverine on carbachol-induced contraction in bovine trachea. Journal of Pharmacological Sciences. 2005;**98**:275-282. DOI: 10.1254/jphs.fpj05013x

[93] Stiefel MF, Spiotta AM, Udoetuk JD, Maloney-Wilensky E, Weigele JB, Hurst RW, et al. Intra-arterial papaverine used to treat cerebral vasospasm reduces brain oxygen. Neurocritical Care. 2006;**4**:113-118. DOI: 10.1385/NCC:4:2:113

[94] Pfizer New Zealand Limited. DBL Papaverine hydrochloride injection product data sheet [İnternet]. 2017. Available from: http://www.medsafe.govt.nz/profs/Datasheet/d/dblPapaverineHydrochlorideinj

[95] National Cancer İnstitute. Papaverine [Internet]. 2019. Available from: https://ncit.nci.nih.gov/ncitbrowser/ConceptReport.jsp?dictionary=NCI_Thesaurus&ns=NCI_Thesaurus&code=C726

[96] Wolf CE, Pierce KL, Goldfine BL, Nanco CR, Poklis JL, Korzun WJ. Using papaverine and its metabolites, 6-desmethyl papaverine and 4',6-didesmethyl papaverine as biomarkers to improve the detection time of heroin use. Journal of Analytical Toxicology. 2019;**43**:600-606. DOI: 10.1093/jat/bkz069

[97] US Natl Inst Health; DailyMed. Current Medication Information for Papaverine hydrochloride injection, solution, 2012. Available from: http://dailymed.nlm.nih.gov/dailymed/lookup.cfm?setid=9e5e2ce4-7a57-4c61-a826-64c8d11d038e

[98] RxList. Papaverine injection [Internet]. 2019. Available from: https://www.rxlist.com/papaverine-injection-drug.htm#overdosage

Chapter 12

The Characteristics of Korean Traditional Post-Fermented Tea (Chungtaejeon)

Doo-Gyung Moon

Abstract

The structure of the bacterial community involved in the production of oriental traditional post-fermented tea (Chungtaejeon) was investigated using 16S rRNA gene analysis. The main microorganisms in fermentation process of Chungtaejeon are identified as *Pantoea sp*. and *Klebsiella oxytoca*. Phylogenetic analysis suggested that the taxonomic affiliation of the dominant species in the Chungtaejeon was γ-proteobacteria. The bacterial community size was higher about 100 times in Chungtaejeon compared with other Korean tea and puer-tea. Also, the fungal community size was higher about seven times in the Chungtaejeon than in the other post-fermented teas. However, the archaeal community size was highest about six times in the Chungtaejeon. Therefore, the bacteria, fungal and archaeal community sizes were highest for Chungtaejeon than in the seven post-fermented teas. As a result, the microbial communities of Chungtaejeon were the largest compared with other teas. The catechin content decreased from 12.10 to 3.80 mg/g, and epicatechin (EC) and gallic acid contents were increased to 28.50 and 8.02 mg/g, respectively, during manufacturing. The *Pantoea sp.* may perform an important role for manufacturing and fermentation to gallic acid from catechins of Chungtaejeon.

Keywords: EGCG (epigallocatechin gallate), dominant bacteria, gallic acid, γ-proteobacteria

1. The history of Korean tea

Tea plant seeds were brought to Korea from China in 828 CE and planted on Jiri Mount in southern Gyeongsang Province. Korean tea history started from the Three Kingdoms, Goryeo and Joseon Dynasties to present for about 1200 year-old tradition. Now, Korean tea culture has been developed to Korean pottery, tea books, and people who like tea and lantern festivals, etc. However, Korean people employ tea as the symbol for communication, reflection, social justice, loyalty, filial piety and manners, etc. Korean Darye translates to "etiquette for tea," which is a way of slowing down and relaxing the mind in everyday life with tea. Almost all of South Korea's tea is grown in the peninsula where people enjoy sea breezes from the Korean strait and the East Sea at the Boseong, Hadong, Jeju and Jangheung regions. Most of the tea produced is green tea picked up from April to May by

IntechOpen

hand and machine. Nowadays, green tea (powder green tea), yellow tea, black tea, post-fermented tea and blending tea are produced in the southern region and its consumption have increased every year.

Byeongcha was popular during the Tang Dynasty (China). It is rapidly spread beyond Korean nobles. A small piece of Byeongcha was dug at an ancient tomb of Goguryeo. It is the representative of Korean tea, which has developed uniquely for long time in terms of Korean climate, custom and preferences. The Borimsa temple served as the main during the Three Kingdoms Dynasty. Historical Borimsa records about tea in Jangheung [1]. The Chungtaejeon is post-fermented tea that is a kind of Byeongcha (mold) and looks similar to the coin that is called Doncha (shape).

2. What is Chungtaejeon?

One of the post-fermented teas Chungtaejeon has been developed and inherited from Jangheung in southern coastal areas in Korea over 1200 years ago. It is an international fermented tea, which has a long history from the Three Kingdoms Dynasty to the early modern period of Korea. Chungtaejeon is a kind of Byeongcha (mold). It looks similar to a coin and is called Doncha (shape, **Figure 1**). Also, it is named since its color changes to blue during the fermentation process.

The historical authenticity of Jangheung Chungtaejeon has been recorded on a tombstone in the Borimsa temple. The Borimsa temple served as the main one during the Three Kingdoms Dynasty. On the other hand, Goryeo ran 19 tea spots, and 13 tea spots were in Jangheung [1]. A tea spot is a special production area for tea. Each province ran a national tea farm in the country where tea is well harvested, and the province governor offered tea as a tribute to the king. Also, Jangheung made the best tea of the country during Joseon Dynasty. Nowadays, Chungtaejeon, which is world recognized as a luxury tea, has been approved as registered one for art of taste (2013), selected as "slow food organization" by international life varieties foundation (2014), Japanese international green tea competition best gold medal (2008), gold medal (2011), best gold medal (2014), gold medal (2015 and 2019) and national important agricultural inheritance (2018) [1].

Figure 1.
The shape of Chungtaejeon of Korean traditional post-fermentation.

3. Manufacturing methods of Chungtaejeon

How to make Chungtaejeon? First of all, collect of tea leaves after picking up for one bud two to three leaves in tea garden on the clear day, and then dry for 24 hours in the room of ventilation and removed after selecting to hard leaves and stems. Second, stem the tea leaves in an pot with water vapor for about 5–15 minutes. Third or fifth one, grind it in the large wood mortar until pasting it. Fourth, it in bamboo mold of coin shape and pre-dry. After that, make holes that are 0.2 cm in diameter by needles with bamboo in the center of the mold and dry in the room. Last, ferment it in the pot in the room and store.

Nowadays, people visit Jangheung for their well-being and healing effect by making Chungtaejeon.

4. The characteristics of Chungtaejeon

The post-fermented tea has softer taste, an increased anti-bacterial effect, antioxidant activity, and lower cholesterol production effects by fermentation of microorganisms due to higher amounts of gallic acid, methoxy phenolic compound, and polyflavonoids [2–7]. The Chungtaejeon changes to blue color during the fermentation process (**Figure 1**).

The raw tea leaves of wild *Camellia sinensis* are dried overnight in a well-ventilated room. The steaming process is carried out for about 15 minutes after removing impurities in the leaves. The steamed tea leaves are pulverized with a motor. After pulverization, the coin-shaped tea balls dry for 2–3 days in a well-ventilated room. The center of coin-shaped tea ball is punched by bamboo stem and connected by a thread to dry it in the sun, ondol room, or shade. Dried Chungtaejeon is kept in a pot to prevent loss of aroma and to avoid moisture. We analyzed bacterial community from raw leaves to products of a 1-year fermented tea (**Figure 2**).

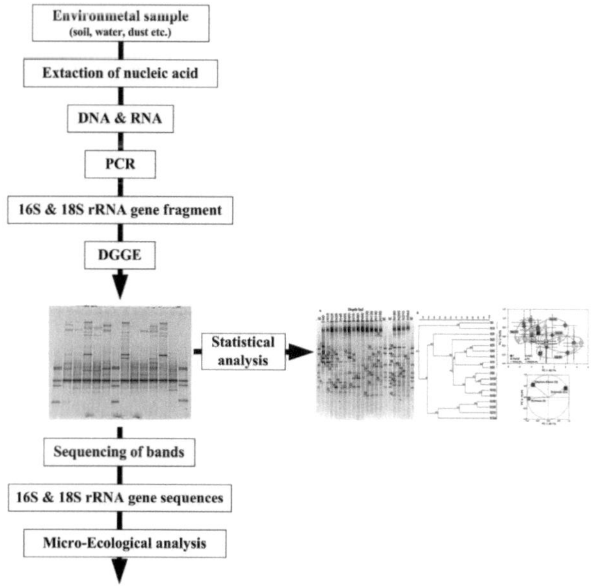

Figure 2.
Flow diagram of PCR-DG-DGGE (double gradient-denaturing gradient gel electrophoresis) analysis for microbial communities.

Item	Before steaming raw leaves	After steaming	After crushing	After forming (0 day)	1 day	3 days	7 days	9 days	11 days	12 days
Acidobacteria	42.86	7.55	2.44	47.6	3.46	7.50	21.4	66.6	13.5	33.3
Alpha-proteobacteria	32.14	1.89	—	—	—	2.50	—	—	—	—
Gamma-proteobacteria	25.0	90.57	97.6	52.4	96.5	90.0	78.5	33.3	86.4	66.6
Total (%)	100	100	100	100	100	100	100	100	100	100

Table 1.
The percentage of the dominant bacteria during manufacturing of Korean traditional post-fermented tea (%).

DNA extractions were carried out in accordance with eukaryotic microalgal nucleic acids extration (EMNE) method. The bacterial universal primers 27F (5′-GAG TTT GAT CMT GGC TCA G-3′) and 518R (5′-WTT ACC GCG GCT GCT GG-3′) were used to amplify the 16S rRNA genes for NGS analysis. The primer sequences and PCR conditions for Roche 454 are described in Pitta et al. [8]. After the PCR reaction, purification was carried out using PCR purification kit (Biofact, Daejeon, Korea). NGS was performed with a Roche 454 GS-FLX plus (454 Life Sciences) according to the procedure described by Galan et al. [9]. To compare modified methods with traditional ones for the overall bacterial communities of the Chungtaejeon, operational taxon unit (OTU)-based and phylogeny-based analyses were performed.

Catechins, gallic acid and caffeine contents were analyzed by HPLC (Agilent 1216 Infinity LC) using column (ZORBAX Eclipse plus C18, 4.6 × 250 mm).

We studied the structure of the fermentation bacterial community of Korean traditional post-fermented tea that was investigated by metagenome analysis using 16S rRNA gene during manufacturing (**Table 1**). The Acidobacteria was changed

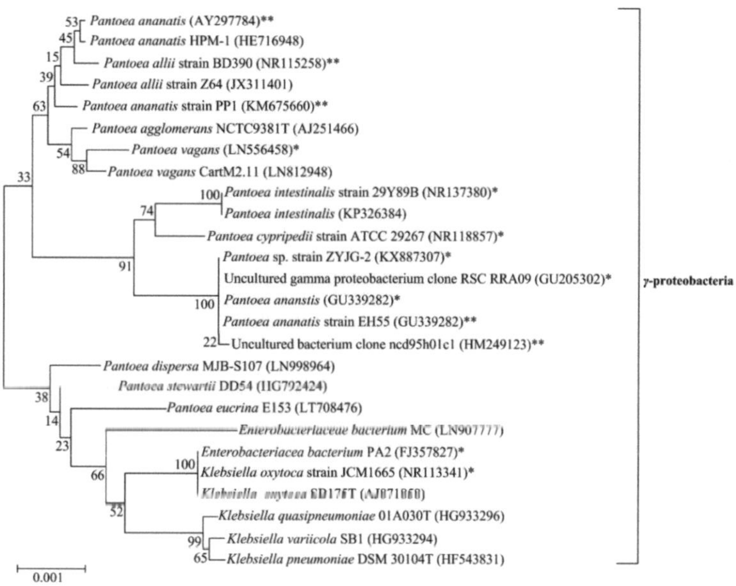

Figure 3.
*Phylogenetic tree on 16S rRNA gene sequence using the neighbor-joining methods showing the Korean traditional post-fermented tea (Chungtaejeon). *: 16S rRNA gene clone library, **:**: DG-DGGE (double gradient-denaturing gradient gel electrophoresis).*

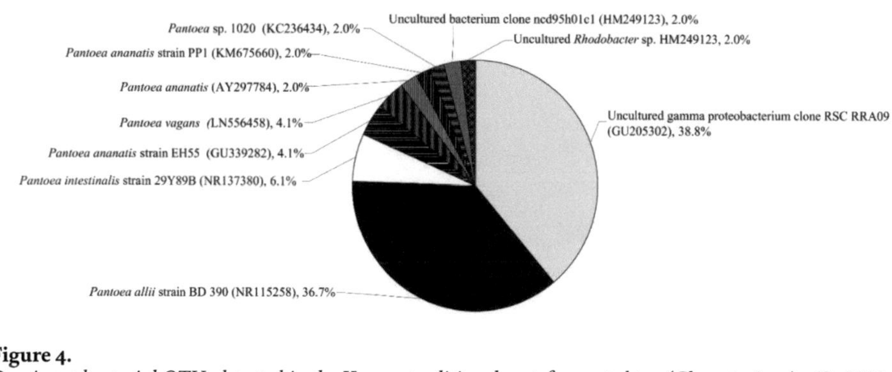

Figure 4.
Dominant bacterial OTUs detected in the Korean traditional post-fermented tea (Chungtaejeon) 16S rRNA gene clone libraries.

from 42.86% before steaming raw leaves to 7.55% after steaming and then ranges from 2.44% after crushing to 66.6% (9 day). The alpha-probacteria range from 1.89 to 32.1%. The gamma-probacteria increased from 25.0% before steaming raw leaves to 97.6% after crushing and then from 33.3 to 66.6% during forming and fermentation by manufacturing process. Therefore, dominant bacteria during manufacturing were gamma-probacteria for Chungtaejeon. The microbial community size was the largest for Chungtaejeon compared with other teas [10, 11]. Also, phylogenetic analysis suggested that the taxonomic affiliation of the dominant species in the post-fermented tea was gamma-proteobacteria (**Figure 3**) for fermentation [10]. Our results were similar to the report of Kim et al. [10, 11] during manufacturing of Korean traditional post-fermented tea (**Table 1**).

However, the structure of the bacterial community involved in the production of oriental traditional post-fermented tea (Chungtaejeon) was investigated using 16S rRNA gene analysis (**Figure 2**). The main microorganisms in fermentation process of Chungtaejeon (**Figure 1**) are identified as *Pantoea sp.* and *Klebsiella oxytoca*, respectively (**Figures 3** and **4**). Phylogenetic analysis suggested that the taxonomic affiliation of the dominant species in the Chungtaejeon was γ-proteobacteria (**Figure 3**). Comparison of bacterial (A), fungal (B), and archaeal (C) community

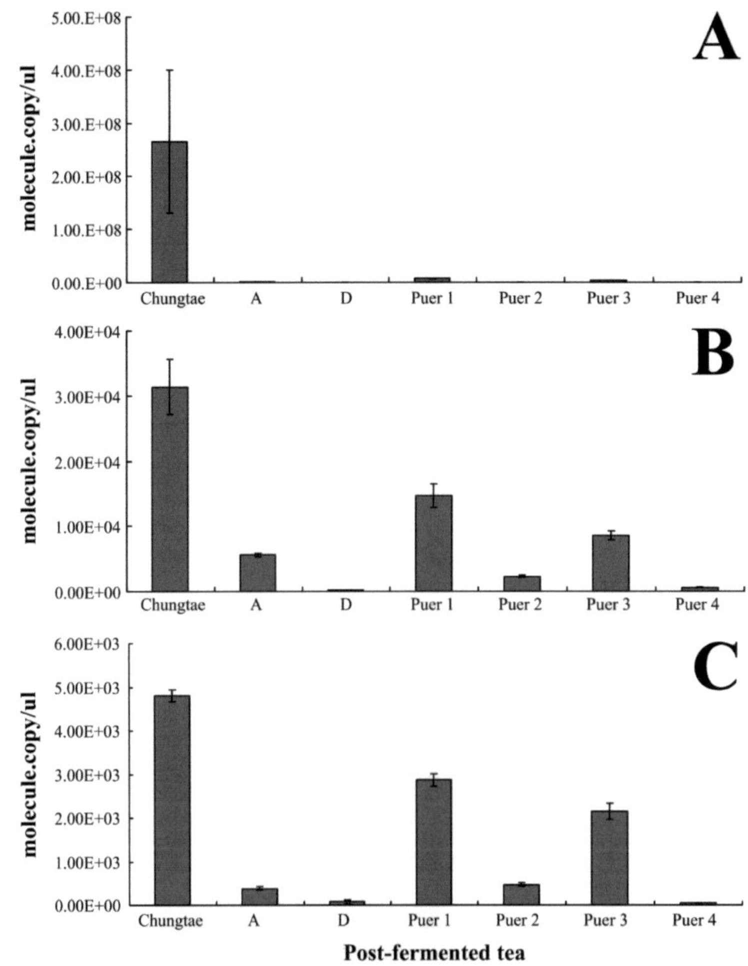

Figure 5.
Comparison of bacterial (a), fungal (B), and archaeal (C) community size using the real-time PCR from the various oriental post-fermented tea (Chungtaejeon). A, D and Puer1 ~ 4 were post-fermented tea for Korea and China, respectively.

Taxonomy	Chungtaejeon	Awabancha	Puer 1	Puer 2
Acidobacteria	12.2	0.2	50.0	34.6
Firmicutes	—	44.7	—	26.6
Alpha-proteobacteria	—	5.2	50.0	17.0
Beta-probacteria	1.1	11.5	—	19.3
Epsilonproteobacteria	—	0.1	—	—
Gamma-proteobacteria	86.7	38.3	—	2.5

Table 2.
Dominant microorganism different of oriental post-fermented tea by metagenome analysis (%).

size using the real-time PCR from the various oriental post-fermented tea (Chungtaejeon), other Korean post-fermented tea and puer-tea from China is shown in **Figure 5**. The bacterial community size was the highest for Chungtaejeon. It is higher about 100 times in Chungtaejeon ($2.65 \times 10^8 \pm 1.35 \times 10^8$ copy/ul) compared with other Korean tea and puer-tea. Also, the fungal community size was higher about seven times in the Chungtaejeon than in the other post-fermented teas. However, the archaeal community size was highest about six times in the Chungtaejeon. Therefore, the bacteria, fungal and archaeal community size were highest for Chungtaejeon in the seven post-fermented teas. As a result, the microbial communities of Chungtaejeon were the largest compared with other teas.

Also, cluster analysis confirmed that microbial population present in both Korean and Chinese post-fermented teas groups into the same class [11]. The dominant microorganism present in Korean post-fermented tea was bacterium, while for the Chinese post-fermented tea, it was fungus [5–7, 11]. However, dominant microorganism different from oriental post-fermented tea by metagenome analysis was a acidobacteria and alpha-proteobacteria for Puer 1 and Puer 2, firmicutes for Awabancha and gamma-proteobacteria for Chungtaejeon (**Table 2**).

Tea is classified as green tea, semi-oxidation tea, oxidation tea and post-fermented tea, depending on its manufacturing methods. Post-fermented tea has softer taste, an increased anti-bacterial effect, antioxidant activity, and lower cholesterol production effects by fermentation of microorganisms due to higher amounts of gallic acid. The gallic acid content was 1.67 mg/g for green tea and 21.98 mg/g for puer-tea [4, 12–15].

5. The beneficial effects of Chungtaejeon

Tea is one of the most popular beverage that is produced from the tea plant, *Camellia sinensis* (L.) O. Kuntze and is consumed as green, black, Oolong and post-fermented tea in different parts of the world. The beneficial effects of green tea such as cancer, heart disease, and liver disease are related to catechins [16]. Black tea has made promising pharmacological effects such as growth promoter, cardioprotector, potent cholesterol-lowering effect, antioxidant and antimicrobial, etc. in humans for various compounds such as flavonoids (Thearubigins and theaflavins and catechins), amino acid (L. theanine), phenolic acids (gallic acid and caffeic acid etc.), vitamins, etc. [17]. The Chungtaejeon is post-fermented tea that is produced in Korea by microorganisms. Chemical analysis of it demonstrated the presence of tannins, flavonoids, glycosides, vitamins, polysaccharides, and volatile oils [18]. Also, Chungtaejeon possessed strong antioxidative effects and effectively inhibited the cytokine that induces proliferation. Furthermore, it prevents migration of human aortic smooth muscle cells (HASMC) by restraining the protein expression and enzymatic action of matrix metalloproteinases (MMP-9) [19].

The dominant bacteria, catechins and gallic acid contents during Chungtaejeon marking process are analyzed from raw leaves to product. The catechin content decreased from 12.10 to 3.80 mg/g, and epicatechin (EC) and gallic acid contents were increased to 28.50 and 8.02 mg/g, respectively, during manufacturing (**Figure 3**). Microorganism oxidizes phenolic compounds of tea and leads to considerable loss of the catechins and formation of the theaflavins, thearubigins, theabrownins and gallic acid [4–7, 13–15]. Our results are similar to decreased catechin content and increased gallic acid content during manufacturing and fermentation. Lee et al. [20] also report that levels of EC, ECG, EGCG, quinate, caffeine and sucrose decreased, whereas gallate and glucose levels increased during tea fermentation. Epigallocatechin-3-gallate (EGCG) content was the highest of 90.2 ± 16.1 mg/g for catechins (**Figure 6**). EGCG was the major catechin among all tea varieties, accounting for about half of the total catechins. Epigallocatechin-3-gallate is the ester of epigallocatechin and gallic acid. EGCG in tea, which is made to green tea, white tea and fermented tea, is a polyphenol under basic research for its potential to affect human health and disease such as dietary supplements. The catechins such as EGCG most changed to gallic acid during fermentation into puer tea [4–7, 12, 14, 15]. The most important pharmacological properties of gallic acid are attributed to its antioxidant and anti-inflammatory potentials [12].

Pantoea agglomerans has been concerned to decompose polyphenol tannic acid and gallic acid [4–6, 15]. It may perform an important role for manufacturing and fermentation into gallic acid from catechins of Chungtaejeon (**Figures 3** and **6**). The gallic acid is the most common member of phenolic acids. Gallic acid or 3,4,5-trihydroxybenzoic acid is one of the most abundant phenolic acids in the plant kingdom such as black tea and post-fermented tea products. The edible use of gallic acid and its ester derivatives is on diverse scientific reports on biological and pharmacological activities of these phytochemicals for antioxidant, antimicrobial, anti-inflammatory, anticancer, cardioprotective, gastroprotective and neuroprotective effects [21]. Also, gallic acid and its derivatives demonstrated a broad range of beneficial effects in prevention and/or management of several disorders, also their acceptable

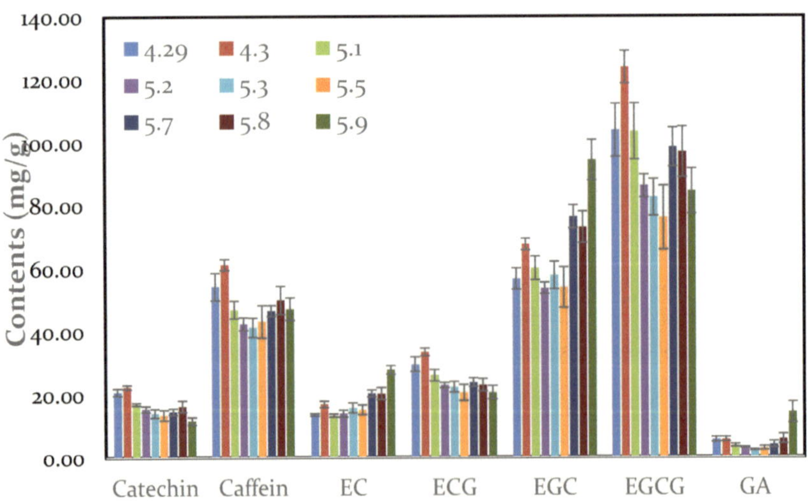

Figure 6.
Catechins, gallic acid and caffeine contents after forming of Korean traditional post-fermented tea (mg/g). Forming time (4.29), 1 day (4.30), 2 days (5.1), 3 days (5.2). 5 days (5.3), 7 days (5.5), 8 days (5.7), 9 days (5.8) and 10 days (5.9). EC (epicatechin), ECG (epicatechin gallate), EGC (epigallocatechin), EGCG (epigallocatechin gallate).

DOI: http://dx.doi.org/10.5772/intechopen.91855

safety and stability profiles, making them significant options to be introduced as dietary supplements [22] like *Camellia sinensis*. Black tea also contains gallic acid.

Black tea has many kinds of biological compounds such as flavonoids [thearubigins (TRs) and theaflavins (TFs) and catechins, amino acid (L. theanine), vitamins (A, C, K), phenolic acids (caffeic acid (CA), gallic acid (GA), chlorogenic acid (CGA) and carbamic acid], lipids, proteins, volatile compounds, carbohydrates, β-carotene and fluoride that are illustrated as having many promising pharmacological effects as growth promoter and cardioprotector, having potent cholesterol-lowering effect, and being antioxidant and antimicrobial for humans [17]. However, Chungtaejeon can be suggested to have beneficial effect in the prevention of atherosclerosis [8, 10, 11, 17–20, 22, 23]. It prevents the risk of atherosclerosis in rats fed a high-fat atherogenic diet *in vivo and vitro* [23]. The chemical composition of tea leaves consists of tanning substances, flavonols, alkaloids, amino acids, enzymes, aroma-forming substances, vitamins, minerals and trace elements contained in theaflavins (TFs), thearubigins (TRs) and theabrownins (TBs) needed to analyze during manufacturing and fermentation of Chungtaejeon and needed to reveal molecular level such as pharmacological value and therapeutic properties for human health in the future.

6. How to drink Chungtaejeon

Tea utensils such as brazier, pipkin, bamboo chopsticks, teacup and cooling bowl are necessary before boiling and brewing. Also, you can add one piece of Chungtaejeon in one pot (about 1 L) before boiling [1]. There are generally two methods for drinking. The first way to drink Chungtaejeon is drinking after boiling. You can turn leaves into roasted ones in gentle heat for 3–4 minutes. It has sterilization effect, and it adds savory flavor and is a unique flavor. And then, you can add Chungtaejeon leaves split in a liter of boiled water. After you boil it for more than 5–4 minutes, you can have it thoroughly infused. Therefore, you pour boiled water in the tea kettle, and you can add Chungtaejeon leaves split and infuse tea sufficiency. Also, you can have it with ginger, yuzu, quince and herbal medicine depending on your preference. The second way to drink Chungtaejeon is drinking after brewing as follows; roast heat pottery, brassware, and soup bowl or roof file slowly and turn them into golden brown so that can be boiled to drink, on the other hand, the green smell of tea will diminish and unique aroma and flavor will deepen if it is lightly roasted. Also, the color of tea gets better and foreign substances can be eliminated. But, tea should be roasted, not burned. Divide it into small pieces (division into 3–4 parts) so that roasted Chungtaejeon can be easily brewed. And then, pour about 500–600 ml of boiling water at 100°C, and then leave it for more than 10 minutes. After that, drink brewed tea using a cup. It can be brewed about 3–5 times more.

Author details

Doo-Gyung Moon
Research Institute of Climate Change and Agriculture, National Institute of Horticultural and Herbal Science, Rural Development Administration, Jeju, Republic of Korea

*Address all correspondence to: dgmoon@korea.kr

References

[1] Jangheung Chungtaejeon Corporation. Tea story. Vol. 2. Jangheung Agricultural Technology Center; 2013. pp. 1-80

[2] Chen YS, Liu BL, Chang YN. Bioactivities and sensory evaluation of Pu-erh teas made from three tea leaves in an improves pile fermentation process. Journal of Bioscience and Bioengineering. 2010;**109**:557-563

[3] Heo BG, Park YS, Chon SU, Lee SY, Cho JY, Gorinstein S. Antioxidant activity and cytoxicity of methol extracts from aerial parts of Korean salad plants. BioFactors. 2007;**30**:79-89

[4] Zhang L, Zhang ZZ, Zhou YB, Ling TJ, Wan XC. Chinese dark teas: Post fermentation, chemistry and biological activities. Food Research International. 2013;**53**:600-607

[5] Zhang W, Yang R, Fang W, Yan L, Lu J, Sheng J, et al. Characterization of thermophic fungal community associated with pile fermentation of Pu-erh tea. International Journal of Food Microbiology. 2016;**227**:29-33

[6] Zhao Y, Zhong GF, Yang XP, Hu XM, Mao DB, Ma YP. Bioconversion of lutein to form aroma compounds by *Pantoea dispersa*. Biochnology Letters. 2015;**37**:1687-1692

[7] Zhu Y, Luo Y, Wang P, Zhao M, Li L, Hu X, et al. Simultaneous determination of free amino acids in Pu-erh tea and their changes during fermentation. Food Chemistry. 2016;**194**:643-649

[8] Pitta DW, Kumar S, Vecchiarelli B, Shirley DJ, Bittinger K, Baker LD, et al. Temporal dynamics in the ruminal microbiome of dairy cows during the transition period. Journal of Animal Science. 2014;**92**:4014-4022

[9] Galan M, Guivier E, Caraux G, Charbonnel N, Cosson JFA. 454 multiplex sequencing method for rapid and reliable genotyping of highly polymorphic genesin large-scale studies. BMC Genomics. 2010;**11**:296

[10] Kim BH, Jang JO, Kang Z, Joa JH, Moon DG. The microbial diversity analysis of the Korea traditional post-fermented tea (Chungtaejeon). Korean Journal of Microbiology. 2017;**53**:170-179

[11] Kim BH, Jang JO, Joa JH, Kim JA, Song SY, Lim CK, et al. A comparison of the microbial diversity in Korean and Chinese post-fermented teas. Microbiology and Biotechnology Letters. 2017;**45**:71-80

[12] Teeradate K, Chiu MT, Huang TC, Hsu JL. Gallic acid content in Taiwanese teas at different degrees of fermentation and its antioxidant activity by inhibiting PKCδ activation: in vitro and in silico studies. Molecules. 2016;**21**:3-11

[13] Wang QP, Peng CX, Gong JS. Effects of enzymatic action on the formation of the brown in during solid state fermentation of Pu-erh tea. Journal of the Science of Food and Agriculture. 2011;**91**:2412-2418

[14] Xie G, Ye M, Wang Y, Ni Y, Su M, Huang H, et al. Characterization of Pu-erh tea using chemical and metabolic profiling approaches. Journal of Agricultural and Food Chemistry. 2009;**57**:3046-3054

[15] Zeida M, Wieser M, Yoshida T, Sugio T, Nagasawa T. Putification and characterization of gallic acid decarboxylase from *pantoea agglomerans* T71. Applied and Environmental Microbiology. 1998;**64**:4743-4747

[16] Sabu MC, Priya TT, Ramadasan K, Ikuo N. Beneficial effects of green tea: A literature review. Chinese Medicine Journal. 2010;**5**:13. DOI: 10.1186/1749-8546-5-13

[17] Muhammad N, Jannat B, Asghar AK, Imaran S, Ihsanullah K, Safaraz AF, et al. Pharmacological value and therapeutic properties of black tea (*Camellia sinensis*): A comprehensive overview. Biomedicine & Pharmacotherapy. 2018;**100**:521-531

[18] Park YS, Lee MK, Heo BG, Ham KS, Kang SG, Cho JY, et al. Comparison of the nutrient and chemical composition of traditional Korean Chungtaejeon and green teas. Plant Foods for Human Nutrition. 2010;**65**:186-191

[19] Rajendra K, Navin S, Jeon ER, Park YS, Kim DW. Chungtaejeon, a Korean fermented tea, scavenges oxidation and inhibits cytokine induces proliferation and migration of human aortic smooth muscle cells. Plant Foods for Human Nutrition. 2011;**66**:27-33

[20] Lee JE, Lee BJ, Chung JO, Shin HJ, Lee SJ, Lee CH, et al. ^{1}H NMR-based metabolomics characterization during green tea (*Camellia sinensis*) fermentation. Foodservice Research International. 2011;**44**:597-604

[21] Choubey S, Varughese L, Kumar V, Beniwai V. Medicinal importance of gallic acid and its ester derivatives: A patent review. Pharmaceutical Patent Analyst. 2015;**4**:305-315

[22] Niloofar K, Fatemeh F, Maryam F, Seyedeh SA, Roodabeh B, Rozita N, et al. Pharmacological effects of gallic acid in health and disease: A mechanistic review. Iranian Journal of Basic Medical Sciences. 2018;**22**(3):225-237

[23] Keshav RP, Lee UW, Kim DW. Chungtaejeon, a Korean fermented tea, prevents the risk of atherosclerosis in rats fed a high-fat atherogenic diet. Journal of integrative medicine. 2016;**14**:134-142

Chapter 13

The Importance of Tea in the Correlation between Caffeine and Health

Burhan Başaran

Abstract

This study aims to examine the studies on the correlation between caffeine intake of individuals and health and to emphasize the importance of tea for health. Caffeine is a substance contained in many foods we frequently consume in our daily diets such as tea, coffee, cola, and energy drinks and is generally known for its stimulant nature. That is why consumers take caffeine into their bodies throughout their lives. The European Food Safety Authority states that daily intake of 400 mg (about 5.7 mg/kg bw for 70 kg) caffeine from all sources does not create any concern for adults. There is no complete consensus about whether caffeine consumption causes various ailments in individuals or whether it has a protective effect against contracting various diseases. The literature review has revealed that coffee and tea in adults and tea, soft, and energy drinks in children and adolescents play an important role in caffeine intake. Tea is a plant that is especially rich in phenolic compounds and has many benefits for human health. Therefore, for conscious consumers, tea is different from other drinks due to caffeine and phenolic compounds and is thought to do more good than harm to people.

Keywords: black tea, green tea, white tea, dietary caffeine intake, health

1. Introduction

Tea is produced from *Camellia sinensis* leaves using various methods and is one of the most commonly used beverages throughout history due to its consumption and health properties [1, 2]. Commercial tea is divided into three main groups: full fermented black tea, semi-fermented oolong tea, and non-fermented green tea according to the production method [3]. Besides these groups, there is non-fermented white tea that is produced and consumed mostly in Asia and has a higher price than other types of tea [4]. Black tea and green tea are the most well-known and consumed types. About 78% of all tea consumed in the world is black tea, 20% is green tea and 2% is oolong tea [5]. The physical and chemical properties of fresh tea leaves vary depending on many factors such as geography, climate, soil, leaf age, and cultivation [6, 7]. Fresh tea leaf consists of polyphenol by 36%, carbohydrate by 25%, protein by 15%, lignin by 6.5%, ash by 5%, amino acid 4%, lipid by 2%, organic acid by 51.5%, chlorophyll by 0.5%, and carotenoid and volatile compounds by less than 0.1% in dry matter [8]. Fresh tea leaf is considered to be one of the plants with the richest amount of phenolic compounds [9]. **Table 1** shows the phenolic compound contents of different types of tea (**Table 1**).

Tea is also an important source of caffeine which is found naturally in coffee, cola, chocolate, cocoa or added as an additive to various energy drinks, medicine and cosmetics [11], is the most consumed drink in the world, whose use is increasing each passing day, and is considered to be a psychoactive substance [12, 13].

Caffeine is a member of a group known as purine alkaloids which also contain 3,7-dimethylxanthine (theobromine), 1,7-dimethylxanthine (paraxanthine), and methyluric acids. Its chemical formula is C8H10N4O2 and its systematic name is known as 1,3,7-trimethylxanthine [14].

Caffeine is largely absorbed by the stomach and small intestine after being taken into the body, quickly moves into body cells and reaches the highest level in 15 to 60 minutes after crossing the blood-brain barrier. The half-life of caffeine varies depending on factors such as puberty, pregnancy, and disease, but it is considered to be 5–6 hours in a healthy individual [15, 16]. Metabolites such as paraxanthine, theobromine, and theophylline are released as a result of caffeine metabolism. Only 10% of caffeine is excreted from the body without being metabolized [17].

Caffeine is available at different levels in many foods that we often consume in our daily lives. Therefore, it can be easily said that individuals consume caffeine regularly every day. However, the prediction that caffeine intake into the body for nutrition and short, medium and long-term exposures can lead to various health problems has led the scientific world to do research in this area. This study aims to examine the studies on the correlation between caffeine intake of individuals and health, and to emphasize the importance of tea for health.

	Green tea (mg/g-km)	Black tea (mg/g-km)	Oolong tea (mg/g-km)
Total phenolic substance	208.80–236.78	221.75–248.31	87.70–195.6
Total catechin	221.94–234.71	187.84–279.43	16.64–282.75
Total flavone and flavonoid glycosides	4.53–5.43	3.03–5.01	4.09–4.68
EGCG (epigallocatechin agallate)	53.14–126.20	2.42–81.93	16.53–132.54
GC (catechin gallate)	5.05–10.52	0.20–10.82	7.5–8.93
EGC (epigallocatechin)	4.40–97.79	0.71–78.82	12.96–19.94
ECG (epicatechin gallate)	14.19–27.80	0.42–13.02	2.09–46.28
EC (epicatechin)	0.20–28.30	0.10–3.59	2.68–2.77
GCG (gallocatechin gallate)	2.60–48.02	0.09–58.89	49.54–60.92
Gallic acid	0.59–5.20	0.57–5.80	1.30–1.37
Caffeine	15.66–77.30	3.14–83.20	2.58–40.84
Theophylline	0.60–0.80	0.10–0.20	—
Theobromine	0.27–6.0	0.41–4.70	0.72–0.99

Table 1.
Phenolic compounds of different types of tea [10].

2. Nutritional caffeine intake

Today, individuals consume low or high levels of caffeine, often knowingly and sometimes unknowingly. Since the production and consumption of caffeine-containing foods vary by country, society and individual, it is quite difficult to accurately calculate individuals' nutritional intake of caffeine. **Table 2** shows caffeine levels in certain foods which are frequently consumed by the general population and considered to be important in terms of caffeine content.

Caffeine is very common in nature, and coffee, tea, energy drinks, chocolate, and cocoa are accepted as sources of nutritional caffeine intake [22]. Caffeine levels in these foods vary according to content, ratio between tea/coffee and water, brewing time and other consumption characteristics [23, 24]. In general, coffee has a higher caffeine level than other foods. As for tea groups, the caffeine level of black tea is higher than other tea types. Caffeine levels are usually at a certain level in soft drinks such as cola and energy drinks as they have standard prescriptions and production techniques. The caffeine level in chocolate varies according to the amount of cocoa it contains. Coffee is also known as the source of nutritional caffeine of adults throughout Europe, especially in Finland, Denmark, Sweden and Switzerland. In all member states of the European Union, there is a "high levels of caffeine" warning on beverage labels containing more than 150 mg/L of caffeine [25].

Many institutions and researchers try to estimate nutritional caffeine intake by examining dietary habits of individuals. The number of studies conducted in this area is quite large and the findings of some studies are given below.

Food group	Food subgroup	Caffeine level (mg/L or mg/kg)
Chocolate	Chocolate milk or chocolate beverages	7–67
	Chocolate snacks	62–418
	Dark chocolate	340–525
Coffee	Coffee drink	320–690
	Cappuccino	250–315
	Espresso coffee	713–1916
	Decaffeinated and imitations	11–29
	Instant coffee, ready to drink	210–690
	Turkish coffee	620–858
Tea	Black tea	181–220
	Green tea	125–320
	White tea	63
	Tea (unspecified)	158–234
	Instant	47–199
Cola beverages (regular)		79–130
Cola beverages (diet)		109–140
Energy drinks		150–335

Table 2.
Caffeine levels in some foods [11, 18–21].

FDA states that a daily intake of 400 mg of caffeine can be considered safe for healthy adults. On the other hand, it has declared that some individuals may be negatively affected by lower doses of caffeine, and studies will begin to investigate the safety of caffeine added to foods, with particular emphasis on children and adolescents [18]. Health Canada has specified safe daily caffeine intake for healthy adults and pregnant women as ≥400 mg/day and <300 mg/day, respectively. The same institution has reported that daily caffeine intake for children of different age groups is in the range of 45 to 85 mg/day (45 mg/day for 4–6 years, 62.5 mg/day for 7–9 years, and 85 mg/day for 10–12 years) [26].

In a study conducted in America with 24,808 individuals between 2001 and 2010, it was reported that more than 85% of adults (≥19 years of age) regularly consumed caffeine and that their average daily caffeine intake was 180 mg/day. In the same study, highest and lowest intake of caffeine was detected in males between the ages of 31 and 50 and females between the ages of 19 and 30, respectively, the caffeine intake of males (211 mg/day) was more than females (161 mg/day), and 98% of the daily caffeine intake came from beverages. The drinks that cause caffeine intake are as follows: coffee by 64%, carbonated soft drinks by 18%, tea by 16%, and energy drinks by less than 1% [27].

In a study examining the daily caffeine intake of adolescents living in the United States between 1999 and 2011, it was reported that more than half of the children at ages ranging from 2.5 to 5 and about 75% of children over the age of 5 consumed caffeine on a daily basis. The mean daily caffeine intakes of children between the ages of 2 and 11 and adolescents between the ages of 12 and 17 were determined as and 50 mg/day, respectively. It was reported that the source of caffeine was carbonated soft drinks for children under the age of 12 and coffee for children that were 12 and older [28].

In another study, daily caffeine intake was calculated as 15, 26, 61, 213, and 135 mg/day for the general population who were 4–8, 9–13, 14–19, and 51–70 years old and those who were 4 years old or younger, respectively. The daily caffeine intake is higher in males (196 mg/day) than females (151 mg/day). Although the distribution of the drinks that cause daily caffeine intake varies by age, the largest contribution is from coffee (64%) and tea (18%) [29].

In another study conducted in the United States, daily caffeine intake was calculated as 120 mg/day (1.73 mg/kg body weight/day) for all age groups. The highest caffeine intake was in individuals in the 35–49 age range (170 mg/day). Daily caffeine intake in pregnant women was estimated to be 58 mg/day. According to the other findings of the study, the daily caffeine intake was calculated as 14 mg/day (0.82 mg/kg body weight/day), 22 mg/day (0.85 mg/kg body weight/day), and 106 mg/day (1.54 mg/kg body weight/day) for ages ranging from 1 to 5, 6 to 9, and 20 to 24, respectively. It was pointed out that the main source of caffeine was coffee in adults and soft drinks in young people. Tea ranked second in both groups [30].

Australian Children's Nutrition and Physical Activity Survey determined the general daily caffeine intake as 18 mg/day in a study on caffeine consumption of 4487 children and adolescents between the ages of 2 and 16 in 2007. The mean daily caffeine intake by age groups was determined as 3, 8, 19, and 42 mg/day for ages ranging from 2 to 3, 4 to 8, 9 to 13, and 14 to 16, respectively. It was stated that the main source of caffeine was drinks (81%) and that the highest contribution was made by soft drinks (31%), coffee (21%) and tea (17%), respectively [31].

In a study conducted in Italy on 1213 adolescents (12–19 years), it was found that 76% of individuals consumed caffeine on a daily basis and the daily caffeine intake was approximately 125 mg/day (2.1 mg/kg body weight/day) [32] while the daily caffeine intake was calculated as 79 mg/day and coffee, tea, and soft drinks were listed as the beverages with the highest contribution to the daily caffeine intake in a study conducted in England on 2008 individuals of varying ages [33].

3. Caffeine and health

Since foods and beverages containing caffeine are common and easy to reach, a very large segment of society regularly consumes caffeine from childhood to old age. Just this information once again reveals the importance of caffeine in our lives. Therefore, systematic and comprehensive studies should be carried out on the effects of nutritional caffeine intake on the health of individuals in the short, medium and long term. A lot of research been conducted in this area.

The fatal dose of caffeine in adults is estimated to be 170 mg/kg body weight/day [34] and there have been reports of some deaths due to caffeine overdose [35]. There is no complete consensus about whether caffeine consumption causes various ailments in individuals or whether it has a protective effect against contracting various diseases [36].

One of the areas in which caffeine's effects on health are most commonly investigated is the cardiovascular system. There are studies showing that caffeine intake by less than 400 mg/day does not have any negative impacts on the cardiovascular system [35], high levels of caffeine consumption leads to an increase in morbidity and mortality in the cardiovascular system by increasing blood pressure and heart rate [37–39], but, despite all this, caffeine has a protective effect [40–42].

Caffeine intake over 300 mg/day has been reported to cause second trimester miscarriages, low birth weight, and an increase in the likelihood of stillbirth [43–45]. Furthermore, the risk of developing childhood obesity increases by 87% in fetuses exposed to caffeine in the womb compared to those not exposed to caffeine [46]. However, there are also studies indicating that there is no correlation between caffeine intake in pregnancy and premature birth and fertility [47–49].

Positive impacts of caffeine consumed at low or medium levels on health such as relieving the airways leading to the lungs of individuals, reducing asthma attacks [50], making people feel healthy, reducing the risk of having type 2 diabetes or Parkinson's disease, and healing liver diseases etc. are also mentioned [51–56].

There is not enough evidence in current studies to prove that caffeine consumption is associated with any type of cancer in the short, medium and long term, it causes an increase in the number of cancer cases, or that it has a protective effect [57]. It has been reported that caffeine intake by less than 400 mg/day does not cause an increase in the cancer risk [11, 58, 59].

One of the most important features of caffeine is that it is a psychoactive compound. While caffeine less than 400 mg/day is generally considered safe, high doses of caffeine taken for a long time lead to caffeine withdrawal syndromes (caffeinism) such as headache, low concentration, restlessness, insomnia, irritability, decreased learning ability and palpitation [60–64].

The prevailing view in the literature is that caffeine consumption has a detrimental effect on an individual's sleep quality. According to current studies, caffeine consumption causes sleep delay, shortening of total sleep time, decreased sleep quality, and, as a result, daytime insomnia. It has also been stated that continued insomnia leads to more caffeine consumption by people [65–67].

It is stated in a study examining the dose-related effects of caffeine that 85–250 mg caffeine per day increases the feeling of alertness and contributes to increased motivation by decreasing fatigue, higher doses between 250 and 500 mg per day may cause restlessness, irritability, insomnia, and anxiety while 15–30 mg/kg body weight/day caffeine may lead to muscle spasms and severe toxic effects on the cardiovascular and central nervous system in healthy adults [68].

It has been found that caffeine accelerates metabolism by causing thermogenesis and lipid oxidation along with other compounds in food and causes weight loss by enabling people to spend more energy [69–72].

Another important feature of caffeine is its negative impact on bone health due to its diuretic effect. Excretion of elements such as calcium, magnesium, sodium and potassium from the body with urine increases due to high levels of caffeine intake. This results in decreased bone mineral density especially in females as well as increasing the risk of osteoporosis [73, 74].

4. Decaffeinated foods

Since caffeine is present in foods such as tea, coffee, and chocolate included in our daily diet, it is not surprising that the daily caffeine intake recommended for both the general society and specific groups is exceeded. Increased caffeine intake is known to cause various health problems. This is why caffeine free or decaffeinated foods are needed. Significant reduction of caffeine content or removal of caffeine is called decaffeination [75]. The first decaffeination process was carried out in 1903 on coffee beans (coffee beans were moistened with salty water and caffeine was removed with benzene) [76]. There are many different methods of caffeine removal that differ depending on the type of food. These are:

Caffeine removal by traditional methods: The main traditional methods for caffeine removal are the removal of caffeine from food by water, organic solvents and supercritical fluids [77].

Caffeine removal by microbial methods: *Pseudomonas, Serratia, Stemphylium, Penicillium,* and *Aspergillus* species are grown on leaf surfaces, and these microorganisms reduce the caffeine level of the food by decomposing the caffeine [78, 79].

5. Conclusion

Caffeine is a compound that is legal, easy to obtain for the general society, socially acceptable to consume, found in many foods in our daily diet, and generally known for its stimulant properties. Caffeine, whose consumption has increased especially due to the changes in dietary habits in recent years, is a compound that still remains popular today. In this sense, both nutritional caffeine intake and the correlation between caffeine and health have been studied from many different angles. The effects of caffeine consumption on health are still a matter of debate. Although caffeine is generally considered safe except for excessive use in adults, studies on this subject are still insufficient for children and adolescents.

Individuals' daily caffeine intake varies depending on many factors such as the source of caffeine, age of the individual, breed, dietary habits and culture. In general, it can be said that the level of caffeine intake increases in proportion to age, and males consume more caffeine than females. It is understood that coffee and tea in adults and tea, soft and energy drinks in children and adolescents play an important role in caffeine intake.

In the light of the data given in **Table 2** and considering the changes in the consumption patterns regarding foods in recent years, it can be easily said that individuals' daily caffeine consumption is much more than 400 mg/day which is usually considered to be safe. But it should be noted that caffeine intolerance can vary from person to person. Even very small amounts of caffeine can negatively affect pregnant women, children, the elderly and individuals who are caffeine intolerant. This is why caffeine intake of these individuals should be limited. For this purpose, popularizing caffeine-reduced products on the market can be considered as a strategy for limiting caffeine intake through nutrition. The literature review reveals that tea ranks second in caffeine intake in adults, children and adolescents. Therefore, tea plays a significant role in caffeine intake. But tea has significant differences distinguishing it from coffee, soft drinks and energy drinks that are other sources of caffeine.

Coffee is a drink that is generally perceived as healthy. However, it is also rich in acrylamide that was found out to be contained in food in 2002 and identified by the International Agency for Research on Cancer (IARC) as a possible carcinogen for humans. Carbonated soft drinks and chocolate foods contain many additives in addition to high levels of sugar. Many researchers and institutions approach energy drinks with suspicion due to very high caffeine content. Early acquaintance of individuals with carbonated soft drinks and energy drinks also increases the likelihood of addiction. In general, it can be said that the dietary quality of individuals who consume this type of drink frequently decreases.

Tea is a plant rich in phenolic compounds, especially catechins, and has many benefits on human health due to its antioxidant, antibacterial, anticarcinogen, antimutagenic, and antiallergic effects. Therefore, for conscious consumers, tea is positively different from other foods and drinks due to caffeine and phenolic compounds and is thought to do more good than harm to people. Green tea stands out among tea types with the lowest content of caffeine and high phenolic content.

Author details

Burhan Başaran
Travel, Tourism and Recreation Department, Ardeşen Vocational School, Recep Tayyip Erdogan University, Rize, Turkey

*Address all correspondence to: burhan.basaran@erdogan.edu.tr

References

[1] Wheeler DS, Wheeler WJ. The medicinal chemistry of tea. Drug Development Research. 2004;**61**(2):45-65

[2] Yang CS, Zhang J, Zhang L, Huang J, Wang Y. Mechanisms of body weight reduction and metabolic syndrome alleviation by tea. Molecular Nutrition & Food Research. 2016;**60**:160e74

[3] Lin JK, Lin CL, Liang YC, Lin-Shiau SY, Juan IM. Survey of catechins, gallic acid, and methylxanthines in green, oolong, pu-erh, and black teas. Journal of Agricultural and Food Chemistry. 1998;**46**(9):3635-3642

[4] Rusak G, Komes D, Likic S, Horzic D, Kovac M. Phenolic content and antioxidative capacity of green and white tea extracts depending on extraction conditions and the solvent used. Food Chemistry. 2007;**110**:852-858

[5] Grove KA, Lambert JD. Laboratory, epidemiological, and human intervention studies show that tea (*Camellia sinensis*) may be useful in the prevention of obesity. The Journal of Nutrition. 2010;**140**:446-453

[6] Salman S, Özdemir F. Beyaz çay: üretimi, bileşimi ve sağlık üzerine etkileri. Akademik Gıda. 2018;**16**(2):218-223

[7] Turkmen N, Sarı F, Velioglu YS. Factors affecting polyphenol content and composition of fresh and processed tea leaves. Akademik Gıda. 2009;7(6):29-40

[8] Łuczaj W, Skrzydlewska E. Antioxidative properties of black tea. Preventive Medicine. 2005;**40**(6):910-918

[9] Tan J, Engelhardt UH, Lin Z, Kaiser N, Maiwald B. Flavonoids, phenolic acids, alkaloids and theanine in different types of authentic Chinese white tea samples. Journal of Food Composition and Analysis. 2017;**57**:8-15

[10] Koca İ, Bostancı Ş. Production, composition, and health effects of oolong tea. Turkish Journal of Agriculture-Food Science and Technology. 2014;**2**(3):154-159

[11] EFSA NDA Panel (EFSA Panel on Dietetic Products, Nutrition and Allergies). Scientific opinion on the safety of caffeine. EFSA Journal. 2015;**13**(5):4102

[12] Baratloo A, Rouhipour A, Forouzanfar MM, Safari S, Amiri M, Negida A. The role of caffeine in pain management: A brief literature review. Anesthesiology and Pain Medicine. 2016;**6**(3):e33193

[13] Diogo JS, Silva LS, Pena A, Lino CM. Risk assessment of additives through soft drinks and nectars consumption on Portuguese population: A 2010 survey. Food and Chemical Toxicology. 2013;**62**:548-553

[14] Heckman MA, Weil J, Mejia D, Gonzalez E. Caffeine (1, 3, 7-trimethylxanthine) in foods: A comprehensive review on consumption, functionality, safety, and regulatory matters. Journal of Food Science. 2010;**75**(3):77-87

[15] Şengül E. Kafein alımının gebelik üzerine olumsuz etkileri. Turkiye Klinikleri. Journal Gynecol Obst. 2009;**19**(1):45-49

[16] Deniz A, Taş F, Tomur A, Koç A. Kafeinin gebelikteki toksik etkileri. İbni Sina Tip Bilimleri Dergisi. 2015;**1**(3):59-63

[17] Donovan JL, DeVane CL. A primer on caffeine pharmacology and ıts drug ınteractions in

clincal psychopharmacology. Psychopharmacology Bulletin. 2001;**35**(3):30-40

[18] Health Canada. Caffeine in food. Available from: https://www.canada.ca/en/health-canada/services/food-nutrition/food-safety/food-additives/caffeine-foods/foods.html [Accessed: 23 March 2020]

[19] Coca-Cola Company UK. 2010. How Much Caffeine Is in Diet Coke, Coca-Cola and Coke Zero? Available from: https://www.coca-cola.ie/faq/how-much-caffeine-is-in-diet-coke-coca-cola-and-coke-zero [Accessed: 23 March 2020]

[20] Hancı M, Bayram S, Karahan S, Ertuğrul K, Bakırcı S. Türk kahvesi ve Türkiye'de satılan bazı içeceklerdeki kafein miktarları. Düzce Tıp Fakültesi Dergisi. 2013;**15**(3):34-38

[21] Mitchell DC, Knight CA, Hockenberry J, Teplansky R, Hartman TJ. Beverage caffeine intakes in the US. Food and Chemical Toxicology. 2014;**63**:136-142

[22] Frary CD, Johnson R, Wang MQ. Food sources and intakes of caffeine in the diets of persons in the United States. Journal of the American Dental Association (Chicago, IL). 2005;**105**(1):110-113

[23] Astill C, Birch MR, Dacombe C, Humphrey PG, Martin PT. Factors affecting the caffeine and polyphenol contents of black and green tea infusions. Journal of Agricultural and Food Chemistry. 2001;**49**:5340-5347

[24] Hogan EH, Hornick BA, Bouchoux A. Communicating the message: Clarifying the controversies about caffeine. Nutrition Today. 2002;**37**(1):28-35

[25] European Commission. Commission Directive 2002/67/EC on the labelling of foodstuffs contaiing quinine, and of foodstuffs containing caffeine. Official Journal of the European Community. 2002;**L**(191):20-21

[26] U.S. Food and Drug Administration. Letter from Jeanne Ireland, Assistant Commissioner for Legislation, FDA, to Senator Richard J. Durbin (August 10, 2012)

[27] Fulgoni VL III, Keast DR, Lieberman HR. Trends in intake and sources of caffeine in the diets of US adults: 2001-2010. The American Journal of Clinical Nutrition. 2015;**101**(5):1081-1087

[28] Ahluwalia N, Herrick K. Caffeine intake from food and beverage sources and trends among children and adolescents in the United States: Review of national quantitative studies from 1999 to 2011. Advances in Nutrition. 2015;**6**(1):102-111

[29] Drewnowski A, Rehm CD. Sources of caffeine in diets of US children and adults: Trends by beverage type and purchase location. Nutrients. 2016;**8**(3):154

[30] Knight CA, Knight I, Mitchell DC, Zepp JE. Beverage caffeine intake in US consumers and subpopulations of interest: Estimates from the share of intake panel survey. Food and Chemical Toxicology. 2004;**42**(12):1923-1930

[31] Beckford K, Grimes CA, Riddell LJ. Australian children's consumption of caffeinated, formulated beverages: A cross-sectional analysis. BMC Public Health. 2015;**15**(1):70

[32] Santangelo B, Lapolla R, Rutigliano I, Pettoello Mantovani M, Campanozzi A. Nearly half of the adolescents in an Italian school-based study exceeded the recommended upper limits for daily caffeine consumption. Acta Paediatrica. 2018;**107**(6):1055-1059

[33] Fitt E, Pell D, Cole D. Assessing caffeine intake in the United Kingdom diet. Food Chemistry. 2013;**140**(3):421-426

[34] Graham DM. Caffeine – It's identity, dietary sources, intake and biological effects. Nutrition Reviews. 1978;**36**(4):97-102

[35] Nawrot P, Jordan S, Eastwood J, Rotstein J, Hugenholtz A, Feeley M. Effects of caffeine on human health. Food Additives and Contaminants. 2003;**20**(1):1-30

[36] Wikoff D, Welsh BT, Henderson R, Brorby GP, Britt J, Myers E, et al. Systematic review of the potential adverse effects of caffeine consumption in healthy adults, pregnant women, adolescents, and children. Food and Chemical Toxicology. 2017;**109**:585-648

[37] Namdar M, Koepfli P, Grathwohl R, Siegrist TP, Klainguti M, Schepis T, et al. Caffeine decreases exercise-induced myocardial flow reserve. Journal of the American College of Cardiology. 2006;**47**(2):405-410

[38] James JE. Critical review of dietary caffeine and blood pressure: A relationship that should be taken more seriously. Psychosomatic Medicine. 2004;**66**(1):63-71

[39] Turnbull D, Rodricks JV, Mariano GF, Chowdhury F. Caffeine and cardiovascular health. Regulatory Toxicology and Pharmacology. 2017;**89**:165-185

[40] Greenberg JA, Dunbar CC, Schnoll R, Kokolis R, Kokolis S, Kassotis J. Caffeinated beverage intake and the risk of heart disease mortality in the elderly: A prospective analysis. The American Journal of Clinical Nutrition. 2007;**85**(2):392-398

[41] Kokubo Y, Iso H, Saito I, Yamagishi K, Yatsuya H, Ishihara J, et al. The impact of green tea and coffee consumption on the reduced risk of stroke incidence in Japanese population: The Japan public health center-based study cohort. Stroke. 2013;**44**(5):1369-1374

[42] Suzuki E, Yorifuji T, Takao S, Komatsu H, Sugiyama M, Ohta T, et al. Green tea consumption and mortality among Japanese elderly people: The prospective Shizuoka elderly cohort. Annals of Epidemiology. 2009;**19**(10):732-739

[43] Wisborg K, Kesmodel U, Bech BH, Hedegaard M, Henriksen TB. Maternal consumption of coffee during pregnancy and stillbirth and infant death in first year of life: Prospective study. BMJ. 2003;**326**(7386):420

[44] Morgan S, Koren G, Bozzo P. Is caffeine consumption safe during pregnancy? Canadian Family Physician. 2013;**59**(4):361-362

[45] Greenwood DC, Thatcher NJ, Ye J, Garrard L, Keogh G, King LG, et al. Caffeine intake during pregnancy and adverse birth outcomes: A systematic review and dose–response meta-analysis. European Journal of Epidemiology. 2014;**29**:725-734

[46] Li DK, Ferber JR, Odouli R. Maternal caffeine intake during pregnancy and risk of obesity in offspring: A prospective cohort study. International Journal of Obesity. 2015;**39**(4):658-664

[47] Maslova E, Bhattacharya S, Lin SW, Michels KB. Caffeine consumption during pregnancy and risk of preterm birth: A meta-analysis. The American Journal of Clinical Nutrition. 2010;**92**(5):1120-1132

[48] Í Soylu L, Jensen A, Juul KE, Kesmodel US, Frederiksen K, Kjaer SK, et al. Coffee, tea and caffeine consumption and risk of primary infertility in women: A Danish cohort

study. Acta Obstetricia et Gynecologica Scandinavica. 2018;**97**(5):570-576

[49] Brent RL, Christian MS, Diener RM. Evaluation of the reproductive and developmental risks of caffeine. Birth Defects Research. Part B, Developmental and Reproductive Toxicology. 2011;**92**(2):152-187

[50] Paganini AH, Kawas CH, Maria M. Corrada non-alcholic beverage and caffeine consumption and mortality: The leisure world cohort study. Preventive Medicine. 2007;**44**:305-310

[51] Ross GW, Abbott RD, Petrovitch H, Morens DM, Grandinetti A, Tung KH, et al. Association of coffee and caffeine intake with the risk of Parkinson disease. Journal of the American Medical Association. 2000;**283**(20):2674-2679

[52] Ruhl CE, Everhart JE. Coffee and tea consumption are associated with a lower incidence of chronic liver disease in the United States. Gastroenterology. 2005;**129**(6):1928-1936

[53] Huxley R, Lee CMY, Barzi F, Timmermeister L, Czernichow S, Perkovic V, et al. Coffee, decaffeinated coffee, and tea consumption in relation to incident type 2 diabetes mellitus: A systematic review with meta-analysis. Archives of Internal Medicine. 2009;**169**(22):2053-2063

[54] Mirmiran P, Carlström M, Bahadoran Z, Azizi F. Long-term effects of coffee and caffeine intake on the risk of pre-diabetes and type 2 diabetes: Findings from a population with low coffee consumption. Nutrition, Metabolism, and Cardiovascular Diseases. 2018;**28**(12):1261-1266

[55] Hu G, Bidel S, Jousilahti P, Antikainen R, Tuomilehto J. Coffee and tea consumption and the risk of Parkinson's disease. Movement Disorders. 2007;**22**(15):2242-2248

[56] Daria P, Sani G, Aromatario M. Caffeine: Cognitive and physical performance enhancer or psychoactive drug? Current Neuropharmacology. 2015;**13**(1):71-88

[57] McGuire S. Institute of Medicine. Caffeine in Food and Dietary Supplements: Examining Safety—Workshop Summary. Washington, DC: The National Academies Press; 2014

[58] Superior Health Council (SHC). 2012. The Use of Caffeine in Foodstuffs. Available from: https://www.health.belgium.be/sites/default/files/uploads/fields/fpshealth_theme_file/19076526/The%20use%20of%20caffeine%20in%20foodstuffs%20%28January%202012%29%20%28SHC%208689%29.pdf [Accessed: 25 March 2020]

[59] Temple JL, Bernard C, Lipshultz SE, Czachor JD, Westphal JA, Mestre MA. The safety of ingested caffeine: A comprehensive review. Frontiers in Psychiatry. 2017;**8**:80

[60] Reissig CJ, Strain EC, Griffiths RR. Caffeinated energy drinks-a growing problem. Drug and Alcohol Dependence. 2009;**99**(1-3):1-10

[61] Hanit RH, Gadoth N. Caffeine-induced heaadache in children and adolescants. Cephalalgia. 2003;**23**:332-335

[62] Chaudhary NS, Grandner MA, Jackson NJ, Chakravorty S. Caffeine consumption, insomnia, and sleep duration: Results from a nationally representative sample. Nutrition. 2016;**32**(11-12):1193-1199

[63] Küçer N. The relationship between daily caffeine consumption and withdrawal symptoms: A questionnaire-based study. Turkish Journal of Medical Sciences. 2010;**40**(1):105-108

[64] Jin MJ, Yoon CH, Ko HJ, Kim HM, Kim AS, Moon HN, et al. The

relationship of caffeine intake with depression, anxiety, stress, and sleep in Korean adolescents. Korean Journal of Family Medicine. 2016;**37**(2):111

[65] Calamaro CJ, Yang K, Ratcliffe S, Chasens ER. Wired at a young age: The effect of caffeine and technology on sleep duration and body mass index in school-aged children. Journal of Pediatric Health Care. 2012;**26**(4):276-282

[66] Drapeau C, Hamel-Hébert I, Robillard R, Selmaoui B, Filipini D, Carrier J. Challenging sleep in aging: The effects of 200 mg of caffeine during the evening in young and middle- aged moderate caffeine consumers. Journal of Sleep Research. 2006;**15**(2):133-141

[67] Rogers PT, Heatherley SV, Hayward RC, Seers HE, Hill J, Kane M. Effects of caffeine and caffeine withdrawal on mood and cognitive performance degraded by sleep restrection. Psychopharmacology. 2005;**179**(4):742-752

[68] Rath M. Energy drinks: What is all the hype? The dangers of energy drink consumption. Journal of the American Academy of Nurse Practitioners. 2012;**24**(2):70-76

[69] Westerterp-Plantenga MS. Green tea catechins, caffeine and body-weight regulation. Physiology & Behavior. 2010;**100**(1):42-46

[70] Westerterp-Plantenga MS, Lejeune MP, Kovacs EM. Body weight loss and weight maintenance in relation to habitual caffeine intake and green tea supplementation. Obesity Research. 2005;**13**(7):1195-1204

[71] Auvichayapat P, Prapochanung M, Tunkamnerdthai O, Sripanidkulchai BO, Auvichayapat N, Thinkhamrop B, et al. Effectiveness of green tea on weight reduction in obese Thais: A randomized, controlled trial. Physiology & Behavior. 2008;**93**(3):486-491

[72] Hudson T. Green tea and women's health. Alternative & Complementary Therapies. 2007;**13**(5):269-272

[73] Rapuri PB, Gallagher JC, Kinyamu HK, Ryschon KL. Caffeine intake increases the rate of bone loss in elderly women and interacts with vitamin D receptor genotypes. The American journal of Clinical Nutrition. 2001;**74**(5):694-700

[74] Hallström H, Byberg L, Glynn A, Lemming EW, Wolk A, Michaëlsson K. Long-term coffee consumption in relation to fracture risk and bone mineral density in women. American Journal of Epidemiology. 2013;**178**(6): 898-909

[75] Miyagishima A, Fujiki S, Okimura A, Arahata S, Inagaki S, Iwao Y, et al. Novel decaffeination of green tea using a special picking method and shortening of the rolling process. Food Chemistry. 2011;**125**(3):878-883

[76] Varman AH, Sutherland JM. Decaffeinated tea. In: Beverages: Technology, Chemistry and Microbiology. Vol. 2. Maryland: Springer Science & Business Media; 1994

[77] Koca İ, Bostancı Ş. Kafeinsiz çay üretim. 2. Rize Kalkınma Sempozyumu. Çay-Lojistik-Turizm. 3-4 Mayıs 2013; Rize, Turkey. pp.156-167

[78] Wang X, Wan X, Hu S, Pan C. Study on the increase mechanism of the caffeine content during the fermentation of tea with microorganisms. Food Chemistry. 2008;**107**(3):1086-1091

[79] Gokulakrishnan S, Chandraraj K, Gummadi SN. Microbial and enzymatic methods for the removal of caffeine. Enzyme and Microbial Technology. 2005;**37**(2):225-232

Chapter 14

Agro-industrial By-Products from Amazonian Fruits: Use for Obtaining Bioproducts

Russany Silva da Costa, Taís Vanessa Gabbay Alves, Raimundo Lopes da Silva, Lindalva Maria de Meneses Costa Ferreira, Roseane Maria Ribeiro Costa and José Otávio Carréra Silva

Abstract

Fruit processing contributes significantly to the agricultural exportation of the Amazonian; however, it generates large amounts of solid waste, despite its high content of bioactive compounds and nutritional properties, and they are discarded in the environment. Therefore, in order to add economic value and potential reuse of agro-industrial by-products from cocoa, cupuassu, pracaxi, and tucumã, we investigated the chemical characteristics of the seed by-product resulting from the industrial extraction of these oils. The investigation of the nutritional and chemical composition of by-product was submitted to green extraction, besides other qualitative and quantitative techniques for the characterization of the main bioactive compounds. The extracts obtained from these by-products had a significant total polyphenol content and antioxidant activity. HPLC analysis identified and quantified some flavonoids present in these by-products (gallic, caffeic and protocatechuic acid, epigallocatechin-3-gallate, epicatechin, catechin, and quercetin). The oil from these species is widely used in the treatment of skin scarring and inflammation and is also used by the cosmetic industry. These results show that these by-products have a great potential for use, since they still have bioactive substances in their composition, which could alternatively be used in the pharmaceutical, cosmetic, or food industries.

Keywords: by-product, antioxidant activity, flavonoids, reuse, value-added applications, Amazonian fruit

1. Introduction

Brazil is currently one of the world's three largest fruit producers, associated with China and India. Brazil makes up 45.9% of world production a year, 2014 accounted for approximately 830.4 million tons of fruit [1]; many of them come from the Amazon region, where there is a diversity in economic fruit species, with huge agro-industrial and nutritional potential in the development of new products [2].

In addition, the Amazon region is home to a large biodiversity of plant species that produce fruit and oilseeds and stand out for their environmental conditions (climate and soil) [3]. Because it has a huge territorial extension, diverse fauna and flora, it is a source of life and income for approximately 200,000 families that collect native fruits, whose commercialization is responsible for 10% of the total income from extractivism [4]. In this sense, countless native species of fruit plants from the Amazon present economic, technological, and nutritional potential.

Given the large production of fruits, mainly for the juice industry, the processing of these fruits generates a huge amount of by-products resulting from seeds, peel, and part of the fruit pulp, resulting in an amount of about 30–40% of the production of these fruits [5]. Given the increase in fruit production, there is an increase in the generation of the so-called agro-industrial by-products, which causes an economic impact, as there is no proper reuse.

Indeed, some studies report nutrient concentrations in fruit by-products even higher than in pulp [6, 7]. The so-called tropical fruit processing by-products have been increasingly used as food additives and sources of bioactive compounds such as polyphenols [8–10]. In addition, the appropriate reuse of these by-products can reduce the environmental impact associated with their disposal, adding value to the entire production chain. Thus, the physicochemical characterization of these by-products and the quantification of their bioactive compounds are of great concern to add value and improve their commercial and industrial reuse, preserving the biome [11, 12].

In the literature, there are several studies related to agro-industrial by-products from fruits of the amazon region with the objective of finding a sustainable destination, among which are worth mentioning those related to the cocoa (*Theobroma cacao*) [13], cupuassu (*Theobroma grandiflorum*) [10], pracaxi (*Pentaclethra macroloba*) [14], and tucumã (*Astrocaryum vulgare* Mart) [15] by-products.

These species are native to tropical forests, originating from the Brazilian Amazon. Cocoa (*Theobroma cacao* L.) belongs to the Malvaceae family, has two varieties, Criollo and Forastero, and is 15–25 cm long and 8–13 cm in diameter, and the pulp is characterized by a thick mass of about from 20 to 40 seeds [16]. In 2015–2016, cocoa production was 3.9 million tons, of which 16.96% came from America, 73.11% from Africa, and 9.93% from Asia and Oceania. In contrast, in the same period, 16 million tons of by-products were generated, with Africa being the largest producer (73.12%), followed by America (16.88%), and Asia and Oceania (10.00%) [16].

Cupuassu [*Theobroma grandiflorum* (Willd. Ex. Spreng.) K. Schum.] belongs to the genus *Theobroma*, the latter being composed of 22 species of tropical plants from the Americas, including cocoa (*Theobroma cacao* L.). Among the Amazonian fruits, it is the one that brings together the best conditions of industrial use, and its pulp has great possibilities of use in the food and cosmetics industry. Due to the various applications in cooking, cupuassu has been arousing economic interest because its pulp is widely used in home and industrial production of various specialties. From the seeds, cupulate® is produced. And also in the cosmetics industry, its fat is considered an important emollient [10].

Pentaclethra macroloba (Willd.) Kuntze is popularly known as pracaxi, paracaxi, or paroá-caxi, belonging to the Fabaceae family. The pracaxi is an oilseed plant from the Amazon region found in Guyana and some parts of Central America [17, 18]. The pracaxi fruit is in the form of a pod of 20–25 cm, curved and dark brown in color, when ripe and contains 4–8 seeds [19]. From the seed is extracted an oil, which is used in the treatment of ulcers and wounds, besides being healing and presenting insecticidal properties against the *Aedes aegypti* mosquito. In the cosmetics industry, it is used in the production of hair products [20, 21].

Fruit abbreviation	Amazonian fruit's species	Parts of plants used	Application	Main phenolic compounds
CA	Cocoa.	Pulp and seed	Food and cosmetics industry	Catechin, epicatechin and gallic acid [13]
CP	Cupuassu	Pulp and seed	Food and cosmetics industry	Epicatechin and glycosylated quercetin [10]
PX	Pracaxi	Tree bark and seed	Ulcers and wounds treatment, inseticidal and cosmetics industry	Catechin [14]
TM	Tucumã	Pulp and seed	Food and cosmetics industry	Gallic acid [15]

Table 1.
Some by-products of Amazonian fruits: application and predominant biocompounds.

Tucumã (*Astrocaryum vulgare* Mart.) is an oleaginous fruit whose mesocarp is fibrous and nutritious, yellow-orange in color, rich in lipids and compounds such as pro-vitamin A [4]. The by-product of tucumã is also an excellent source of carotenoids [4]. The food industry uses its pulp to produce creams and ice creams. After obtaining the pulp, the tucumã seed is discarded (tons each year) [22]. The cosmetic industry uses tucumã pulp for oil extraction, which is used in skin moisturizing cosmetics, body lotions, and hair care products.

Since the waste of Amazonian fruit by-products, and consequently their antioxidant potential, due to the presence of bioactive compounds (e.g., polyphenols) (**Table 1**), it is possible to highlight some alternatives for better use of these by-products. In this sense, its use as enriched ingredients in food formulations with nutritional and functional properties [10] stands out, for supplementation/complementation in cookies, bread, cereal bars, cakes, and pastes.

Given the high nutritional and economic (underutilized) value of by-products, many studies have been conducted with a common goal, their reuse [10, 13, 15, 23–28]. In this perspective, the valorization of agri-food by-products is presented not only as a necessity, but as an opportunity to obtain new products with added value and a great impact on the economy of industries. Thus, several authors have demonstrated that the vast diversity of fruits found in the Brazilian territory, especially the Amazon, presents nutritional richness and can be better utilized directly by the population and also by the food or cosmetic industries [10, 25, 28, 29]. To this end, further studies are needed to better understand the nutritional, functional, and economic potential of fruit by-products, especially those found in the Brazilian Amazon.

2. Nutritional composition

Brazilian fruit by-products, particularly those from the Amazon region, need further investigation in order to obtain more information on their nutritional composition. Relevant data on the chemical composition of those traditionally inedible parts such as peel and seeds are even rarer. In addition, large amounts of by-products from these fruits are not consumed regularly; among them are seeds that are generally wasted in the environment [8]. Therefore, this chapter aimed to gather information on the nutritional potential of cocoa (CA), cupuassu (CP), pracaxi (PX), and tucumã (TM) seed by-products, with the objective

Parameter (g/100 g)	CA*	CP*	PX*	TM*
Lipids	33.5 ± 0.5	24.4 ± 0.8	14.9 ± 0.1	15.5 ± 0.4
Proteins	17.3 ± 0.4	14.2 ± 0.3	21.5 ± 0.6	11.1 ± 0.2
Total fibers	15.0 ± 0.4	22.3 ± 0.3	20.9 ± 0.7	41.4 ± 0.7
Carbohydrates	42.9 ± 0.4	26.6 ± 0.5	32.9 ± 0.0	63.1 ± 0.8
VET[a]	539.8 ± 2.3	382.0 ± 2.0	352.5 ± 3.3	436.4 ± 1.8

*VET means energetic value.*Results expressed as mean of triplicates ± standard deviation, expressed on a dry basis.*
[a]Values expressed in (kcal/100 g).

Table 2.
Nutritional composition of Amazonian fruit by-products: Cocoa (CA) [13], cupuassu (CP) [10], pracaxi (PX) [14], and tucumã (TM) [15].

(depending on the results) of encouraging its consumption by the population, taking advantage of its use as ingredients in animal feed and even human food formulations.

The results of the nutritional composition of the CA, CP, PX, and TM by-products are presented in **Table 2**.

Amazonian fruit by-products: CA, CP, PX, and TM showed important, up-to-date, and reliable nutritional values on the macronutrient composition of these by-products. As expected, because it is organic matter, carbohydrates were the most abundant macronutrients in the by-product studied. The CA, CP, PX, and TM also presented values of lipids, total fibers, and protein, and considerable energy value (**Table 2**). Thus, the contents of these macronutrients were higher than those reported for cupuassu pulp [5]. Protein content was 42% higher than fermented or roasted cupuassu seeds and 56% higher than cocoa seeds [7]. Compared to the studied by-products, pracaxi presented almost twice the protein content. And the by-product of the tucumã seed had the highest total fiber content, which may be related to its seed size (up to 22.9 mm in diameter) [30].

Compared to the cupuassu seeds, the cocoa seeds presented 17.74% higher protein content, 37.16% carbohydrate content, and 27.25% lipid content, while total fiber content was lower 33.41% (**Table 2**) [10, 13].

Given the above, the CA, CP, PX, and TM seed by-products presented significant nutritional values of macronutrients (carbohydrates, proteins, lipids, and crude fibers) (**Table 2**), suggesting the possibility of their reuse by the food industry, as a possible food supplement, as it is a great alternative for food product enrichment, by increasing its nutritional value with a low-cost raw material and its importance as a source of human and animal food.

3. By-product processing

The industrial processing to obtain the industrial by-products (**Figure 1**) is similar; usually this process is performed from the fruit, where the first step is the separation of the pulp (**Figure 1A**), and the seeds are then subjected to a cooking process at 65°C for 45 min (**Figure 1B,C**) and then pressed (**Figure 1D**) to remove crude oil or butter (raw material for the cosmetic industry). Therefore, the resulting by-product (residual cake) is usually discarded by the industry, but it can be used as a raw material and proceeds to the standardization stage, being added in an oven with air circulation (40 ± 2°C) until obtaining of constant weight (**Figure 1E**). After dehydration, it was pulverized and from this extraction is performed (**Figure 1F**), obtaining the biocompounds (**Figure 1G**) [31–33].

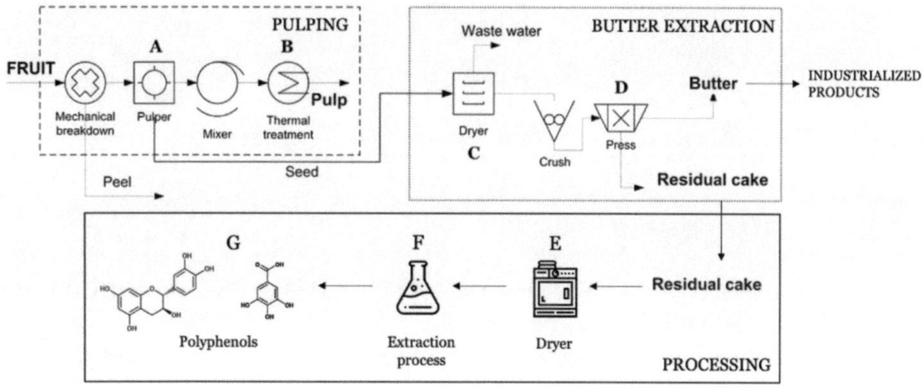

Figure 1.
Residue processing and obtaining biocompounds (adapted from González et al. [31]).

4. Extractive process

4.1 Green extraction

Green extraction is based on efficient and conscientious use of plant raw materials to ensure an extraction process with better yield conditions. In this sense, this model aims to optimize the extractive process and reduce extraction time, number of operations, energy consumption, and amount of waste generated and processing costs [34].

Good manufacturing practices are in this context to improve the elective parameters during the extraction process, in order to optimize the steps during the plant cultivation process, ensuring the lowest water consumption and the reduction of pesticide and fertilizer use. In addition, it is possible to develop genetic improvement protocols to obtain extracts with the highest concentration of biocompounds of industrial interest. Techniques that allow efficient production and a reduction in the generation of environmental waste should be pointed out. The use of natural, less toxic, easy to degrade solvents or with a lower risk of environmental contamination is one of the most recent bets on the production of new products [34].

4.2 Obtaining the extract

Extracts are preparations obtained from medicinal plant derivatives (powder), which may be in liquid (fluid extract), semi-solid (soft extract), or solid (dry extract) form. They can be obtained by various methods, as shown in **Figure 2** (Adapted from Silva Junior et al. [35]).

Maceration is an extractive process in which the proportion of vegetable drug powder and solvent volume influences the efficiency of the method; generally, the ratio 1:5 or 1:10 (plant drug/extract) is used. The plant material will be in contact with the solvent at rest in a closed container; at certain time intervals, the mixture should be agitated, and the final process time is variable [35, 36].

Percolation is an extractive methodology in which it is obtained by exhaustion. Prior to this process, maceration of the plant drug should be performed. To perform the technique, in a percolator, the vegetable drug and the solvent must be added. The volume should be 1:10 (plant drug/extract). The percolator faucet should be opened, and the liquid flow rate varies according to the velocity and can be classified as slow (1.0 mL/min), moderate (1.0-3.0 mL/min), and fast (3.0-5.0 mL/min) [36].

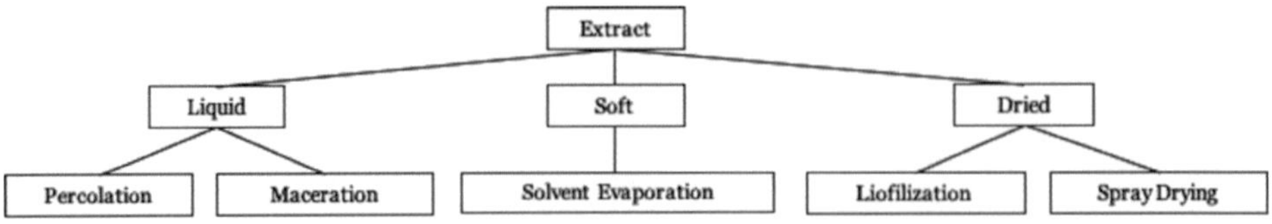

Figure 2.
Types of extracts and their methods of obtaining.

To obtain the extracts of the by-products of CA [32], CP [33], PX [14], and TM [15], the percolation methodology was performed.

The standardization of plant material for extracts and their fractions requires applying techniques that characterize it to ensure correct use according to quality parameters, among which stand out the identification and quantification of the main classes of secondary metabolites and chemical markers, as well as investigation of pharmacological activities of industrial interest [35].

5. Chemical composition

Polyphenols (**Figure 3**) are the secondary metabolites found mostly in both fruit and by-products. Many factors may alter the contents of polyphenols, as well as other bio-compounds in a plant species, such as the area under cultivation, the maturation time, climatic conditions during the cultivation stage, the season of the harvest year, and the storage of the crop raw material. In addition, the seeds of CA, CP, PX, and TM during their processing go through a heating step, and the temperature used can contribute negatively in reducing the levels of the assets [37]. They are derived from phenylpropanoids, where their main representatives are phenols, lignins, and flavonoids [6].

Flavonoids are low molecular weight compounds and are derived from benzo-g-pyrone, are responsible for the pigmentation of flowers, and protect against damage caused by light, fungi, or parasites. They have a chemical structure consisting of 15 carbon atoms, characterized by the presence of a diphenylpropan (C_6-C_3-C_6) attached to two benzene rings (A and B), and depending on the oxidation of the pyran ring, the ring A is derived from the acetate/malonate route, and ring B is derived from phenylalanine, and its name is variable according to ring C [6].

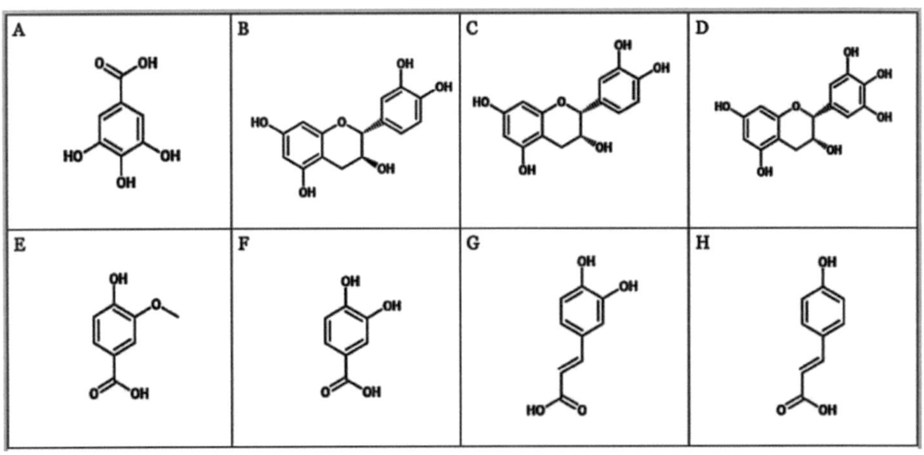

Figure 3.
Polyphenol structural formula: A) gallic acid; B) (+)-catechin; C) (−)-epicatechin; D) (−)-epigallocatechin; E) vanillic acid; F) protocatechuic acid; G) caffeic acid; H) p*-coumaric acid (adapted from Alves [32]).*

5.1 Qualitative analysis

5.1.1 Determination of biocompound contents by ultraviolet spectrophotometry (UV: Vis)

Pre-formulation studies are of fundamental importance in relation to quality control tests, which are based on analyzes that allow the characterization of the raw material for the evaluation of its potential to be used as a finished product asset [27]. The characterization of an extract implies the definition of the contents of its main chemical constituents, as well as moisture, color, particle size distribution, viscosity, and technological properties, ensuring the safety, efficacy and quality of a product. When formulating products containing natural actives, the standard identification of markers or the development of methods that allow the quantification of purified chemical groups before, during, and after the obtaining process is essential [38].

There are many qualitative analytical methods used to identify total polyphenol levels (**Table 3**) such as UV–vis spectrophotometry, where one of the main colorimetric methods used is the Folin–Ciocalteu reagent. For the analysis, in a 25 mL volumetric flask, 4.8 ml of deionized water, 0.2 ml of sample, and 0.5 ml of Folin–Ciocalteu reagent are added. To this mixture is added 1.0 mL of 20% sodium carbonate solution, and then deionized water should be added to complete the volume of 10 mL and homogenize. The reaction system should be kept for 1 hour at room temperature and protected from light. After time, aliquots were collected and analyzed by UV–vis spectrophotometer (Perkin Elmer, Wellesley, MA, USA) at a wavelength of 725 nm. TP was standardized against gallic acid and expressed as micrograms of gallic acid equivalents per gram of dried extract (mgGAE/g) [39].

The total flavonoid contents (**Table 3**) can be determined by spectrophotometric analysis using aluminum chloride and for the extracts as described by [39, 40] for the analysis of CA, CP, PX, and TM, the UV–vis spectrophotometer was used (Perkin Elmer, Wellesley, MA, USA) to a wavelength of 510 nm.

5.1.2 Fourier transform infrared spectroscopic profile identification (FT-IR)

Infrared absorption spectroscopy is used in the literature to identify possible characteristic functional groups in organic compounds, providing important information on the chemical structure of the sample [38].

Samples	TP (mg/g)	TF (mg/g)	Source
CA	229.6 ± 3.24[a]	0.68 ± 0.02[b]	[13]
CP	50.1 ± 5.30[a]	5.92 ± 3.40[c]	[10]
PX	2.66 ± 0.01[a]	0.11 ± 0.25[d]	[14]
TM	1.35 ± 0.08[a]	0.33 ± 0.01[b]	[15]

Results are expressed as mean triplicate ± standard deviation.
[a]*GAE = gallic acid equivalent.*
[b]*EQE = quercetin equivalent.*
[c]*ECA = catechin equivalent.*
[d]*ERT = rutin equivalent.*

Table 3.
Total polyphenol (TP) and total flavonoid (TF) content of crude extracts of CA, CP, PX, and TM by-products.

The identification of the functional bands using the Fourier transform infrared absorption spectroscopy (**Figure 4**) was performed in the CA by-product extract, where it was possible to observe bands at 1600, 2920, and 3331 cm^{-1}, which correspond to C-C stretching of aromatic ring by phenol group, C-H stretching of aromatic ring by phenol group, and O-H stretching of phenol group, respectively [27, 41]. In the CP by-product extract, the bands 1037 cm^{-1} corresponding to acid vibrations, 1120 cm^{-1} alcohol vibrations, 1668 cm^{-1} esters and sulfonic vibrations, and 2931 cm^{-1} O-H axial deformation of alcohol groups were displayed [33]. In PX, bands were shown at 1037, 1384, 1635, and 3448 cm^{-1} for C-O axial deformation of alcohol and phenols, C-H axial deformation vibration of methyl, C=C axial deformation of carbonyl ring, and –OH deformation vibration of carbohydrates and carboxylic acids, respectively [38]. And the TM by-product extract exhibited bands at 1403 cm^{-1} aromatic ring stretching vibration, 1609 cm^{-1} ketone C=O stretching vibration, and 3381 cm^{-1} free –OH stretching vibration [42]. All bands observed in CA, CP, PX, and TM were correlated to chemical structures present in polyphenols [27, 33, 38, 42].

5.1.3 Thermal analysis

Thermogravimetric analysis (TGA) is a technique that assesses the loss in mass of a substance as a function of temperature, it allows a variability of results to be observed, such as the temperature range at which the sample is degraded, until the temperature of the sample remains stable, at which temperature a change in physical state such as melting, among others, occurs. In addition, it is possible to plot a derivative on top of the TGA curve; this analysis can show which temperature range the greatest loss of

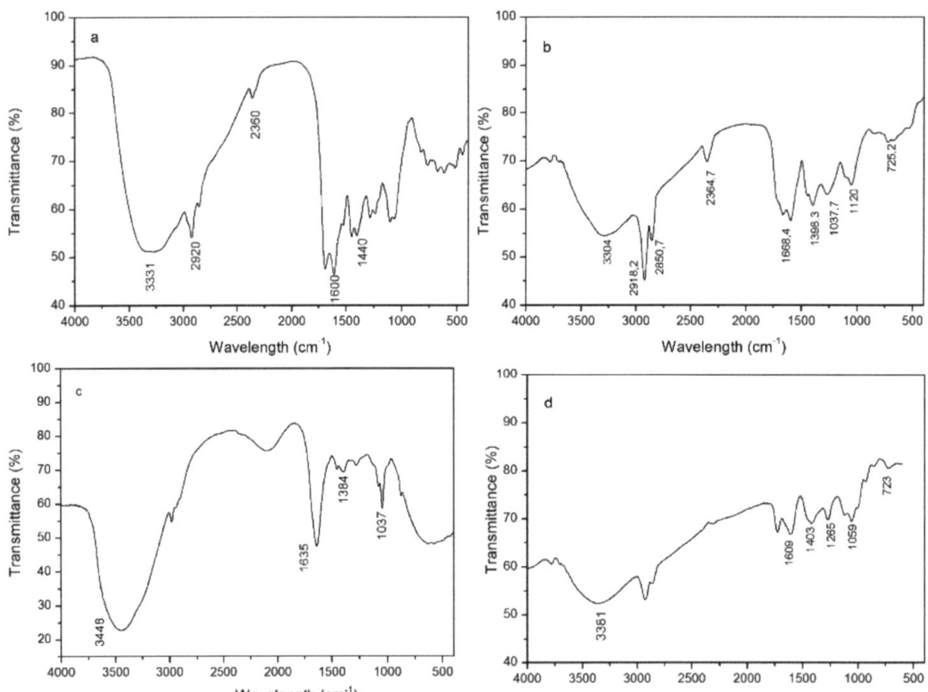

Figure 4.
Identification of the functional bands using FT-IR absorption spectroscopy. The extracts of the by-products were compressed into Kbr and scanned in the 4000-400 cm^{-1} wavelength absorption range, with a resolution of 2.0 cm^{-1} and a scan number of 20 scans. a) CA [27], b) CP [33], c) PX [14], and d) TM [15].

mass occurs in [27]. This technique has been used for the evaluation of by-products and their extracts, such as CA [27], CP [10], PX [14], and TM [15].

The analysis of behavior and thermal stability can be employed in the quality control of raw materials and evaluation in the development of herbal medicines [43]. The thermogravimetric analysis (**Figure 5**) of the CA, CP, PX, and TM extracts presented on average three events of mass loss. The first corresponds to the evaporation of solvents, such as water. The second event represents successive reactions and may be related to loss of sugars. And the latter confers with the degradation and carbonization of organic matter from biocompounds [14, 15, 27, 28].

Such analysis is important to evaluate the thermal stability of the extracts, bearing in mind that prior knowledge has the purpose of guaranteeing the physical–chemical stability of the thermal constituents present in the extracts [38, 44]. The thermogravimetric study makes it possible to obtain information on the relationship between humidity and the maximum temperature of stability of the extract [44].

5.2 Quantitative analysis

5.2.1 High-performance liquid chromatography (HPLC)

High-performance liquid chromatography is a separation technique that is among the main techniques used to determine polyphenol and flavonoid levels. Natsume et al. [45] point out that it is possible to carry out the identification and quantification of those elements in plant extracts and their derivatives; the analysis applied may be: reverse-phase HPLC (RP-LC) and reverse-phase HPLC-mass spectrometry (RP-LC/MS), and in particular, when it comes to the genus *Theobroma*, the majority of flavonoids observed in the species were catechin and epicatechin (**Figure 6**).

The literature points to the activity of certain flavonoids (**Table 4**), such as those identified and quantified in CA and CP a possible correlation with interesting

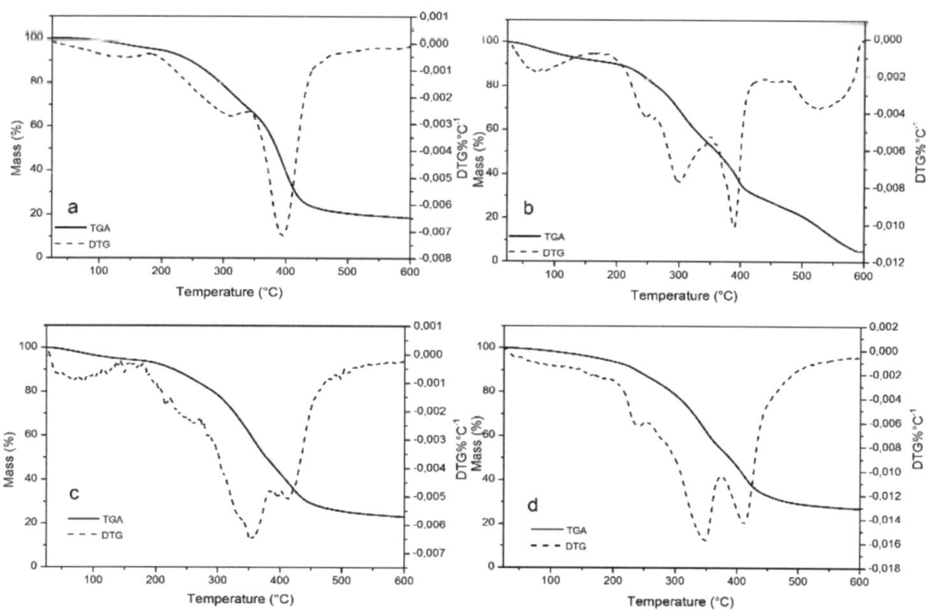

Figure 5.
TGA and DTG curves of the by-product extracts obtained at 25 to 600° C at 10° C/ min under N2 atmosphere and flow of 50 mL/min: a) CA [28], b) CP [10], c) PX [14], and d) TM [15].

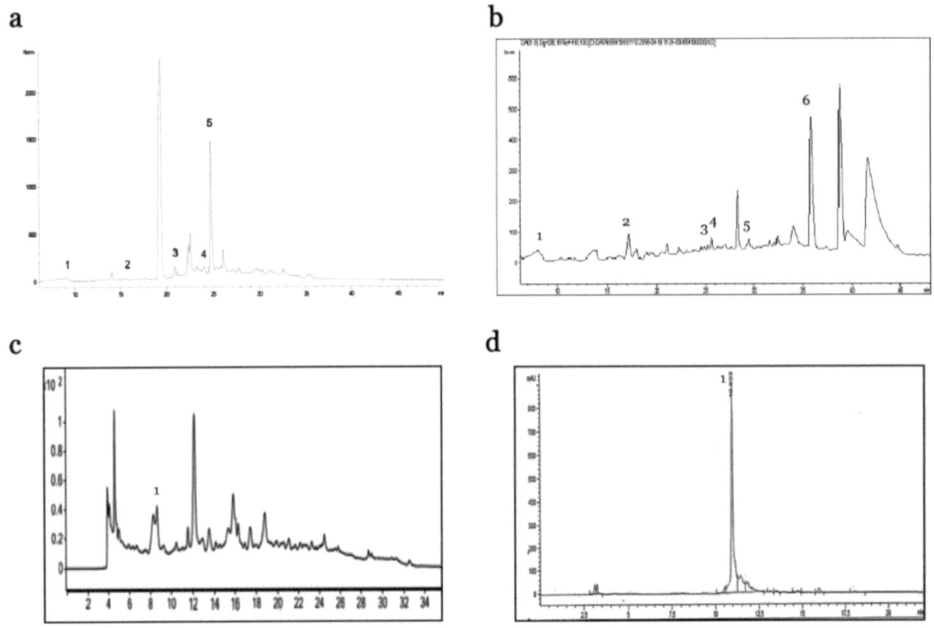

Figure 6.
Chromatograms about the phenolic compounds (280 nm) identified in the extracts of the by-products of a) CA [28], where 1—gallic acid, 2—protocatechuic acid, 3—catechin, 4—epigallocatechin-3-gallate, and 5—epicatechin; b) CP [33], where 1—gallic acid, 2—protocatechuic acid, 3—epigallocatechin-3-gallate, 4—epicatechin, 5—p-coumaric acid, and 6—glycosylated quercetin; c) PX [14], where 1—catechin; d) TM [15], where 1—gallic acid.

Compounds	CA[a]	CA[b]	CP[a]	CP[b]	PX[a]	PX[b]	TM[a]	TM[b]
Gallic acid	9.05	1.10	8.80	0.06	—		10.03	20.0
Protocatechuic acid	17.08	1.12	17.32	0.33	—		—	
p-coumaric acid	—	—	29.5	0.01	—		—	
Catechin	23.20[a]	11.00	—	—	9.0	0.21	—	
Epigallocatechin-3-gallate	24.80	0.66	25.45	0.07	—		—	
Epicatechin	24.80[a]	24.00	25.9	0.21	—		—	
Glycosylated quercetin	—	—	35.6	0.28	—		—	
Quercetin[c]	—	—	40.78	0.58	—		—	

[a]*Retention time (min).*
[b]*Concentration (mg/g).*
[c]*Viewed at 369 nm.*

Table 4.
Polyphenolic compounds detected at 280 nm by HPLC in the extract of CA, CP, and PX by-products and UHPLC in the extract of TM by-product [10, 13–15].

pharmacological activities, such as cardiac protector [46] reducing oxidative stress [47]. The phenolic acids, such as gallic acid, found in CA, CP, and TM, confers antioxidant properties both to food and to the body, so they are indicated for the treatment and prevention of cancer, cardiovascular diseases, and other illnesses [48]. According to Natsume et al. [45], some flavonoids may be related to the reduction of the probability of developing atherosclerosis, since the ingestion of these biocomposites contribute to the inhibition of oxidation of low-density lipoprotein (LDL).

6. Antioxidant activity

The compounds containing antioxidant activity inhibit and/or diminish the effects caused by free radicals, protecting the cells against the harmful effects of oxygenated and nitrogenous free radicals, formed during the oxidative process of the cells [49]. The evaluation of the antioxidant activity of plant material shall be determined by at least two methods, in order to obtain a slightly more complete context of this activity [50]. The use of ABTS colorimetric methods, together with DPPH, bring better reproducibility and sensitivity [51]. Given this, the antioxidant capacity of the raw extracts of the cocoa, cupuassu, pracaxi, and tucumã by-products was evaluated by the $ABTS^+$ and DPPH methods, being performed in triplicate.

Table 5 shows the antioxidant activities of CA, CP, PX, and TM extracts. According to the results, CP extract was the one that presented the best antioxidant activity, both by the $ABTS^+$ and by the DPPH methods, while the CA presented the lowest values, determined by both methods. The four by-product extracts showed good results for the $ABTS^+$ method.

These results were higher than those reported in the literature by Leong and Shui [52], who achieved values of the radical $ABTS^+$ for some fruits, such as mango, passion fruit, pineapple, and guava that were 38.0; 5.5; 7.7; and 20.9 μM Trolox/g, respectively.

Studies have demonstrated that the $ABTS^+$ assay can be used to evaluate the antioxidant activity of a wide variety of substances [53], being commonly applied to determine the antioxidant activity in plants and based on the antioxidant capacity to neutralize the ABTS cation radical [54].

The antioxidant capacity values presented by different by-products CA, CP, PX, and TM, in addition to the vegetable sample difference, may also be related to different types of plant samples, which lead to different levels of phenolic compounds in extracts [55].

The results presented by the different by-products bring perspectives on the use of their antioxidant capabilities and may be better used in the development of new products.

Extract	$ABTS^{+*}$ μM Trolox/g	$DPPH^*$ μM Trolox/g
CA	225.0 ± 3.46	6.74 ± 0.20
CP	1497.82 ± 5.78	1717.73 ± 5.54
PX	597.23 ± 0.37	599.54 ± 0.01
TM	1247.88 ± 3.60	326.0 ± 1.21

*Results expressed as a mean of triplicate ± standard deviation.

Table 5.
Values of antioxidant activity, by ABTS and DPPH methods, of raw extracts of CA [13], CP [10], PX [14], and TM [15].

7. Drying by spray drying

Microencapsulation is a technique that aims to protect the assets, from possible causes that produce their instability, such as oxidation, humidity, and photolysis, among others. For this purpose, the sample is surrounded by a polymeric layer that was denominated an encapsulating agent [13]. Therefore, there are several works in this area, which aim to guarantee and/or increase the stability of phenolic compounds present in tropical fruit by-products, including cocoa, cupuassu, pracaxi, and tucumã [13, 28, 56, 57].

Thus, the spray drying technique was chosen for drying and obtaining microencapsulated extract of CA, CP, PX, and TM by-products, in order to protect the phenolic compounds against oxidation and environmental factors. For the drying of the materials, different conditions of inlet temperature (IT) and feed flow (FF) were used, in addition to the encapsulating agents.

To confirm the encapsulation, from the microencapsulated dry extract, the total polyphenol (TP) and total flavonoid (TF) contents were determined, besides their polyphenol (Y_{TP}) and flavonoid (Y_{TF}) microencapsulation yields, as well as the antioxidant activity by the $ABTS^+$ method. Analysis of scanning electron microscopy (SEM) of microparticles from microencapsulated extracts of Amazonian fruit by-products was performed.

Table 6 shows the results obtained in the drying of cocoa, cupuassu, pracaxi, and tucumã extracts. Maltodextrin was the encapsulating agent for drying all extracts of fruit by-products, and a percentage of 5.0% was used. For CA extract, chitosan was also used in a percentage of 0.5%. Chitosan was used as a polymer for encapsulation as the microparticles formed with the cocoa extract by-product because it was used in pisciculture, since maltodextrin is readily soluble in water [32]. For the CA and CP extracts, FF = 2.5 and 5.0 mL/min were used, respectively, while for both, IT was 170°C. For PX and TM extracts, the FF value was 10 and 7.5 mL/min, and the IT was 160 and 100°C, respectively.

The TM extract showed the highest values for all methods used in relation to the other extracts. It was observed that the higher TP content influenced a higher antioxidant activity for all extracts, where the TM extract had the highest content of polyphenols and, consequently, higher antioxidant activity by the $ABTS^+$ method. Despite the difference in the TP contents for the extracts of CA (80.44 ± 2.84 mgGAE/g) and CP (38.93 ± 1.24 mgGAE/g), the two presented values of Y_{PT} close to 64.87 ± 0.16% and 67.20 ± 1.90%, respectively.

The CP and TM extracts showed better microencapsulation yields (Y_{PT} and Y_{FT}), being above 50%. The Y_{PT} of the CP and TM extracts were close, where 96.50 ± 0.10 and 93.95 ± 2.62% of the polyphenols were present in the microparticles of the extracts, respectively, suggesting that they were not affected by the high temperatures used.

All extracts presented higher antioxidant activities in relation to the studies performed by Rezende et al. (129.16–155.24 μM Trolox/g) in the drying of acerola by-product extract; this may be due to use of gum arabic, along with maltodextrin, as an encapsulation agent, besides the plant material used and drying parameters [28, 59].

ME	TP* (mgGAE/g)	TF* (**)	Y_{PT}* (%)	Y_{FT}* (%)	$ABTS^{+}$* (μMTrolox/g)
CA	80.44 ± 2.84	22.52 ± 2.34	31.98 ± 2.08	64.87 ± 0.16	623.76 ± 20.06
CP	38.93 ± 1.24	11.28 ± 0.37	93.95 ± 2.62	67.20 ± 1.90	435.13 ± 4.1
PX	19.06 ± 0.32	12.46 ± 0.21	31.88 ± 0.02	21.90 ± 0.13	163.28 ± 0.32
TM	130.00 ± 0.024	27.17 ± 0.002	96.50 ± 0.10	83.02 ± 0.01	956.01 ± 7.63

**Results expressed as a mean of triplicate ± standard deviation.*
***mgQE/g for microencapsulated CA, PX, and TM extracts; mgCE/g for microencapsulated CP extracts.*
ME = microencapsulated extracts; TP = total polyphenols; TF = total flavonoids; YPT and YFT = microencapsulation yields of polyphenols and flavonoids, respectively.

Table 6.
Values of microencapsulated extracts of CA [13], CP [10], PX [14], and TM [15] obtained by spray drying.

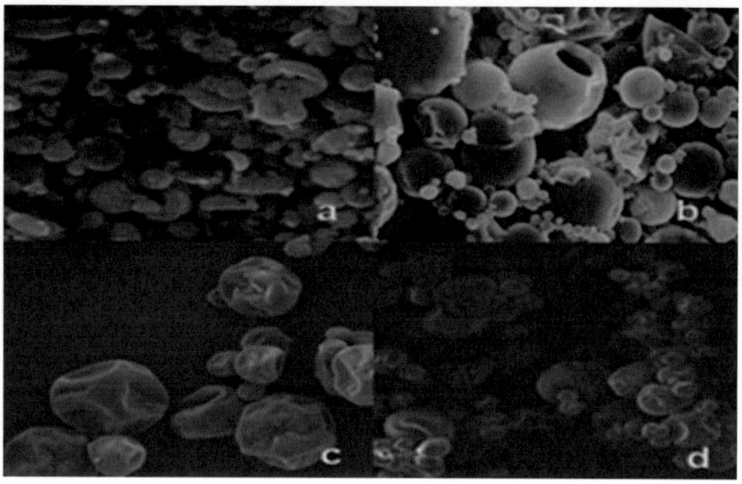

Figure 7.
Photomicrograph (SEM) of the microencapsulated extract of Amazonian fruit by-products. a) CA [13]; b) CP [10]; c) PX [14], magnification in 1.000x; d) TM, magnification in 5.000x [15].

With the drying results of the raw extracts of the Amazonian fruit by-products, it was observed that all of them maintained the presence of bioactive compounds and also showed antioxidant activity, using maltodextrin as an encapsulating agent.

There are a variety of polymers that can be used in microencapsulation; maltodextrin is of natural origin and is used as a wall material for encapsulation of various plant extracts because it has advantages such as biocompatibility, biodegradability, low toxicity, and reduced moisture in the wall of the microparticle [60].

The SEM analysis was performed for all four by-products in order to verify the formation of microparticles after spray drying. In the microencapsulated extract of the cocoa by-product (**Figure 7a**), the microparticles did not present any type of cracks and were not very grouped, besides not being rough. For the microencapsulated extract of the cupuassu by-product, the microparticles exhibited a very regular spherical structure, with low agglomeration, few ruptures, and heterogeneous size (**Figure 7b**), indicating poor structure deformation.

Figure 7c shows the photomicrographs of the microparticles present in the extract of the pracaxi by-product, where it is possible to verify that in general, the particles presented a spherical shape and rough surface characteristic of particles obtained by the spray drying method, which occurred probably during the drying and cooling process [56]. The presence of heterogeneous sizes and aggregate formation was also observed.

The microparticles of the microencapsulated extract of the tucumã by-product (**Figure 7d**) showed morphologies with low deformation in their structures and heterogeneity. Its external surfaces were without cracks and thus does not lead to rupture, which is essential to ensure greater protection of the asset. In general, they exhibit regular spherical shapes, although some are rounded, without strong agglomeration that may be due to the repulsion of loads, with varied size and presence of roughness.

8. Industrial application

The reuse of vegetable waste (by-products) is a viable alternative to contribute to the production chain segment, reduce costs, and contribute to the reduction of environmental contamination [32]. In this context, they have a wide spectrum of use in the food, pharmaceutical, cosmetic and veterinary industries.

The use of residues of certain fruits as raw material in the food industry in place of synthetic antioxidants and in the production of food that can be included in human food, such as biscuits, breads, cereal bars, cakes, and pastes among other products is of great economic interest and has represented an important segment in industries [26]. By-products have been used in innovative biotechnological processes to obtain enzymes with proteolytic and keratinolytic properties [57].

The exploration of by-products of fruit and vegetable processing, as a source of functional compounds and their application in cosmetics, is a promising field, used in personal, perfumery, and cosmetics hygiene products [25]. Animal feed supplementation is one of the most frequent applications for plant by-products. Its indications on the market are pisciculture, poultry, cattle, and pigs [23, 58]. The use of nutraceuticals in diets is adopted by improving the development, performance, and immunity of the animal. In this segment, enzymes, nucleotides, chitin, chitosan, vitamins, antioxidants, and plant extracts stand out in this segment [59].

The application is possible, thanks to the levels of nutrients that they possess, because they have an expressive amount of protein, nutrients, minerals, and bioactive compounds and ensure good digestibility [33, 60] contributing to generate a low-cost product with promising characteristics for its use.

The use of elements with a low-cost and easy access enables the use of by-products from industrial processing as a strategy to optimize the entire course of the productive stage [61]. Several studies use by-products of vegetable origin as a raw material for industrial reuse [39, 62]. These matrices present nutrient contents significantly interesting for total or partial use in fish feed supplementation [63]. Within this perspective, the cocoa and pracaxi extracts from the by-product present all the prerequisites to be applied in this market [13, 14].

In this perspective, the elaboration and characterization of flours, from fruit by-products, have been the object of numerous studies, which point to good nutritional characteristics and potential for their application as ingredients in food [33]. Due to their nutritional characteristics, cupuassu and tucumã flours emerge as a highly desirable food ingredient to enrich other foods [33, 64].

The market is made up of niches of consumers of natural foods (energy products), such as athletes, sportsmen, children, and workers who need to eat caloric foods [65]. As a result of the growing interest of consumers for more nutritious natural foods, with good intake of carbohydrates, proteins, vitamins, minerals, fibers, and an adequate balance of calories, the market for cereal bars has been increasing [64].

In view of the above, the extracts obtained from the by-products of cupuassu and tucumã can be used in this segment. The by-product of cupuassu can be used in the enrichment of multimixture flour, which is incorporated in the feeding of children in a state of infant malnutrition, a project already applied by the Sociedade Bíblica do Brasil in partnership with the Pastoral da Criança [33] and the by-product of tucumã in the preparation of bakery products (in the form of bread and cookies) and pasta in the production of cereal bars explored as functional food [15, 64].

9. Conclusions

The cocoa, cupuassu, pracaxi, and tucumã seed by-products presented concentrations of macronutrients such as proteins, fibers, total fats, and carbohydrates as ingredients potentially to be used as food or animal feed. In addition, these extracts showed significant antioxidant activity, and phenolic compounds (including protocatechuic acid, gallic acid, caffeic acid, and *p*-coumaric acid) and flavonoids (quercetin, glucosylated quercetin, epicatechin, catechin, and epigallocatechin-3-gallate) were the most abundant compounds in these extracts. By means of the response surface methodology, it was determined that the optimal conditions for the microencapsulation of cocoa, cupuassu, and pracaxi seeds by-product extract are: IT = 170°C, FF = 2.5 mL/min, and MD = 5.0%; IT = 170°C, FF = 5.0 mL/min, and MD = 5.0%; IT = 160°C, FF = 10.0 mL/min and MD = 5.0%, respectively, and optimal conditions for microencapsulation of tucumã seeds IT = 100°C, FF = 7.5 mL/min, and MD = 5.0%. Under these conditions, the microparticles were obtained with good stability and some heterogeneity, with spherical structure, confirming the efficiency of the microencapsulation process with the use of maltodextrin as a drying adjuvant.

Therefore, it is suggested that the microencapsulated extracts of cocoa, cupuassu, pracaxi, and tucumã seed by-products can be used in the food, cosmetic, pharmaceutical, and veterinary industries And used as a potential source of nutrients to be deployed as an integrator of alternative human food or animal feed, an opportunity in which they acquire economic value and at the same time reducing the environmental impact related to their disposal in the environment.

Acknowledgements

The authors would like to thank the CNPq, FAPESP, and CAPES for financial support.

Conflict of interest

The authors declare that there is no conflict of interest.

Author details

Russany Silva da Costa[1], Taís Vanessa Gabbay Alves[2], Raimundo Lopes da Silva[2], Lindalva Maria de Meneses Costa Ferreira[1], Roseane Maria Ribeiro Costa[2] and José Otávio Carréra Silva[1*]

1 Laboratory R&D Pharmaceutical and Cosmetic, Faculty of Pharmaceutical Sciences, Federal University of Pará, Belém, PA, Brazil

2 Laboratory of Pharmaceutical Nanotechnology, Faculty of Pharmaceutical Sciences, Federal University of Pará, Belém, PA, Brazil

*Address all correspondence to: carrera@ufpa.br

References

[1] SEAB, Secretaria de Estado da Agricultura e do Abastecimento; DERAL-Departamento de Economia Rural, Fruticultura; Res. Téc. Eng. Agr. Paulo Fernando de Souza Andrade; 2017

[2] Matos AT. In: Viçosa, editor. Tratamento e Aproveitamento Agrícola de Resíduos Sólidos. UFV; 2014

[3] Chisté RC, Fernandes E. Bioactive compounds from Amazonian fruits and peach palm fruits, β-carotene was the major carotenoid, followed by γ- their antioxidant properties. In: Silva LR, Silva BM, editors. Natural Bioactive Carotene and δ-Carotene. Compounds from Fruits and Vegetables as Health Promoters – Part I. United Arab Emirates: Bentham Science Publishers; 2016. pp. 244-264

[4] Noronha Matos KA, Praia Lima D, Pereira Barbosa AP, Zerlotti Mercadante A, Campos Chisté R. Peels of tucumã (*Astrocaryum vulgare*) and peach palm (*Bactris gasipaes*) are by-products classified as very high carotenoid sources. Food Chemistry. 2019;**272**:216-221

[5] Sousa MSB, Vieira LM, Da Silva MJM, Lima A. Nutritional characterization and antioxidant compounds in pulp residues of tropical fruits. Ciência Agrotecnica. 2011;**35**(3):554-559

[6] Ignat I, Volf I, Popa VI. A critical review of methods for characterization of polyphenolic compounds in fruits and vegetables. Food Chemistry. 2011;**126**:1821-1835

[7] Pugliese AG, Tomas-Barberan FA, Truchado P, Genovese M. Flavonoids, proanthocyanidins, vitamin C, and antioxidant activity of *Theobroma grandiflorum* (Cupuassu) pulp and seeds. Journal Agriculture and. Food Chemistry. 2013;**61**:2720-2728

[8] Ayala-Zavala JF, Vega-Vega V, Rosas-Domínguez C, Palafox-Carlos H, Villa-Rodriguez JA, Wasim Siddiqui MD, et al. Agro-industrial potential of exotic fruit Byproducts as a source of food additives. Foodservice Research International. 2011;**44**(7):1866-1874

[9] Pantaleón-Velasco MR, Ruiz-López II, Pérez-Silva A, Bravo-Clemente L, Mateos R, Ruiz-Espinosa H, et al. Antioxidant and functional properties of a high dietary fiber powder from carambola (*Averrhoa carambola* L.) pomace. International Journal of Food Science and Technology. 2014;**49**:2101-2110

[10] Costa RS, Dos Santos OV, Lannes SCS, Casazza AA, Aliakbarian B, Perego P, et al. Bioactive compounds and value-added applications of cupuassu (*Theobroma grandiflorum* Schum.) agro-industrial by-product. Food Science and Technology. 2019. DOI: 10.1590/fst.01119

[11] Ballesteros LF, Teixeira JA, Mussatto SI. Chemical, functional, and structural properties of spent coffee grounds and coffee silverskin. Food and Bioprocess Technology. 2014;**7**:3493-3503

[12] Janissen B, Huynh T. Chemical composition and value-adding applications of coffee industry byproducts: A review. Resources, Conservation and Recycling. 2018;**128**:110-117

[13] Gabbay Alves TV, Costa RS, Aliakbarian B, Casazza AA, Perego P, Arruda MSP, et al. Bioactive compounds and antioxidante potential for polyphenol-rich cocoa extract obtained by agro-industrial residue. Natural Product Research. 2017. DOI: 10.1080/14786419.2017.1399381

[14] Bezerra HM. Avaliação da atividade antioxidante e otimização das condições de microencapsulação por spray drying do extrato seco do subproduto agro-industrial do pracaxi (Pentaclethra macroloba Willd.) [dissertação]. Belém: Universidade Federal do Pará; 2018

[15] Ferreira LMMC. Obtenção, caracterização e microencapsulação de extratos ricos em antioxidantes naturais a partir do co-produto das amêndoas do tucumã (*Astrocaryum vulgare* Mart.) [dissertação]. Belém: Universidade Federal do Pará; 2019

[16] Vasquez ZS, Neto DPde C, Pereira GVM, Vandenberghe LPS, Oliveira PZ, Tiburcio HLGR, et al. Biotechnological approaches for cocoa waste management: A review. Waste Management. 2019;**90**:72-83

[17] Lorenzi H. Árvores Brasileiras: Manual de Identificação e Cultivo de Plantas Arbóreas no Brasil. 2nd ed. Vol. 2. Nova Odessa: Instituto Plantarum; 2002. p. 368

[18] Pesce C. Oleaginosas da Amazônia. 2nd ed. Belém: Museu Paraense Emílio Goeldi. Núcleo de Estudos Agrários e Desenvolvimento Rural; 2009. p. 334

[19] Cruz ED, Barros HSD. Germinação de sementes de espécies amazônicas: pracaxi [*Pentaclethra macroloba* (Willd.) Kuntze]. Embrapa Amazônia Oriental-Comunicado Técnico (INFOTECA-E); 2015

[20] Santiago GMP, Viana FA, Pessoa ODL, Santos RP, Pouliquen YBM, Arriaga AMC, et al. Avaliação da atividade larvicida de saponinas triterpênicas isoladas de *Pentaclethra macroloba* (Willd.) Kuntze (Fabaceae) e Cordia Piauhiensis Fresen (Boraginaceae) sobre Aedes Aegypti. Brazilian Journal of Pharmacognosy. 2005;**15**:187-190

[21] Silva JO, Fernandes RS, Ticli FK, Oliveira CZ, Mazzi MV, Franco J, et al. Triterpenoid saponins, new metalloprotease snake venom inhibitors isolated from *Pentaclethra macroloba*. Toxicon. 2007;**50**:283-291

[22] Bergmann JC, Tupinambá DD, Costa OYA, Almeida JRM, Barreto CC, Quirino BF. Biodiesel production in Brazil and alternative biomass feedstocks. Renewable and Sustainable Energy Reviews. 2013;**21**:411-420. DOI: 10.1016/j.rser.2012.12.058

[23] Almeida JS, Santos Neto DS, Paiva KSL, et al. Utilização de subprodutos de frutas na alimentação animal. Revista Eletrônica Nutritime. 2014;**11**(3):3430-3443

[24] Teixeira UHG, Simioni TA, Pina DS, Gomes FJ, Paula DC, Botini LA. Potencial de utilização de coprodutos agroindustriais para suplementos. Revista Eletrônica Nutritime. 2014;**11**(2):3363-3386

[25] Barbosa NP, Conceição EC. Alimentos com Potencial Aplicação em Cosméticos Naturais. Revista de Processos Químicos. 2016:127-131

[26] de CR VB, Moreira YR, Alfaiate MB, Souza MH, Mendonça PP, Deminicis BB. Utilização de subprodutos e resíduos de frutas na suplementação de ovinos (*Ovis aries*). Archives of Veterinary Science. 2017;**22**(2):08-17

[27] Gabbay Alves TV, Silva da Costa R, Aguiar Gomes AT, Ferreira da Costa CE, Perego P, Silva Júnior JOC, et al. Quality control of Amazonian cocoa (*Theobroma cacao* L.) by-products and microencapsulated extract by thermal analysis. Journal of Thermal Analysis and Calorimetry. 2018;**134**(2):993-1000

[28] Da Costa RS, Teixeira CB, Gabbay Alves TV, Ribeiro-Costa RM, Casazza AA, Aliakbarian B, et al. Optimization of spray drying conditions to microencapsulate

cupuassu (*Theobroma grandiflorum*) seed by-product extract. Natural Product Research. 2018. DOI: 10.1080/14786419.2018.1462178

[29] Arruda HS, Pereira GA, Pastore GM. Oligosaccharide profile in Brazilian Cerrado fruit araticum (*Annona crassiflora* Mart.). LWT–Food Science and Technology. 2016

[30] Lima RR, Trassato LC, Coelho V. O tucumã (*Astrocaryum vulgare* Mart.): Principais características e potencialidade agroindustrial. (EMBRAPA – CPATU). Boletim de Pesquisa. 1986;**75**:27

[31] González AA, Moncada J, Idarraga A, Rosenberg M, Cardona CA. Potential of the amazonian exotic fruit for biorefineries: The Theobroma bicolor (Makambo) case. Industrial Crops and Products. 2016;**86**:58-67. DOI: 10.1016/j.indcrop.2016.02.015

[32] Alves TVG. Aproveitamento de co-produto agroindustrial: micropartículas contendo extrato de cacau (*Theobroma cacao* L.) Padronizadas em antioxidantes naturais para suplementação da dieta de tambaqui (*Colossoma macropomum*) [Tese]. Belém: Universidade Federal do Pará; 2017

[33] Costa da RS. Otimização do processo de microencapsulação do extrato do coproduto agroindustrial do cupuassu (*Theobroma grandiflorum*) - Padronizado em antioxidantes naturais visando o enriquecimento nutricional e funcional da multimistura [Tese]. Belém: Universidade Federal do Pará; 2017

[34] Chemat F, Abert-Vian M, Fabiano-Tixier AS, Strube J, Uhlenbrock L, Gunjevic V, et al. Green extraction of natural products. Origins, current status, and future challenges. Trends in Analytical Chemistry. 2019;**118**:248-263

[35] Silva Júnior JOC, Ribeiro Costa RM, Teixeira FM, Barbosa WLR. Processing and quality control of herbal drugs and their derivatives. In: Shoyama Y, editor. Quality Control of Herbal Medicines and Related Areas. Rijeka: IntechOpen; 2011. DOI: 10.5772/20346

[36] Brasil. Farmacopéia Brasileira. 6 Edição. Vol. 1; 2019

[37] Oracz J, Zyzelewicz D, Nebesny E. The content of polyphenolic compounds in cocoa beans (*Theobroma cacao* L.), depending on variety, growing region and processing operations: A review. Critical Reviews in Food Science and Nutrition. 2013. DOI: 10.1080/10408398.2012.686934

[38] Silva-Júnior JOC, Vieira JLF, Barbosa WLR, Pereira NL. Caracterização físico-química do extrato fluido e seco por nebulização de Symphytum officinale L. Brazilian Journal of Pharmacognosy. 2006;**16**:671-677

[39] Aliakbarian B, Casazza AA, Perego P. Valorization of olive oil solid waste using high pressure–high temperature reactor. Food Chemistry. 2011;**128**:704-710. DOI: 10.1016/j.foodchem.2011.03.092

[40] Jemai H, Bouaziz M, Sayadi S. Phenolic composition, sugar contents and antioxidant activity of Tunisian sweet olive cultivar with regard to fruit ripening. Journal of Agriculture and Food Chemistry. 2009;**57**:2961-2968

[41] Batista NN, de Andrade DP, Ramos CL, Dias DR, Schwan RF. Antioxidant capacity of cocoa beans and chocolate assessed by FTIR. Food Research International. 2016;**90**:313-319

[42] Silverstein RM, Webster FX. Identificação espectrométrica de compostos orgânicos. 7th ed. Rio de Janeiro, RJ: LTC; 2007

[43] Zhang LM, Zhao X, Ji JJ, Dai YJ. TG-DTG as an effective method for the

characterization of rutin extracted from the buds of Sophora japonica L. Journal of Thermal Analysis and Calorimetry. 2008:1-6

[44] Costa RS, Negrão CAB, Camelo SRP, Ribeiro-Costa RM, Barbosa WLR, Costa CEF, et al. Investigation of thermal behavior of *Heliotropium indicum* L. lyophilized extract by TG and DSC. Journal of Thermal Analysis and Calorimetry. 2013;**111**(3):1959-1964

[45] Natsume M, Osakabe N, Yamagishi M, Takizawa T, Nakamura T, Miyatake H, et al. Analyses of polyphenols in cacao liquor, cocoa, and chocolate by normal-phase and reversed-phase HPLC. Bioscience, Biotechnology, and Biochemistry. 2000;**64**(12):2581-2587. DOI: 10.1271/bbb.64.2581

[46] Kris-Etherton PM, Kari RD, Hecker D, Bonanome A, Coval MS, Binkoski AS, et al. Bioactive compounds in foods: Their role in the prevention of cardiovascular disease and cancer. American Journal of Medicine. 2002;**113**:71-88

[47] Prince PSM. A biochemical, electrocardiographic, electrophoretic, histopathological and in vitro study on the protective effects of (−)epicatechin in isoproterenol-induced myocardial infarcted rats. European Journal of Pharmacology. 2011;**671**:95-101

[48] Ferguso LR, Harris PJ. Protection against cancer by wheat bran. European Journal of Cancer Prevention. 2006;**8**(1):17-26

[49] Angelo PM, Jorge N. Phenolic compounds in foods – A brief review. Revista do Instituto Adolfo Lutz. 2007;**66**(1):1-92

[50] Esparza-Martínez FJ, Miranda-López R, Guzman-Maldonado SH. Extractable and non-extractable phenolics and antioxidant capacity of mandarin waste dried at different temperatures. Plant Foods for Human Nutrition. 2016;**71**:294-300

[51] Agourram A, Ghirardello D, Rantsiou K, Zeppa G, Belviso S, Romane A, et al. Phenolic content, antioxidant potential, and antimicrobial activities of fruit and vegetable by-product extracts. International Journal of Food Properties. 2013;**16**:1092-1104

[52] Martínez R, Torres P, Meneses MA, Figueroa JG, Perés-Alvarez JA, Viuda-Matos M. Chemical, technological and in vitro antioxidant properties of mango, guava, pineapple and passion fruit dietary fiber concentrate. Food Chemistry. 2012;**135**:1520-1526

[53] Leong LP, Shui G. An investigation of antioxidant capacity of fruits in Singapore markets. Food Chemistry. 2002;**76**:69-75

[54] Zhang Y, Sun Y, Xi W, Shen Y, Qiao L, Zhong L, et al. Phenolic compositions and antioxidant capacities of Chinese wild mandarin (*Citrus reticulata* Blanco) fruits. Food Chemistry. 2014;**145**:674-680

[55] Contreras-Calderón JA, Calderón-Jaimes LB, Guerra-Corrêa JC, Benites VM, Rebellatto A. O uso dos resíduos animais como fertilizantes. II Simpósio Internacional sobre Gerenciamento de Resíduos Agropecuários e Agroindustriais – II SIGERA 15 a 17 de março de; 2011

[56] Saénz C, Tapia S, Chávez J, Robert P. Microencapsulation by spray drying of bioactive compounds from cactus pear (Opuntia ficus-indica). Food Chemistry. 2009;**114**:616-622

[57] Lemes AC, Álvares GT, Egea MB, Brandelli A, Kalil SJ. Simultaneous production of proteases and antioxidant

compounds from agro-industrial by-products. Bioresource Technology. 2016;**222**:210-216. DOI: 10.1016/j.biortech.2016.10.001

[58] Lousada Junior JE, Da JMC C, JNM N, Rodriguez NM. Physicochemical characterization of tropical fruit byproducts for use in animal feed. Revista Ciência Agronômica. 2006;**37**:70

[59] Aklakur M, Rather MA, Kumar N. Nano delivery: An emerging avenue for nutraceuticals and drug delivery. Critical Reviews in Food Science and Nutrition. 2015;**8398**. DOI: 10.1080/10408398.2013.839543

[60] Chaves BN, Stefanello FS, Burin AP, Ritt LA, Nornberg JL. Utilização de resíduos industriais na dieta de bovinos leiteiros. REGET. 2014;**18**:150-156

[61] Pereira-Junior GP, Pereira EMDO, Pereira-filho M, Barbosa PDS, Eduardo S, Brandão LV. Desempenho produtivo de juvenis de tambaqui (*Colossoma macropomum* CUVIER, 1818) alimentados com rações contendo farinha de crueira de mandioca (*Manihot esculenta*, CRANTZ) em substituição ao milho (Zea mays). Acta Amaz. 2013;**43**:217-226. DOI: 10.1590/S0044-59672013000200013

[62] Jahurul MHA, Zaidul ISM, Ghafoor K, Al-Juhaimi FY, Nyam KL, Norulaini NAN, et al. Mango (*Mangifera indica* L.) by-products and their valuable components: A review. Food Chemistry. 2015;**183**:173-180. DOI: 10.1016/j.foodchem.2015.03.046

[63] Garcıa AJ, Esteban MB, Márquez MC, Ramos P. Biodegradable municipal solid waste: Characterization and potential use as animal feedstuffs. Waste Management. 2005;**25**:780-787. DOI: 10.1016/j.wasman.2005.01.006

[64] Simões DLV. Composição nutricional e elaboração do biscoito e da barra de cereal do fruto de Tucumã (*Astrocaryum vulgare* Mart.) [Dissertação]. Lisboa: Universidade Nova de Lisboa; 2010

[65] Guerra I. Importância da alimentação do atleta visando a melhora da performance. Nutrição em Pauta. 2002;**4**(55):63-66

Chapter 15

Thermodynamics and Kinetics of *Camellia sinensis* Extracts and Constituents: An Untamed Antioxidant Potential

Douglas Vieira Thomaz

Abstract

Given the relevance of impairing reactive oxygen species (ROS) buildup in tissues, the use of exogenous antioxidants is highly regarded as a valid prophylaxis against oxidative stress and its deleterious effects on organisms. In this regard, *Camelia sinensis* uses as a remarkable antioxidant source have been reported in various folk and standard medicine systems around the world. In this chapter, the thermodynamics and kinetics of *Camelia sinensis* constituents is concisely discussed focusing on the implications of its redox profiling toward antioxidant capacity. Notwithstanding, the biological repercussion of ROS reduction as well as its therapeutic potential is also addressed to provide readers a basic background on the relevance of investigating the physicochemical features of medicinal plants.

Keywords: redox, black tea, electrochemistry, free radicals, reactive oxygen species

1. Introduction

1.1 The principles of antioxidant thermodynamics in biological systems

Antioxidant capacity has been described by many authors as an important property of phytomedicines due to the extent of biological effects linked to it [1–3]. This key feature in plant-based drugs and nutraceuticals is widely exploited by industry to increase product appeal as well as to improve the health benefits allegedly promised by their consumption [3–5]. The main premise behind the antioxidant lore is that it counteracts the oxidative stress generated by biological systems [6, 7]. In this sense, the regular intake of free radical scavengers would safeguard the proper attunement of homeostatic balance by diminishing oxidative stress [8, 9].

The natural workings of biological systems, although quite complex at first glance, are based on two simple reactions which quite often occur in close connection to each other [8]. These reactions are reduction and oxidation, commonly shortened to redox reactions. The oxidation is based on electrons (e) leaving the chemical species (A), as represented in Eq. (1), while reduction is based on a chemical species (A) accepting electrons (e), as represented in Eq. (2). These reactions can occur simultaneously, leading to a redox couple, as showcased in Eq. (3).

$$A^{0} \rightarrow A^{+1} + 1e^{-1} \tag{1}$$

IntechOpen

$$A^{+1} + 1e^{-1} \rightarrow A^{0} \tag{2}$$

$$A^{0} \leftrightarrow A^{+1} + 1e^{-1} \tag{3}$$

Regarding all processes which are related to the occurrence of life, redox reactions are deeply important since biological systems feed on the chemical energy produced by oxidation [8]. To a certain extent, cells behave quite like thermal power plants, thereby "burning" certain nutrients to obtain energy for self-sustainability. This process is however imperfect, as oxidation byproducts may be highly reactive, therefore degrading functional biomolecules which come in contact with them [5, 9, 10]. These reactive byproducts are the often-demonized reactive oxygen species (ROS), which include oxygen-bearing free radicals and other reactive compounds [11, 12]. An overview of ROS and other reactive chemicals is showcased in **Table 1**.

Albeit ROS are widely known for their negative effects on organisms, these compounds do have remarkable importance to sustain life. Although such statement might be seemingly paradoxical, oxidative stress and ROS buildup are essential to proper inflammatory response as well as immunologic defense [8, 11, 12]. Moreover, the relevance of ROS in the natural workings of cell physiology is still being discovered, since these compounds have been linked to many biochemical pathways and cell signaling processes. For instance, superoxide anion showcases affinity to sulfur residues bound to iron coordination complexes as the heme in cytochromes, while some non-radical ROS may showcase affinity to exposed sulfur-bearing amino acids such as cysteine, thence selectively targeting these moieties [11–13]. Therefore, the true concept behind antioxidants is not the full disruption of oxidative stress, but the balance of this natural phenomenon toward proper homeostasis.

Non-radical ROS	Radical ROS
O_2 (oxygen)	$-{}^{\cdot}O_2$ (superoxide anion)
H_2O_2 (hydrogen peroxide)	${}^{\cdot}OH$ (hydroxyl)
O_3 (ozone)	${}^{\cdot}HO_2$ (perhydroxyl)
Obtainable through excitation	
${}^{1}O_2$ (singlet oxygen)	

Obs.: Radical ROS are obtainable by sequential single-electron oxidation of water and its oxidation products, namely, hydroxyl radical, hydrogen peroxide, superoxide anion, and molecular oxygen. Singlet oxygen is obtainable through excitation of molecular oxygen.

Table 1.
Overview of ROS and other reactive chemicals.

Enzymes	Nonenzymatic	Proteins
Catalase	Bilirubin	Lactoferrin
Superoxide dismutase	Uric acid	Metallothionein
Glutathione peroxidase	Glutathione	Transferrin
Glutathione reductase	Lipoic acid	Ceruloplasmin
Thioredoxin reductase	Melatonin	Ferritin

Table 2.
Endogenous antioxidants reported in literature.

DOI: http://dx.doi.org/10.5772/intechopen.92813

Considering that all chemicals showcase intrinsic physicochemical features regarding their proneness to undergo redox reactions, the feasibility of antioxidant capacity can be linked to the thermodynamics of selected compounds in comparison to endogenous antioxidants of both chemical and enzymatic nature [8, 11, 14–16]. Thus, the Gibbs free energy, enthalpy and entropy of alleged antioxidants can be compared to those of endogenous free radical scavengers and oxireductive enzymes in order to assess the thermodynamic feasibility of antioxidant capacity [9, 17, 18]. **Table 2** showcases the main endogenous antioxidants reported in literature.

The activity of endogenous antioxidants is remarkably high in order to counteract the effect of the oxidative stress promoted by cell physiology; however, the presence of compounds bearing higher thermodynamic proneness to reduce ROS could safeguard the whole biological material in site. This antioxidant capacity can be assessed by many methods, such as thermodynamic evaluation through spectrophotometry, electrochemistry, and other approaches. However, redox reactions may be dependent on mass transfer through solution or other processes which reduce reaction speed, which thence raises the importance of also evaluating the kinetics of antioxidant capacity.

2. Basic physicochemical features of *Camellia sinensis* antioxidants

Camellia sinensis, also known as green or black tea according to its production method, is a widely commercialized herb whose therapeutic applications are highly

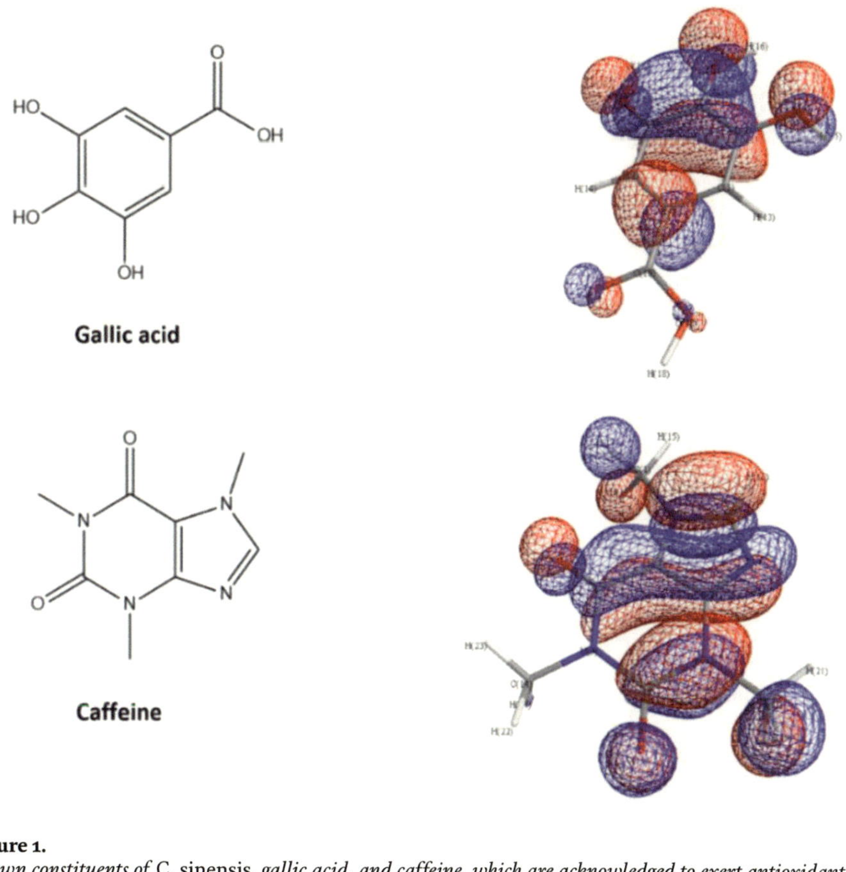

Figure 1.
Known constituents of C. sinensis, *gallic acid, and caffeine, which are acknowledged to exert antioxidant power, and graphical rendering of HOMO-0 for each constituent. Negative charges are rendered in blue, and positive charges are rendered in red.*

regarded in the folk medicine of many regions around the globe [19–23]. This plant is acknowledged to harbor a chemically diverse metabolism, and its constituents are known to promote strong antioxidant action in biological systems [18, 24]. **Figures 1** and **2** showcase some major constituents of *C. sinensis* which are known to exert antioxidant action as well as the surface rendering of the first state of their highest occupied molecular orbital (HOMO-0) using standard Hückel molecular orbital theory [25, 26].

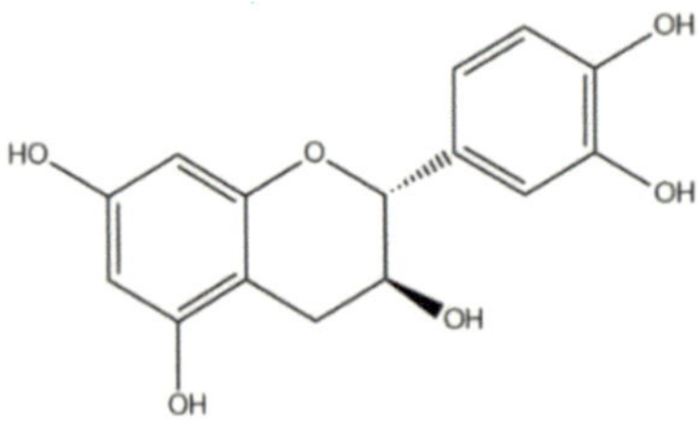

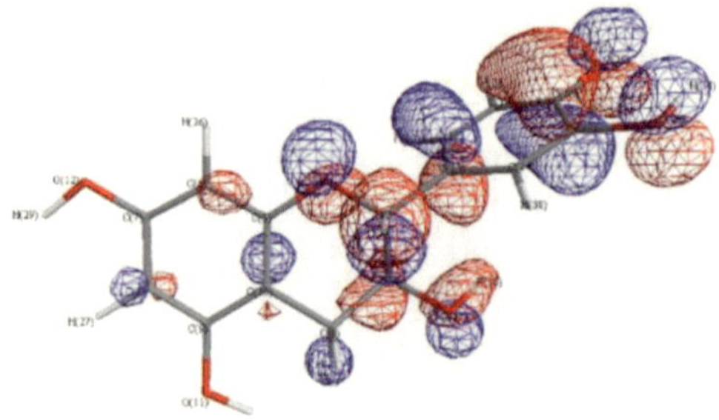

Catechin

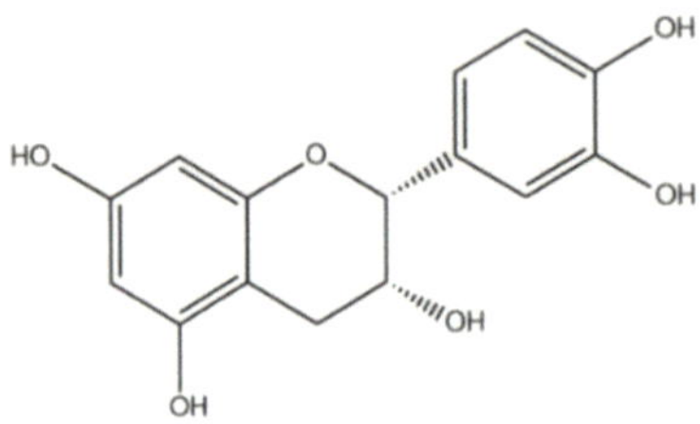

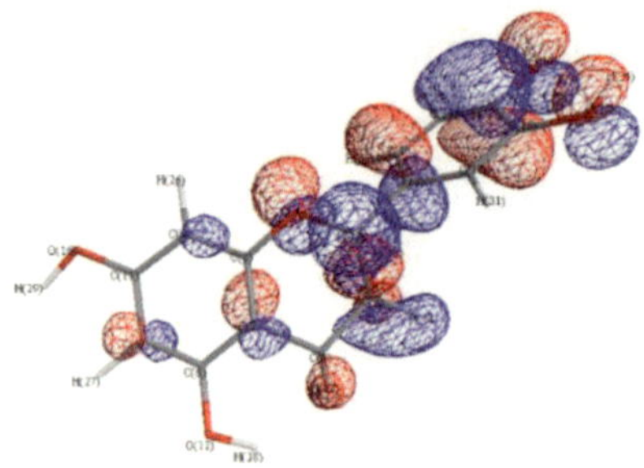

Epicatechin

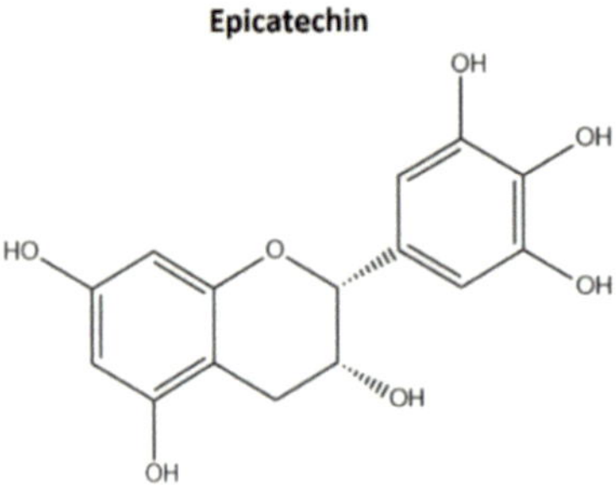

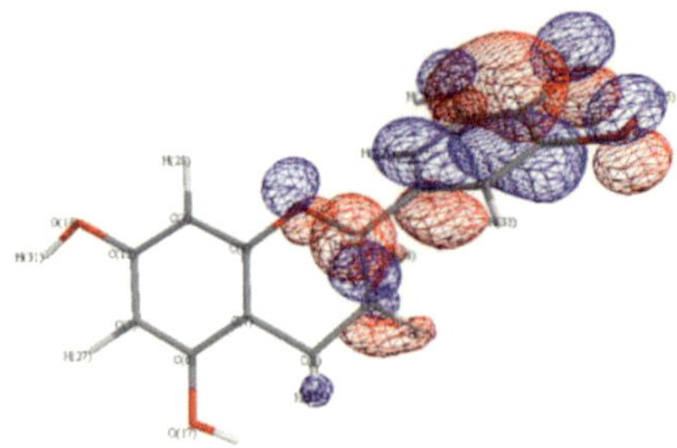

Epigallocatechin

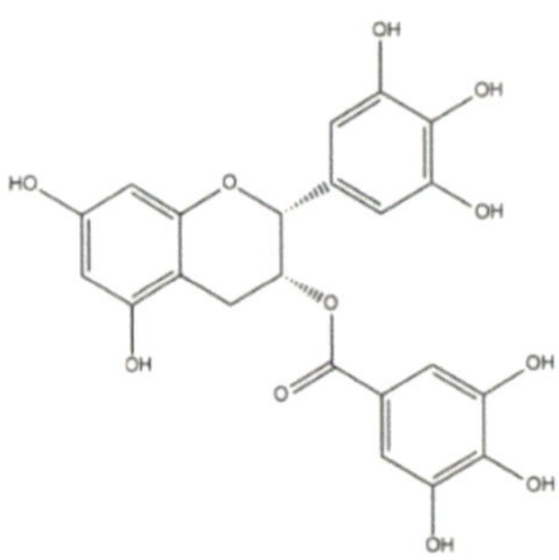

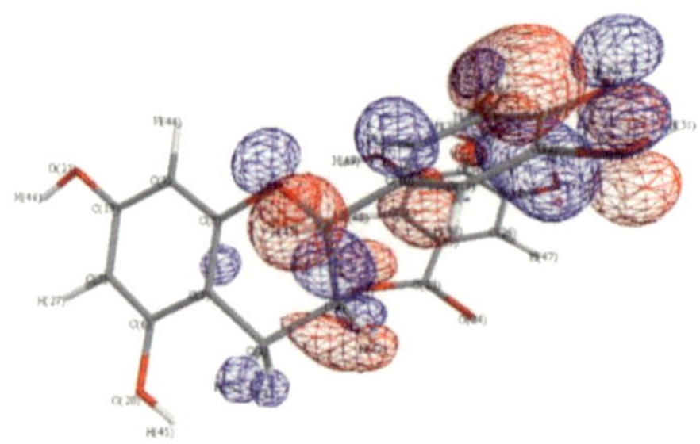

Epigallocatechin-gallate

Figure 2.
Known flavonoid constituents of C. sinensis, *which are acknowledged to exert antioxidant power, and graphical rendering of HOMO-0 for each constituent. Negative charges are rendered in blue and positive charges are rendered in red.*

Constituent	HOMO-0 (eV)	LUMO-0 (eV)	ΔE (|LUMO-HOMO|)
Catechin	−10.885	1.293	12.178
Epicatechin	−10.938	1.254	12.192
Epigallocatechin	−10.528	2.382	12.91
Epigallocatechin gallate	−10.549	−3.117	7.432
Gallic acid	−11.026	−3.01	8.016
Caffeine	−9.26	2.047	11.307

Obs.: ΔE also displayed in eV.

Table 3.
HOMO-0, LUMO-0, and energy gaps (ΔE) for each major C. sinensis *constituent.*

As showcased in **Figures 1** and **2**, the resonance systems promoted by aromatic rings in both the flavonoids, as well as gallic acid and caffeine, allow the inference that the inductive effect of functional groups may be transmitted along the molecule [25, 26]. This assumption is supported by both Hückel molecular orbital theory and its implications in medicinal chemistry through the vinylogy principle [27]. In this sense, all these molecules could undergo redox reactions due to their easily excitable electrons in the conjugated aromatic systems (flavonoids) and single aromatic ring (gallic acid and caffeine) [25]. Notwithstanding, this interpretation is further corroborated by experimental data as well as by the evaluation of their energy gap (ΔE) in comparison to that of other known antioxidants such as alpha-tocopherol, which showcase ΔE of 10.2011 eV when analyzed under the same method [25, 26, 28, 29]. **Table 3** showcases the ΔE of major *C. sinensis* constituents.

The ΔE of chemical compounds is an important parameter to evaluate the thermodynamic feasibility of redox processes; hence lower values suggest easier electron transfer and therefore higher possibility of occurrence [25, 26]. Nonetheless, complex approaches such as density functional theory (DFT) are often employed to gather thermodynamic data from ab initio computational models in order to investigate the proneness of chemicals to undergo oxidation [28].

3. Thermodynamics of redox systems and ways to explore it

Thermodynamic data obtained through ab initio or empiric/semi-empiric approaches such as DFT and Hückel molecular orbital theory can also be combined to other physicochemical investigations in order to render more reproducible models regarding energy shifts in redox reactions. Among the most associated techniques are quantum chemistry and electrochemistry [28], which provide relevant information concerning energy parameters and the kinetics of chemical processes.

When energy levels are concerned, the overall behavior of a system can be investigated by the fundamental laws of thermodynamics, as summarized in Eq. (4).

$$\Delta G = \Delta H - T\Delta S \tag{4}$$

wherein ΔG stands for Gibbs free energy, ΔH stands for enthalpy, T stands for temperature, and ΔS stands for entropy. Note that all energy parameters are approached as variations between different states of the system.

Considering that Gibbs free energy can also be expressed according to Eq. (5), we can thence declare the dependence of energy to the equilibrium constant of reactions, as showcased in Eqs. (6) and (7).

$$\Delta G = -RTLn\,K_{eq} \tag{5}$$

$$A_{red} \leftrightarrow A_{ox} + 1e^{-1} \tag{6}$$

$$K_{eq} = \frac{A_{ox}}{A_{red}} \tag{7}$$

wherein R stands for the universal gas constant.

From these relations, both Van't Hoff and Nernst equation are achievable. Taking Nernst equation for instance, one can clearly see how electric potential is a thermodynamic parameter, as showcased in Eqs. (8) and (9).

$$\Delta G = -nF\,\Delta E \tag{8}$$

$$\Delta E = -\frac{RT}{nF}Ln\frac{A_{ox}}{A_{red}} \tag{9}$$

wherein n stands for the number transferred electrons in the system and F stands for the Faraday constant.

Considering that electric potential is a thermodynamic parameter, the antioxidant capacity is therefore conditioned by both energy levels of oxidized and reduced forms and the reaction rate thereby associated. In this context, electrochemical tests such as chronoamperometry and voltammetry can be used to investigate the electric potentials associated to redox reactions, as well as the kinetic profile they follow [8, 16].

4. Investigating the basics of *Camellia sinensis* antioxidant thermodynamics and kinetics

Many authors reported that *C. sinensis* antioxidant power is remarkable due to the capacity of its constituents to promptly reduce standard free radicals such as 2,2-diphenyl-1-picrylhydrazyl (DPPH) and 2,2′-azino-bis(3-ethylbenzothiazoline-6-sulfonic acid) (ABTS) [18, 24, 30, 31]. Although results are often noteworthy, colorimetric tests tend to be biased due to the strong color of *C. sinensis* extracts, which may lead to imprecise results. In this regard, many authors pursued the exploration of the redox behavior of this plant using electrochemical methods, which may provide less color-biased results due to their unique dependence on electron transfer [17, 32, 33].

The electroanalytical investigation of plantstuff is a growing field on science due to its promising perspectives regarding the quality control, authenticity, and physicochemical characterization of plant secondary metabolites [32–34]. Nonetheless, most voltammetric assays can be applied to plants without strenuous pretreatment of the vegetal sample [32]. However, given that plant secondary metabolites of phenolic origin such as those of *C. sinensis* showcase electrochemical processes which are mainly controlled by mass transfer in the bulk solution [17], as well as proneness of oxidation products to undergo adsorption on electrode surface, a careful electrode surface renewal protocol needs to be adopted.

Regarding the basic background on voltammetric studies, these tests involve the interpolation of two functions, namely, electric potential *versus* time ($E \times t$) and electric current *versus* time ($I \times t$). When graphically displayed, a voltammogram is the plot of electric current *versus* the applied electric potential following a specific signal pattern during a defined time interval ($I \times E$). Therefore, any change in electric current which is non-capacitive by nature can be attributed to redox processes taking place in the electrochemical cell [35–37].

During voltammetric investigation, the scanning of electric potential toward positive values, aka anodic scan, leads to the visualization of oxidative processes, while the reverse scan, aka cathodic scan, leads to the visualization of reduction processes [35, 37]. Taking these concepts into account, the redox profiling and electrochemical characterization of both isolated plant constituents and vegetal extracts can be elucidated by varying the kind of scan which is being performed [10, 17, 33, 34]. **Figure 3** showcases an example of a cyclic voltammogram presenting a response which could be attributed to a reversible redox reaction, while **Figure 4** depicts an overview of the main mechanisms which are involved in the electrooxidation of *C. sinensis* constituents [7, 27, 38–40].

Literature reports that *C. sinensis* extracts showcase anodic peaks at electric potentials bellow 0.5 V when analyzed under voltammetry [17, 18, 30, 31, 41], which is nonetheless a remarkable feature. Considering that most of the endogenous antioxidant arsenal operates at electric potentials close to this value, the reductive power of *C. sinensis* constituents is noteworthy, since they could undergo oxidation thereby stabilizing ROS or restituting endogenous antioxidants [8, 10, 27, 42].

Notwithstanding, many authors showcased evidence of redox reversibility in the processes which take place at 0.5 V in *C. sinensis* extracts, which suggests that the antioxidant compounds could undergo followed redox reactions to promote the reduction of ROS in biological systems [32, 33]. When compared to voltammetric profiles of isolated compounds, *C. sinensis* voltammograms evidence the richness of electroactive compounds which are present in this plant, which further corroborates to the appeal of this plant in the development of therapeutic and nutraceutical products to balance oxidative stress in biological systems.

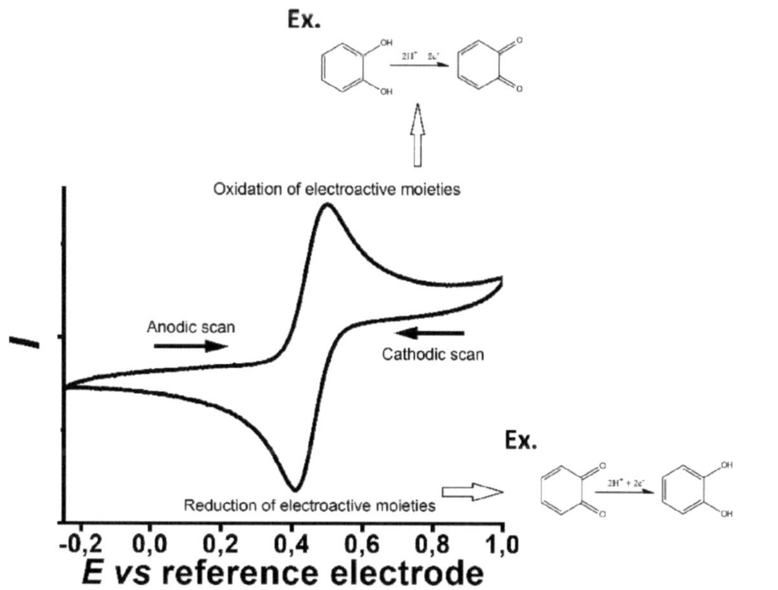

Figure 3.
Example of a cyclic voltammogram presenting a response which could be attributed to a reversible reaction.

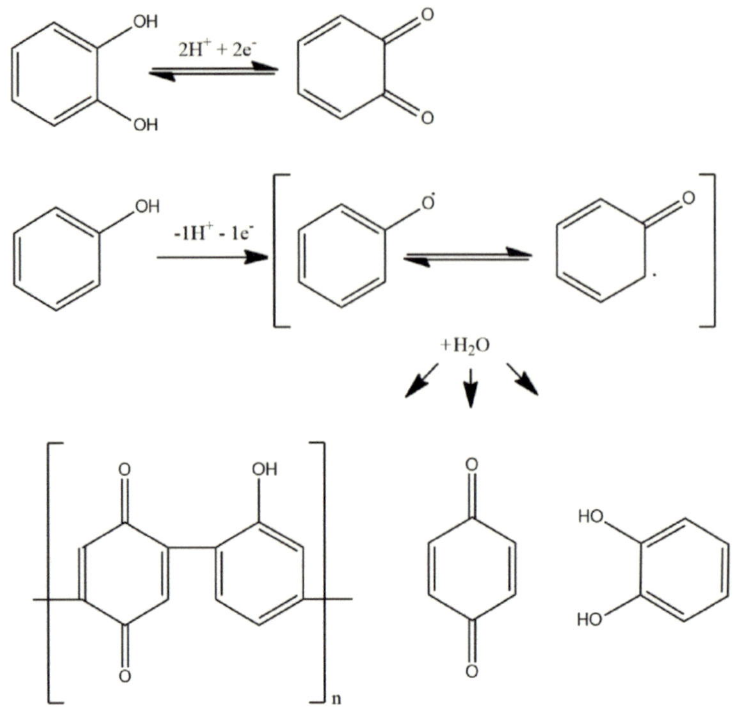

Figure 4.
Main mechanisms involved in the electrooxidation of phenolic compounds which are occurrent in C. sinensis. *Note that catechol oxidation is reversible, while phenol undergoes irreversible oxidation, leading to both 1,2 and 1,4 catechol and an electro-polymerized product.*

5. Conclusions

This chapter aimed to provide readers with a basic background on the relevance of investigating the physicochemical features of *Camelia sinensis*, as well as concisely discuss the implications of the redox profiling in the understanding of antioxidant capacity. It was observed that several methods can be used to investigate the underlying thermodynamic and kinetic features which are intrinsically linked to the antioxidant power of phytomedicines. Moreover, literature extensively reports the remarkable antioxidant power of *C. sinensis* extracts and constituents, therefore highlighting the relevance of this plant as an important asset for the development of therapeutic and nutraceutical formulations.

Author details

Douglas Vieira Thomaz
Faculty of Pharmacy, Federal University of Goias, Goiânia, Goias, Brazil

*Address all correspondence to: douglasvthomaz@gmail.com

References

[1] Boots AW, Haenen GRMM, Bast A. Health effects of quercetin: From antioxidant to nutraceutical. European Journal of Pharmacology. 2008;**585**(2-3):325-337. DOI: 10.1016/j.ejphar.2008.03.008

[2] Krishnamurthy P, Wadhwani A. Antioxidant enzymes and human health. Antioxidant Enzyme. 2012; **32**(8):595-603. DOI: 10.5772/48109

[3] Devasagayam TPA, Tilak JC, Boloor KK, Sane KS, Ghaskadbi SS, Lele RD. Free radicals and antioxidants in human health: Current status and future prospects. The Journal of the Association of Physicians of India. 2004;**52**:794-804

[4] Pham-Huy LA, He H, Pham-Huy C. Free radicals, antioxidants in disease and health. International Journal of Biomedical Sciences. 2008;**4**(2):89-96

[5] Pandey KB, Rizvi SI. Plant polyphenols as dietary antioxidants in human health and disease. Oxidative Medicine and Cellular Longevity. 2009;**2**(5):270-278. DOI: 10.4161/oxim.2.5.9498

[6] da Cunha CEP, Rodrigues ESB, Fernandes Alecrim M, Thomaz DV, Macêdo IYL, Garcia LF, et al. Voltammetric evaluation of diclofenac tablets samples through carbon black-based electrodes. Pharmaceuticals. 2019;**12**(2):83-94. DOI: 10.3390/ph12020083

[7] Procházková D, Boušová I, Wilhelmová N. Antioxidant and prooxidant properties of flavonoids. Fitoterapia. 2011;**82**(4):513-523. DOI: 10.1016/j.fitote.2011.01.018

[8] Jomova K, Valko M. Thermodynamics of free radical reactions and the redox environment of a cell. ACS Symposium Series. 2011;**3**:71-82. DOI: 10.1021/bk-2011-1083.ch003

[9] Thomaz DV. How phytocomponents may be valuable against oxidative stress in brain tissue? Global Drugs and Therapeutics. 2020;**1**:1-3. DOI: 10.31487/j.gdt.2020.01.04

[10] Thomaz DV, Peixoto LF, de Oliveira TS, Fajemiroye JO, da Silva Neri HF, Xavier CH, et al. Antioxidant and neuroprotective properties of *Eugenia dysenterica* leaves. Oxidative Medicine and Cellular Longevity. 2018;**1**:1-11. DOI: 10.1155/2018/3250908

[11] Kapoor D, Singh S, Kumar V, Romero R, Prasad R, Singh J. Antioxidant enzymes regulation in plants in reference to reactive oxygen species (ROS) and reactive nitrogen species (RNS). Plant Gene. 2019;**19**:100182. DOI: 10.1016/J.PLGENE.2019.100182

[12] Lu H, Hu H, Yang Y, Li S. The inhibition of reactive oxygen species (ROS) by antioxidants inhibits the release of an autophagy marker in ectopic endometrial cells. Taiwanese Journal of Obstetrics & Gynecology. 2020;**59**:256-261. DOI: 10.1016/J.TJOG.2020.01.014

[13] Huihui Z, Xin L, Yupeng G, Mabo L, Yue W, Meijun A, et al. Physiological and proteomic responses of reactive oxygen species metabolism and antioxidant machinery in mulberry (*Morus alba* L.) seedling leaves to NaCl and $NaHCO_3$ stress. Ecotoxicology and Environmental Safety. 2020;**193**:110259. DOI: 10.1016/J.ECOENV.2020.110259

[14] Jabeen H, Saleemi S, Razzaq H, Yaqub A, Shakoor S, Qureshi R. Investigating the scavenging of reactive oxygen species by antioxidants via theoretical and experimental methods. Journal of Photochemistry and Photobiology B: Biology. 2018;**180**:268-275. DOI: 10.1016/J.JPHOTOBIOL.2018.02.006

[15] Aqeel S, Naheda A, Raza A, Khan K, Khan W. Differential status and significance of non-enzymatic antioxidants (reactive oxygen species scavengers) in malaria and dengue patients. Acta Tropica. 2019;**195**:127-134. DOI: 10.1016/J.ACTATROPICA.2019.04.033

[16] Chen Y, Xiao H, Zheng J, Liang G. Structure-thermodynamics-antioxidant activity relationships of selected natural phenolic acids and derivatives: An experimental and theoretical evaluation. PLOS One. 2015;**10**(3):e0121276. DOI: 10.1371/journal.pone.0121276

[17] Thomaz DV, Leite KC de S, Moreno EKG, Garcia LF, Alecrim MF, Macêdo IYL, et al. Electrochemical study of commercial black tea samples. International Journal of Electrochemical Science. 2018;**13**(2018):5433-5439. DOI: 10.20964/2018.06.55

[18] Novak I, Šeruga M, Komorsky-Lovrić Š. Characterisation of catechins in green and black teas using square-wave voltammetry and RP-HPLC-ECD. Food Chemistry. 2010;**122**(4):1283-1289. DOI: 10.1016/j.foodchem.2010.03.084

[19] Gramza A, Korczak J. Tea constituents (*Camellia sinensis* L.) as antioxidants in lipid systems. Trends in Food Science and Technology. 2005;**16**(8)351-358. DOI: 10.1016/j.tifs.2005.02.004

[20] Chan EWC, Lim YY, Chew YL. Antioxidant activity of *Camellia sinensis* leaves and tea from a lowland plantation in Malaysia. Food Chemistry. 2007;**102**(4):1214-1222. DOI: 10.1016/j.foodchem.2006.07.009

[21] Boehm K, Borrelli F, Ernst E, Habacher G, Hung SK, Milazzo S, et al. Green tea (*Camellia sinensis*) for the prevention of cancer. Cochrane Database of Systematic Reviews. 2009;**1**:14651858. DOI: 10.1002/14651858.CD005004.pub2

[22] Namita P, Mukesh R, Vijay KJ. *Camellia sinensis* (green tea): A review. Global Journal of Pharmacology. 2012;**6**(2):52-59

[23] Patel SH. *Camellia sinensis*. Journal of Agromedicine. 2005;**10**(2):57-64. DOI: 10.1300/j096v10n02_08

[24] Veljković JN, Pavlović AN, Mitić SS, Tošić SB, Stojanović GS, Kaličanin BM, et al. Evaluation of individual phenolic compounds and antioxidant properties of black, green, herbal and fruit tea infusions consumed in Serbia: Spectrophotometrical and electrochemical approaches. Journal of Food and Nutrition Research. 2013;**52**(1):12-24

[25] Matito E, Feixas F, Solà M. Electron delocalization and aromaticity measures within the Hückel molecular orbital method. Journal of Molecular Structure: THEOCHEM. 2007;**811**(1-3):3-11. DOI: 10.1016/j.theochem.2007.01.015

[26] Yates K. Hückel molecular orbital theory. In: Hückel Molecular Orbital Theory. Vol. 1. Academic Press; 1978. pp. 1-384. DOI: 10.1016/b978-0-12-768850-3.50005-5

[27] Thomaz DV. Flavonoid chemistry and neuroprotection. Frontiers in Drug, Chemistry and Clinical Research. 2020;**3**:1-3. DOI: 10.15761/FDCCR.1000140

[28] Rodrigues ESB, de Macêdo IYL, da Silva Lima LL, Thomaz DV, da Cunha CEP, de Oliveira MT, et al. Electrochemical characterization of central action tricyclic drugs by voltammetric techniques and density functional theory calculations. Pharmaceuticals. 2019;**2**(3):116. DOI: 10.3390/ph12030116

[29] Zinola CF. Density functional theory. In: Electrocatalysis: Computational, Experimental, and Industrial Aspects. Vol. 1. CRC Press; 2010. p. 664. DOI: 10.1201/9781420045451

[30] Bhattacharyya R, Tudu B, Das SC, Bhattacharyya N, Bandyopadhyay R, Pramanik P. Classification of black tea liquor using cyclic voltammetry. Journal of Food Engineering. 2012;**109**(1):120-126. DOI: 10.1016/j.jfoodeng.2011.09.026

[31] Kilmartin PA, Hsu CF. Characterisation of polyphenols in green, oolong, and black teas, and in coffee, using cyclic voltammetry. Food Chemistry. 2003;**82**(4):501-512. DOI: 10.1016/S0308-8146(03)00066-9

[32] Thomaz DV, Couto RO, Roberth A de O, Oliveira LAR, Leite KC de S, Bara MT de F, et al. Assessment of noni (*Morinda citrifolia* L.) product authenticity by solid state voltammetry. International Journal of Electrochemical Science. 2018;**13**(2018):8983-8994. DOI: 10.20964/2018.09.390

[33] Moreno EKG, Thomaz DV, Machado FB, Leite KCS, Rodrigues ESB, Fernandes MA, et al. Antioxidant study and electroanalytical investigation of selected herbal samples used in folk medicine. International Journal of Electrochemical Science. 2019;**14**(2019):838-847. DOI: 10.20964/2019.01.82

[34] Leite KC de S, Garcia LF, Lobón GS, Thomaz DV, Moreno EKG, de CMF, et al. Antioxidant activity evaluation of dried herbal extracts: An electroanalytical approach. Brazilian Journal of Pharmacognosy. 2018;**28**(3):325-332. DOI: 10.1016/j.bjp.2018.04.004

[35] Tien HT. 715—Cyclic voltammetry of electron-conducting bilayer lipid membranes. Bioelectrochemistry and Bioenergetics. 1984;**13**:299-316. DOI: 10.1016/0302-4598(84)87033-6

[36] Popov BN, Popov BN. Basics of corrosion measurements. Corrosion Engineering. 2015:181-237. DOI: 10.1016/B978-0-444-62722-3.00005-7

[37] Brainina KH, Bond AM. In: Smyth MR, Vos JG, editors. Analytical Voltammetry. Elsevier, Amsterdam; 1992. (xxvi + 578 pages), Dfl. 495.00. ISBN: 0-444-88938. TrAC Trends in Analytical Chemistry 1993;**12**(6):13-14. DOI: 10.1016/0165-9936(93)87068-9

[38] Hodnick WF, MllosavljeviĆ EB, Nelson JH, Pardini RS. Electrochemistry of flavonoids: Relationships between redox potentials, inhibition of mitochondrial respiration, and production of oxygen radicals by flavonoids. Biochemical Pharmacology. 1988;**37**:2607-2611. DOI: 10.1016/0006-2952(88)90253-5

[39] Hendrickson HP, Kaufman AD, Lunte CE. Electrochemistry of catechol-containing flavonoids. Journal of Pharmaceutical and Biomedical Analysis. 1994;**12**:325-334. DOI: 10.1016/0731-7085(94)90007-8

[40] Rice-Evans CA, Miller NJ, Paganga G. Structure-antioxidant activity relationships of flavonoids and phenolic acids. Free Radical Biology & Medicine. 1996;**20**(7):933-956. DOI: 10.1016/0891-5849(95)02227-9

[41] Nagles E, García-Beltrán O. Determination of rutin in black tea by adsorption voltammetry (AdV) in the presence of Morin and quercetin. Food Analytical Methods. 2016. DOI: 10.1007/s12161-016-0538-y

[42] de Oliveira TS, Thomaz DV, da Silva Neri HF, Cerqueira LB, Garcia LF, Gil HPV, et al. Neuroprotective effect of *Caryocar brasiliense* Camb. Leaves is associated with anticholinesterase and antioxidant properties. Oxidative Medicine and Cellular Longevity. 2018;**2018**:1-12. DOI: 10.1155/2018/9842908

Chapter 16

Microorganisms as Alternative Sources of New Natural Products

Lucía Ortega Cabello

Abstract

Microbial natural products have become important over the last decades due to the ability of bacteria and fungi to subsist in different habitats such as marine and extreme environments. Microorganisms are able to synthesize new compounds with diverse therapeutic activity equal to or better than the activity of compounds already known, thus being promising for the treatment of different diseases such as cancer or the solution to health problems such as antibiotic resistance. The production of microbial natural compounds can be improved by modifying culture media, growing conditions, amplifying gene expression or by co-cultivation techniques, which are the major challenges in the industrial production of such compounds.

Keywords: microorganisms, antioxidants, antibiotics, antitumor, polymers

1. Introduction

The lack of effectiveness in current therapeutics using already known compounds has made necessary the rediscovery of natural products, either for obtaining new compounds or modifying their structure to improve their activity, where plants are the most popular sources.

However, due to seasonal and environmental conditions that influence their production, alternative sources have been searched for. Microorganisms have been considered as good alternative sources due to the self-sustainability and controllable growth conditions such as carbon source, nitrogen source, pH and temperature [1, 2], thus leading to the possibility of discovering new compounds.

In this chapter, we will focus on the uses of microbial secondary metabolites as antioxidants, antibiotics, antitumor and polymers from mainly *Streptomyces* genus, which have been important in soil bioremediation and biocatalysis for the obtention of enantiopure compounds [3–5].

2. Microorganisms as sources of natural products

Since the discovery of penicillin and streptomycin in 1928 and 1943 respectively [6, 7], microorganisms have become fascinating alternative sources because of the diversity of natural products with new structures to be elucidated and studied for biological activity.

Microorganisms can be found in very extreme environments (soil/marine, high/low temperature, acid/alkaline) [8], with the isolation of these microorganisms being a major challenge to date because there are uncultivable microbes, complicating natural product discovery. To overcome this problem, different techniques have been applied such as co-cultivation, as well as exploration of isolation techniques on natural habitats [9]. Co-cultivation has attracted attention because it can induce the biosynthesis of new compounds [10] such as libertellenone A, B, C and D from co-cultivating α-proteobacterium and *Libertella* sp. [11] and stearidonic acid from *Rhizobium* strain 10II and *Ankistrodesmus* sp. [12].

Terrestrial fungus and actinobacteria are the most important sources of antimicrobials, cytotoxic compounds and antioxidants, among others [13]. However, in the last few years, marine environment has attracted attention due to the diversity and effectivity of natural products [14], such as apratoxins from cyanobacteria from the *Lyngbya* genus used as cytotoxic agents to induce apoptosis [15], as well as salinisporamides isolated from *Salinispora tropica* with activity against human colon carcinoma [16].

3. Antibiotics

The inadequate use of current antibiotics has led to antibiotic resistance, which is a global threat because of the adaptation rate of microorganisms [17]. Natural product discovery as a potential solution to antibiotic resistance has been important if we recall the discovery of penicillin and streptomycin. Nevertheless, actinobacteria isolated from soil have already been widely exploited, limiting the search of new antibiotics [18].

Due to the latter, the need to search new microorganisms associated with higher life forms or from unknown environments such as marine and extreme ecosystems [19, 20], as well as co-cultivation techniques between antagonists strains have been useful [21, 22], as the case of the co-cultivation of a *Micromonospora* sp. with a *Rhodococcus* strain to enhance the production of keyicin [23], as well as the co-cultivation of a marine *Pestalotia* sp. with an unidentified bacteria to obtain pestalone which resulted in high activity against *Staphylococcus aureus* and *Enterococcus faecium* [24].

Among the examples of marine microbial sources is a *Streptomyces* strain isolated from a marine sediment in India that produced ala-geninthiocin along with val-geninthiocin, geninthiocin and staurosporine; all compounds were found to be effective against *Staphylococcus aureus* and *Candida albicans* [25]. Another example is tetrahydroanthra-γ-pyrone from marine *Streptomyces* sp. (isolated from Binzhou shell island), which presented activity against *Bacillus subtilis*, *Staphylococcus aureus* and *Enterococcus faecalis* with a minimum inhibitory concentration (MIC) from 3 to 46 μg/mL [26].

The presence of metals has been explored to increase the production of antibiotics, such as the presence of nickel chloride in the cultivation of *Streptomyces pratensis* (isolated from the east coast of China), which enhanced the production of angucycline-type antibiotics, with moderate antimicrobial activity against *Pseudomonas aeruginosa*, *Klebsiella pneumonia*, *Escherichia coli* and *Staphylococcus aureus* with a MIC of 16 μg/mL [27]. Other minerals that have been tested to increase antimicrobial production on other *Streptomyces* strains are magnesium, calcium, manganese, cobalt, copper and iron salts, where copper sulfate and iron chloride resulted in the best induction of antimicrobial biosynthesis [28].

The *Micromonospora* is *a* genus of actinobacteria known to produce other antibiotics such as aminorifamycins and sporalactams with a good antimicrobial activity against *Mycobacterium tuberculosis* [29], as well as phocoenamicins with a potent activity against *Staphylococcus aureus* and *Mycobacterium tuberculosis* (MIC 32–64 μg/mL); the differences in their activity are attributed to different functional groups in the macrocyclic core [30].

Marine fungi have been considered as antibiotic sources such as *Penicillum* sp. (isolated from the coast of China), which produced four new compounds (neocitreoviridin, 10z-isocitreoviridinol, penicillstresseol and isopencillstressol) in the presence of cobalt. Penicillstresseol and isopencillstressol presented a MIC of 0.5 μg/mL against *Staphylococcus aureus*, followed by 10z-isocitroviridinol with a MIC value of 1–4 μg/mL, while neocitreoviridin exhibited a strong activity against *Pseudomonas aeruginosa* with a MIC around 4 μg/mL [31].

Marine *Engyodontium album* (isolated from a sponge) produced six new polyketides, where engyodontochone A and engyodontochone B were the ones that exhibited the best antimicrobial activity against *Staphylococcus aureus*, which was better than that of chloramphenicol [32].

Emerimicin IV extracted from *Emericellopsis minima* (isolated from a bay in Chile) exhibited a strong antimicrobial activity against *Enterococcus faecalis* and moderate to low activity against *Staphylococcus aureus* with a MIC value of 12.5 and 100 μg/mL respectively [33].

Extremophiles have also been useful in the discovery of new antibiotics due to the extreme growth conditions such as salinity (>1.0 M NaCl), pH (<5.0, >8.0), temperature (1–15°C and >45°C) and pressure (380 atm and >500–1200 atm); such conditions can be found on oceans, hypersaline lakes, hot springs and hydrothermal vents, among other places [34]. Actinobacteria are known to survive a range of the conditions previously reviewed such as the ones isolated from Kazakhstan where screening for antagonistic strains against *Escherichia coli* and *Aspergillus niger* [35].

Co-cultivation techniques have also been used for antibiotic synthesis such as *Penicillium fuscum* with *Penicillium camemberti/clavigerum*, whose co-culture allowed the extraction and purification of new macrolides named berkeleylactones. Berkeleylactone A was the one that exhibited the best activity against *Staphylococcus aureus, Bacillus anthracis, Streptococcus pyogenes, Candida albicans* and *Candida glabrata* [36].

4. Antioxidants

Antioxidants are molecules capable of counteracting at low concentrations the damage of mainly reactive oxygen and nitrogen species (ROS and RNS), which are generated from metabolic pathways such as mitochondrial respiratory chain and lipid β-oxidation among others [37, 38]; depending on the ROS/RNS, they can attack different targets [39, 40] whether biomolecules such as proteins, lipids and nucleic acids or cell organelles [22, 41]. Usually ROS and RNS at moderate concentration are useful for defense, signaling mechanisms and cellular maturation [42–45]; however when ROS and RNS concentration are in excess, different pathologies can be caused due to oxidative stress by causing tissue damage [41, 43, 45, 46].

In this regard actinobacteria have played their role as potential sources of antioxidants where [47] isolated *Streptomyces* strains in the Oman sea presented an inhibitory concentration 50 (IC_{50}) that ranges from 356.8 to 566.4 μg/mL against 2,2-diphenyl-1-picrylhydrazyl (DPPH) radical inhibition.

Growth media is important for the production of antioxidants such as the case reported by [48] on *Streptomyces variabilis* (isolated from the Gulf of Khambhat) using six different media: starch casein agar, yeast malt extract agar (ISP2), glycerol asparagine agar (ISP5), inorganic salt agar (ISP4), tyrosine agar (ISP7) and gause' synthetic agar (GSA), and incubated at 30°C for 7–9 days. GSA medium was selected because there was a larger quantity of cell mass compared to other media;

its metabolites were extracted with ethyl acetate and antioxidant activity was tested against DPPH, metal and hydrogen peroxide (H_2O_2) radical in a concentration range from 0.5 to 2.0 mg/mL. The best radical scavenging activity was against H_2O_2 radical (64% of antioxidant activity) at a concentration of 0.5 mg/mL.

Specific radical scavengers can be obtained depending on the microorganism such as the strain of *Streptomyces antioxidans* (isolated in the forest of Tanjung Lumpur), in a research reported by [49], which exhibited 79.84% of antioxidant activity against superoxide radical at an extract concentration of 1.5 mg/mL; most compounds present in the extract were pyrazines, fatty acids and a phenolic compound. Similar compounds have been found by [50] in a strain of *Streptomyces monachensis* isolated from a mangrove in Malaysia with an antioxidant activity against superoxide radical as well as metal chelating activity of 83.80 and 75.50% respectively.

Among other antioxidants found on microorganisms extracted due to their possible coloring properties are carotenoid pigments mainly used as vitamins in the case of carotenes and xanthophylls, which can be found on bacteria (*Gordonia rubropertincta*), yeast (*Blakeslea trispora)* and microalgae (*Haematococcus pluvialis*) [51].

In this regard, 50 carbon atom carotenoids identified as bacterioruberin derivatives have been detected as main pigments of *Haloterrigena turkmenica* grown in halobacterium medium, which were tested with DPPH and ferric reducing antioxidant power (FRAP) assays [52].

As mentioned earlier, growth media can influence in the production of antioxidants. Three yeasts isolated from Brazil were tested in different media. The highest carotenoid producer was *Rhodotorula mucilaginosa* in malt and yeast extract medium (MYM) followed by glycerol and corn steep liquor (GCSLM) with a biomass production of 13.5 and 7.9 g/L and a carotenoid content of 1068.5 and 224.8 μg/L respectively.

The authors noticed changes in the carotenoid profile with a higher content of β-carotene followed by astaxanthin and lutein in MYM (91.8, 6.9 and 1.3% respectively). With GCSLM, astaxanthin and lutein content increased (23.3 and 71.2% respectively) and β-carotene content decreased (71.2%).

This change in the carotenoid profile influenced greatly in the antioxidant activity where the pigments presented antioxidant activity against DPPH, 2.7 and 14.7% for MYM and GCSLM respectively. A similar, yet higher behavior was observed with 2,2′-azino-bis(3-ethylbenzothiazoline-6-sulfonic acid) (ABTS) and FRAP [53]. The increase in antioxidant activity could be due to the increase of xanthophyll content since the presence of oxygenated moieties in the carotenoid structures increases the antioxidant activity [54].

Similar experiments have been carried out by adding bivalent ions such as ferrous, calcium, copper and zinc among others as initiators of the Fenton reaction or as cofactors for carotenoid biosynthesis [55–57]. However, in a research reported by [58], such behavior was not observed on carotenoid pigments from marine strains of *Rhodococcus* and *Gordonia* genera (isolated from the Gulf of Mexico). However, a change in the carotenoid profile was observed on *Rhodococcus* sp., which may improve the antioxidant activity for two reasons:

1. The increase on the selective carotenoids that may present the best antioxidant activity [58].

2. The possible formation of carotenoid-metal complexes, mainly in the oxygenated groups [59].

These carotenoids were identified as glycosidic carotenoids; such carotenoid extracts demonstrated a better antioxidant activity against DPPH radical (IC_{50} of 1.07 and 0.09 μg/mL for *Rhodococcus* sp. and *Gordonia* sp. respectively) than β-carotene (IC_{50} of 19.59 μg/mL) [60]. Furthermore, these extracts were compared against those reported by [61] where the authors calculated an IC_{50} of 11.6 and 9.1 μg/mL for carotenoid extracts from two varieties of *Bactris gasipae*, presenting a better antioxidant activity than the bacterial extracts.

Some of these microbial carotenoid pigments are already commercially available for their use as supplement like Lycogen™, which is a carotenoid pigment from a mutant strain of *Rhodobacter sphaeroides* [62], which contains spheroidenone, bixin (a carotenoid found on *Bixa orellana* L.) and hydroxyspheroidenone [63, 64]. Another pigment that is already available is astaxanthin from microalgae *Haematococcus pluvialis*, whose production cost is estimated at $552/Kg, being competitive with synthetic carotenoids ($1000/Kg) [65].

5. Antitumor

Tumoral cells are submitted to high levels of ROS and RNS, manifesting uncontrolled proliferation, death evasion, angiogenesis, invasiveness and metastasis, causing loss of cellular function due to changes in the DNA [66].

The action of ROS and RNS may trigger different factors that stimulate angiogenic processes such as the vascular endothelial growth factor inducing proliferation, migration and tubule formation [67], as well as the induction of epithelial-mesenchymal transition by upregulation of transforming growth factor β [68].

In this regard, antioxidants serve as chemopreventive agents on healthy tissue while increasing the damage on cancerous cells; this phenomenon has been studied on secondary metabolites of plant origin such as soy isoflavone, and polyphenols such as resveratrol and hydroxychalcones [69].

Among microbial compounds that presented a correlation between antitumor and antioxidant activity were in extracts of *Streptomyces malaysiense* with compounds identified as pyrrolizidines and deferoxamine, exhibiting antioxidant as well as antitumor activities. Deferoxamine, which is listed in the World Health Organization's List of Essential Medicines, presents antioxidant activity by chelating iron and antitumoral activities [70].

Another interesting example of antitumoral compounds is an already known compound that is widely used for breast cancer stage III and IV treatment, which is doxorubicin [71], isolated from a mutant strain of *Streptomyces peuceticus*. Doxorubicin works as a DNA intercalating agent by inhibiting the activity of topoisomerase II in DNA replication [72].

Since the 1950s there has been an increase in the interest of studying marine microbial sources for drug discovery in the area of anticancer drugs such as tetracenoquinocin and 5-iminoarianciamicina, extracted from *Streptomyces* sp. in 2010, which were effective against human cervical carcinoma HeLa cells and myelogenous leukemia LH-60 [73].

A similar case is the research reported by [74], where they isolated 32 strains from lagoon sediment in Lagos. The strains were identified mostly as *Streptomyces* and *Micromonospora.* Nine isolates from *Streptomyces* genus presented cytotoxic activity against human acute myelocytic and promyelocytic leukemia, cervical carcinoma, human gastric, breast adenocarcinoma cell lines varying their effectiveness at a concentration below 1 mg/mL. The compounds present in the extracts were identified as kigamicin and staurosporine analogues.

Other kind of compounds found in *Streptomyces* strains are pyrrolopyrazines (found on *Streptomyces colonosanans*), which presented anticancer activity against human colon cancer cell lines [75].

Diketopiperazines from *Streptomyces nigra* (isolated from a mangrove soil) were effective against several human cancer-derived cell lines, while with normal cell lines they were inactive at a concentration range of 50–100 μg/mL. Other compound found on *Streptomyces nigra* was β-carboline, which is a compound widely found in plants with anticancer activity against a variety of cancer cell lines that act inhibitis DNA topoisomerase as well as intercalates in the DNA strands, changing the DNA structure; and tamoxifen [76] which is commonly used to control breast cancer after chemo and radiotherapy have been applied to the patient.

Another actinobacteria with discovered antitumor activity is *Rhodococcus*, where [77] a *Rhodococcus* strain was isolated from a contaminated soil. The extract exhibited cytotoxic activity against HepG2 and Hela cell line with an IC_{50} of 33 and 73 μg/mL respectively.

Another class of compounds that exhibit antitumor properties are polysaccharides, which inhibit cell growth and induce apoptosis as well as exert a synergistic effect with other chemotherapeutical agents such as doxorubicin [78], such as that reported on resveratrol [79, 80]. Examples of these kind of compounds are exopolysaccharides (EPs) produced by *Bacillus mycoides* composed of a sugar mixture containing galactose, mannose, glucose and glucuronic acid; such EPs exhibited antitumor activity by observing morphological abnormalities in HepG2 and Caco-2 cancer cell lines with an IC_{50} of 138 and 159 μg/mL respectively, while on normal cells the IC_{50} was 245 μg/mL [81]. A similar activity was observed with *Bacillus licheniformis* EP constituted by glucose, galactose, fructose, mannose and galacturonic acid on MCF cancer cell lines with an IC_{50} value of 840 μg/mL [82].

It can be observed from both *Bacillus* species that changes in the polysaccharide composition may influence the antitumor activity; as observed by [83] in three EPs of *Streptococcus thermophilus*, two of them were mainly composed by mannose, while the other contained mainly glucose with a protein moiety. The latter exhibited a higher antitumor activity on HepG2 cells with an IC_{50} of 313.75 μg/mL, while for the other two compounds the antitumor activity was below 50%.

Some *Trichoderma* species are also able to synthesize EP constituted by mannose, glucose, galacturonic acid and glucuronic acid with a mannan core, where the antitumor activity was more effective on HeLa cells than on MCF-7 cells by arresting G2/M phase and inducing apoptosis [84].

Fungal endophytes are another kind of microorganisms that could be used as alternative sources of bioactive compounds found in plants. Such as taxol (a chemotherapeutic), pestalactams and penicestorids. Taxol was discovered initially on *Taxus brevifolia*, and it presents a similar activity as doxorubicin [85]. Another example, camphotecin, found commonly on *Campotheca acuminata*, was also found on *Fusarium solani*. Camphotecin from *Fusarium solani* was proved to induce apoptosis on Vero cells at a concentration of 30 μg/mL for 24 h with a maximum apoptosis of 15% [86].

Endophytic fungi are also able to produce EP with antitumor activity. An example is *Bionectria ochroleuca* whose activity was proved to be effective against liver, gastric and colon cancer cell lines in a concentration range from 100 to 450 μg/mL without exhibiting toxicity in healthy cells [87].

Fungal co-cultivation techniques have also been used in the obtention of antitumor compounds. For example, *Isaria felina* with *Aspergillus sulphureus* was used for obtaining oxirapentyn L, which exhibited antitumor activity at IC_{50} greater than 100 μg/mL [88].

6. Polymers

Biopolymers such as lipopolysaccharides (LPSs), EP and extracellular polymeric substances (EPSs) are high-molecular weight substances secreted by microorganisms [89]. In the case of EP, their antitumor properties have been observed in some bacteria as well as in endophytic fungi. EPSs are exopolymers, constituted by polysaccharides, lipids, proteins and nucleic acids; the composition provides these biopolymers unique properties that can be manipulated for a variety of technological applications [90].

LPSs from Gram negative bacteria possess a lipid moiety and a glucosamine fraction with phosphate groups to improve membrane stability [91, 92]. Some of these LPSs have been studied as flocculating and emulsifying agents; for example, the one produced by *Trichosporon mycotoxinivorans* at a concentration of 8.6 mg/mL was able to flocculate kaolin and charcoal with 80 and 78% of efficiency respectively, while the emulsifying activity by mixing water and kerosene presented an emulsification efficiency of 81% [93].

Another application of LPSs is to enhance the immune response by accelerating the maturation of dendritic cells using immobilized LPS nanostructures; compared to LPS solutions and LPS monolayers, such structures could be useful in HIV patients [94]. In a similar manner, inactivated LPSs from non-sulfur photosynthetic bacteria have been used to stimulate immune response [95].

EPs have become important in material science, being useful as storage molecules, protective capsular layers and as matrix components of biofilms due to their water-binding capacity because of hydroxyl and carboxyl groups. EPs can be used in drug delivery, enzyme immobilization, tissue engineering, among other uses [96], their production depends on composition and growth conditions applied on the culture media [97].

EPs from lactic acid bacteria have been used as emulsifiers and viscosifiers because of their pseudoplastic rheological behavior; the sugar identified have been dextran, reuteran, levan and insulin, pullulan (homopolysaccharides), kefiran and hyaluronic acid (heteropolysaccharides) among others depending on the strain used to produce EP [98, 99].

An example of this kind of EP is levan produced by *Bacillus licheniformis* reported by [100] where the authors studied its physicochemical properties and concluded its utility in stabilizing topical formulations. Other uses that have been studied of levan but from *Halomonas smyrnensis* were on tissue engineering and prosthetics [101].

Hyaluronic acid from *Streptococcus equi* was compared against kefiran isolated from kefir grains (also produced on lactic acid bacteria) demonstrating antioxidant and immunostimulatory activities [102].

Marine EPs are mainly heteropolysaccharides composed of pentoses, hexoses, aminosugars or uronic acids [103]. The EPs of *Pantoea* sp. [97] presented wound healing activity by facilitating cell migration on fibroblasts. The EPs of *Bacterium polaribacter* increased 1.42-fold the wound closure at an EP concentration of 1 mg/mL.

EPS in microbial cells aids in the fixation to marine surfaces, thus forming biofilm communities though a three-dimensional arrangement in which the cells can localize extracellular activities and conduct agonist/antagonist interactions. In marine bacteria, EPSs generally contain higher levels of glucuronic and galacturonic acids. Among the sugars found on EPSs are glucose, galactose, mannose, fructose, rhamnose, uronic acids, N-acetyl-glucosamine and N-acetyl-galactosamine; the protein moiety can occur as peptides, aminosugars, glycoproteins, proteoglycans and amyloid proteins. Proteins can occur as peptides, aminosugars, glycoproteins, proteoglycans and amyloid proteins. Extracellular DNA and extracellular nucleases can be found, thus influencing on the physical consistency [90].

EPS production depend on the presence of divalent cations [90], as it is in the case of *Bacillus vallismortis;* which EPS was better in composition in the presence of zinc enhancing the adsorption capacity [104]; while in the presence of the ferric ion the EPS production is limited [90].

An application of EPS is in microencapsulation of vitamins to formulate functional foods as demonstrated on *Cyanoteche* sp. The authors extracted its EPSs and made encapsulation tests of vitamin B12 either alone or in the presence of arabic gum by spray-drying technique. EPS alone presented a particle diameter of 8 μm and when combined with arabic gum the particle diameter was smaller than that of EPS alone; both microcapsules presented different release kinetics due to the different swelling mechanisms of the EPS [105]. EPS from another *Cyanoteche* strain was found to be useful for controlled delivery of small molecules such as procainamide as well as proteins. The authors found out that adding bivalent cations such as Ca^{2+}, as well as considering the protein charge, the release kinetics could improve [106].

Other encapsulation studies were performed on the EPSs of *Bacillus subtilis* in the preservation of *Lactobacillus plantarum* as probiotic, facilitating its survival in gastric conditions during co-cultivation of both strains [107].

EPSs have been widely used in sludge treatment for pharmaceutically active ingredients removal such as ciprofloxacin as well as sulfonamides. EPS from *Klebsiella* sp. was tested against sulfonamides; the high protein content of EPS (mainly tryptophan and tyrosine) is a critical factor in the adsorption of sulfonamides for hydrophobic interactions with sulfonamides [108]. The same thing happens with ciprofloxacin being important to reach the isoelectric point of the protein moiety as well as the use of iron salts to enhance the adsorption of ciprofloxacin [109–111].

The latter ability of EPSs to adsorb antibiotics needs further studies in order to model and improve the kinetics of controlled release dosage forms giving us a natural and possible biocompatible alternative material for design of molecular pharmaceutical forms.

7. Challenges and trends in the discovery and development of microbial natural products

Even though the plethora of natural products of microbial origin mainly isolated from marine and extreme environments is a large field of research, developments in technological aspects such as the increase in natural product production for industrial scale-up or overcoming the difficulty in isolating microorganisms are needed [112].

Genome mining focused on the activation of silent genes, to search gene clusters serving as molecular markers, with complementary informatic tools has been a solution [113, 114]. This technique can be used in metabolic engineering, producing an heterologous host through genetic engineering, using plasmids or recombinant systems using interspaced palindromic repeat, one of the most recent techniques applied in genetic engineering [115].

Another trend also used on unculturable microorganisms is the discovery of environmental DNA coupled to cosmids for gene expression, which have also been used for the selective isolation of biologically active natural products [116].

Search of ideal media and culture conditions have also been a major challenge in optimizing the amount of metabolite present on the microorganism, which have been developed by either trial and error or statistical design [116]; an example is the presence of metallic salts to activate enzymes involved in the biosynthetic route or by manipulating temperature, light, aeration and pH [117] as we saw in the obtention throughout the chapter [117].

Another technique widely observed along the chapter was co-cultivation technique between bacteria, fungi or in combination to improve metabolite production.

Conventionally organic solvents are often used for natural product extraction, which is an important step for industrial scale-up [118, 119]. However, due to the health and environmental hazards, alternative extraction techniques have been searched for with the purpose of reducing residues and thus environmental impact [119, 120]. Among the alternative extraction techniques are ultrasound, microwave, enzyme and pulse electric field. The latter techniques have been widely explored on plants; nevertheless, on microorganisms, they have been poorly explored, thus representing a critical challenge in natural product research [121, 122].

The possibility to expand the research in this regard, is also the search of alternative solvents such as supercritical fluids, [119, 120], ionic liquids, gas-expanded liquids and vegetable oils [121–123].

8. Conclusions

Microbial natural products are a wide research field with much potential to be explored, the main goals being:

a. Isolating new microorganisms, being successful in marine as well as extreme environments, including genetic diversity studies for unculturable microorganisms.

b. Screening of isolates with potential biological activity, by performing extractions of different polarity to begin the selectivity of compounds.

c. Extraction and purification for identification of active compounds, where new extraction techniques can be explored such as supercritical fluids, microwave, enzyme, among others to make the discovery process more eco-friendly.

d. Elucidation of action mechanisms of new active compounds through in silico studies, for considering the possibility of improving the activity.

e. Improvement of natural compounds production for industrial scale-up, as it has already been seen that these are the main challenges and trends through alternative techniques such as co-cultivation, genome mining and media formulation, the last one being the first approach for production enhancement.

f. Preclinical and clinical trials of microbial natural products with already discovered potential activity, to determine biocompatibility and innocuousness of compounds such as EPs and EPSs for antitumoral activity as well as tissue engineering.

As it can be seen, there is a long way ahead in natural product discovery that could solve many health and environmental issues such as antibiotic resistance, cancer, soil and water contamination, tissue engineering, among other contributions.

Acknowledgements

The author thanks the Department of Biological Systems for their support.

Author details

Lucía Ortega Cabello
Biological Systems Department, Metropolitan Autonomous University-Xochimilco Unit, Mexico City, Mexico

*Address all correspondence to: lortegac@correo.xoc.uam.mx

References

[1] Sen T, Barrow CJ, Deshmukh SK. Microbial pigments in the food industry—Challenges and the way forward. Frontiers in Nutrition. 2019;**6**:1-14. DOI: 10.3389/fnut.2019.00007

[2] Saini RK, Keum YS. Microbial platforms to produce commercially vital carotenoids at industrial scale: An updated review of critical issues. Journal of Industrial Microbiology & Biotechnology. 2019;**46**(5):657-674. DOI: 10.1007/s10295-018-2104-7

[3] Guimarães TC, Gomes TS, Fernandes CD, Barros FD, Oliveira KV, Bilal M, et al. Antitumor microbial products by actinomycetes isolated from different environments. In: Arora PK, editor. Microbial Technology for Health and Environment. Singapore: Springer; 2020. pp. 113-160. DOI: 10.1007/978-981-15-2679-4_5

[4] Chaturvedi S, Khurana SMP. Importance of actinobacteria for bioremediation. In: Khurana S, Gaur R, editors. Plant Biotechnology: Progress in Genomic Era. Singapore: Springer; 2019. pp. 277-307. DOI: 10.1007/978-981-13-8499-8_13

[5] Anteneh YS, Franco C. Whole cell actinobacteria as biocatalysts. Frontiers in Microbiology. 2019;**10**:77. DOI: 10.3389/fmicb.2019.00077

[6] Gaynes R. The discovery of penicillin—New insights after more than 75 years of clinical use. Emerging Infectious Diseases. 2017;**23**(5):849-853. DOI: 10.3201/eid2305.161556

[7] Woodruff HB. Selman A. Waksman, winner of the 1952 Nobel prize for physiology or medicine. Applied and Environmental Microbiology. 2014;**80**(1):2-8. DOI: 10.1128/AEM.01143-13

[8] Huang T, Lin S. Microbial natural products: A promising source for drug discovery. Journal of Applied Microbiology and Biochemistry. 2017;**1**(2):5. DOI: 10.21767/2576-1412.100005

[9] Ma K, Kim SD. Development of techniques for isolating microorganisms. Enliven Archive. 2018;**5**(1):1-5

[10] Arora D, Gupta P, Jaglan S, Roullier C, Grovel O, Bertrand S. Expanding the chemical diversity through microorganisms co-culture: Current status and outlook. Biotechnology Advances. 2020;**40**:107521. DOI: 10.1016/j.biotechadv.2020.107521

[11] Oh D-C, Jensen PR, Kauffman CA, Fenical W. Libertellenones A–D: Induction of cytotoxic diterpenoid biosynthesis by marine microbial competition. Bioorganic & Medicinal Chemistry. 2005;**13**(17):5267-5273. DOI: 10.1016/j.bmc.2005.05.068

[12] Do Nascimento M, Dublan MLA, Ortiz-Marquez JCF, Curatti L. High lipid productivity of an Ankistrodesmus-Rhizobium artificial consortium. Bioresource Technology. 2013;**146**:400-407. DOI: 10.1016/j.biortech.2013.07.085

[13] Demain AL. Importance of microbial natural products and the need to revitalize their discovery. Journal of Industrial Microbiology & Biotechnology. 2014;**41**:185-201. DOI: 10.1007/s10295-013-1325-z

[14] Torregrosa-Crespo J, Montero Z, Fuentes JL, Reig García-Galbis M, Garbayo I, Vílchez C, et al. Exploring the valuable carotenoids for the large-scale production by marine microorganisms. Marine Drugs. 2018;**16**(6):203. DOI: 10.3390/md16060203

[15] Russo P, Cesario A. New anticancer drugs from marine cyanobacteria. Current Drug Targets. 2012;**13**:1048-1053. DOI: 10.2174/138945012802009035

[16] Williams PG, Buchanan GO, Feling RH, Kauffman CA, Jensen PR, Fenical W. New cytotoxic salinosporamides from the marine actinomycete *Salinispora tropica*. The Journal of Organic Chemistry. 2005;**70**(16):6196-6203. DOI: 10.1021/jo050511+

[17] Ferri M, Ranucci E, Romagnoli P, Giaccone V. Antimicrobial resistance: A global emerging threat to public health systems. Critical Reviews in Food Science and Nutrition. 2017;**57**(13):2857-2876. DOI: 10.1080/10408398.2015.1077192

[18] Moore BS, Carter GT, Brönstrup M. Are natural products the solution to antimicrobial resistance? Natural Product Reports. 2017;**34**(7):685-686. DOI: 10.1039/C7NP90026K

[19] Niu G, Li W. Next-generation drug discovery to combat antimicrobial resistance. Trends in Biochemical Sciences. 2019;**44**(11):961-972. DOI: 10.1016/j.tibs.2019.05.005

[20] McSorley FR, Johnson JW, Wright GD. Natural products in antibiotic discovery. In: Fong IW, Shlaes D, Drlica K, editors. Antimicrobial Resistance in the 21st Century. Cham: Springer; 2018. pp. 533-562. DOI: 10.1007/978-3-319-78538-7_17

[21] Ueda K, Beppu T. Antibiotics in microbial coculture. The Journal of Antibiotics. 2017;**70**(4):361-365. DOI: 10.1038/ja.2016.127

[22] Netzker T, Flak M, Krespach MK, Stroe MC, Weber J, Schroeckh V, et al. Microbial interactions trigger the production of antibiotics. Current Opinion in Microbiology. 2018;**45**:117-123. DOI: 10.1016/j.mib.2018.04.002

[23] Adnani N, Chevrette MG, Adibhatla SN, Zhang F, Yu Q, Braun DR, et al. Coculture of marine invertebrate-associated bacteria and interdisciplinary technologies enable biosynthesis and discovery of a new antibiotic, keyicin. ACS Chemical Biology. 2017;**12**(12):3093-3102. DOI: 10.1021/acschembio.7b00688

[24] Cueto M, Jensen PR, Kauffman C, Fenical W, Lobkovsky E, Clardy J. Pestalone, a new antibiotic produced by a marine fungus in response to bacterial challenge. Journal of Natural Products. 2001;**64**:1444-1446. DOI: 10.1021/np0102713

[25] Iniyan AM, Sudarman E, Wink J, Kannan RR, Vincent SGP. Ala-geninthiocin, a new broad spectrum thiopeptide antibiotic, produced by a marine *Streptomyces* sp. ICN19. The Journal of Antibiotics. 2019;**72**(2):99-105. DOI: 10.1038/s41429-018-0115-2

[26] Han Y, Wang Y, Yang Y, Chen H. Shellmycin A-D, novel bioactive tetrahydroanthra-γ-pyrone antibiotics from marine Streptomyces sp. shell-016 Marine Drugs. 2020;**18**(1):58. DOI: 10.3390/md18010058

[27] Akhter N, Liu Y, Auckloo BN, Shi Y, Wang K, Chen J, et al. Stress-driven discovery of new angucycline-type antibiotics from a marine *Streptomyces pratensis* NA-ZhouS1. Marine Drugs. 2018;**16**(9):331. DOI: 10.3390/md16090331

[28] Al Farraj DA, Varghese R, Vágvölgyi C, Elshikh MS, Alokda AM, Mahmoud AH. Antibiotics production in optimized culture condition using low cost substrates from *Streptomyces* sp. AS4 isolated from mangrove soil sediment. Journal of King Saud University-Science. 2020;**32**(2):1528-1535. DOI: 10.1016/j.jksus.2019.12.008

[29] Williams DE, Dalisay DS, Chen J, Polishchuck EA, Patrick BO, Narula G, et al. Aminorifamycins and sporalactams produced in culture by a *Micromonospora* sp. isolated from a northeastern-pacific marine sediment are potent antibiotics. Organic Letters. 2017;**19**(4):766-769. DOI: 10.1021/acs.orglett.6b03619

[30] Pérez-Bonilla M, Oves-Costales D, De la Cruz M, Kokkini M, Martín J, Vicente F, et al. Phocoenamicins B and C, new antibacterial spirotetronates isolated from a marine *Micromonospora* sp. Marine Drugs. 2018;**16**(3):95. DOI: 10.3390/md16030095

[31] Auckloo BN, Pan C, Akhter N, Wu B, Wu X, He S. Stress-driven discovery of novel cryptic antibiotics from a marine fungus *Penicillium* sp. BB1122. Frontiers in Microbiology. 2017;**8**:1450. DOI: 10.3389/fmicb.2017.01450

[32] Wu B, Wiese J, Wenzel-Storjohann A, Malien S, Schmaljohann R, Imhoff JF. Engyodontochones, antibiotic polyketides from the marine fungus *Engyodontium album* strain LF069. Chemistry-A European Journal. 2016;**22**(22):7452-7462. DOI: 10.1002/chem.201600430

[33] Inostroza A, Lara L, Paz C, Perez A, Galleguillos F, Hernandez V, et al. Antibiotic activity of emerimicin IV isolated from *Emericellopsis minima* from Talcahuano Bay, Chile. Natural Product Research. 2018;**32**(11):1361-1364. DOI: 10.1080/14786419.2017.1344655

[34] Rekadwad B, Khobragade C. Marine polyextremophiles and their biotechnological applications. In: Kalia VC, Kumar P, editors. Microbial Applications. Cham: Springer; 2017. pp. 319-331. DOI: 10.1007/978-3-319-52666-9_15

[35] Trenozhnikova L, Azizan A. Discovery of actinomycetes from extreme environments with potential to produce novel antibiotics. Central Asian Journal of Global Health. 2018;7(1):337. DOI: 10.5195/cajgh.2018.337

[36] Stierle AA, Stierle DB, Decato D, Priestley ND, Alverson JB, Hoody J, et al. The berkeleylactones, antibiotic macrolides from fungal coculture. Journal of Natural Products. 2017;**80**(4):1150-1160. DOI: 10.1021/acs.jnatprod.7b00133

[37] Nathan C, Cunningham-Bussel A. Beyond oxidative stress: An immunologist's guide to reactive oxygen species. Nature Reviews. Immunology. 2013;**13**(5):349-361. DOI: 10.1038/nri3423

[38] Lushchak VI. Free radicals, reactive oxygen species, oxidative stress and its classification. Chemico-Biological Interactions. 2014;**224**:164-175. DOI: 10.1016/j.cbi.2014.10.016

[39] Schieber M, Chandel NS. ROS function in redox signaling and oxidative stress. Current Biology. 2014;**24**:R453-R462. DOI: 10.1016/j.cub.2014.03.034

[40] Krumova K, Cosa G. Overview of reactive oxygen species. In: Nonell S, Flors C, editors. Singlet Oxygen: Applications in Bioscience and Nanoscience. EU: Royal Society of Chemistry; 2016. p. 3

[41] Nimse SB, Pal D. Free radicals, natural antioxidants, and their reaction mechanisms. RSC Advances. 2015;**5**:27986-28006. DOI: 10.1039/C4RA13315C

[42] Radi R. Oxygen radicals, nitric oxide, and peroxynitrite: Redox pathways in molecular medicine. PNAS. 2018;**115**(23):5839-5848. DOI: 10.1073/pnas.1804932115

[43] Chehue Romero A, Olvera Hernández EG, Flores Cerón T, Álvarez-Chávez A. The exogenous antioxidants. In: Morales-González JA, editor. Oxidative Stress and Chronic Degenerative Diseases—A Role for Antioxidants. Croatia: InTechOpen; 2013. p. 33. DOI: 10.5772/52490

[44] Pham-Huy LA, He H, Pham-Huy C. Free radicals, antioxidants in disease and health. International Journal of Biomedical Sciences. 2008;**4**(2):89-96

[45] Phaniendra A, Jestadi DB, Periyasami L. Free radicals: Properties, sources, targets and their implications in various diseases. Indian Journal of Clinical Biochemistry. 2015;**30**(1):11-26. DOI: 10.1007/s12291-014-0446-0

[46] Xu J, Leeuwenburgh C. Free radicals and oxidative stress: Basic concepts and misconceptions. In: Miller J, Le Prell CG, Rybak L, editors. Free Radicals in ENT Pathology. SWT: Humana Press; 2015. pp. 10-15

[47] Gozari M, Bahador N, Jassbi AR, Mortazavi MS, Hamzehei S, Eftekhar E. Isolation, distribution and evaluation of cytotoxic and antioxidant activity of cultivable actinobacteria from the Oman Sea sediments. Acta Oceanologica Sinica. 2019;**38**(12):84-90. DOI: 10.1007/s13131-019-1515-2

[48] Dholakiya RN, Kumar R, Mishra A, Mody KH, Jha B. Antibacterial and antioxidant activities of novel actinobacteria strain isolated from Gulf of Khambhat, Gujarat. Frontiers in Microbiology. 2017;**8**:2420. DOI: 10.3389/fmicb.2017.02420

[49] Ser HL, Tan LTH, Palanisamy UD, Abd Malek SN, Yin WF, Chan KG, et al. *Streptomyces antioxidans* sp. nov., a novel mangrove soil actinobacterium with antioxidative and neuroprotective potentials. Frontiers in Microbiology. 2016;7:899. DOI: 10.3389/fmicb.2016.00899

[50] Law JWF, Ser HL, Ab Mutalib NS, Saokaew S, Duangjai A, Khan TM, et al. *Streptomyces monashensis* sp. nov., a novel mangrove soil actinobacterium from East Malaysia with antioxidative potential. Scientific Reports-UK. 2019;**9**(1):1-18. DOI: 10.1038/s41598-019-39592-6

[51] De Carvalho J, Cardoso L, Ghiggi V, Woiciechowski A, Vandenberghe L, Soccol C. Microbial pigments. In: Brar SK, Dhillon GS, Soccol CR, editors. Biotransformation of Waste Biomass into High Value Biochemicals. NY: Springer; 2014. pp. 73-97

[52] Squillaci G, Parrella R, Carbone V, Minasi P, La Cara F, Morana A. Carotenoids from the extreme halophilic archaeon *Haloterrigena turkmenica*: Identification and antioxidant activity. Extremophiles. 2017;**21**:933-945. DOI: 10.1007/s00792-017-0954-y

[53] Pereira Cipolatti E, Diaz Remedi R, dos Santos SC, Bueno Rodrigues A, Markowiski Gonçalves Ramos J, Veiga Burkert CA, et al. Use of agroindustrial byproducts as substrate for production of carotenoids with antioxidant potential by wild yeasts. Biocatalysis and Agricultural Biotechnology. 2019;**20**:101208. DOI: 10.1016/j.bcab.2019.101208

[54] Young AJ, Lowe GL. Carotenoids-antioxidant properties. Antioxidants. 2018;**7**(2):28. DOI: 10.3390/antiox7020028

[55] Elfeky N, Elmahmoudy M, Zhang Y, Guo J, Bao Y. Lipid and carotenoid production by *Rhodotorula glutinis* with a combined cultivation mode of nitrogen, sulfur, and aluminium stress. Applied Sciences. 2019;**9**(12):2444. DOI: 10.3390/app9122444

[56] Massoud R, Khosravi-Darani K. A review on the impacts of process variables on microbial production of carotenoid pigments. In:

Grumezescu AM, Holban AM, editors. Food Biosynthesis. UK: Academic Press; 2017. pp. 183-211. DOI: 10.1016/B978-0-12-811372-1.00006-3

[57] Mayasari E, Raya I, Natsir H. Effect of Fe^{2+} and Mn^{2+} addition on growth and β-carotene production of *Dunaliella salina*. Journal of Physics: Conference Series. 2018;**979**(1):012012. DOI: 10.1088/1742-6596/979/1/012012

[58] Ortega-Cabello L, Pérez-Méndez HI, Manjarrez-Alvarez N, Solís-Oba A, López-Luna A. Effect of iron salts on *Rhodococcus* sp. and *Gordonia* sp. on carotenoid production. Revista Mexicana de Ingeniería Química. 2017;**16**(1):1-10

[59] Chatterjee T, Das A. Natural agents with iron chelation potential: A new hope and promise for cheaper supplement of costly iron chelating medicines for transfusion dependent thalassaemics. International Journal of Advanced Research. 2016;**4**(3):1258 1268

[60] Ortega Cabello L, Pérez Méndez HI, Manjarrez Alvarez N, López-Luna A, Solís Oba A, Solís OM. Comparación de la actividad antioxidante de carotenoides en extractos crudos y pre-purificados de actinobacterias marinas. Revista Mexicana de Ciencias Farmaceuticas. 2016;**47**(2):60-65

[61] Quesada S, Azofeifa G, Jatunov S, Jiménez G, Navarro L, Gómez G. Carotenoids composition, antioxidant activity and glycemic index of two varieties of *Bactris gasipaes*. Emirates Journal of Food and Agriculture. 2011;**23**(6):482-489

[62] Wang CC, Ding S, Chiu KH, Liu WS, Lin TJ, Wen ZH. Extract from a mutant *Rhodobacter sphaeroides* as an enriched carotenoid source. Food & Nutrition Research. 2016;**60**(1):29580. DOI: 10.3402/fnr.v60.29580

[63] Li Z, Kong L, Hui B, Shang X, Gao L, Luan N, et al. Identification and antioxidant activity of carotenoids from superfine powder of *Rhodobacter sphaeroides*. Emirates Journal of Food and Agriculture. 2017;**29**(11):833-845. DOI: 10.9755/ejfa.2017.v29.i11.1479

[64] Raddatz-Mota D, Pérez-Flores LJ, Carrari F, Mendoza-Espinoza JA, de León-Sánchez FD, Pinzón-López LL, et al. Achiote (*Bixa orellana* L.): A natural source of pigment and vitamin E. Journal of Food Science and Technology. 2017;**54**(6):1729-1741. DOI: 10.1007/s13197-017-2579-7

[65] Saini RK, Keum YS. Progress in microbial carotenoids production. Indian Journal of Microbiology. 2017;**57**(1):129-130. DOI: 10.1007/s12088-016-0637-x

[66] Hecht F, Pessoa CF, Gentile LB, Rosenthal D, Carvalho DP, Fortunato RS. The role of oxidative stress on breast cancer development and therapy. Tumor Biology. 2016;**37**(4):4281-4291. DOI: 10.1007/s13277-016-4873-9

[67] Auyeung KK, Ko JK. Angiogenesis and oxidative stress in metastatic tumor progression: Pathogenesis and novel therapeutic approach of colon cancer. Current Pharmaceutical Design. 2017;**23**(27):3952-3961. DOI: 10.2174/1381612823666170228124105

[68] Ma-on C, Sanpavat A, Whongsiri P, Suwannasin S, Hirankarn N, Tangkijvanich P, et al. Oxidative stress indicated by elevated expression of Nrf2 and 8-OHdG promotes hepatocellular carcinoma progression. Medical Oncology. 2017;**34**(4):57. DOI: 10.1007/s12032-017-0914-5

[69] Thyagarajan A, Sahu RP. Potential contributions of antioxidants to cancer therapy: Immunomodulation and radiosensitization. Integrative Cancer

Therapies. 2018;**17**(2):210-216. DOI: 10.1177/1534735416681639

[70] Ser HL, Palanisamy U, Yin W-F, Chan K-G, Goh BH, Lee LH. *Streptomyces malaysiense* sp. nov.: A novel malaysian mangrove soil actinobacterium with antioxidative activity and cytotoxic potential against human cancer cell lines. Scientific Reports-UK. 2016;**6**:24247. DOI: 10.1038/srep24247

[71] Bauermeister A, Zucchi TD, Moraes LAB. Mass spectrometric approaches for the identification of anthracycline analogs produced by actinobacteria. Journal of Mass Spectrometry. 2016;**51**(6):437-445. DOI: 10.1002/jms.3772

[72] Malla S, Niraula NP, Singh B, Liou K, Sohng JK. Limitations in doxorubicin production from *Streptomyces peucetius*. Microbiological Research. 2010;**165**(5):427-435. DOI: 10.1016/j.micres.2009.11.006

[73] Busi S, Pattnaik SS. Current status and applications of actinobacteria in the production of anticancerous compounds. In: Singh BP, Gupta VK, Passari AK, editors. New and Future Developments in Microbial Biotechnology and Bioengineering. Vol. 2018. Netherlands: Elsevier; 2018. pp. 137-153

[74] Davies-Bolorunduro OF, Adeleye IA, Akinleye MO, Wang PG. Anticancer potential of metabolic compounds from marine actinomycetes isolated from Lagos Lagoon sediment. Journal of Pharmaceutical Analysis. 2019;**9**(3):201-208. DOI: 10.1016/j.jpha.2019.03.004

[75] Law JWF, Ser HL, Duangjai A, Saokaew S, Bukhari SI, Khan TM, et al. *Streptomyces colonosanans* sp. nov., a novel actinobacterium isolated from Malaysia mangrove soil exhibiting antioxidative activity and cytotoxic potential against human colon cancer cell lines. Frontiers in Microbiology. 2017;**8**:877. DOI: 10.3389/fmicb.2017.00877

[76] Chen C, Ye Y, Wang R, Zhang Y, Wu C, Debnath SC, et al. *Streptomyces nigra* sp. nov. is a novel actinobacterium isolated from mangrove soil and exerts a potent antitumor activity in vitro. Frontiers in Microbiology. 2018;**9**:1587. DOI: 10.3389/fmicb.2018.01587

[77] Zhang XG, Liu ZY, Liu JW, Zeng YL, Guo GJ, Sun QY. Antitumor activity of a *Rhodococcus* sp. Lut0910 isolated from polluted soil. Tumour Biology. 2017;**39**(6):1-9. DOI: 1010428317711661

[78] Chen L, Huang G. Antitumor activity of polysaccharides: An overview. Current Drug Targets. 2018;**19**(1):89-96. DOI: 10.2174/1389450118666170704143018

[79] Washington KE, Kularatne RN, Biewer MC, Stefan MC. Combination loading of doxorubicin and resveratrol in polymeric micelles for increased loading efficiency and efficacy. ACS Biomaterials Science & Engineering. 2018;**4**(3):997-1004. DOI: 10.1021/acsbiomaterials.7b00972

[80] Gu J, Hu W, Zhang D-D. Resveratrol, a polyphenol phytoalexin, protects against doxorubicin-induced cardiotoxicity. Journal of Cellular and Molecular Medicine. 2015;**19**(10):2324-2328. DOI: 10.1111/jcmm.12633

[81] Farag MMS, Moghannem SAM, Shehabeldine AM, Azab MS. Antitumor effect of exopolysaccharide produced by *Bacillus mycoides*. Microbial Pathogenesis. 2020;**140**:103947. DOI: 10.1016/j.micpath.2019.103947

[82] Insulkar P, Kerkar S, Lele SS. Purification and structural-functional characterization of an exopolysaccharide from *Bacillus licheniformis* PASS26 with in-vitro

antitumor and wound healing activities. International Journal of Biological Macromolecules. 2018;**120**:1441-1450. DOI: 10.1016/j.ijbiomac.2018.09.147

[83] Sun N, Liu H, Liu S, Zhang X, Chen P, Li W, et al. Purification, preliminary structure and antitumor activity of exopolysaccharide produced by *Streptococcus thermophilus* CH9. Molecules. 2018;**23**(11):2898. DOI: 10.3390/molecules23112898

[84] Li H, Yu H, Zhu H. Structure studies of the extracellular polysaccharide from *Trichoderma* sp. KK19L1 and its antitumor effect via cell cycle arrest and apoptosis. Applied Biochemistry and Biotechnology. 2017;**182**(1):128-141. DOI: 10.1007/s12010-016-2315-1

[85] Chen L, Zhang QY, Jia M, Ming QL, Yue W, Rahman K, et al. Endophytic fungi with antitumor activities: Their occurrence and anticancer compounds. Critical Reviews in Microbiology. 2016;**42**(3):454-473. DOI: 10.3109/1040841X.2014.959892

[86] Ran X, Zhang G, Li S, Wang J. Characterization and antitumor activity of camptothecin from endophytic fungus *Fusarium solani* isolated from *Camptotheca acuminate*. African Health Sciences. 2017;**17**(2):566-574. DOI: 10.4314/ahs.v17i2.34

[87] Li Y, Guo S, Zhu H. Statistical optimization of culture medium for production of exopolysaccharide from endophytic fungus *Bionectria ochroleuca* and its antitumor effect in vitro. EXCLI Journal. 2016;**15**:211. DOI: 10.17179/excli2016-154

[88] Smetanina OF, Yurchenko AN, Ivanets EV, Kalinovsky AI, Khudyakova YV, Dyshlovoy SA, et al. Unique prostate cancer-toxic polyketides from marine sediment-derived fungus *Isaria felina*. The Journal of Antibiotics. 2017;**70**(7):856-858. DOI: 10.1038/ja.2017.53

[89] Singh S, Katoch A, Kaur R, Sran KS, Kumar B, Choudhury AR. Fermentative production and application of marine microbial exopolysaccharides. In: Trincone A, editor. Enzymatic Technologies for Marine Polysaccharides. USA: CRC Press; 2019. pp. 189-218

[90] Decho AW, Gutierrez T. Microbial extracellular polymeric substances (EPSs) in ocean systems. Frontiers in Microbiology. 2017;**8**:922. DOI: 10.3389/fmicb.2017.00922

[91] Caroff M, Karibian D. Structure of bacterial lipopolysaccharides. Carbohydrate Research. 2003;**338**(23): 2431-2447. DOI: 10.1016/j.carres.2003.07.010

[92] Bonev B. Lipopolysaccharide, structure and assembly of bacterial outer membranes. Biophysical Journal. 2018;**114**(3):270a

[93] Domingues VS, Monteiro AS, Ferreira GF, Santos VL. Solid flocculation and emulsifying activities of the lipopolysaccharide produced by *Trichosporon mycotoxinivorans* CLA2. Applied Biochemistry and Biotechnology. 2017;**182**(1):367-381. DOI: 10.1007/s12010-016-2332-0

[94] Liu Y, Wang KH, Chen HY, Li JR, Laurence TA, Ly S, et al. Periodic arrangement of lipopolysaccharides nanostructures accelerates and enhances the maturation processes of dendritic cells. ACS Applied Nano Materials. 2018;**1**(2):839-850. DOI: 10.1021/acsanm.7b00254

[95] Masoud H. Novel adjuvants derived from attenuated lipopolysaccharides and lipid As of purple non-sulfur photosynthetic bacteria. Vaccine. 2019;**37**(26):3472-3477. DOI: 10.1016/j.vaccine.2019.04.097

[96] Moradali MF, Rehm BH. Bacterial biopolymers: From pathogenesis to

advanced materials. Nature Reviews. Microbiology. 2020;**18**:195-210. DOI: 10.1038/s41579-019-0313-3

[97] Sahana TG, Rekha PD. A novel exopolysaccharide from marine bacterium *Pantoea* sp. YU16-S3 accelerates cutaneous wound healing through Wnt/β-catenin pathway. Carbohydrate Polymers. 2020;**238**:116191. DOI: 10.1016/j.carbpol.2020.116191

[98] Ahmed Z, Ahmad A. Biopolymer produced by the lactic acid bacteria: Production and practical application. In: Holban AM, Mihai A, editors. Microbial Production of Food Ingredients and Additives. UK: Academic Press; 2017. pp. 217-257

[99] Verma ML, Kumar S, Jeslin J, Dubey NK. Microbial production of biopolymers with potential biotechnological applications. In: Pal K, Banerjee I, Sarkar P, Kim D, Deng WP, Dubey NK, Majumder K, editors. Biopolymer-Based Formulations. Netherlands: Elsevier; 2020. pp. 105-137

[100] Pantelić I, Lukić M, Gojgić-Cvijović G, Jakovljević D, Nikolić I, Lunter DJ, et al. *Bacillus licheniformis* Levan as a functional biopolymer in topical drug dosage forms: From basic colloidal considerations to actual pharmaceutical application. European Journal of Pharmaceutical Sciences. 2020;**142**:105109. DOI: 10.1016/j.ejps.2019.105109

[101] Avsar G, Agirbasli D, Agirbasli MA, Gunduz O, Oner ET. Levan based fibrous scaffolds electrospun via co-axial and single-needle techniques for tissue engineering applications. Carbohydrate Polymers. 2018;**193**:316-325. DOI: 10.1016/j.carbpol.2018.03.075

[102] Radhouani H, Gonçalves C, Maia FR, Oliveira JM, Reis RL. Biological performance of a promising Kefiran-biopolymer with potential in regenerative medicine applications: A comparative study with hyaluronic acid. Journal of Materials Science: Materials in Medicine. 2018;**29**(8):124. DOI: 10.1007/s10856-018-6132-7

[103] Sun ML, Zhao F, Chen XL, Zhang XY, Zhang YZ, Song XY, et al. Promotion of wound healing and prevention of frostbite injury in rat skin by exopolysaccharide from the arctic marine bacterium *Polaribacter* sp. SM1127. Marine Drugs. 2020;**18**(1):48. DOI: 10.3390/md18010048

[104] Ding P, Song W, Yang Z, Jian J. Influence of Zn(II) stress-induction on component variation and sorption performance of extracellular polymeric substances (EPS) from *Bacillus vallismortis*. Bioprocess and Biosystems Engineering. 2018;**41**(6):781-791. DOI: 10.1007/s00449-018-1911-6

[105] Estevinho BN, Mota R, Leite JP, Tamagnini P, Gales L, Rocha F. Application of a cyanobacterial extracellular polymeric substance in the microencapsulation of vitamin B12. Powder Technology. 2019;**343**:644-651. DOI: 10.1016/j.powtec.2018.11.079

[106] Leite JP, Mota R, Durão J, Neves SC, Barrias CC, Tamagnini P, et al. Cyanobacterium-derived extracellular carbohydrate polymer for the controlled delivery of functional proteins. Macromolecular Bioscience. 2017;**17**(2):1600206. DOI: 10.1002/mabi.201600206

[107] Yahav S, Berkovich Z, Ostrov I, Reifen R, Shemesh M. Encapsulation of beneficial probiotic bacteria in extracellular matrix from biofilm-forming *Bacillus subtilis*. Artificial Cells, Nanomedicine and Biotechnology. 2018;**46**(2):974-982. DOI: 10.1080/21691401.2018.1476373

[108] Pi S, Li A, Cui D, Su Z, Feng L, Ma F, et al. Biosorption behavior and

mechanism of sulfonamide antibiotics in aqueous solution on extracellular polymeric substances extracted from *Klebsiella* sp. J1. Bioresource Technology. 2019;**272**:346-350. DOI: 10.1016/j.biortech.2018.10.054

[109] Polesel F, Lehnberg K, Dott W, Trapp S, Thomas KV, Plósz BG. Factors influencing sorption of ciprofloxacin onto activated sludge: Experimental assessment and modelling implications. Chemosphere. 2015;**119**:105-111. DOI: 10.1016/j.chemosphere.2014.05.048

[110] Gu C, Gao P, Yang F, An D, Munir M, Jia H, et al. Characterization of extracellular polymeric substances in biofilms under long-term exposure to ciprofloxacin antibiotic using fluorescence excitation-emission matrix and parallel factor analysis. Environemental Science and Pollution Research. 2017;**24**(15):13536-13545. DOI: 10.1007/s11356-017-8986-5

[111] Zhang H, Jia Y, Khanal SK, Lu H, Fang H, Zhao Q. Understanding the role of extracellular polymeric substances on ciprofloxacin adsorption in aerobic sludge, anaerobic sludge, and sulfate-reducing bacteria sludge systems. Environmental Science & Technology. 2018;**52**(11):6476-6486. DOI: 10.1021/acs.est.8b00568

[112] Singh BP, Rateb M, Rodriguez-Couto S, MDLTD P, Li WJ. Microbial secondary metabolites: Recent developments and technological challenges. Frontiers in Microbiology. 2019;**10**:914. DOI: 10.3389/fmicb.2019.00914

[113] Baltz RH. Gifted microbes for genome mining and natural product discovery. Journal of Industrial Microbiology & Biotechnology. 2017;**44**(4-5):573-588. DOI: 10.1007/s10295-016-1815-x

[114] Kang HS. Phylogeny-guided (meta) genome mining approach for the targeted discovery of new microbial natural products. Journal of Industrial Microbiology & Biotechnology. 2017;**44**(2):285-293. DOI: 10.1007/s10295-016-1874-z

[115] Zhang MM, Wang Y, Ang EL, Zhao H. Engineering microbial hosts for production of bacterial natural products. Natural Product Reports. 2016;**33**(8):963-987. DOI: 10.1039/C6NP00017G

[116] Nybo SE, Kharel MK. Recent developments in the quest for novel microbial natural products. In: Rahman A, editor. Studies in Natural Products Chemistry. USA: Elsevier; 2018. pp. 109-152

[117] Tomm HA, Ucciferri L, Ross AC. Advances in microbial culturing conditions to activate silent biosynthetic gene clusters for novel metabolite production. Journal of Industrial Microbiology & Biotechnology. 2019;**46**(9-10):1381-1400. DOI: 10.1007/s10295-019-02198-y

[118] Pagliaro M. Green and economically viable extraction of natural products: From lab to marketplace. Chimica Oggi. 2017;**35**(3):71-72

[119] Abas F, Intan SI, Nordin HL. Generalized likelihood uncertainty estimation (GLUE) methodology for optimization of extraction in natural products. Food Chemistry. 2018;**250**:37-45. DOI: 10.1016/j.foodchem.2018.01.023

[120] Zhang Q-W, Li-Gen L, Wen-Cai Y. Techniques for extraction and isolation of natural products: A comprehensive review. Chinese Medicine-UK. 2018;**13**(1):20. DOI: 10.1186/s13020-018-0177-x

[121] Ahmad I. Ionic liquid-based microwave-assisted extraction: Fast and green extraction method of

secondary metabolites on medicinal plant. Pharmacognosy Reviews. 2018;**12**(23):20-26

[122] Li Y, Fabiano-Tixier AS, Chemat F. Vegetable oils as alternative solvents for green extraction of natural products. In: Chemat S, editor. Edible Oils: Extraction, Processing and Applications. United States: CRC Press; 2017. p. 25

[123] Herrero M, Ibañez E. Green extraction processes, biorefineries and sustainability: Recovery of high added-value products from natural sources. The Journal of Supercritical Fluids. 2018;**134**:252-259. DOI: 10.1016/j.supflu.2017.12.002

Chapter 17

Cannabis Effect on Female Reproductive Health

Somenath Ghosh

Abstract

Cannabis sativa is a cheap hallucinating agent used in different parts of the world from time unknown as a part of various religious as well as social practices. *Cannabis* which is a special type of Marijuana can provide temporary relief from analgesia, body pain, and in some other clinical conditions. But impacts of *Cannabis* on reproductive health of males and females are multi-faceted and differentially fatal. In males, *Cannabis* can cause changes in testicular morphology, sperm parameters (in terms of semen quality, sperm morphology, sperm mortality, and sperm motility), male reproductive hormones and finally causing reduced libido. In females, *Cannabis* can reduce female fertility by disrupting hypothalamic release of gonadotropin releasing hormone (GnRH), leading to reduced estrogen and progesterone production and anovulatory menstrual cycles. Current research suggest that *Cannabis* may negatively impact on male and female fertility conditions. However, male sterility considering the *Cannabis* impact is totally lacking in human as well as in sub-human primates. However, very limited studies are available on *Cannabis* effect on primate female reproduction considering Rhesus monkeys. Hence, further studies are needed to validate that robust findings in animal models will carry over into human experience.

Keywords: *Cannabis*, CB1, female mice, impairment, reproduction, stress

1. Introduction

Cannabis which is a type of marijuana has been used by the people of Indian subcontinent from time unknown [1]. They not only use this herb as a part of holy practice but also use it for recreational purposes [2]. Irrespective of sex, this hallucinogenic agent is used by most part of the world particularly by the populations of South America, India, Bangladesh, and Pakistan from a long time ago [3]. Reports suggesting the roles of *Cannabis* causing systemic neuropathy [4], neuronal disability [5], impaired fetal development [6], and mal-functioning of male reproductive system [7–10] are documented. But no reports available are depicting the effects of marijuana in female reproductive system.

The main causative agent of marijuana/cannabinoids is the endocannabinoid. This is a neutral lipid and highly conserved molecule throughout evolutionary history [11]. They are having different derivatives like anandamide [12], 2-arachidonoylglycerol [13] and Δ^9-tetrahydrocannabinol (THC) [14]. However, among all of the fatty acid derivatives of cannabinods or endocannabinoids (eCBs) the Δ^9-tetrahydrocannabinol (THC) has now been established as the most important hallucinogenic agent of this molecule [15]. There are literatures suggesting the role of this Δ^9-tetrahydrocannabinol (THC) in regulation of functions of central nervous system and thus regulating the reproductive functions by affecting/ modulating hypothalamo-pituitary-gonadal axis (HPG-axis) [16] *via* its receptor CB1 and CB2 [17]. Now it has been reported that CB1 receptors are localized mostly in whole vertebrate central nervous system (CNS) and some peripheral tissues, whereas CB2 receptors are mostly expressed in peripheral tissues and immune cells, however, they have recently been found also in the CNS [18]. But, with all the advancement in psycho-neuro-endocrine research, till date it is a matter of debate how THC is going to regulate reproductive system at peripheral level. Some literatures suggest that, there is a general agreement on the inhibitory effect exerted by cannabinoids and eCBs on GnRH release [19] Thus, it is affecting the subsequent FSH and LH release in females and impairing female reproduction [20].

But, all the above mentioned reported phenomenon are occurring in the central nervous system and no definitive proof has been reported till date how the enocannabinoids are affecting peripheral reproductive performances in females (in terms of gonadal activity, steroidogenesis, receptor expressions, free radical generations). Thus, aim of the present study was to note the cannabinoid (particularly endocannabinoid) induced oxidative stress and reproductive impairments in female mice specifically taking peripheral reproductive organs (ovary) in consideration.

2. Subjects and methods

2.1 Animals and maintenance

In bred adult (12–15 weeks of age), female Parkes strain mice were used for this study. Mice were maintained under hygienic conditions in a well-ventilated room with 12-h photoperiod (8 AM to 8 PM, light) with 50 ± 20% relative humidity, 25 ± 2°C temperature and were fed pelleted food (Mona Laboratory Animal Feeds, Varanasi, India); drinking water was available *ad libitum*. Five mice in each group were housed in polypropylene cages (430 mm × 270 mm × 300 mm), with dry rice husk as the bedding material. General health condition and body weight of the animals were monitored regularly during the entire tenure of the experiment. All experiments were conducted in accordance with principles and procedures approved by Departmental Research Committee under supervision of Committee for the Purpose of Control and Supervision of Experiments on Animals, (CPCSEA), Govt. of India (2007).

Preparations of different doses of *Cannabis* extracts:

Leaves and flowers of fresh *Cannabis* plant (100 g cannabis plant) were extensively ground in mortar and pestle with 1 ml autoclaved double distilled water. From the 1 g/ml paste, 12 mg was weighed and further dissolved in 1 ml autoclaved double-distilled water to make a stock solution of 12 mg/ml. This solution was filtered to get a clear solution. Finally, the mice were gavaged *Cannabis* by means of a 100 μl micro-pipette using the 12 mg/ml stock.

2.2 Purity assessment of *Cannabis* preparations

The dry-weight ratio of D9-tetrahydrocannabinol (THC) to cannabidiol (CBD) and the percent CBD and THC in the cannabis variant found in this region of the world has been previously reported [21]. The proportion of high THC/CBD chemotype plants in most accessions assigned to *C. sativa* was of 25% (Hillig and Mahlberg) [21].

2.3 Experimental design

Mice were randomly allocated into three groups (groups 1–3). Each group comprised of five female mice (n = 5/group). Group 1 was treated with distilled water (vehicle treated; controls); group 2 was gavaged with 6 mg/100 g body weight/day aqueous *Cannabis* preparation; group 3 was gavaged with 12 mg/100 g bodyweight/day aqueous *Cannabis* preparation. The mode of oral delivery of extracts were following the protocol published previously [21]. The tips used for this purpose to deliver the dose from the micro-pipette had the pointed surface cut to avoid any injury in the mouth of the mouse. The micro-pipette was used to deliver a small volume of (~20 or 40 μl) dose. The study was continued for 30 days.

2.4 Collection of desired tissues

Mice were weighed before the start of experiment as well as before killing. The animals were etherized to death and blood was collected from heart. Subsequently serum was separated and was stored at −20°C until biochemical estimations of total serum cholesterol and estradiol by ELISA. Both the ovaries and uterine horns were excised, blotted free of blood and fat tissues and were weighed. The ovary on one side of the animal was fixed in Bouin's fluid for histology and immunohistochemical localization of CB1 receptor. The contra-lateral ovary of each mouse was stored at −20°C until used for enzyme assays (for steroidogenesis, Caspase-3 and free radical parameters) and western blot analysis of CB1 receptor.

2.5 Antibodies and reagents

All of the chemicals used for the present study were of analytical grade and were purchased either from Sigma Aldrich (St. Louis, MO, USA) or from Merck (Germany). For western blot analysis, polyclonal primary antibody against CB1 receptor was purchased from Affinity BioReagents (Rockford, IL, USA, Cat No. RQ1287) and horseradish peroxidase (HRP)-linked secondary antibody was purchased from Bangalore Genei Pvt. Ltd. (Bangalore, India). For immunohistochemistry (IHC), ABC Kit was purchased from ABC staining kit (Universal Elite, Vector Laboratories, Burlingame, CA). For 3β HSD and 17β HSD assays, pregnenelone was purchased from Sigma Aldrich (St. Louis, MO, USA).

3. Experimental approaches

3.1 Histological preparations

Ovaries were embedded in paraffin wax and serially sectioned of 6 μm using a microtome (Leica, Germany). One set of slide was prepared and was further processed for hematoxylin and eosin staining following the protocol published elsewhere [22]. The permanent slides were prepared by mounting with DPX (Distyrene Plasticizer Xylene, SRL, India), after 24 h were observed under microscope (Leitz MPV3 with photo-automat software) and were documented for general histology.

3.2 Immunohistochemistry of CB1 receptor

Immunohistochemistry for CB1 receptor was performed following the protocol published elsewhere [21]. Ovaries of both treated and untreated adult mice were paraffin embedded, and 6 mm sections were analyzed by immunohistochemistry, for CB1receptor to show where, CB1, receptor is localized in mice ovaries and to have a generalized idea about the receptor expression pattern. For the secondary antibody and enzyme conjugates, ABC staining was used. Briefly after deparaffinization and hydration, and blocking of endogenous peroxidase with 3% H_2O_2 in methanol, sections were incubated with blocking serum for 1 h, followed by incubation with primary antibody (CB1 at a dilution of 1:50) for 1 h at room temperature. The sections were then washed and incubated with the biotinylated secondary antibody for 30 min at room temperature, followed by another 30 min with horse radish avidin-peroxidase conjugated. After washing, sections were incubated with the chromagen substrate (0.1% 3,3- diaminobenzidine tetrahydrochloride, DAB, Sigma-Aldrich, USA) in 0.05 M Tris buffer, pH 7.6, and 0.01% H_2O_2 for 10 min and then counterstained with Elrich's hematoxylin. The permanent slides were prepared by mounting with DPX (Distyrene Plasticizer Xylene, SRL, India), after 24 h were observed under microscope (Leitz MPV3 with photo-automat software) and were documented.

3.3 Estimation of total serum cholesterol

The total serum cholesterol was estimated by commercial cholesterol estimation kit following manufacturer's protocol (Span Diagnostics, Surat, Gujarat, India).

3β hydroxy steroid dehydrogenase enzyme activity:

3β HSD (EC 1.1.1.145) enzyme was assayed according to the protocol of Shivanandappa and Venkatesh [23] using ovarian homogenate. Ten percent tissue homogenate was prepared in 0.1 M Tris-Cl buffer (pH 7.8). The homogenate was centrifuged at 12,000 × g at 4°C and the supernatant was used as the source of enzyme. The enzyme was assayed in 0.1 M Tris-Cl buffer (pH 7.8) containing 500 mM NAD, 100 mM pregnenolone as substrate and enzyme (50 ml) in a total volume of 3.0 ml and incubated at 37°C for 1 h. The reaction was stopped by the addition of 2.0 ml of phthalate buffer (pH 3.0) and the absorbance was noted at 490 nm. The enzyme activity was calculated from the standard curve of NADH and expressed as nmol NADH formed/h/mg protein.

3.4 17β hydroxy steroid dehydorgenase enzyme activity

17β HSD (EC 1.1.1.62) activity was measured by following the protocol of Jarabek et al. [24]. In brief, 10% homogenate of the ovarian tissues were prepared in normal Phosphate Buffered Saline (PBS; pH 7.4) and 250 μl of the supernatant was mixed with 250 μl of 440 μM sodium pyrophosphate buffer (pH 10.2), 10 μl ethanol containing 0.3 μM estradiol (Sigma, St. Louis, USA) and 240 μl of 25 mg% BSA. Enzyme activity was measured after addition of 50 μl of 0.5 μM NAD to the mixture in a spectrophotometer at 340 nm against a blank (without NAD). One unit of enzyme activity was the amount causing a change in absorbance of 0.001/min at 340 nm.

3.5 Evaluation of SOD activity in ovary

Superoxide dismutase (SOD; EC 1.15.1.1) activity was assayed following the method of Das et al. [25]. Just after sacrifice, 10% homogenates of all ovarian tissues from group 1 and set-III mice were prepared in 150 mM phosphate buffered saline (PBS, pH 7.4) and centrifuged for 30 min at 12,000 g at 4°C. The supernatant was again centrifuged for 60 min at 12,000 g at 4°C and then processed for enzymatic activity based on a modified spectrophotometric method using nitrite formation by superoxide radicals. A 0.5 ml of homogenate was added to 1.4 ml of reaction mixture comprised of 50 mM phosphate buffer (pH 7.4), 20 mM L-methionine, 1% (v/v) Triton X- 100, 10 mM hydroxylamine hydrochloride, 50 mM ethylene diamine tetraacetic acid (EDTA) followed by a brief pre-incubation at 37°C for 5 min. Next, 0.8 ml of riboflavin was added to all samples along with a control containing buffer instead of sample and then exposed to two 20 W fluorescent lamps fitted parallel to each other in an aluminum foil coated wooden box. After 10 min of exposure, 1 ml of Greiss reagent was added and absorbance of the color formed was measured at 543 nm. One unit of enzyme activity is defined as the amount of SOD inhibiting 50% of nitrite formation under assay conditions.

3.6 Estimation of catalase activity in ovary

Catalase (CAT; EC 1.11.1.6) activity was measured following the procedure of Sinha [26]. This method is based on the fact that dichromate in acetic acid is reduced to chromic acetate when heated in the presence of H_2O_2 with the formation of perchromic acid as an unstable intermediate. The chromic acetate thus produced is measured calorimetrically. The catalase preparation is allowed to split H_2O_2 for different periods of time. The reaction is stopped at a particular time by the addition of dichromate/acetic acid mixture and the remaining H_2O_2 is determined by measuring chromic acetate calorimetrically after heating the reaction mixture. There is production of green color at the end of the process. Immediately after sacrifice, 20% homogenate of ovarian tissues from groups 1 to 3 were prepared in PBS (10 mM; pH = 7.0) and then centrifuged at 12,000 g for 20 min at 4°C. Supernatant was taken for enzyme estimation. About 5 ml of PBS was added to 4 ml of H_2O_2 (200 mM) and then 1 ml of enzyme extract was added. After 1 min 1 ml of this solution was taken in a tube and 2 ml of $K_2Cr_2O_7$ (5%) solution was added. Then, it was boiled for 10 min and absorbance was measured at 570 nm. The activity of CAT was expressed as amount of H_2O_2 degraded per minute.

3.7 Estimation of lipid peroxidation (LPO) assay by thiobarbituric acid reactive substances (TBARS) level estimation in ovary

After sacrifice of the mice of all the groups, the ovarian tissues were dissected out on a sterile watch glass placed in ice box, cleaned from adherent tissues and processed immediately for estimation of lipid peroxidation. Ovarian tissues of groups 1–3 experimental mice were weighed and homogenized in a tenfold excess of 20 mM Tris-HCl buffer (pH 7.4) and the 10% homogenates were centrifuged for 15 min at 3000 × g at 4°C. The supernatant was subjected to thiobarbituric acid (TBA) assay by mixing with 8.1% sodium dodecyl sulfate (SDS), 20% acetic acid, 0.8% TBA and then digested it for 1 h at 95°C. The reaction mixture was immediately cooled in running water, vigorously shaken with 2.5 ml of n-butanol and pyridine reagent (15:1) and centrifuged for 10 min at 1500 × g (Ohkawa et al.) [27]. The absorbance of the upper phase was measured at 534 nm. Total thiobarbituric acid reactive substances (TBARS) were expressed as malondialdehyde (MDA; nmol/g tissue weight) taking 1,1,1,1-tetraethoxy propane (TEP) as standard. The standard curve was calibrated using 10 nM TEP.

3.8 Glutathione peroxidase (GPx) estimation in ovary

Glutathione peroxidase (GPx; EC 1.11.1.9) activity was assayed as described by Mantha et al. [28]. The reaction mixture (1 ml) contained 50 μl sample, 398 μl of

50 mM phosphate buffer (pH 7.0), 2 μl of 1 mM EDTA, 10 μl of 1 mM sodium azide, 500 μl of 0.5 mM NADPH, 40 μl of 0.2 mM GSH, and 1 U glutathione reductase. The reaction mixture was allowed to equilibrate for 1 min at room temperature. After this, the reaction was initiated by addition of 100 mMH_2O_2. The absorbance measured kinetically at 340 nm for 3 min. The GPx activity was expressed as nmol of oxidized NADPH oxidized to NADP+ per min per mg of protein using an extinction coefficient (6.22 mM^{-1} cm^{-1}) for NADPH.

3.9 Caspase 3 activity assay

Thecal cell suspension was prepared following the protocol of Sharma et al., [29]. In brief, thecal cell suspensions from all the groups were prepared by mincing the entire ovary in ice-cold 1 × PBS, at 4°C. After washing, cell pellets were collected by centrifugation at 500 g for 10 min at 4°C and the supernatant was gently removed. Cell pellets were lysed by the addition of 50 ml of cold lysis buffer (5 mM Tris, 20 mM EDTA, 0.5% Triton-X 100, pH 6.0) per 2 × 6 10^6 cells and incubated on ice for 10 min. Lysates were centrifuged at 10, 000 g for 1 min at 4°C, and the supernatant was transferred to a fresh tube and processed for caspase-3 (EC 3.4.22.xx) activity using a caspase-3 colorimetric assay kit, according to manufacturer's instructions (R&D Systems, Inc. MN). Each enzymatic reaction, carried out in a 96-well flat bottom microplate, required 50 ml cell lysate, 50 ml reaction buffer, and 5 ml caspase-3 colorimetric substrate (DEVD-pNA). The plate was incubated at 37°C for 2 h with a substrate blank and sample blank. At the end of the incubation period, the absorbance of enzymatically released chromophore p-nitroanilide (pNA) was read at 405 nm in a microplate reader (Tecan, Spectra II-micro-ELISA plate reader, Austria). Caspase-3 activity was determined by comparing the absorbance or optical density (OD) of pNA from apoptotic samples with the untreated control and expressed as fold increase in $OD_{405}/10^6$ cells per ml [29].

3.10 Serum level of estradiol

Estradiol was assayed using ELISA kit (Biotron Diagnostics Inc., USA) according to manufacturer's protocol. The coefficient of intra- and inter-assay variation was less than 4.1 and 6.4%, respectively. The analytical sensitivity was 10 pg/ml.

3.11 Western blot analysis of Cannabinoid receptor 1 (CB1) analysis

The ovarian tissue protein pooled from six mice was extracted as described earlier [30]. For western blot analysis, 10% ovarian homogenate was prepared. Equal amounts of proteins (50 mg) determined by Bradford's method were loaded on SDS PAGE (10%) for electrophoresis. Thereafter, proteins were transferred electrophoretically to nitrocellulose membrane (NC; Sigma-Aldrich, USA) overnight at 4°.C NC was then blocked for 60 min with Tris-buffered saline (TBS; Tris 50 mM, pH 7.6) and then incubated with primary antiserum (CB1 at a dilution of 1:250) for 1 h. Then, membranes were washed for 10 min each (three washes) in TBS-Tween 20. Then, NC membrane was incubated with secondary conjugated with serum immunoglobulin (1:500) for 30 min and then washed in TBS for 10 min (three times). Signals were detected using an ECL kit (Bio-Rad, Hercules, CA). Blot for each protein was repeated for three times. The densitometry analysis of blots was performed by scanning and quantifying the bands for density value by using computer-assisted image analysis (Image J 1,38X, NIH). The densitometry data were presented as the mean of the integrated density value ± SEM. A pre-stained multicolor broad range marker (Spectra TM multicolor broad range marker; 10 to 260 kDa x SM-1841; Fermentas, MD, USA) was also run along with sample proteins to clarify the position of band obtained as published elsewhere previously to detect the specificity of the bands [30].

3.12 Statistical analyses

The data were analyzed on Microsoft Office Excel worksheet followed by one way ANOVA. All data are expressed as mean ± SEM. The data were considered significant if $P < 0.05$. Further, to note the level of significance between the experimental groups Duncan's multiple range post-hoc test was applied. All of the estimations were done in single lot using replicates and were repeated thrice. Analyses were done using Statistical Package for Social Sciences software version 16 for windows (SPSS, 16.0, IBM, Chicago, IL, USA) and in accordance to Brunning and Knitz [31].

4. Results

4.1 Histomorphology of ovary

The ovarian sections of both 6 mg/100 g of body weight and 12 mg/100 g of body weight showed degeneration of ovarian micro-architecture in comparison to control. There was absence of corpora-lutea in the ovaries of *Cannabis* treated mice. The ovaries of 12 mg/100 g of body weight showed highest number of degenerating follicles.

4.2 Immunohistochemistry of CB1 receptor in ovary

CB1 receptors protein was demonstrated immunohistochemically in the ovaries of the control and *Cannabis* treated groups of mice. The immunoreactivity of CB1 receptors was mainly observed in the granulose cells of secondary follicles in the control group. There was a dose-dependent increase in the expression of the CB1 receptor in the ovarian sections. Intense staining was also observed in the degenerating follicles and oocyte (group 3). However, negative control did not show any immunostaining.

4.3 Body weight

We noted a significant ($P < 0.05$) decrease in body weight in a dose dependent manner following *Cannabis* treatment in comparison to control. However, the differences of body weight between two experimental groups were not significant ($P > 0.05$).

4.4 Ovarian weight

We recorded the ovarian weight upon *Cannabis* treatment. It was observed that upon *Cannabis* treatment the ovarian weight was significantly low ($P < 0.01$) in dose dependent manner as compared to control. Among two experimental groups, the difference in weight was also statistically significant ($P < 0.05$).

4.5 Uterine weight

We recorded the same result in uterine weight also where *Cannabis* treatment profoundly ($P < 0.01$) decreased uterine weight as compared to control. However, the difference in uterine weight between two experimental groups was not significant ($P > 0.05$).

4.6 Total serum cholesterol

Serum cholesterol also showed significant dose dependent decrease ($P < 0.01$) in serum cholesterol level upon *Cannabis* treatment being lowest in 12 mg/100 g of body weight group as compared to control. However, the difference between two experimental groups was statistically non-significant ($P > 0.05$).

4.7 3β HSD enzyme activity in ovary

Significant decrease in 3β HSD enzyme activity ($P < 0.01$) was noted in a dose dependent manner in *Cannabis* treated ovaries as compared to control. However, the difference between two experimental groups was statistically non-significant ($P > 0.05$).

4.8 17β HSD enzyme activity in ovary

Significant decrease ($P < 0.01$) in 17β HSD enzyme activity was noted in *Cannabis* treated ovaries in comparison to control. The difference in decreased activity between two experimental groups was also statistically significant ($P < 0.05$).

4.9 SOD activity in ovary

Significant increase in SOD activity was noted in *Cannabis* treated groups in dose dependent manner being significantly high ($P < 0.01$) in both the groups of 6 mg/100 g of body weight and 12 mg/100 g of body weight as compared to control. The level was highest in the latter group in comparison to 6 mg/100 g of body weight ($P < 0.05$).

4.10 Catalase activity in ovary

Significant increase in catalase activity was noted in *Cannabis* treated groups in dose-dependent manner ($P < 0.01$) as compared to control. But, among the treated groups the level was not significant ($P > 0.05$).

4.11 Malondialdehyde level in ovary

Significant decrease in ovarian malondialdehyde levels were noted in a dose dependent manner following *Cannabis* treatment being lowest in 12 mg/100 g of body weight dose ($P < 0.01$). The level in the 6 mg/100 g of body weight dose was intermediate with significantly lower level ($P < 0.05$) than control. Among the treated groups, group 3 showed least level of MDA activity ($P < 0.05$).

4.12 GPx level in ovary

Glutathione peroxide (GPx) level was found to be significantly high ($P < 0.01$) in both the treatment groups when compared to control. Among 6 mg/100 g body weight and 12 mg/100 g body weight groups, the latter showed significantly high level ($P < 0.01$).

4.13 Caspase 3 activities in ovarian thecal cells

Caspase 3 activity was assayed in the ovarian thecal cells upon cannabis treatment. We noted a significant increase of caspase 3 in the thecal cells in dose dependent manner being highest in 12 mg/100 g of body weight dose ($P < 0.01$) in comparison to control. Further, among the treated groups, group 3 presented the highest level of caspase 3 activity ($P < 0.01$).

4.14 Serum level of estradiol

Serum level of estradiol was found to be significantly low ($P < 0.05$) in 6 mg/100 g of body weight dose; however, the level was further significantly low ($P < 0.01$) in 12 mg/100 g of body weight dose as compared to control which was recorded to be significantly low among the treated groups ($P < 0.05$).

4.15 Western Blot analysis of CB1 receptor in ovaries of mice

We noted a significant increase ($P < 0.05$) in Cannabinoid receptor type 1 (CB1) in 6 mg/100 g of body weight treatment group. The level was further significantly high ($P < 0.01$) in 12 mg/100 g of body weight group as compared to control group. Further, the level of expression was highest in group 3 ($P < 0.05$) as compared among the treated groups.

5. Discussions

The present study was confined on the role of chronic *Cannabis* induced oxidative stress and reproductive impairment in female mice. In the recent years, there are several literatures available depicting the role of *Cannabis* in neuro-degeneration [32], neuro-myopathy [33] and different other neurological disorders [34]. But, till date there are no data or reports are available depicting the role of *Cannabis* treatment in regulating/modulating the female reproduction, however, it had been predicted from prolonged time that *Cannabis* is potent enough to interfere in reproduction in males [35] and females.

Our study, in relation to the dose dependent effect of *Cannabis* treatment in the female reproduction is the preliminary and elaborated study depicting the deleterious and detrimental effects of *Cannabis* in female reproduction. Our study is divided into two different parts addressing the role of *Cannabis* in reproductive impairments in female mice due to oxidative stress and loss in the functions of steroidogenesis.

We noted a significant decrease in body weight, ovarian and uterine weight upon *Cannabis* treatment suggesting the first clue in reproductive impairment upon *Cannabis* treatment. The results were further supported by degeneration in ovarian histomormhology and increase in expressions of CB1 receptors in ovaries of different treatment groups. Cumulatively, the histological and immunohistochemical data suggest a dose dependent impairment in ovarian as well as reproductive functions which are in agreement with previous reports in where *Cannabis* causes reproductive impairment in males [36]. We have also studied the different aspects of free radical as well as reproductive enzyme activities (3β HSD and 17β HSD). The SOD, catalase and GPx levels were significantly high in ovary tissues where as MDA level was significantly low. The increased results of free radical scavenging enzyme activities suggest the reproductive impairment in mice is may be due to high generation of free radicals and also due to different physiological malfunctions which are yet to be traced out [37–40].

Further, significant decrease in total serum cholesterol levels, estradiol levels in circulation, 3β HSD and 17β HSD enzyme activities in ovarian tissues upon *Cannabis* treatment were noted. Thus, we may suggest that upon *Cannabis* treatment reproduction in females was impaired by *Cannabis* treatment by generation of free radicals in female reproductive tissues. The results were also discussed in light of apoptosis in thecal cells by Caspase 3 activity assays and it was found to be significantly high in different doses of *Cannabis* treatments. To delineate the possible molecular mechanism of *Cannabis* function in ovary, we checked the CB1 receptor expression in ovarian tissues and we also found that the CB1 receptor expressions were significantly high in both the 6 mg/ 100 g of body weight and 12 mg/100 g of body weight groups which are in agreement with the reports published earlier [41–43].

Thus, we may suggest that *Cannabis* treatments were not only impairing the reproduction in females but also chronic duration of doses is responsible for high fecundity in terms of reproductive malfunctions.

6. Conclusion

This study, for the first time showed the effect of administration of *Cannabis*, in controlling the reproductive process in female mice. It also showed the interrelation between the exogenous administrations of *Cannabis*, the possible mechanism that was not dealt by earlier workers showing anti-fertility effect of cannabis for females in particular.

Acknowledgements

Financial support to SG by Council of Scientific and Industrial Research (CSIR), New Delhi, India is gratefully acknowledged.

Author details

Somenath Ghosh[1,2]

1 Pineal Research Laboratory, Department of Zoology, Banaras Hindu University, Varanasi, India

2 Department of Zoology, Rajendra College, Chapra, India

*Address all correspondence to: somenath9936439419@gmail.com

References

[1] Ashton CH, Moore PB, Gallagher P, Young AH. Cannabinoids in bipolar affective disorder: A review and discussion of their therapeutic potential. Psychopharmacology. 2005;**9**:293-300

[2] Aversa A, Rossi F, Francomano D, et al. Early endothelial dysfunction as a marker of vasculogenic erectile dysfunction in young habitual cannabis users. International Journal of Impotence Research. 2008;**20**:566-573

[3] Battista N, Pasquariello N, Tommaso D, Maccarrone M. Interplay between endocannabinoids, steroids and cytokines in the control of human reproduction. Journal of Neuroendocrinology. 2008;**20**:82-89

[4] Bayewitch M, Rhee MH, Avidor-Reiss T, et al. D^9-Tetrahydrocannabinol antagonizes the peripheral cannabinoid receptor-mediated inhibition of adenylyl cyclase. The Journal of Biological Chemistry. 1996;**1271**:9902-9905

[5] Brown TT, Dobs AS. Endocrine effects of marijuana. Journal of Clinical Pharmacology. 2002;**42**:S90-S96

[6] Cacciola G, Chioccarelli T, Mackie K, et al. Expression of type-1 cannabinoid receptor during rat postnatal testicular development: Possible involvement in adult Leydig cell differentiation. Biology of Reproduction. 2008;**79**:758-765

[7] Cobellis G, Cacciola G, Scarpa D, et al. Endocannabinoid system in frog and rodent testis: Type-1 cannabinoid receptor and fatty acid amide hydrolase activity in male germ cells. Biology of Reproduction. 2006;**75**:82-89

[8] Chakravarty I, Shah PG, Sheth AR, Ghosh JJ. Mode of action of delta-9-tetrahydrocannibinol on hypothalamo–pituitary function in adult female rats. Journal of Reproduction and Fertility. 1979;**57**:113-115

[9] Cravatt BF, Giang DK, Mayfield SP, et al. Molecular characterization of an enzyme that degrades neuromodulatory fatty-acid amides. Nature. 1996;**384**: 83-87

[10] Campbell VA. Tetrahydrocannabinol-induced apoptosis of cultured cortical neurones is associated with cytochrome c release and caspase-3 activation. Neuropharmacology. 2001;**40**:702-709

[11] Dalterio S, Bartke A, Burstein S. Cannabinoids inhibit testosterone secretion by mouse testes *in vitro*. Science. 1977;**196**:1472-1473

[12] Singh SK, Chakravarty S. Antispermatogenic and antifertility effects of 20, 25-diazacholesterol dihydrochloride in mice. Reproductive Toxicology. 2003;**17**:37-44

[13] Devane WA, Hanus L, Breuer A. Isolation and structure of a brain constituent that binds to the cannabinoid receptor. Science. 1992;**258**: 1946-1949

[14] Dhawan K, Kumar S, Sharma A. Reversal of cannabinoids (delta9- THC) by the benzoflavone moiety from methanol extract of *Passiflora incarnata* Linneaus in mice: A possible therapy for cannabinoid addiction. The Journal of Pharmacy and Pharmacology. 2002;**54**: 875-881

[15] Dixit VP, Lohiya NK. Effects of Cannabis extract on the response of accessory sex organs of adult male mice to testosterone. Indian Journal of Physiology and Pharmacology. 1975;**19**:98-100

[16] Dixit VP, Gupta CL, Agarwal M. Testicular degeneration and necrosis induced by chronic administration of cannabis extract in dogs. Endokrinology. 1977;**69**:299-305

[17] EL-Gohary M, Eid MA. Effect of cannabinoid ingestion (in the form of bhang) on the immune system of high school and university students. Human & Experimental Toxicology. 2004;**23**: 149-156

[18] Gammon CM, Freeman GM Jr, Xie W, et al. Regulation of gonadotropin-releasing hormone secretion by cannabinoids. Endocrinologie. 2005;**146**:4491-4499

[19] Gaoni Y, Mechoulamm R. Isolation, structure and partial synthesis of an active constituent of hashish. Journal of the American Chemical Society. 1964; **86**:1646-1647

[20] Gye MC, Kang HH, Kang HJ. Expression of cannabinoid receptor 1 in mouse testes. Archives of Andrology. 2005;**51**:247-255

[21] Hillig KW, Mahlberg PG. A chemotaxonomic analysis of cannabinoid variation in *Cannabis* (Cannabaceae). American Journal of Botany. 2004;**91**:966-975

[22] Ghosh S, Haldar C. Histology and immunohistochemical localization of different receptors (MT1, MT2, AR, GR, ERα) in various organs (spleen, thymus, testis, ovary, uterus) of Indian goat *C. hircus*. International Journal of Research Studies in Biosciences. 2015;**7**:1-13

[23] Shivanandappa T, Venkatesh S. A colorimetric assay method for 3beta-hydroxy-delta5-steroid dehydrogenase. Analytical Biochemistry. 1997;**254**:57-61

[24] Blomquist CH, Lindemann NJ, Hakanson EY. 17β-Hydroxysteroid and 20α-hydroxysteroid dehydrogenase activities of human placental microsomes: Kinetic evidence for two enzymes differing in substrate specificity. Archives of Biochemistry and Biophysics. 1985;**239**:206-215

[25] Das K, Samanta L, Chainy GBN. A modified spectrophotometric assay of superoxide dismutase using nitrite formation by superoxide radicals. Indian Journal of Biochemistry & Biophysics. 2000;**37**:201-204

[26] Sinha AK. Colorimetric assay of catalase. Analytical Biochemistry. 1972; **47**:389-394

[27] Ohkawa H, Ohishi N, Yagi K. Reaction of linoleic acid hydroperoxide with thiobarbituric acid. Journal of Lipid Research. 1978;**19**:1053-1057

[28] Mantha SV, Prasad M, Kalra J, Prasad K. Antioxidant enzymes in hypercholesterolemia and effect of vitamin E in rabbits. Atherosclerosis. 1993;**101**:135-144

[29] Sharma S, Haldar C, Chaube SK. Effect of exogenous melatonin on X-ray induced cellular toxicity in lymphatic tissue of Indian tropical male squirrel, *Funambulus pennantI*. International Journal of Radiation Biology. 2008;**84**: 363-374

[30] Ghosh S, Singh AK, Haldar C. Seasonal modulation of immunity by melatonin and gonadal steroids in a short day breeder goat *Capra hircus*. Theriogenology. 2013;**82**:1121-1130

[31] Bruning JL, Knitz BL. Computational Handbook of Statistics. 2nd ed. Illinois: Scott, Foresman and Company; 1977

[32] Hall W, Solowij N. Adverse effects of cannabis. Lancet. 1998;**14**:1611-1616

[33] Harmon J, Aliapoulios MA. Gynecomastia in marijuana users. The New England Journal of Medicine. 1972; **287**:936

[34] Jackson AL, Murphy LL. Role of hypothalamic-pituitary-adrenal axis in the suppression of luteinizing hormone release by Delta-9-THC. Neuroendocrinology. 1977;**65**:446-452

[35] Lewis SE, Maccarrone M. Endocannabinoids, sperm biology and human fertility. Pharmacological Research. 2009;**60**:126-131

[36] Maccarrone M, Finazzi-Agro A. The endocannabinoid system, anandamide and the regulation of mammalian cell apoptosis. Cell Death and Differentiation. 2003;**10**:946-955

[37] Matsuda LA, Lolait SJ, Brownstein MJ, et al. Structure of a cannabinoid receptor and functional expression of the cloned cDNA. Nature. 1990;**346**:561-564

[38] Meccariello R, Franzoni MF, Chianese R, et al. Interplay between the endocannabinoid system and GnRH-I in the forebrain of the anuran amphibian *Rana esculenta*. Endocrinologie. 2008; **149**:2149-2158

[39] Moore TH, Zammit S, Lingford-Hughes A, et al. Cannabis use and risk of psychotic or affective mental health outcomes: A systematic review. Lancet. 2007;**370**:319-328

[40] Munro S, Thomas KL, Abu-Shaar M. Molecular characterization of a peripheral receptor for cannabinoids. Nature. 1993;**365**:61-65

[41] Murphy LL, Munoz RM, Adrian BA, Villanua M. Function of cannabinoid receptors in the neuroendocrine regulation of hormone secretion. Neurobiology of Disease. 1998;**5**:432-446

[42] Pertwee RG. The therapeutic potential of drugs that target cannabinoid receptors or modulate the tissue levels or actions of endocannabinoids. The AAPS Journal. 2005;7:554-625

[43] Pertwee RG, Ross RA, Craib SJ, Thomas A. Cannabidiol antagonizes cannabinoid receptor agonists and noradrenaline in the mouse vas deferens. European Journal of Pharmacology. 2002;**456**:99-106

Chapter 18

Antioxidant Activity of Areca Nut to Human Health: Effect on Oral Cancer Cell Lines and Immunomodulatory Activity

Liza Meutia Sari

Abstract

Many herbs have been discovered to be potential sources of the antitumor and immunomodulatory drug. Areca nut (*Areca catechu* L.) has a high content of phenols and flavonoids and is highly related to antioxidant activity. Areca nut is a traditional herbal medicine that is popular around Indonesia, India, Thailand, and Taiwan. However, data on its effect on human health showed various results. This chapter's aim to review the phytochemical and polyphenolic content, the molecular structure of bioactive compounds, the side effect of the crude extract, the role of catechin in cancer mechanism, the antioxidant activity, the cytotoxicity, and immunomodulatory activity of the areca nut. Areca nuts from Aceh province in Indonesia, contain flavonoids, phenolics, catechin, quercetin, and a small percentage of tannins which contribute to antioxidant activity. The areca nut has anticancer potential activity so it can be used in combination with chemotherapeutic agents to enhance the effect at lower doses and thus minimize chemotherapy-induced toxicity. Areca nuts also show immunomodulatory activity which can increase the body's immune system.

Keywords: Areca nut, phytochemical content, anticancer, immunomodulatory

1. Introduction

One of the plants with the potential to be developed as herbal medicine is the *pinang* plant (*Areca catechu* Linn; *areca*, *Palmaceae*). The term *Areca catechu* refers to the largest amount of the phytochemical compounds in areca nut that is arecoline and catechin. Areca nut is a traditional herbal medicine that is popular around Indonesia, India, Thailand, and Taiwan. This plant spreads widely in South East Asia, South Africa, and the Pacific Ocean Islands. It is usually used in betel chewing common among the Indians and Malays as a breath freshener, digestive aid, anthelmintic, aphrodisiac, and to maintain stamina [1]. Several kinds of research studied the functions of the areca plant showed that the stem can be used as antibacterial and antifungal [2]. Epidemiologic study and laboratory research found a large number of anticancer compounds obtained from the nuts. Some of them contained flavonoid, polyphenol, carotenoid, selenium, vitamin C, and E which showed chemotherapy and preventive activity. An antioxidant is one of the solutions to minimize oxidative stress levels caused by cancer treatment. The antioxidant is a compound that is capable to cleanse, dispose, and repel free radical or reactive oxygen species forming in the body [3]. The previous study showed that methanol and water extract of areca nut contains tannin and high total phenolic [4–6]. This compound has stronger free radical scavenging activity than ascorbic acid [7].

2. Phytochemical content of areca nut

Determination of areca plant as follows:

Kingdom	: *Plantae*
Division	: *Magnoliophyta*
Subdivision	: *Angiospermae*
Class	: *Liliopsida*
Nation	: *Arecales*
Family	: *Arecacea/palmae*
Subfamily	: *Arecoideae*
Genus	: *Areca*
Species	: *Areca catechu* Linn.

Characteristics of areca nut	
Stems	Slender, grow upright, reach 10 to 30 m high, 15 to 20 cm in diameter, and unbranched [8]. The new stem formation occurs after two years (**Figure 1**).
Leaves	The pinnate compound leaves grow together at the tip of the stem, almost like coconut trees. Leaf midrib tubular has 80 cm long and short petiole. It has 1–1.8 cm long, 5 cm wide, with torn and jagged tips.
Flowers	Flower cobs with long, easy-to-fall spathes appear under the leaves. They appear at the beginning and the end of the rainy season.
Nuts	• Oval elongated has 3.5 to 7 cm long, fibrous mesocarp and a thin woody endocarp enveloping one nut. • The color of the ripe nut is red-orange (**Figure 2**). • Sweet, fresh, and it gives a sense of addiction. • The nuts appear at the age of 5 to 8 years depending on the soil condition.
Roots	Fiber roots and very similar to the roots of the coconut plants.
Life span	25 to 30 years.
Grow locations	Areca plants can produce optimally when planted in locations with an altitude of 0–1,400 masl. The required rainfall to grow areca nut optimally is between 2,000–3,000 mm/year which is evenly distributed throughout the year or rainy days around 100–150 days. The desired temperature is 20 °C–32 °C, and the humidity is between 50–90%. The soil acidity which is good for plant growth around pH 4–8 [9].
Constituents	• **Polyphenols**: **Phenolics, flavonoids**, syringic acid, procyanidin (dimeric, trimeric, and tetrameric), tannins, isorhamnetin 3-O-rutinoside, **catechin**, jacareubin, ent-catechin, epicatechin, **quercetin**, luteolin, and chrysoeriol [4, 5]. • **Alkaloids**: Arecoline, arecaidine, guvacoline, and guvacine. Arecoline is colorless volatile resembling nicotine. • **Fatty acids**: Myristic acid, palmitic acid, oleic acid, linoleic acid, eicosanoic acid, arachidonic acid, docosanoic acid, tetracosanoic acid, hexacosanoic • Carbohydrate (19.13%), protein (10.22%), fats (12.84%), crude fibers (14.40%),

Characteristics of areca nut	
	• Minerals: Calcium, phosphorus, iron, vitamin B6, and C.
Actions	• Antioxidant, anti-aging, antihelminthic, antimicrobial, analgesic, and anti-inflammatory [2, 10–13]. • It can reduce the risk of dental caries. • Apoptosis and cell cycle arrest in carcinoma cell lines [14]. • Anti-inflammatory/Anti-melanogenesis • Hypolipidemic • Hypoglycemic activity • Antidepressant • A-glucosidase inhibitory • Antihypertensive • Immunomodulatory activity • Vascular relaxation • Anticonvulsant activity • Anti-allergic
Side effects	• Chewing habits can induce the risk of developing oral squamous cell carcinoma. • The arecolines can enhance Alzheimer's type of dementia. • Dental attrition, oral leukoplakia, areca staining, lichenoid lesions.

Areca nut contains several alkaloids that belong to the pyridine group, especially arecoline. Arecoline affects the oculomotor nerve which can cause mydriasis, mild

Figure 1.
Areca plants [8].

Figure 2.
The ripe areca nuts.

paralysis, and pupillary dilatation [15]. Several studies showed that arecoline can induce neurotoxicity through its action in oxidative stress and generating Reactive Oxygen Species (ROS), hepatotoxicity, testicular toxicity, oropharyngeal cancers, and oral submucous fibrosis [16, 17]. Besides alkaloids, ent-catechin, and jacareubin are the major compound in areca nut [7]. Catechin and its analogs are antioxidant, anti-allergic, anticancer, anti mutation, and anti-inflammatory. It can improve liver function and scavenge free radicals [18]. The beneficial effect of catechins is reported in the treatment of cancer, cardiovascular diseases, diabetes, neurodegenerative diseases, and liver diseases [19]. Apparently, catechins have function not only as a powerful antioxidant [20], preventing oxidative damage in healthy cells [21], but also as an antiangiogenic, antitumor agent [22], and a modulator of tumor cell response to chemotherapy [23]. It can induce apoptosis by increasing caspases [24] and promotes cell growth arrest by altering the expression of cell cycle regulatory proteins [25]. Jacareubin is reported to have an anti-inflammatory property and increases the activity of H^+, K^+-ATPase which is needed to gain the energy of the cells [26, 27].

The majority of phytochemical ingredients in areca nut extract are phenolic compounds such as flavonoids, tannins, and alkaloids [28]. It was established that a great quantity of tannins has been found in dark dry fruits such as tea. The areca nut also has a reddish-brown color. The levels of tannin found in areca nut have different contents in different areas or regions [29]. It could be caused by climate conditions and environmental factors where it grows. Climatic factors such as temperature, weather, and rainfall. The environmental factors such as soil type and fertility, the height of growing, and plant maintenance.

The presence of catechin and quercetin mostly were identified through High-Performance Liquid Chromatography (HPLC) analysis. These two compounds are well-known antioxidants and could have contributed to the observed antioxidant activity [30]. Previous studies have identified several phenolic compounds in the areca nuts including trimer procyanidin, dimer procyanidin (B1 dan B2), catechin, and isorhamnetin 3-O-rutinoside [31]. Many of these compounds have antioxidants activities [20, 32]. Catechin was proved to have strong antioxidant activity which could contribute towards the anti-cancer effects [33–36]. It has also been shown that polymerized catechin suppresses the activity of *Staphylococcus aureus* α-toxin and as an effective urease inhibitor in *Staphylococcus saprophyticus* strains. Although the level of quercetin is not much in areca nut extract, it also has been proved as an antioxidant. It possesses an anti-inflammatory potential that can be expressed in different cell types, both in animal and human models [37]. Quercetin is also able to inhibit the growth of cancer cells through induction of apoptosis and inhibitory proliferation in gastrointestinal, breast, esophagus, and ovary cancer [38].

3. Quantitative data of polyphenolic content in areca nut

Various studies have been conducted to calculate the content of polyphenolic areca nut from various regions around the world. The results also showed various results. This difference shows that the variation in polyphenolic content depends on the geographical conditions in which the plant grows so that it affects the quality and quantity of the phytochemical composition in areca nut. This variation is also influenced by the species and type of the areca nuts used such as freshness, maturity, and form of the nut drying process. One of the polyphenolic content derived from areca nut on the island of Sumatra in Indonesia is shown in **Table 1**. The total phenolic and flavonoid contents of areca nut are expressed as Tannin Acid Equivalents (TAE) and Catechin Equivalents (CE).

Methanolic extract	Compounds	Estimated amount
Ripe areca nut	Total phenolics	80.3 mg TAE/gr[a]
	Total flavonoids	238.5 mg CE/gr[b]
	Catechin	2.79 mg/gr[c]
	Quercetin	0.14 mg/gr[c]
	Tannin	0.007%[d]
Unripe areca nut	Total phenolics	56.6 mg TAE/gr[a]
	Total flavonoids	18.13 mg CE/gr[b]
	Catechin	—
	Quercetin	—
	Tannin	8.7%[d]

[a]*Folin–Ciocalteu method.*
[b]*Dowd method.*
[c]*HPLC.*
[d]*Titrimetri assay.*

Table 1.
Quantitative data of polyphenolic content in areca nut from Aceh, Indonesia.

4. Molecular structure of bioactive compounds in areca nut

4.1 Phenolic compounds

Phenolic compounds are the largest part of the phytochemicals in plants, especially fruit, seeds, and leaves. This collection of compounds provides many health benefits because they contain many antioxidants that can scavenge oxidative stress due to reactive oxygen species. Intake of fruit, vegetables, and whole grains that are rich in phenolics can lower the risk of cardiovascular disease, chronic inflammation, cancer, and neurodegenerative diseases. Polyphenolics contain an aromatic hydroxyl ring derived from L-phenylalanine. Phenolic compounds comprise one (phenolic acids) or more (polyphenols) aromatic rings with attached hydroxyl groups in their structures. Several types of polyphenols that are classified as phenolic acids are hydrolyzed tannins, lignans, stilbene, and flavonoids. The intact polyphenolic form is usually absorbed directly from the digestive tract.

Microorganisms can cause the degradation of polyphenolics into aglycones and aromatic acids. Polyphenolics are detected in all tissues, especially the digestive tract and oral mucosa. All types of polyphenolics are excreted in the urine and bile ducts. The bioavailability of polyphenolics is low, and the remaining at urinary excretion is 0.3% for anthocyanins and 43% for isoflavones. The absorption of phenolic compounds in food is determined by the chemical structure which influences the degree of glycosylation and acylation, basic structure, conjugation with other phenolic compounds, molecular size, polymerization, and solubility. The maximum concentration in plasma rarely exceeds 1 mM after ingestion of 10–100 mg of a single phenolic compound. Polyphenolics that are mostly absorbed by the body are isoflavones, gallic acid, flavanones, catechins, and quercetin glycosides, while proanthocyanidins, anthocyanins, and gallolylated catechins are less widely absorbed.

4.2 Flavonoid

The total flavonoids in ripe areca nuts were found to be the highest in levels. Flavonoids belong to a group of natural substances with variable phenolic structures and are found in vegetables, tea, flowers, fruit, grains, vegetables, grains, and wine. This natural ingredient is known to have many health benefits. Flavonoids are a group of more than 4000 polyphenolic compounds that occur naturally in foods of plant origin. These compounds possess a common phenylbenzopyrone structure (C6-C3-C6), and they are categorized according to the saturation level and opening of the central pyran ring, mainly into flavones, flavanols, isoflavones, flavonols, flavanones, and flavanonols. Flavonoids are polyphenolic compounds that consist of several types based on their structure, namely flavonols (quercetin and kaempferol), flavones (luteolin and wogonin), flavanols (catechin, gallocatechin), isoflavones (genistein), flavanones, and flavanonols.

Flavonoids have an antibacterial, anti-inflammatory, antioxidant, allergy, antimutagenic, antiviral, antineoplastic, anti-thrombotic, vasodilatory, and anti-hepatotoxic activities. Flavonoids can cause cell cycle arrest, thereby inhibiting the activity of Cyclin-Dependent Kinases (Cdks), Phosphorylation Kinases (PKs), and signal transduction of cell proliferation. Flavonoids are also able to modulate Mitogen-Activated PKs (MAPKs). The Cdk enzyme is an enzyme that functions to control the activity of the phosphorylase enzyme which regulates every phase of the cell cycle, especially during DNA replication and chromosome formation. The anti-tumor properties of flavonoids can inhibit the release of prooxidant enzymes, modulate carcinogen metabolism, suppress protein tyrosine kinase activity, anti-proliferation, anti-metastasis, inhibit some drug resistance, are antioxidant and anti-angiogenesis, induce apoptosis and cell cycle retention. The chemical structures of flavonoids in areca nut can be seen in **Figure 3**.

4.3 Catechin

Catechins are flavonoids (flavanols) that are included in the polyphenol group, which have high concentrations in vegetables, fruits, and beverages. Catechins contain two aromatic rings and several hydroxyl groups. Catechins are divided into two groups, namely free catechins and esterified catechins. Catechin, gallocatechin, epicatechin, epigallocatechin are non-esterified catechins, whereas epigallocatechin gallate, epicatechin gallate, gallocatechin gallate are classified as esterified. The chemical structure of catechins can be seen in **Figure 4**.

Several *in vitro* studies have proven the role of catechins as protection for several diseases such as degenerative, heart disease, and bacterial infection. The catechins in green tea are able to inhibit carcinogenesis of the skin, lungs, esophagus, stomach, liver, small intestine, colon, and mammary glands in animal experiments.

5. The side effect of crude extract of areca nut

Among the benefits of consuming areca nut, other studies have shown that these nuts are carcinogens so that they can cause oral malignancy lesions. Arecoline is the most common alkaloid found in areca nut which is known to cause cytotoxicity in mammalian cells *in vivo* and cause carcinogenicity. Some of the oral lesions that can be caused by crude areca nuts include:

5.1 Oral submucous fibrosis

Oral submucous fibrosis (OSF) is a chronic disease that produces hyperkeratosis, epithelial atrophy, tissue fibrosis, and precancerous lesions. Pathological characteristics include chronic inflammation, excessive collagen deposition in the connective tissues below the oral mucosal epithelium, local inflammation in the lamina propria

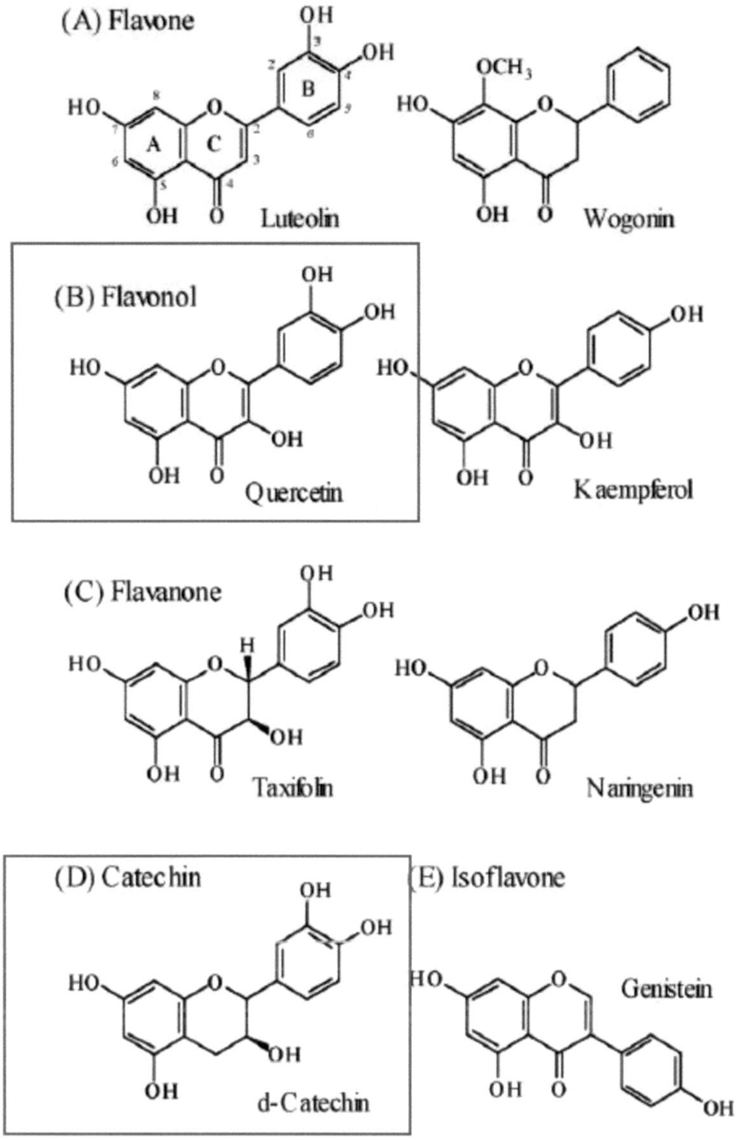

Figure 3.
Chemical structures of the flavonoid family. The areca nut contains flavonol (quercetin) and flavanol (catechin).

or deep connective tissues, and degenerative changes in the muscles. OSF patients experience a severe burning sensation in the mouth after ingesting spicy foods. Other symptoms of OSF include dry mouth, pain, taste disorders, restricted tongue mobility, trismus, dysphagia, and altered tone. This disease contributes significantly to mortality because of its high malignant transformation rate (1.5–15%). Previous research has shown that the arecoline in areca nuts can induce COX-2 expression in humans. It also suggests contributing to pathogenesis of OSF in betel chewers. Consistently, the elevated expression level of COX-2 has also observed in arecoline-treated HGF-1 cells and primarily cultured HGF cells in the study, suggesting these cells of different origins derived from oral cavity might have similar inflammatory responses upon exposure to arecoline, which in turn promote oral lesions and tumorigenesis.

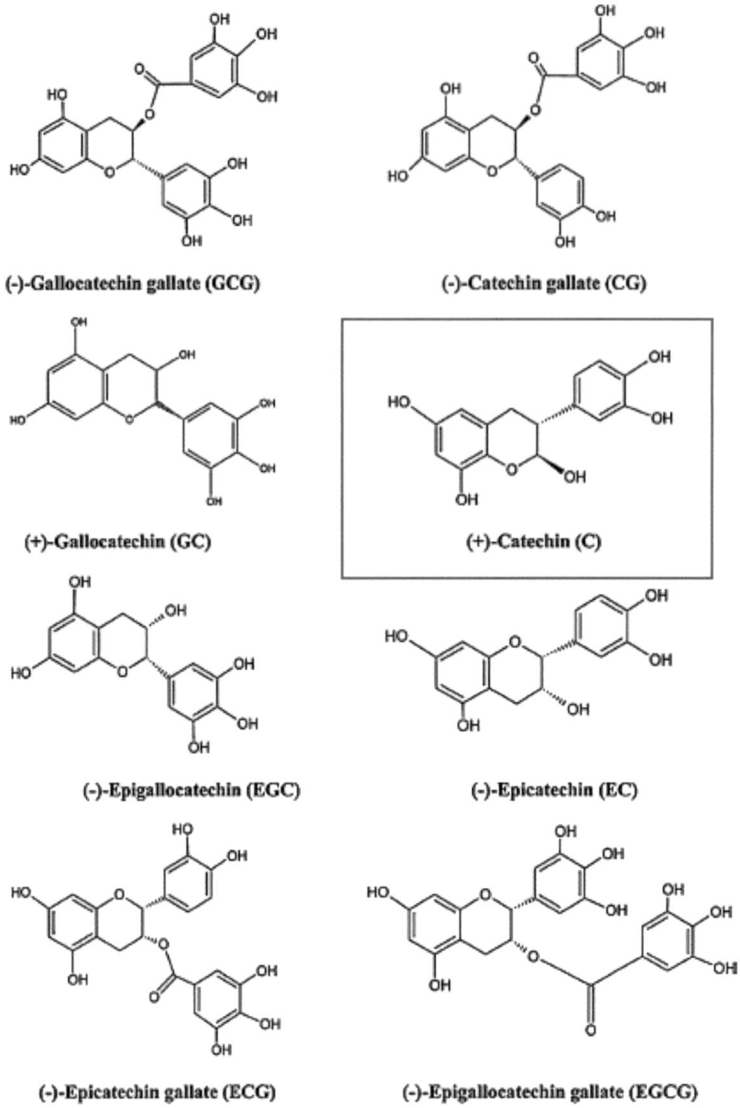

Figure 4.
Chemical structures of catechin. The areca nut contains (+)-Catechin.

5.2 Oral squamous cell carcinoma (OSCC)

Betel (areca nut) quid chewing is a widely prevalent habit correlated with a high incidence of oral cancer. A Previous study showed that the carcinogenic mechanism of action of betel quid chewing is caused by DNA damage induced by arecaidine and Cu (II). The Cu (II) is an important structural metal ion in chromatin; however, it is reported to be an important factor in DNA damage induced by many organic compounds. It was found that arecaidine alone had no significant effect on inducing DNA damage, but it caused significant DNA double strand breaks in the presence of Cu (II) ions under alkaline conditions. Further studies showed that reactive oxygen species were generated and Cu (II) was formed in the reaction. Areca nuts contain arecoline and arecaidine, both of which can cause mutation. When chewing areca nut, arecoline dissolves in saliva and its concentration can reach 140 g/mL. arecoline can cause cytotoxicity and genotoxicity for multicultured human cell lines and inhibits oral mucosal and gingival fibroblast which can cause epithelial damage and delayed wound healing.

The areca nut does not contain carcinogenic substances, but this carcinogenic effect is caused by nitrosamines product in the long term and uncontrollable. Nitrosamines are produced by nitrosation of the alkaloids in dried stored nuts, when in the mouth, and especially in the acid conditions of the stomach, in the presence of nitric oxide generated by bacteria [39]. The combination with nitric oxide, produced by bacteria, causes the production of methylnitrosaminoproprionitrile which is proved to cause carcinogenesis in animal studies [39]. This endogen nitrosation is higher significantly in a patient with bad oral hygiene [40]. If areca nut is combined with tobacco, then chewed by people with bad oral hygiene, there will be a very high accumulation of nitrosamines product [41]. This process usually occurs continuously for the long term because the seeds have addictive properties. Some studies also reported increased reactive oxygen species such as hydroxyl oxide in the oral cavity caused by a combination of polyphenol auto-oxidation from areca nut with high alkaline pH of slaked lime (Calcium hydroxide paste) [42]. If the areca nut is chewed with *Piper betel* and slaked lime, these two materials will cause mucous membrane erosion, so that carcinogenic substance could easily penetrate cells through the mucous membrane [43]. Some part of the community in India and Pakistan use industrial packaging areca nut called *gutka* and *pan masala*. *Gutka* contains areca nut, *Piper betel*, tobacco, and slaked lime, while *pan masala* was prepared without tobacco. Approximately 40% of *gutka* and *pan masala* packaging are contaminated with aflatoxin which has carcinogenic properties from *Aspergillus flavus*, *Aspergillus niger,* and *Rhizopus spp.* fungi [44]. The occurrence of the OSCC's risk depends greatly on the composition of the compound which determines the quality of the seeds, the method of using seeds which are associated with oral hygiene, duration of use, the presence or absence of toxin caused by fungi contamination in the seeds, and the presence or absence of other carcinogenic substances such as tobacco and slaked lime. The occurrence of OSCC could also be influenced by several factors such as intrinsic factors (tumor suppressor gene abnormality or mutation and oncogene) and extrinsic factors (tobacco smoking, vitamin A and iron deficiency, candidiasis, viral infection, and immunosuppression).

The dried areca nut, gambir, and calcium hydroxide wrapped in betel leaf are usually used as symbol of respect for guests in traditional ceremonies in Aceh, Indonesia.

6. Role of catechin in cancer mechanisms

The term catechins are commonly used to refer to the family of flavonoids and the subgroup flavan-3-ols or simply, flavanol. Catechins are differentiated from the ketone-containing flavonoids such as quercetin and rutin, which are called flavonols. High concentrations of catechin can be found in fresh tea leaves, red wine, broad beans, black grapes, apricots, and strawberries [45].

Role of catechin in cancer mechanisms	
Initiation Stage	• Catechins neutralize the procarcinogens by inhibiting the activity of cytochrome P450 enzyme and modulating free radicals [46]. • Catechins inhibit the activity of Nicotinamide adenine dinucleotide phosphate (NADPH)-cytochrome c reductase. • Epigallocatechin gallate (EGCG) could interact with hepatic cytochrome P450 and inhibit the P450-dependent mixed-function oxidase enzymes in the skin and liver [47]. • The epicatechin derivative structure is capable of inhibiting microsomal enzyme system derived from the catechin hydroxyl group. The pyrogallol structure causes catechin molecules to have a strong metal-chelating ability that can bind to metal transition ions and act as preventive antioxidants [48]. • The EGCG reduces cell proliferation and induces apoptosis in low-dose H_2O_2 (10 M)-treated colon carcinoma cells and downregulates 12-O-tetradecanoylphorbol-13-acetate-mediated oxidative stress in cervical carcinoma cells [49]. • The EGCG can suppress the growth of HepG2 human HCC cells. The oral administration of green tea for 1 year can inhibit the progression of high-grade prostate intraepithelial neoplasia to prostate cancer [50].
Promotion Stage	• The inhibitory mechanism of the promotion stage is divided into three processes, namely the intervention of intracellular signaling pathways, increases the caspase activity, and cell cycle modulation. • Catechins inhibit phosphorylation of extracellular signal-regulated protein kinases (ERK)-1 and 2 and suppress the activity of p38 MAPKs in human fibrosarcoma cells [51]. The ERK enzymes are important transducers of proliferation signals. • Catechin hydrate exhibits anticancer effects by blocking the proliferation of MCF7 cells and inducing apoptosis in part by modulating expression levels of caspase-3, −8, and − 9 and p53 [52]. • (−)-Epigallocatechin-3-gallate has potential as a novel therapeutic agent for patients with B-cell malignancies including multiple myeloma via induction of apoptosis mediated by modification of the redox system. • Catechins can also inhibit the cell cycle. The cell cycle is controlled by numerous mechanisms ensuring correct cell division [53]. The mechanisms are regulation of cyclin-dependent kinases (CDK) by cyclins, CDK inhibitors, and phosphorylating events [53]. • **Cell-cycle dysregulation** is a hallmark of tumor cells. The ability of normal cells to undergo cell-cycle arrest after damage to DNA is crucial for the maintenance of genomic integrity [54]. • The biochemical pathways that stop the cell cycle in response to cellular stressors are called **checkpoints.** • Defective checkpoint function results in genetic modifications that contribute to carcinogenesis. The regulation of checkpoint signaling also has important clinical implications because the abrogation of checkpoint function can alter the sensitivity of tumor cells to chemotherapeutics.
Progression Stage	• Metastasis requires down-regulation of cell adhesion receptors necessary for tissue-specific, cell–cell attachment, as well as up-regulation of receptors that enhance cell motility. • Inhibition of migration and invasion of tumor cells could be a target of anticancer therapy. • Catechins can inhibit MMP-2 and MMP-9 in endothelial cells [55]. MMP-2 and MMP-9 secretions are elevated in several types of human cancers and their elevated expression has been associated with poor prognosis. • **Angiogenesis**, the development of new capillaries from preexisting blood vessels, is required in physiological processes such as wound healing and pathological conditions including tumor growth and metastases.

Role of catechin in cancer mechanisms
• **Tumor angiogenesis** is a complex process that consists of several steps including the secretion of angiogenic factors by tumor and host cells, activation of proteolytic enzymes, endothelial cell migration, invasion, endothelial cell proliferation, and capillary formation. • Vascular endothelial growth factor (VEGF) and its receptors have been known as important angiogenic factors and are commonly overexpressed in several types of human cancers. • Catechins especially EGCG is proved to inhibit tumor growth and angiogenesis by the down-regulation of VEGF expression in serum-deprived HT29 human colon cancer cells [56]. • Catechin in green tea extract inhibits angiogenesis partly through the disruption of STAT3-mediated transcription of genes, including VEGF [57]. • Several members of the signal transducers and activators of transcription (STAT) family play a role in tumorigenesis. The STAT3 activity is commonly upregulated in breast cancer and regulates the expression of angiogenic genes including VEGF and MMP9 [57].

7. Antioxidant activity of areca nut

Antioxidants protect cells from deleterious effects of oxidation and are also employed as dietary supplements to neutralize the adverse effects of oxidative stress. Many of the natural antioxidants of interest are of plant origin and belong to bioactive compounds in the phenolic and polyphenolic class. The phytochemical contents of areca nut mostly come from phenolic and flavonoid content which produce antioxidant activity. Total phenolic content test using the Folin–Ciocalteu method was performed based on oxidation–reduction mechanism. Methanol is the best solvent for areca nut compared to petroleum ether, ethyl acetate, and water, and was used in this study [58]. The value of total phenolic concentration in 1 gram extract is 80.3 mg TAE/gr extract. The highest phenolic concentration was usually found in methanol, acetonide, and water solvent. This concentration is depended on how big the solvent's polarity is used at extraction. The high phenolic solubility in the polar solvent showed a high concentration of phenolic content in the extract. When compared with literature using the same method, the phenolic content of areca nut in the study is lower than that of a study conducted in Assam, India (146.7 mgTAE/g extracts) and Hainan province, Taiwan (167.70 mgTAE/g extracts) [5, 58]. Total flavonoid test using Dowd modification method also revealed a higher content of flavonoid (238.5 mg CEmg/gr extract). This number is higher than that reported by Zhang et al. and Wang et al., which was 77.36 and 10.45–142.65 mg CE mg/gr extract, respectively. This indicates that the variation of polyphenol content depends on the geographical locations where the plant grows. This variation is also affected by species and the characteristics of the nut used in the study, including freshness, maturity, and methods of drying.

The flavonoid in areca nut extract has antioxidant and also pro-oxidant activity. These two activities were also possessed by other herbal plants, *curcumin* [59, 60]. Lower doses of *curcumin* (12.5 μM) has the properties of reactive oxygen species scavenging, anti-inflammatory, apoptosis induction, and proliferation inhibition in myeloid leukemic cells, but in a higher dose and long term, the metabolite contents of *curcumin,* which is lipophilic or water-insoluble, could increase the level of cellular reactive oxygen species that causes carcinogenic potential through oxidative DNA damage or metal-mediated *DNA* damage at P450 cytochrome [61, 62]. This damage occurred because of the presence of Cu(II)-CYP2D6 which caused the damage of the DNA, especially 5-TG-3, 5-GC-3, and GG sequences [61]. *Curcumin* could induce lung cancer by increasing reactive oxygen species resulting in disarray between mitogen-activated protein kinase, NF-jb, and p53, causing genetic mutation and oxidative stress [63]. This finding concluded that the antioxidant effect which was started by an oxidative stimulus, depending on time, dose, and certain cancer type, could also cause unwanted side effects [63].

Catechins are phytochemical compounds found in high concentrations in a variety of plant-based foods and beverages. Catechins are classified as flavanols. Catechin is the highest phenolic compound in the areca nut extract (2.79 mg/g). In comparison with catechin, quercetin is found in much smaller amounts in the areca nut extract (0.14 mg/g). The areca nut extract from Aceh, Indonesia, has a low level of tannin (0.007%) through the titrimetric analysis. This finding showed different results with less amount than areca nut from various districts in Karnataka, India, showed 1.13%–3.39% tannin in areca nut extract with the same technique [29]. Tannin is a polyphenol compound of plant origin, bitter in taste, which reacts with and coagulates protein, or various other compounds including amino acids and alkaloids.

The main characteristic of an antioxidant is its ability to trap free radicals. The antioxidant activity of an extract can be measured by 2,2-diphenyl-1-picryl-hydrazyl-hydrate (DPPH) free radical method. DPPH is a stable nitrogen-centered free radical, the color of which changes from violet to yellow upon reduction by either the process of hydrogen- or electron- donation. The substance which is performs this reaction can be considered as antioxidants and therefore radical scavengers [64]. The DPPH test is a direct and reliable method for determining radical scavenging action. The reducing properties are generally associated with the presence of reductones, which have been shown to exert antioxidant action by breaking the free radical chain by donating a hydrogen atom. The result of the antioxidant activity curve showed linear line formula so the EC_{50} value was acquired. The EC_{50} value is the extract concentration which was able to catch 50% free radical. The EC_{50} value was measured from the association curve between the percentages of radical catcher DPPH against the concentration of the treatment's solution. This value is inversely proportional to antioxidant extract capability. The higher the antioxidant activity, the lower EC_{50} would be. The study showed that the EC_{50} value of areca nut extract was 15.95 μg/mL (**Figure 5**). The polyphenol could

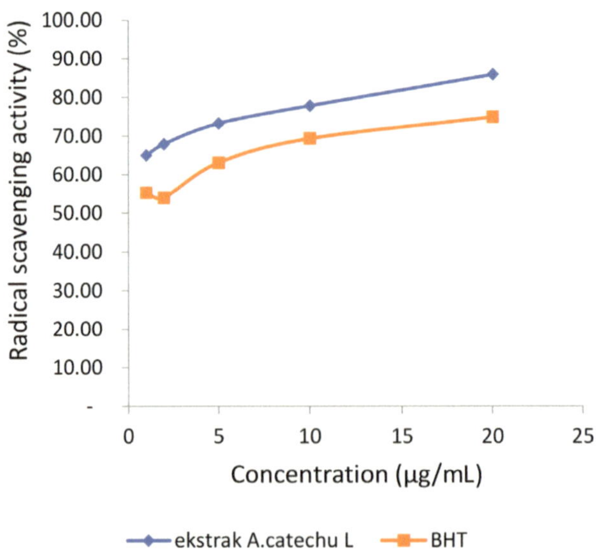

Figure 5.
Antioxidant activity estimated by DPPH method. The linear curve of areca nut extract has similar linearity with BHT control. The dosage used was 1–20 μg/mL.

dispose of free radicals by becoming a hydrogen donor so the free radical chain reaction was broken. The EC_{50} value of the extract was smaller than Butylated Hydroxytoluene (BHT); synthetic antioxidant posing as control. The result of the study showed stronger antioxidant activity of areca nut compared to BHT control, so it could be concluded that the activity potency of the combination of several phenolic compounds in the extract could work synergic and resulted in more potent antioxidant than the activity of one isolate in the extract [38]. Polyphenolics had a stimulation effect on mitochondria's activity, so it could be more efficient in creating energy and preventing free radicals. It can increase the cell viability graph over 100% in human keratinocyte (HaCat) cell line [28].

8. The cytotoxicity and immunomodulatory activity of areca nut to the human health

8.1 Cytotoxicity activity of areca nut

The majority of the community prefer to choose herbal medicine because the natural ingredient is considered to be safer and cheaper than chemical drugs. The precipitating factors of the increasing use of herbal medicine in developed countries are the side effects of chemical drugs, the high cost of modern medicine, and the increasing life expectancy when the prevalence of chronic disease had increased, so herbal medicine becomes an alternative treatment that is believed to cover all classes of the community especially in Indonesia [65]. In Taiwan and South-Eastern Asia, areca nut chewing has been associated with the development of oral squamous cell carcinoma (OSCC) through epidemiological studies and has been classified as a human carcinogen by the IARC (2004) [6].

MTS assay was done to observe the cytotoxicity effect of areca nut extract in HSC-2, HSC-3, and HaCat cells. Areca nut has a stimulation effect on mitochondria's activity, so it could be more efficient in creating energy and preventing free radicals. It can increase the cell viability graph over 100% in HaCat cell line by MTS assay. The areca nut extract in a certain dose could increase respiration and metabolism in the HaCat cell line [28]. Cytotoxicity of the extract was displayed with the viability percentage. The result showed that IC_{50} of areca nut extract in HSC-2 cells was 629.50 μg/mL and the IC_{50} value in HSC-3 occurred in lower concentration which was 164.06 μg/mL. The cytotoxicity effect started from the smallest concentration which was between 160–640 μg/mL, but at a concentration higher than 1280 μg/mL, the extract started to show proliferative activity. Areca nut extract provided a stronger cytotoxicity effect in HSC-3 cells than in HSC-2 cells. The cytotoxicity graph of the three cell lines is shown in **Figure 6.**

MTS assay showed increased cellular viability in doses higher than IC_{50}, which was 16.2% followed by 4.5% in HSC-3 cells and an increase of 13.8% and 4% in HSC-2 cells. This was caused by strong antioxidant activity in the extract. Polyphenol had a direct stimulation effect on mitochondria's activity, so it was more efficient in producing energy and free radical scavenging. This was probably caused by the high extracellular formazan intensity from tetrazolium reduction by dehydrogenase succinate enzyme in cells performing respiration and metabolism. The more energy and cellular respiration product, the more formazan would be formed. Areca nut extract did not cause cytotoxicity in HaCat cell lines. This study showed that areca nut could induce formazan in large numbers so it showed a large increase in cell number. This was not caused by the high HaCat cells viability, but because of the high absorbance value by following the dense formazan color which was influenced by cellular respiration and metabolism. This absorbance value was read

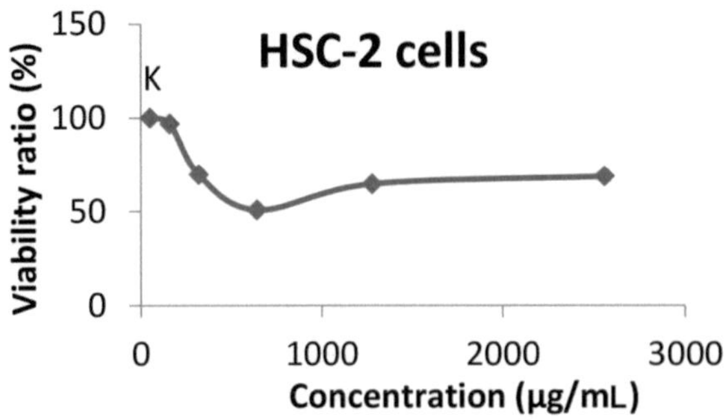

a.

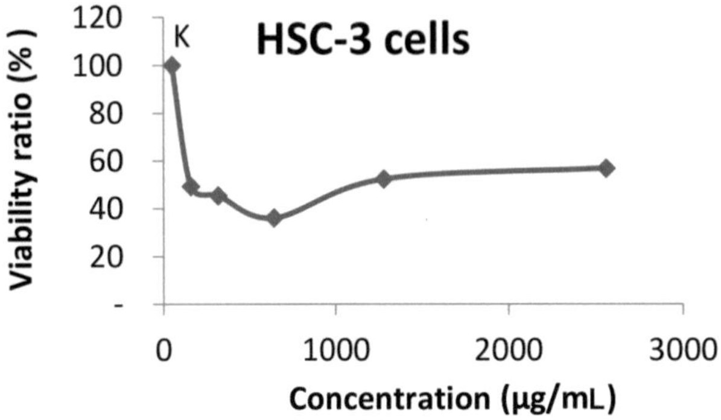

b.

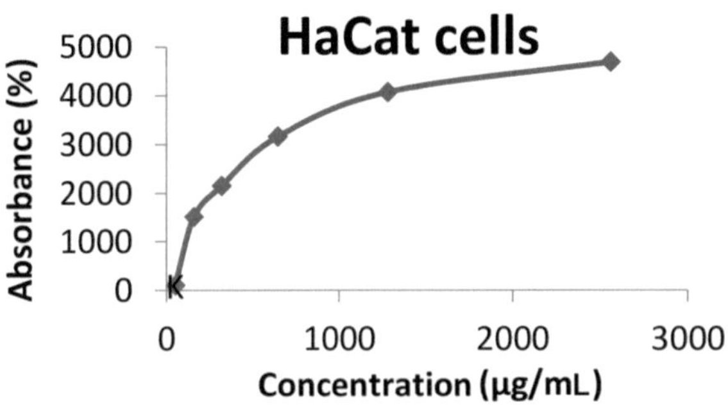

c.

Figure 6.
The result of MTS assay in HSC-2, HSC-3, and HaCat cell lines. Areca nut extract has a stronger cytotoxicity effect in HSC-3 cells (b) than in HSC-2 cells (a). The extract does not show a cytotoxicity effect in HaCat cells (c).

based on 490 nm wavelength. The denser the color of formazan's product, the higher the absorbance value would be. Although this study did not perform the test for Nicotinamide Adenosine Dinucleotide Hydrogen (NADH) content in the extract, it is possible that areca nut extract might contain NADH which could provide additional energy for the cells. NADH is an active coenzyme form of vitamin B3 (Niacin) which has an important role in the nervous system. This vitamin is found in all living cells of plants, animals, and humans. Tharakan et al. showed that *Tricopus zeylanicus* containing NADH had antioxidant activity by inducing lipid peroxidase and inhibiting lipoxygenase activity [66]. Further studies are needed to detect NADH content in areca nut.

The flavonoid in areca nut extract has antioxidant and also pro-oxidant activity. These two activities were also possessed by other herbal plants, *curcumin* [59, 60]. Lower doses of *curcumin* (12.5 μM) has the properties of reactive oxygen species scavenging, anti-inflammatory, apoptosis induction, and proliferation inhibition in myeloid leukemic cells, but in a higher dose and long term, the metabolite contents of *curcumin,* which is lipophilic or water-insoluble, could increase the level of cellular reactive oxygen species that causes carcinogenic potential through oxidative DNA damage or metal-mediated DNA damage at P450 cytochrome [61, 62]. This damage occurred because of the presence of Cu(II)-CYP2D6 which caused the damage of the DNA, especially 5-TG-3, 5-GC-3, and GG sequences [61]. *Curcumin* could induce lung cancer by increasing reactive oxygen species resulting in disarray between mitogen-activated protein kinase, NF-jb, and p53, causing genetic mutation and oxidative stress [63]. This finding concluded that the antioxidant effect which was started by an oxidative stimulus, depending on time, dose, and certain cancer type, could also cause unwanted side effects [63].

The principle of the MTS assay method was determined by a reduction–oxidation reaction occurring in cells. MTT/MTS reagent was reduced into formazan salts by succinate dehydrogenase enzyme found in living mitochondria cells. This reaction was allowed to take place for 4 hours then stopper reagent was added in MTT assay. The stopper reagent will lysate the cell's membrane so that formazan salts could get outside of the cell, and it could be dissolved. The MTS assay does not need a stopper reagent because formazan could dissolve in the culture medium with the addition of PMS. The formazan salt, that was in extracellular, was quantified with a spectrophotometer and measured in form of absorbance (**Figure** 7).

The higher the absorbance, the higher cell viability would be. The IC_{50} value between HSC-2 and HSC-3 cells had quite a large difference range. This difference showed the selective toxicity difference and type of cell death in some OSCC cell lines during exposure by natural anticancer or synthetic agent. The flavonoid compound showed weak cytotoxicity activity against HSC-2 cells so the IC_{50} value was higher [67, 68]. The factors of the substance in plants that influenced the cytotoxicity capability were the presence or absence of hydrophilic and hydrophobic groups in one same molecule, the presence or absence of isoprenyl groups, the presence or absence of polycyclic and/or halogen structure, the most condensed structure (low molecular weight is more cytotoxic) and lipophilicity. The factors from inside the cells which could probably cause this cytotoxicity was the difference

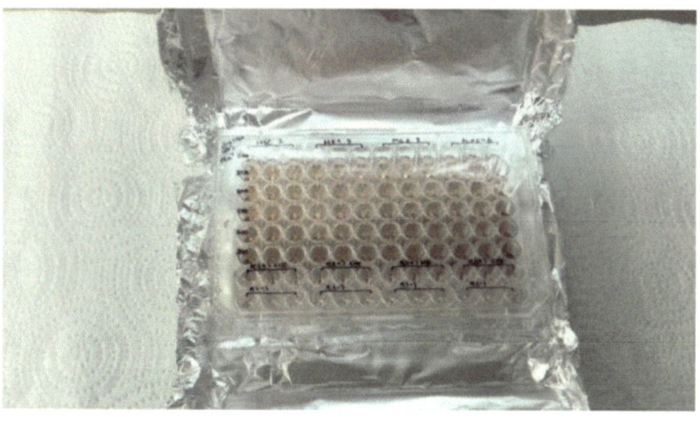

Figure 7.
The cells preparation in MTS assay.

of protein expression which was resistant against some anticancer agent and expression of the drug's metabolism enzyme or natural substances [68]. Environmental factors were serum type, the presence of metal ion, oxygen concentration, and external pressure [68]. The success of the MTS assay depended greatly on a cell's metabolism and respiration capability. This study also showed the morphology of HSC-2 cells which was different from HSC-3 cells. The HSC-2 cells had bigger, wider cytoplasm and tighter cell junction than HSC-3 cells that probably could cause weaker extract's cytotoxicity activity than in HSC-2 cells so the IC_{50} value of HSC-2 was higher than HSC-3 cells.

Analysis of apoptotic cells using flow cytometry demonstrated that areca nut can induce apoptosis in oral cancer cell lines, HSC-2 and HSC-3 cell lines [14]. The caspase-3 activity as an effector caspase is shown to be related to late apoptosis activity because of the increase of caspase-3 with increasing late apoptotic cells percentage in both cells after areca nut treatment. Analysis of caspase activity confirmed that apoptosis might be the major mechanism of cell death induced by areca nut. As far as we know, there is no similar report regarding the caspase-3 activity induced by areca nut extract, but this result is similar to several previous studies that used plants containing flavonoids to increase caspase activity in cancer cells [41–43]. This finding may have biological implications in cancer treatment. The caspases inside cells are in an inactive form (procaspase), but activation induces the production of other caspases leading to cell death through proteolytic activity [44, 45]. Caspase-3 activation is a crucial component in the apoptotic signaling cascade. The apoptosis pathway involved in areca nut-induced cell death in both cancer cell lines may be through the extrinsic and intrinsic pathways. The areca nut also caused the cell-cycle arrest in HSC-3 and HSC-2 cell lines. The areca nut inhibited cell proliferation by the enhancement of Ki-67 after 24 hours of extract treatment in both cells [69].

8.2 Immunomodulatory activity

The areca nut extract was found to increase the white blood cell count post-challenge with *Staphylococcus aureus* induction significantly indicating that the extract could stimulate the hemopoietic system. Another study found that areca nut extract can induce calcium signals in at least three immune cell lines and human primary immune cells (PBMCs), inducing the production of pro-inflammatory cytokines. Further separation of the PBMCs into T cells, B cells, and monocytes would potentially be of interest in elucidating specific responses to each cell subtype [70]. However, areca nut extract can also induce antigen-specific immune responses and promote inflammatory reactions *in vivo*, which may contribute to immune deregulation associated with areca-related diseases [71]. In fact, the study of areca nut as an immunomodulatory drug still causes various results. It depends on the content of polyphenolic compound in the areca nut extract which greatly affects the efficacy of the nut.

The areca nut can also increase the activity and capacity of macrophages. The process by which a cell ingests and disposes of foreign material, including microorganisms is called phagocytosis. In normal conditions, most phagocytes are circulating in the blood and when there is an inflammation, the phagocytes will leave the bloodstream and migrate to the site of inflammation. The circulation in the capillaries and venules is rapidly moving with the red blood cells in the mainstream and neutrophils and other leukocytes tending to flow more slowly along the vessel's periphery. Monocytes and macrophages have the same functions as neutrophils but for a longer time and in a later stage of the inflammatory response. Monocytes are produced in the bone marrow, enter the circulation, and migrate to the inflammatory site, where they develop into macrophages. Macrophages are more active as phagocytes than their monocytic precursors. Macrophages, particularly those residing in the tissue, are often important cellular initiators of the inflammatory response.

Several bacteria are resistant to killing by granulocytes and can even survive inside macrophages. Microorganisms such as *Mycobacterium tuberculosis*, *Salmonella typhi*, and *Mycobacterium leprae* can remain dormant or even multiply inside the phagolysosomes of macrophages. However, the bactericidal activity of macrophages can be markedly increased with the help of inflammatory cytokines produced by cells of the acquired immune system (a subset of T lymphocyte) or cells activated through Toll-like receptors. Macrophage activation results in increased phagocytic activity, the size of itself, plasma membrane area, glucose metabolism, and a number of lysosomes. The activation of leukocytes, monocytes, and macrophages which is induced by *S. aureus* and treated with areca nut is shown in **Figure 8**.

The areca nut extract probably stimulates the proliferation of macrophages, which in turn leads to the activation of macrophage activity. Further study is needed to find out how extract can increase macrophage activity and capacity.

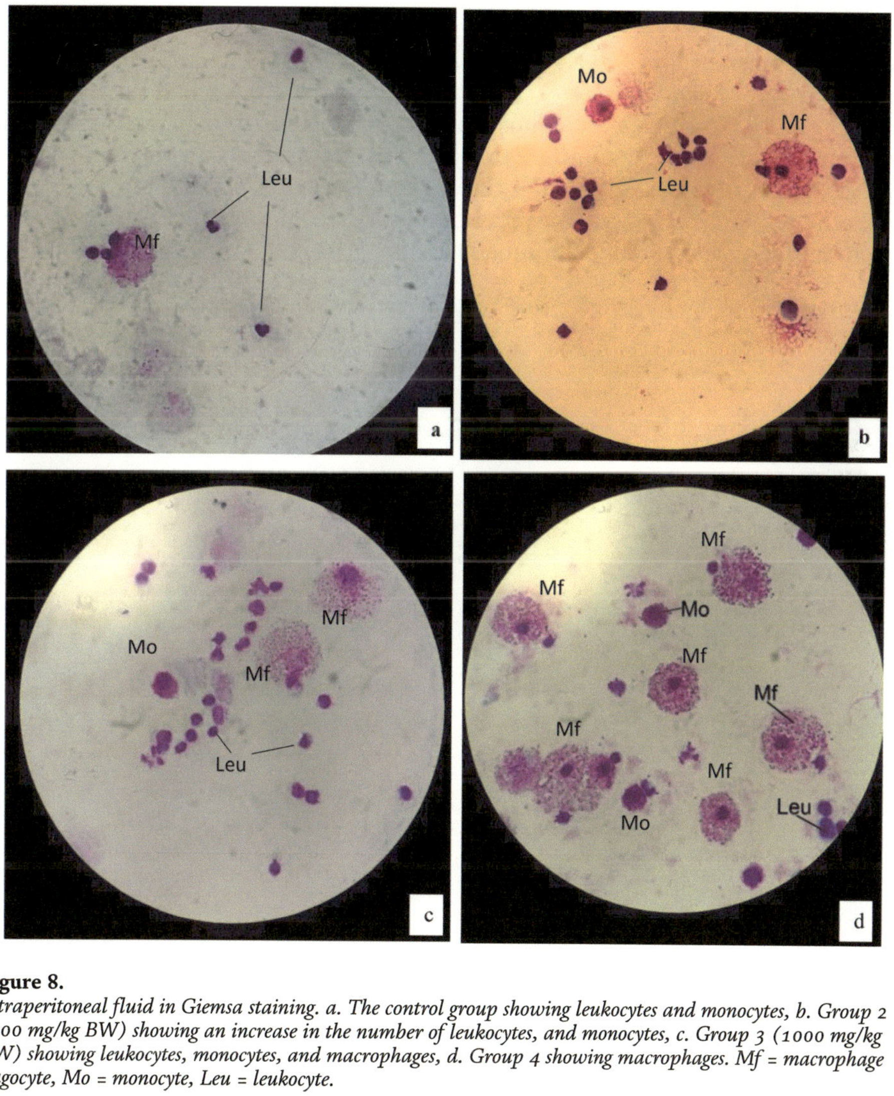

Figure 8.
Intraperitoneal fluid in Giemsa staining. a. The control group showing leukocytes and monocytes, b. Group 2 (500 mg/kg BW) showing an increase in the number of leukocytes, and monocytes, c. Group 3 (1000 mg/kg BW) showing leukocytes, monocytes, and macrophages, d. Group 4 showing macrophages. Mf = macrophage Fagocyte, Mo = monocyte, Leu = leukocyte.

However, not only is the effect of the treatment given, but the increase in macrophage activity against *S. aureus* infection might also be caused by the internal factor of the macrophage itself. The previous study reported the role of macrophage transmembrane expression that suppressed the *S. aureus*-induced production of nitric oxide and proinflammatory cytokines in mouse macrophages [72]. Moreover, it has been reported by several groups that this bacterium can invade and survive within a variety of cells such as neutrophils, macrophages, T-lymphocytes, epithelial cells, endothelial cells, fibroblasts, and osteoblasts which may be related to the intracellular persistence of bacteria within host cells [73–75]. The markers of biochemical examination (ureum, creatinine, SGOT, and SGPT) did not show changes in liver and kidney in mice after two weeks of treatment and one hour before and post-challenge with *S. aureus*. This study is in line with previous studies which revealed that the areca nut consumed in the long term in humans does not cause hepatotoxicity [76, 77]. However, another study showed that raw areca nut given for 28 days caused mild hepatotoxicity and nephrotoxicity in mice [78].

9. Conclusions

This chapter provides an overview of the characteristics of the areca nut and its impact on human health, especially in an anticancer and immunomodulatory effect. Areca nuts from Aceh, Indonesia, contain flavonoids, phenolics, catechin, quercetin, and a small percentage of tannins which contribute to antioxidant activity. The areca nut has anticancer potential activity so it can be used in combination with chemotherapeutic agents to enhance the effect at lower doses and thus minimize chemotherapy-induced toxicity. Areca nuts also show immunomodulatory activity which can increase the body's immune system.

Author details

Liza Meutia Sari
Oral Medicine Department, Syiah Kuala University, Banda Aceh, Indonesia

*Address all correspondence to: lizameutiasari@unsyiah.ac.id

References

[1] Hamsar MN, Ismail S, Mordi M, Ramanathan S, et al. Antioxidant capacity and the effect of different parts of areca catechu extracts on gluthatione-S-Transferase activity invitro. Free Rad Antiox 2011;1(1):28–33.

[2] Jaiswal P, Kumar P, Singh VK, Singh DK. *Areca catechu* L.: A valuable herbal medicine againts different health problems. Res J Med Plant 2011;5(2): 145–152.

[3] Velioglu YS, Mazza G, Gao L, Oomah BD. Antioxidant activity and total phenolics in selected fruits, vegetables, and grain products. J Agri Food Chem 1998;46(9):4113–4117.

[4] Ahsan H, Ali A, Ali R. Oxygen free radicals and systemic autoimmunity. Clin Exp Immunol 2003;31:398–404.

[5] Zhang WM, Wei J, Chen WX, Zhang HD. The chemical composition and phenolic antioxidants of *areca* (*areca catechu* L.) seeds. ICABE 2011;1-2:16–22.

[6] Huang PL, Chi CW, Liu TY. Effects of Areca catechu L. containing procyanidins on cyclooxygenase-2 expression in vitro and in vivo. Food Chem Toxicol 2010;48(1):306–313.

[7] Xing Z, Jiao W, Zhuang H, Li MW, Fu DH. Antioxidant and cytotoxic phenolic compounds of areca nut (*Areca catechu L.*). Chem Res Chinese Universities 2010;26(1):161–164.

[8] Staples GW, Bevacqua RF. Spesies profiles for pacific island agroforestry. Areca catechu (betel nut palm): www.traditionaltree.org; 2006. p. 1–24.

[9] IARC. WHO-biennial report. International agency for research on cancer, part I. 2004:1–192.

[10] Kim BJ, Kim JH, Kim HP, Heo MY. Biological screening of 100 plant extracts for cosmetic use (II): Anti-oxidative activity and free radical scavenging activity. Int J Cosmetic Sci 1997;19:299–307.

[11] Kook K, Lee JJ, Cho J, Park, Choi JD. The effects of *Areca Catechu* L Extract on Anti-Inflammation and AntiMelanogenesis. Int J Cosmetic Sci 1999;21(4):275–280.

[12] Lee KK, Choi JD. The effects of *Areca catechu* L extract on antiaging. Int J Cosmetic Sci 1999;21(4):285–295.

[13] Bhandare AM, Kshirsagar AD, Vyawahare NS, Hadambar AA, Thorve VS. Potential analgesic, anti-inflammatory and antioxidant activities of hydroalcoholic extract of Areca catechu L. nut. Food Chem. Toxicol 2010;48:3412–3417.

[14] Sari LM, Subita GP, Auerkari EI. Areca nut extract demonstrated apoptosis-inducing mechanism by increased caspase-3 activities on oral squamous cell carcinoma F1000 Res 2019;7(723):1–10.

[15] Giri S, Idle JR, Chen C, Zabriskie TM, Krausz KW, et al. A metabolic approach to the metabolism of the areca nut alkaloids arecoline and arecaidine in the mouse. Chem Res Toxicol 2006;19:818–827.

[16] Zhou J, Sun Q, Yang Z, Zhang J. The hepatotoxicity and testicular toxicity induced by arecoline in mice and protective effects of vitamins C and E. Korean J Physiol Pharmacol 2014;18(2): 143–148.

[17] Chuerduangphui J, Ekalaksananan T, Chaiyarit P, Patarapadungkit N, Chotiyano A, Kongyingyoes B, et al. Effects of arecoline on proliferation of oral squamous cell carcinoma cells by

dysregulating c-Myc and miR-22, directly targeting oncostatin M. PLoS ONE 2018;13(1):e0192009.

[18] Liu R. Potential Synergy of Phytochemicals in Cancer Prevention: Mechanism of Action. J. Nutr 2004;134 (12):3479S–3485S.

[19] Min KJ, and Kwon TK. Anticancer effects and molecular mechanisms of epigallocatechin-3-gallate. IMR 2014;3: 16–24.

[20] Lambert JD, and Elias RJ. The antioxidant and pro-oxidant activities of green tea polyphenols: A role in cancer prevention. Arch Biochem Biophys 2010;501:65–72.

[21] Roychoudhury S, Agarwal A, Virk G, and Cho CL. Potential role of green tea catechins in the management of oxidative stress-associated infertility. Reprod Biomed online 2017;34(5):487–498.

[22] Ren W, Qiao Z, Wang H, Zhu L, and Zhang L. Flavonoids: Promising Anticancer Agents. Med Res Rev 2003; 23(4):519–534.

[23] Shimizu M, Shirakami Y, Sakai H, Kubota M, Kochi T, et al. Chemopreventive Potential of Green Tea Catechins in Hepatocellular Carcinoma. Int J Mol Sci 2015;16(3): 6124–6139.

[24] Iwasaki R, Ito K, Ishida T, Hamanoue M, Adachi S, et al. Catechin, green tea component, causes caspase independent necrosis-like cell death in chronic myelogenous leukemia. Cancer Sci 2009;100(2):349–356.

[25] Khiewkamrop P, Phunsomboon P, Richert L, Pekthong D, and Srisawang P. Epistructured catechins, EGCG and EC facilitate apoptosis induction through targeting de novo lipogenesis pathway in HepG2 cells. Cancer Cell Int 2018;18 (46):1–13.

[26] Gopalakrishnan C, Shankaranarayanan D, Nazimudeen SK, Viswanathan S, Kameswaran L. Anti-inflammatory and C.N.S. depressant activities of xanthones from Calophyllum inophyllum and Mesua ferrea. Ind J Pharmac 1980;12(3):181–191.

[27] Chilpa RR, Baggio CH, Solano DA, Muniz EE, Kaufmann FC, et al. Inhibition of gastric H+, K+,-ATPase activity by flavonoids coumarins and xanthones isolated from Mexican medicinal. J Ethnopharm 2006;105:167–172.

[28] Sari LM, Subita GP, Auerkari EI. Potential antioxidant and cytotoxic activity of areca nut (*Areca catechu* L.) extract in human oral squamous cell carcinoma and keratinocyte cells. Asian J Pharm Clin Res 2017;10(10):286–291.

[29] Gurumurthy B. Diversity in Tannin and Fiber Content in Areca Nut (*Areca catechu*) Samples of Karnataka, India. Int J Curr Microbiol App Sci 2018;7(1): 2899–2906.

[30] Wolfe KL, Liu RH. Structure-activity relationships of flavonoids in the cellular antioxidant activity assay J Agric Food Chem 2008;56:8404–8411.

[31] Sazwi NN, Nalina T, Rahim AZH. Antioxidant and cytoprotective activities of *Piper betle, Areca catechu, Uncaria gambir* and betel quid with and witouth calcium hydoxide. BMC Complem Altern M 2013;13(351):1–12.

[32] Wetwitayaklung P, Paechamud T, Limmatvapirat C, et al. The study of antioxidant capacity in various parts of *Areca catechu* L. Naresuan Univ J 2006; 14(1):1–14.

[33] Khiewkamrop P, Phunsomboon P, Richert L, Pekthong D, Srisawang P. Epistructured catechins, EGCG, and EC facilitate apoptosis induction through targeting de novo lipogenesis pathway

in HepG2 cells. Cancer Cell Int 2018;18 (46):1–13.

[34] Zaveri NT. Green tea and its polyphenolic catechins: medicinal uses in cancer and noncancer applications. Life Sci 2006;78:2073–2080.

[35] Singh BN, Shankar S, Srivastava RK. Green tea catechin, epigallocatechin-3-gallate (EGCG): mechanisms, perspectives, and clinical applications. Biochem Pharmacol 2011;82(12):1807–1821.

[36] Mayr C, Wagner A, Neureiter D, Pichler M, Jakab M, et al. The green tea catechin epigallocatechin gallate induces cell cycle arrest and shows potential synergism with cisplatin in biliary tract cancer cells. BMC Complement Altern Med 2015;15(194):1–7.

[37] Gerates L, Moonen HJJ, Brauers K, Wouters EFM, Bast A, et al. Dietary flavones and flavonols are inhibitor of poly (ADP-ribose) polymerase-1 in pulmonary epithelial cells J Nutr 2007; 137: 2190–2195.

[38] Ramamoorthy P, Bono A. Antioxidant activity, total phenolic, and flavonoid content of *Morinda citrifolia* fruit extracts from various extraction processes. IJEST 2007;2(1):70–80.

[39] Bhisey RA, Boucher BJ, Hsi-Chen TH, Gajalakshmi V, Gupta PC, et al. Monographs on the Evaluation of Carcinogenic Risks to Humans.Betel-Quid and Areca-Nut Chewing and some Areca-Nut Related Nitrosamines. IARCS Monographs 2004;85:1–293.

[40] Nagao T, Ikeda N, Warnakulasuriya S, Fukano H, Yuasa H, et al. Serum antioxidant micronutrients and the risk of oral leukoplakia among Japanese. Oral Oncol 2000;36: 466–470.

[41] Carossa S, Pera P, Doglio P, Lombardo S, Colagrande P, et al. Oral nitric oxide during plaque deposition. Eur J Clin Invest 2001;31:876–879.

[42] Nair U, Bartsch H, Nair J. Alert for an epidemic of oral cancer due to use of the betel quid substitutes gutkha and pan masala: a review of agents and causative mechanisms. Mutagenesis 2004;19:251–262.

[43] Johnson NW, Jayasekara P, dan Amarasinghe HK. Squamous cell carcinoma and precursor lesions of the oral cavity: epidemiology and aetiology. Periodontology 2000. Singapura: John Wiley and Sons; 2011.

[44] Anttila A, Bhat RV, Bhond JA, Bhorgoff SJ, Bosch FX, et al. Monographs on the Evaluation of Carcinogenic Risks to Humans. Some Traditional Herbal Medicines, Some Mycotoxins, Naphthalene and Styrene. IARCS Monographs 2002;82:171–174.

[45] Gadkari PV, Balaramana M Catechins: Sources, extraction and encapsulation: A review Food Bioprod Process 2014;xxx:1–17.

[46] Chen L, Zhang HY Cancer Preventive Mechanisms of the Green Tea Polyphenol (−)-Epigallocatechin-3-gallate. Molecules 2007;12:946–957.

[47] Mukhtar H, Wang ZY, Katiqan SK, Agarwal R. Tea components: antimutagenic and antigagenic effects. Prev Med 1992;21:351–360.

[48] Misaka S, Kawabe K, Onoue S, Werba JP, Giroli M, et al. Effects of green tea catechins on cytochrome p450 2B6, 2C8, 2C19, 2D6, and 3A activities in human liver and intestinal microsomes. Drug Metab Pharmacokinet 2013;28(3):244–249.

[49] Park IJ, Lee YK, Hwang JT, Kwon DY, Ha J, Park OJ. Green tea catechin controls apoptosis in colon cancer cells by attenuation of H_2O_2-stimulated COX-2 expression via the AMPK signaling

pathway at low-dose H_2O_2. Ann N Y Acad Sci 2009;1171:538–544.

[50] Bettuzzi S, Brausi M, Rizzi F, Castagnetti G, Peracchia G, et al. Chemoprevention of human prostate cancer by oral administration of green tea catechins in volunteers with high-grade prostate intraepithelial neoplasia: A preliminary report from a one-year proof-of-principle study. Cancer Res 2006;66:1234–1240.

[51] Maeda-Yamamoto M, Suzuki N, Sawai Y, Miyase T, Sano M, et al. Association of suppression of extracellular signal-regulated kinase phosphorylation by epigallocatechin gallate with the reduction of matrix metalloproteinase activities in human fibrosarcoma HT1080 cells. J Agric Food Chem 2003;51:1858–1863.

[52] Alshatwi AA. Catechin hydrate suppresses MCF-7 proliferation through TP53/Caspase-mediated apoptosis. J Exp Clin Canc Res 2010;29(167):1–9.

[53] Vermeulen K, Berneman ZN, Van Bockstaele DR. Cell Cycle and Apoptosis. Cell Prolif 2003;36(3):165–175.

[54] Stewart ZA, Westfall MD, Pietenpol JA. Cell-cycle dysregulation and anticancer therapy. Trends Pharmacol Sci 2003;2(3):139–145.

[55] Fassina G, Vene R, Morini M Mechanisms of inhibition of tumor angiogenesis and vascular tumor growth by epigallocatechin-3-gallate. Clin Cancer Res 2004;10:4865–4873.

[56] Jung YD, Kim MS, Shin BA, Chay KO, Ahn BW, Liu W, et al. EGCG, a major component of green tea, inhibits tumour growth by inhibiting VEGF induction in human colon carcinoma cells. Br J Cancer 2001;84:844–850.

[57] Leong H, Mathur PS, Greene GL. Green tea catechins inhibit angiogenesis through suppression of STAT3 activation. Breast Cancer Res Treat 2009;117(3):505–515.

[58] Hannan A, Karan S, Chatterjee TP. A comparative study of invitro antioxidant capacity of different extract of *areca* seed collected from *Areca catechu* L plant grown in Assam. Int J Pharm Pharm 2012;4(2):420–427.

[59] Kong Y, Ma W, Liu X, Zu Y, Fu Y, et al. Cytotoxic Activity of Curcumin towards CCRF-CEM Leukemia Cells and Its Effect on DNA Damage. Molecules 2009;14:5328–5338.

[60] Gibellini L, Pinti M, Nasi M, Biasi SD, Roat E, et al. Interfering with ROS Metabolism in Cancer Cells: The Potential Role of Quercetin. Cancers 2010;2:1288–1311.

[61] Sakano K, Kawanishi S. Metal-mediated DNA damage induced by curcumin in the presence of human cytochrome p450 isozymes Arch Biochem Biophys 2002;405:223–230.

[62] Gupta SC, Hevia D, Patchva S, Park B, Koh W, et al. Upsides and downsides of reactive oxygen species for cancer: The roles of reactive oxygen species in tumorigenesis, prevention, and therapy. Antioxid Redox Signal 2012;16(11):1295–1322.

[63] Dance-Barnes ST, Kock ND, Moore JE, Lin EY, Mosley J, et al. Lung tumor promotion by *curcumin*. Carcinogenesis 2009;30(6):1016–1023.

[64] Roy S, Choudhury MD, Paul SB. Antioxidant potential of rhiome of alocasia decipiens schott. Asian J Pharm Clin Res 2013;6(2):120–122.

[65] Dewoto HR. Pengembangan Obat Tradisional Indonesia Menjadi Fitofarmaka. Maj Kedokt Indon 2007;57 (7):205–211.

[66] Tharakan B, Dhanasekaran M, Manyam BV. Antioxidant and DNA

Protecting Properties of Anti-fatigue Herb Trichopus zeylanicus. Phytother Res 2005;19:669–673.

[67] Okamura M, Shimada J, Sakagami H. Comparative Analysis of Cell Death Induction by Cisplatin and 5-FU in Human Oral Squamous and Hepatocellular Carcinoma Cell Lines. Anticancer Res 2008;28:253–260.

[68] Sakagami H, Kobayashi H, Chien CH, Kanegae H, Kawase M. Selective Toxicity and Type of Cell Death Induced by Various Natural and Synthetic Compounds in Oral Squamous Cell Carcinoma. In vivo 2007;21:311–320.

[69] LM Sari, Wulandari D, Bustami A, Subita GP, Auerkari EI. Tannin Screening, Phenolic Compounds Analysis, and Antiproliferative Activity of Areca Nut Extract by Decreasing Ki-67 Protein in Oral Squamous Carcinoma Cell Lines. Trop J Nat Prod Res 2020;4 (9):563–570.

[70] Faouzi M, Neupane RP, Yang J, Williams P, Penner R. Areca nut extracts mobilize calcium and release pro-infammatory cytokines from various immune cells. Sci Rep 2018; 1075:1–13.

[71] Wang CC, Lin HL, Wey SP, Jan TR. Areca nut extract modulates antigen-specific immunity and augments inflammation in ovalbumin-sensitized mice. Immunopharmacol Immunotoxicol 2011;33(2):15–22.

[72] Zhu B, Yu Y, Liu X, Han Q, Kang Y, et al. CD200 Modulates *S. aureus*-Induced Innate Immune Responses Through Suppressing p38 Signaling. Int J Mol Sci 2019;20(659):1–15.

[73] Hamza T, Li B. Differential responses of osteoblasts and macrophages upon *Staphylococcus aureus* infection. BMC Microbiol 2014;14(207): 1–11.

[74] Kubica M, Guzik K, Koziel J, Zarebski M, Richter W, et al. A potential new pathway for *Staphylococcus aureus* dissemination: the silent survival of *S. aureus* phagocytosed by human monocyte-derived macrophages. PLoS One 2008;3(1):e1409.

[75] Gresham HD, Lowrance JH, Caver TE, Wilson BS, Cheung AL, Lindberg FP. Survival of *Staphylococcus aureus* inside neutrophils contributes to infection J Immunol 2000;164(7):3713–3722.

[76] Singroha K, Kamath VV. Liver function tests as a measure of hepatotoxicity in areca nut chewers. J Dent Res Rev 2016;3(2):1–5.

[77] Pai SR, Shirke AJ, Gothoskar SV. Longterm feeding study in C17 mice administered saccharin coated betel nut and 1,4dinitrosopiperazine in combination. Carcinogenesis 1981;2: 175–177.

[78] Lohith TS, Shridhar NB, Dilip SM, Jayashree P, Suhasini K. Repeated dose 28-day oral toxicity study of raw areca nut extract in rats. Int Res J Pharm 2013; 4(5):238–240.

Chapter 19

Health Benefits of Aqueous Extract of Black and Green Tea Leaves

Zainab Khudhur Ahmad Al-Mahdi, Ruqaya M.J. Ewadh and Nada Khazal Kadhim Hindi

Abstract

Tea, next to water, is the beverage humans consume. Drinking the beverage tea is great for joining and collecting family members and public communities since ancient times. Tea plant *Camellia sinensis* has been cultivated for thousands of years, and its leaves have been used for medicinal purposes. Various studies suggest that polyphenolic compounds present in green and black tea are associated with beneficial effects in prevention of cardiovascular diseases, particularly of atherosclerosis and coronary heart disease. Anti-ageing, antidiabetic and many other health beneficial effects associated with tea consumption are described. Evidence is accumulating that catechins and theaflavins, which are the main polyphenolic compounds of green and black tea, respectively, are responsible for most of the physiological effects of tea. This review describes the evidence from clinical and epidemiological studies in the prevention of chronic diseases like cancer and cardiovascular diseases and inhibits pathogenic bacteria and general health promotion associated with tea consumption.

Keywords: green tea, black tea, health benefit, antibacterial activity

1. Green tea

Green tea is a popular drink, especially in Asian countries, although its popularity continues to spread across the globe. The health benefits of green tea, derived from the leaves of the *Camellia sinensis* [*Camellia sinensis* is a species of evergreen shrub or small tree whose leaves and leaf buds are used to produce tea. It is of the genus *Camellia* (Chinese: 茶花; pinyin: *cháhuā*, literally: "tea flower")] plant, have been studied for many years. Fairly recently, researchers have begun to look at the possibility of using green tea in antimicrobial therapy, and the potential prevention of infections. The particular properties of catechins found in the tea have shown promise for having antimicrobial effects. There are four main catechins (polyphenols) found in green tea: (-)-epicatechin (EC), (-)-epicatechin-3-gallate (ECG), (-)-epigallocatechin (EGC) and (-)-epigallocatechin-3-gallate (EGCG). Three of these, ECG, EGC and EGCG have been shown to have antimicrobial effects against a variety of organisms [1, 2].

Among the health benefits that have been studied using green tea are: as an antioxidant, anti-inflammatory, anticarcinogenic, in cardiovascular health, oral

IntechOpen

health, and as an antimicrobial. Antioxidant effects come from the ability of green tea to limit the amount of free radicals by binding to reactive oxygen species (ROS) [3, 4]:

1. **Anti-inflammatory** effects may be a result of increased production of IL-10, an anti-inflammatory cytokine [5]

2. **Inflammation** is involved in, among other conditions, arthritis, cardiovascular disease, ageing, and cancer [5, 6].

3. **The anticarcinogenic** effects of green tea have been seen in many types of cancer, and the mechanisms may include inhibiting angiogenesis and cell growth, and inducing apoptosis in cancer cells [5–7].

4. **Cardiovascular** effects include the antioxidant and anti-inflammatory effects, and consumption of green tea has been shown to inhibit atherosclerosis, reduce lipid levels overall, and improve the ratio of LDL to HDL [6]. The effects for oral health are related to both teeth and gums. The main cause of dental caries is the bacteria *Streptococcus mutans*.

5. **Antioxidants** in Green Tea May Lower Your Risk of Some Types of Cancer, Cancer is caused by uncontrolled growth of cells. It is one of the world's leading causes of death. It is known that oxidative damage contributes to the development of cancer and that antioxidants may have a protective effect [8].

    - Green tea is an excellent source of powerful antioxidants, so it makes sense that it could reduce your risk of cancer, which it appears to do; **Breast cancer:** A meta-analysis of observational studies found that women who drank the most green tea had a 20–30% lower risk of developing breast cancer, the most common cancer in women [9].

    - **Prostate cancer:** One study found that men drinking green tea had a 48% lower risk of developing prostate cancer, which is the most common cancer in men [10].

    - **Colorectal cancer:** An analysis of 29 studies showed that those drinking green tea were up to 42% less likely to develop colorectal cancer [11]. Many observational studies have shown that green tea drinkers are less likely to develop several types of cancer. However, more high-quality research is needed to confirm these effects [12].

1.1 Green tea can kill bacteria

- Which Improves Dental Health and Lowers Your Risk of Infection [13]

- The catechins in green tea also have other biological effects.

- Some studies show that they can kill bacteria and inhibit viruses like the influenza virus, potentially lowering your risk of infections.

- *Streptococcus mutans* is the primary harmful bacteria in the mouth. It causes plaque formation and is a leading contributor to cavities and tooth decay.

- Studies show that the catechins in green tea can inhibit the growth of *Streptococcus mutans*. Green tea consumption is associated with improved dental health and a lower risk of caries [14].

- **Green tea** has a direct **antimicrobial** effect on this bacteria; plus, it seems to inhibit the attachment of the bacteria to oral surfaces. In addition, green tea is a natural source of fluoride [1, 6]. Green tea has been shown to have antimicrobial effects against a variety of gram positive and gram negative bacteria (e.g., *Escherichia coli*, *Salmonella* spp., *Staphylococcus aureus*, *Enterococcus* spp.), some fungi (e.g., *Candida albicans*) and a variety of viruses (e.g., HIV, herpes simplex, influenza) [5, 7].

1.2 Green tea composition

The medically important components of green tea are the polyphenols, most importantly the flavonoids. The main flavonoids in tea are the catechins, making up 30–40% of the water-soluble solids in green tea [3, 11]. The different types of tea vary in the amount of catechins that they contain, with green tea containing the most, then Oolong tea, then black tea. The initial steaming process in the production of green tea destroys the enzyme polyphenol oxidase, thus protecting the polyphenol content. There are four main catechins in tea: (-)-epicatechin (EC), (-)-epicatechin-3-gallate (ECG), (-)-epigallocatechin (EGC) and (-)-epigallocatechin-3-gallate (EGCG). In green tea, EGCG is the most abundant, representing approximately 59% of the total catechins. EGC is next, making up approximately 19%, then ECG, at 13.6% and EC, at 6.4% [6, 10].

Green tea contains less caffeine than coffee, but enough to produce an effect. It also contains the amino acid L-theanine, which can work synergistically with caffeine to improve brain function [8].

Multiple studies show that the catechin compounds in green tea can have various protective effects on neurons in test tubes and animal models, potentially lowering the risk of Alzheimer's and Parkinson's [15].

1.3 Black, green and white

Black, green and white teas are the most popular beverages worldwide [16].

Tea leaves are known for its antibacterial activity against any microorganisms. It is one of the most popular beverages worldwide.

Black tea has a long history of use dating back to China approximately 5000 years ago. It is made from the dried leaves of *Camellia sinensis*, a perennial evergreen shrub formerly known as *Thea sinensis*. It is native to southeastern Asia. Green tea, black tea, and oolong tea are all derived from the same plant. Black tea results from the oxidation of *Camellia sinensis* leaves.

The chemical components in tea include alkaloids (theobromine, caffeine, theophylline), polyphenols, amino acids, polysaccharides, volatile acids, vitamins, lipids as well as inorganic elements [17–19]. Black tea is used for treating headaches, low blood pressure, preventing heart disease, including atherosclerosis and heart attack, preventing Parkinson's disease, reducing the risk of stomach and colon cancer, lung, ovarian and breast cancers [20].

Currently, a growing consumption of tea is observed in western countries, where it has been considered as functional food.

Nutritional Value of Black Tea Like other types of tea, black tea contains:

- Caffeine

- Amino acids

- Carbohydrates
- Proteins
- Potassium
- Major minerals and trace minerals
- Manganese
- Fluoride
- Polyphenols

Black tea also contains catechins (the powerful antioxidants in tea that fight cancer-causing cells and help prevent heart disease), tannins (the naturally occurring chemical compounds that give black tea and red wine their astringency), guanine (a natural stimulant) and xanthine (another natural stimulant, similar to caffeine).

The many antioxidants and polyphenols in black tea are associated with a number of health benefits. Specifically, black tea contains complex flavonoids, which are polyphenols that aid in disease prevention. A single cup of black tea contains an average of 200 mg of flavonoids. Many doctors now recommend getting 600 mg of flavonoids per day for a range of health benefits. The flavonoid polyphenols in black tea known as thearubigin and theaflavin act as especially powerful antioxidants. Interestingly, these two flavonoids are more concentrated in black tea than in green tea [21].

Additionally, black tea is low in sodium and calories (if you do not add a sweetener). Plus, black tea has a bold flavor, making it a good substitute for those accustomed to soft drinks other unhealthy beverages (which also tend to have bold flavors) [22].

2. The biological properties of tea

The biological properties of tea include effects on the Central System (CNS) and antioxidant effects, attributed to the presence of methylxanthines, such as caffeine and phenolic compounds, especially catechins [23].

Black tea is more oxidised than all other types of teas. It contains antioxidants and other substances that might help protect the heart and blood vessels. It is also used for treating headache and low blood pressure; preventing heart disease, including "hardening of the arteries" (atherosclerosis) and heart attack [24].

3. Black tea fights diseases and infections; how black tea benefits

The tannins in black tea do not just give it its characteristic taste. Several studies have shown that tannins help fight viruses such as influenza ("the flu"), dysentery, and hepatitis. Black tea also contains alkylamine antigens, which help boost immune response [25]. Both Iranian non fermented (green tea) and fermented (black tea) have anti *Streptococcus mutans* activity in vitro. The anti-*Streptococcus mutans* activity of black tea appears on a lower concentration than green tea [18].

Black and green tea have antibacterial activity against many pathogens.

Effectiveness of aqueous extract of green, black and red tea leaves against some types of Gram positive and negative bacteria [19].

An *in vitro* study recorded that the black tea has antibacterial activity to the following pathogens:

- *Escherichia coli*
- *Klebsiella pneumonia*
- *Bacillus subtilis*
- *Micrococcus luteus*
- *Staphylococcus aureus*
- *Salmonella typhi*
- *Pseudomonas aeruginosa*

All tea extracts have shown significant antibacterial activity against *S. aureus* ATCC 25922 with Aqueous extract of Green tea exhibiting highest activity. All Green tea extracts exhibited significant activity against *E. coli* ATCC 25923 higher than Black tea extracts. As compared to Green tea extracts, Black tea extracts showed much lower activity against *P. aeruginosa* ATCC 27853. *S. aureus* was found to be most susceptible to tea extracts followed by *E. coli* and *P. aeruginosa*. Green tea and Black tea extracts have shown significant antibacterial activity with former being more effective than later. In future there is immense potential of clinical application of polyphenolic contents of tea extracts as adjuvant therapeutic agents to tackle the menace of growing antibiotic resistance [20, 23].

Black tea extract also had the ability to completely inhibit *Pseudomonas* growth on blood agar and inhibited protease activity and adhesion. There were also differences in Congo red binding seen in bacterial cell suspensions cultured in growth media that contained tea extract. The synergistic activity of tea extract with antibiotics has changed the resistance of *P. aeruginosa* (without the tea) to sensitive (in the presence of tea extract) [22].

4. Anticancer activity of black tea

Black tea is recorded to have anticolon cancer, lung, ovarian and breast cancers [21, 25]; green or black tea has polyphenols as prophylactic and therapeutic agents. Theaflavins and catechins seem to act on cancer cells largely through different pathways, so utilisation of both could offer synergistic anticancer effects, but so far no work has been done on the cumulative effects of EGCG and TF on prostate cancer. Therefore, in this study, we have investigated if EGCG in combination with TF can reduce the rate of prostate cancer growth, and we have observed greater cell death compared to application of either TF or EGCG alone.

Consumption of black tea, rich in polyphenols, has been found to reduce ovarian cancer risk. Theaflavin (TF1), theaflavin-3-gallate (TF2a), theaflavin-3′-gallate (TF2b) and theaflavin-3, 3′-digallate (TF3) are four main theaflavin derivatives found in black tea.

All four theaflavin derivatives inhibited ovarian cancer cells. Some of the effects and mechanisms of TF1 are different from those of the other three theaflavin derivatives [17, 21].

4.1 Black tea and skin health

Drinking black tea benefits the skin in three ways. First, it nourishes the skin with vitamins B2, C, and E, with minerals such as magnesium, potassium, and zinc, and essential polyphenols and tannins. Second, its caffeine and some of its other chemical components can kill oral viruses, which helps prevent skin infections (and pimples). Third, black tea has been shown to reduce "mimic wrinkles" and signs of premature ageing.

Black tea can also benefit your skin with direct contact/application. For example, placing black tea bags under your eyes helps reduce puffiness and dark circles. And using black tea for herbal baths can provide an antioxidant boost for your skin and may even provide low levels of sun protection [22, 26].

5. Black tea benefits

The healthiest tea for you is the one you will want to drink every day. By that definition, if you live in the West, the tea that is the healthiest for you is probably black tea. Over 90% of all tea sold in the West is black tea (or red tea, as it is known in the East).

Like green tea, oolong tea and white tea, black tea is made from leaves of the *Camellia sinensis* plant, so it shares many tea health benefits with other tea types. However, black tea is unique, and it is known to be especially beneficial for certain health purposes [8, 25].

The major health benefits of black tea include its nutritional value, anti-cancer benefits, digestive benefits, beneficial effects on skin and hair health, and much more. Grab a cup of tea (preferably organic tea) and learn more.

Black tea also contains catechins (the powerful antioxidants in tea that fight cancer-causing cells and help prevent heart disease), tannins (the naturally occurring chemical compounds that give black tea and red wine their astringency), guanine (a natural stimulant) and xanthine (another natural stimulant, similar to caffeine).

The many antioxidants and polyphenols in black tea are associated with a number of health benefits. Specifically, black tea contains complex flavonoids, which are polyphenols that aid in disease prevention. A single cup of black tea contains an average of 200 mg of flavonoids. Many doctors now recommend getting 600 mg of flavonoids per day for a range of health benefits. The flavonoid polyphenols in black tea known as thearubigin and theaflavin act as especially powerful antioxidants. Interestingly, these two flavonoids are more concentrated in black tea than in green tea [22, 23, 27].

5.1 Black tea and cardiovascular health

Black tea is abundant in antioxidants, such as flavonoids. These antioxidants have been demonstrated to lower the risk of heart disease. They do this by preventing the oxidation of LDL cholesterol and preventing damage in both the bloodstream and at artery walls. Additionally, black tea flavonoids can both improve coronary vasodilation and reduce clots, and its manganese may reduce the risk of coronary heart disease by helping cardiac muscle function. Studies have shown that as few as three cups of tea per day can improve heart health [25].

5.2 Cancer prevention

Perhaps the most-studied tea health benefit is its anti-cancer benefit. While most of the study has been on green tea, there is a growing body of evidence that black tea also plays a role in cancer prevention [25].

It appears that the polyphenols in tea help prevent the formation of potential carcinogens in the body. This is particularly true with certain types of cancer, such as ovarian cancer, lung cancer, prostate cancer, colorectal cancer, and bladder cancer. Some studies also show that black tea may help prevent stomach cancer, prostate cancer, breast cancer and oral cancers (especially for those who use tobacco products) [25–28].

The mechanism with which black tea prevents cancer is an interesting one. Black tea contains a compound called TF-2. This chemical causes apoptosis ("programmed death") of cancer cells without harming normal, healthy cells. This helps to stop cancer growth before it even becomes noticeable, and may help in cases where cancer has already been diagnosed. Additionally, black tea also prevents cancer by inhibiting the formation and growth of malignant tumors [25].

5.3 Black tea benefits the immune system

The tannins in black tea do not just give it its characteristic taste. Several studies have shown that tannins help fight viruses such as influenza ("the flu"), dysentery, and hepatitis. Black tea also contains alkylamine antigens, which help boost immune response [25].

5.4 Black tea and oral health

There are many folk tales about tea, freshening the breath and cleansing the mouth. It turns out that they are true. Research has found that black tea may reduce oral cancers. Additionally, tea's polyphenols and tannin's kill and prevent the bacteria that cause tooth decay, and to drastically reduce the oral bacteria that cause bad breath [25].

5.5 Black tea's digestive benefits

The tannin in tea in general (and black tea in particular—it has more of them than other tea types) offer digestive benefits. They soothe gastric and intestinal illnesses, generally aid in digestion and decrease intestinal activity (making them useful for those with diarrhea) [24, 25].

5.6 Black tea and hair health

Although it may seem rather vain compared to some of the other more life-altering health benefits of black tea, black tea is fantastic for your hair.

The high levels of antioxidants and caffeine in black tea both benefit hair health. The caffeine decreases a hormone that causes hair loss (known as DHT or dihydrotestosterone), while the antioxidants promote healthy hair growth. However, excess caffeine may stunt hair growth, so be careful not to overdo it. Black tea can also add shine, luster, and darkness to your hair if you incorporate it into your hair care regimen [25].

5.7 Bone and connective tissue benefits of black tea

If you drink tea regularly, you are more likely to have stronger bones and connective tissue than someone who does not drink tea regularly. Scientists believe this may be due to tea's phytochemicals [25].

5.8 The effect of black tea on brain (and nervous system)

The caffeine in black tea has been shown to improve mental focus and concentration by promoting blood flow in the brain. Unlike drinks with higher levels of caffeine and other stimulants (i.e., coffee and energy drinks), the caffeine in black tea is less likely to over-stimulate the heart and cause other unpleasant side effects.

Caffeine aside, studies show that L-theanine (an amino acid found in black tea) balances the effects of caffeine in a unique way, helping you concentrate more fully on tasks and act in a focused but relaxed manner. Furthermore, studies show that 1 month of four cups of black tea a day reduces levels of the stress hormone cortisol enough to boost your memory function, and some studies suggest that regular black tea consumption may prevent Parkinson's disease [8].

5.9 Black tea increases your energy level

Moderate caffeine consumption not only stimulates metabolism, but it also increases alertness and overall brain function. The caffeine in tea is mitigated by the naturally occurring chemical L-theophylline, which makes the effects of tea on energy level more smooth and continuous than the sometimes jarring effects of coffee and caffeinated sodas. Additionally, while caffeine mainly stimulates the muscles, L-theophylline targets the heart, kidneys and respiratory system, so the overall impact on the body is more evenly distributed and balanced [25].

6. Conclusion

Our review shows that tea have great health benefit for human body, the nutritional value of tea components shows great health benefit for heart and vascular system, fight cancer, inhibit pathogens, healthy for skin, hair, bone regulate metabolism and more.

Author details

Zainab Khudhur Ahmad Al-Mahdi[1]*, Ruqaya M.J. Ewadh[2] and Nada Khazal Kadhim Hindi[3]

1 Department of Basic and Medical Science, College of Dentistry, University of Babylon, Iraq

2 Department of Clinical Laboratory Science, College of Pharmacy, University of Babylon, Iraq

3 Department of Basic and Medical Science, College of Nursing, University of Babylon, Iraq

*Address all correspondence to: nadakhazal@yahoo.com

References

[1] Kamel GJ. History of Tea: Botanics, whose name in Latin was Camellus was missionary to the Philippines, died in Manilla in 1706. [...] Camellias were named in posthumous honor of George Joseph Kamel by Carolus Linnæus; 2003

[2] Botten D, Fugallo G, Fraternali F, Molteni C. Structural properties of green tea catechins. The Journal of Physical Chemistry. B. 2015;119(40):12860-12867

[3] Golender L. History of Tea: Botanics, the first edition of Linnaeus's Species Plantarum published in 1753 suggested calling the tea plant Thea sinensis; 2003

[4] Chacko SM, Thambi PT, Kuttan R, Nishigaki I. Beneficial effects of green tea: A literature review. Chinese Medicine. 2010;5:13

[5] Serafini M, Del Rio D, Yao DN, Bettuzzi S, Peluso I. Chapter 12: Health benefits of tea. In: IFF B, Wachtel-Galor S, editors. Herbal Medicine: Biomolecular and Clinical Aspects. 2nd ed. Boca Rotan, FL: CRC Press; 2011. pp. 239-262

[6] Jigisha A, Nishant R, Navin K, Pankaj G. Green tea: A magical herb with miraculous outcomes. International Research Journal of Pharmacy. 2012;3:139-148. Available online at: http://www.irjponline.com/admin/php/uploads/1083

[7] Subramani C, Natesh RK. Molecular mechanisms and biological implications of green tea polyphenol, (–)-epigallocatechin -3- gallate. International Journal of Pharma Bioscience and Technology. 2013;1:54-63

[8] Reuter S, Gupta SC, Chaturvedi MM, Aggarwal BB. Oxidative stress, inflammation, and cancer: How are they linked? Free Radical Biology and Medicine. December 2010;49(11):1603-1616

[9] Ogunleye AA, Xue F, Michels KB. Green tea consumption and breast cancer risk or recurrence: A metaanalysis. Breast Cancer Research and Treatment. 2010;119(2):477-484. DOI: 10.1007/s10549-009-0415-0

[10] Kurahashi N, Sasazuki S, Iwasaki M, Inoue M, Tsugane S, JPHC. Green tea consumption and prostate cancer risk in Japanese men: A prospective study. American Journal of Epidemiology. 2008;167(1):71-77

[11] Chen Y, Wu Y, Du M, Chu H, Zhu L, Tong N, et al. An inverse association between tea consumption and colorectal cancer risk. Oncotarget. 2017;8(23):37367-37376

[12] Yuan JM. Cancer prevention by green tea: Evidence from epidemiologic studies. The American Journal of Clinical Nutrition. 2013;98(6 Suppl):1676S-1681S

[13] Steinmann J, Buer J, Pietschmann T, Steinmann E. Anti-infective properties of epigallocatechin-3-gallate (EGCG), a component of green tea. British Journal of Pharmacology. 2013;168(5):1059-1073

[14] Lee HJ, Lee YN, Youn HN, Lee DH, Kwak JH, Seong BL, et al. Antiinfluenza virus activity of green tea by-products in vitro and efficacy against influenza virus infection in chickens. Poultry Science. 2012;91(1):66-73

[15] Caruana M, Vassallo N. Tea polyphenols in Parkinson's disease. Advances in Experimental Medicine and Biology. 2015;863:117-137

[16] Shagana JA, Geetha R. Comparative analysis of antimicrobial activity of black tea, green tea and white tea extracts on Streptococcus mutans by

tube dilution method. International Journal of Pharmaceutical Sciences and Research. 2017;9(9):1581-1582

[17] Jalayer NJ, Niakan M, Kharazi Fard MJ, Zardi S. Antibacterial activity of Iranian green and black tea on *Streptococcus mutans*: An in vitro study. Journal of Dentistry (Tehran). 2011;**8**(2):55-59

[18] Fatima A, Malik F, Shafiq A, Jawaid S, Hakim ST, Nadeem SG. Evaluation antibacterial activity of three most consumed tea extracts against pathogenic bacteria. International Journal of Current Microbiology and Applied Sciences. 2016;**5**(11):824-827

[19] Mehta A, Saxena G, Mani A. Comparative analysis of antibacterial activity of aqueous, ethanolic, methanolic and acetone extracts of commercial green tea and black tea against standard bacterial strains. International Journal of Current Microbiology and Applied Sciences. 2016;**5**(11):145-152

[20] Koňariková K, Ježovičová M, Keresteš J, Gbelcová H, Ďuračková Z, Žitňanová I. Anticancer effect of black tea extract in human cancer cell lines. Springerplus. 2015;**4**:127. DOI: 10.1186/s40064-015-0871-4

[21] Naji SA, Abbas AF, AL-Hushemi EH. Antibacterial activities of black tea alcohol ethanolic extract and some antibiotics against isolates *Staphylococcus aureus* and *Staphylococcus epidermidis* isolated from conjunctivitis. Diyala Agricultural Sciences Journal. 2009;**1**(2):52-60

[22] Gao Y, Rankin GO, Tu Y, Chen YC. Inhibitory effects of the four main theaflavin derivatives found in black tea on ovarian cancer cells. Anticancer Research. 2016;**36**(2):643-651

[23] Flayyih MT, Yousif HS, Subhi IM. Antimicrobial effects of black tea (Camellia sinensis) on Pseudomonas aeruginosa isolated from eye infection. Iraqi Journal of Science. 2013;**54**(2):255-265

[24] Bagyalakshmi B, Balamurugan A. Inhibitory activity of fresh green tea and black tea extracts (*Camellia sinensis*) on intestinal pathogens isolated from diarrheal sample. International Research Journal of Pharmacy. 2017;**8**(10):93-98. DOI: 10.7897/2230-8407.0810188

[25] Kobalka AJ, Keck RW, Jankun J. Synergistic anticancer activity of biologicals from green and black tea on DU 145 human prostate cancer cells. Central-European Journal of Immunology. 2015;**40**(1):1-4. DOI: 10.5114/ceji.2015.50825

[26] Hindi NKK, Abdul-Husin IF, Al-Mahdi ZKA, Ewadh RMJ, Hossain AO, Kadhim MJ, et al. RJPT - effectiveness of aqueous extract of green, black and red tea leaves against some types of gram positive and negative bacteria. Research Journal of Pharmacy and Technology. 2017;**10**(6):1957-1962

[27] Goodwin L. Black tea benefits. cardiovascular health, cancer prevention, digestive benefits and more. The Spruce Eat. 2019. Available form: https://www.thespruceeats.com/black-tea-benefits-765048

[28] Archana S, Abraham J. Comparative analysis of antimicrobial activity of leaf extracts from fresh green tea, commercial green tea and black tea on pathogens. Journal of Applied Pharmaceutical Science. 2011;**01**(08):149-152